T0302058

An Introduction to Condensed Matter Physics for the Nanosciences

The book provides an accessible introduction to the principles of condensed matter physics with a focus on the nanosciences and device technologies. The basics of electronic, phononic, photonic, superconducting, optics, quantum optics, and magnetic properties are explored, and nanoscience and device materials are incorporated throughout the chapters. Many examples of the fundamental principles of condensed matter physics are taken directly from nanoscience and device applications.

This book requires a background in electrodynamics, quantum mechanics, and statistical mechanics at the undergraduate level. It will be a valuable reference for advanced undergraduates and graduate students of physics, engineering, and applied mathematics.

Features

- Contains discussions of the basic principles of quantum optics and its importance to lasers, quantum information, and quantum computation.
- Provides references and a further reading list to additional scientific literature so that readers can use the book as a starting point to then follow up with a more advanced treatment of the topics covered.
- Requires only a basic background in undergraduate electrodynamics, quantum mechanics, and statistical mechanics.

Professor Emeritus Arthur McGurn, CPhys, FInstP, is a Fellow of the Institute of Physics, a Fellow of the American Physical Society, a Fellow of Optica (formerly Optical Society of America), a Fellow of the Electromagnetics Academy, and an Outstanding Referee for the journals of the American Physical Society. He received his PhD in Physics in 1975 from the University of California, Santa Barbara, followed by postdoctoral studies at Temple University, Michigan State University, and George Washington University (NASA Langley Research Center). Prof. McGurn's research interests include the theory of: magnetism in disorder materials, electron conductivity, the properties of phonons, ferroelectrics and their nonlinear dynamics, Anderson localization, amorphous materials, the scattering of light from disordered media and rough surfaces, the properties of speckle correlations of light, quantum optics, nonlinear optics, the dynamical properties of nonlinear systems, photonic crystals, and meta-materials. He has over 150 publications spread amongst these various topics. Since 1981, he has taught physics for 38 years at Western Michigan University, where he is currently a Professor Emeritus of Physics and a WMU Distinguished Faculty Scholar. A number of PhD students have graduated from Western Michigan University under his supervision.

An Introduction to Condensed Matter Physics for the Nanosciences

Arthur McGurn

CRC Press is an imprint of the
Taylor & Francis Group, an **informa** business

First edition published 2023
by CRC Press
4 Park Square, Milton Park, Abingdon, Oxon, OX14 4RN

and by CRC Press
6000 Broken Sound Parkway NW, Suite 300, Boca Raton, FL 33487-2742

CRC Press is an imprint of Informa UK Limited

ISBN: 9780367466473 (hbk)
ISBN: 9780367517090 (pbk)
ISBN: 9781003031987 (ebk)

DOI: 10.1201/9781003031987

Typeset in Times
by codeMantra

Contents

Preface........xi

Chapter 1 Introduction........1

1.1 Electrical Properties........2
1.2 Optical Transport and the Interaction of Light with Matter........4
1.3 Electrons in a Variety of Dimensions........4
1.4 Semiconductors........5
1.5 The Landauer Approach to Conductivity........6
1.6 Photonic Crystals and Metamaterials........7
1.7 Quantum Optics........7
1.8 Anderson Localization and Mott Localization........8
1.9 Quantum Hall Effect........9
1.10 Phenomena Related to the Hall Effect........10
1.11 Correlated Electron Systems........10
1.11.1 Superconductivity........10
1.12 Josephson Junctions........11
1.13 Fractional Quantum Hall Effect........12
1.14 Coulomb Blockade........12
1.15 Resonance........13
1.16 Scaling and Renormalization Group........13
References........15

Chapter 2 Conductivity........17

2.1 Basic Ideas of Conductivity........17
2.2 Quantum Effects........20
2.3 Magnetic Field Effects........25
2.3.1 Classical Treatment of a Two-Dimensional Gas in a Harmonic Confining Potential........28
2.3.2 Orbital and Landau Diamagnetism: Magnetism of a Two-Dimensional Fermi Gas........29
2.3.3 Magnetic Properties and Landau Diamagnetism of a Slab of Fermi Gas........33
2.3.4 Pauli Paramagnetism........34
2.4 Stoner Theory of Permanent Magnetism of Metals........36
2.5 Localization Properties of the Fermi Gas Model........40
2.5.1 Periodic Potential........45
2.5.2 Modes in the Anderson Localized Gas........46
2.5.3 Phase Coherence in Localization........48
2.5.4 Dependence of Localization on the System Dimension........50
2.5.5 Hopping and Variable-Range Hopping Conductivity........52
2.5.6 Ioffe–Regel Criterion and Minimum Metallic Conductivity........56
2.6 Mott Transition........57
2.7 Wigner Crystal........58
2.8 Superconductivity........58
2.8.1 Planar Interface between a Superconductor and Normal Metal........61

2.8.2 Flux Quantization ... 63
2.9 Ahronov–Bohm Effect in Normal Metals ... 65
References ... 66

Chapter 3 Conductivity: Another View ... 69

3.1 The Landauer Formulation ... 69
3.2 Scattering within the Waveguide: Ohmic Limit ... 73
3.3 Landauer–Buttiker Formula ... 77
3.4 Universal Conductance Fluctuations ... 82
3.5 Nonzero Temperature ... 85
References ... 87

Chapter 4 Properties of Periodic Media ... 89

4.1 Tight-Binding Model ... 91
4.1.1 A One-Dimensional Model of a Chain of Atoms ... 93
4.1.2 A Two-Dimensional Tight-Binding Model of Graphene ... 97
4.1.2.1 The Tight-Binding Hamiltonian of Graphene ... 103
4.1.2.2 Dispersive Properties of the Excitations in Graphene ... 106
4.1.3 Graphene Conductivity ... 107
4.1.4 Graphene Nanotubes ... 108
4.1.5 Some of the Interesting Properties of Graphene ... 109
4.2 Quantum Dots, Quantum Wells, and Quantum Wires ... 110
4.2.1 Properties of $GaAs$ and $Ga_{1-x}Al_xAs$ ... 111
4.2.2 Effective Mass Approximation ... 112
4.2.3 Envelope Function Approximation ... 113
4.2.4 Quantum Wells and Heterostructures ... 115
4.2.4.1 Quantum Wells ... 115
4.2.4.2 Quantum Heterostructures ... 118
4.2.5 Quantum Wires and Dots ... 121
4.3 Excitons ... 123
4.4 Photonic and Phononic Crystals ... 125
4.4.1 Properties of Waves in Periodic Media ... 126
4.4.2 Examples of Periodic Media in One and Two Dimensions ... 128
4.4.2.1 Examples of Two-Dimensional Photonic Crystals ... 129
4.4.2.2 Two-Dimensional Semiconducting Photonic Crystal ... 129
4.4.3 Applications of Photonic and Phononic Crystals ... 134
References ... 135

Chapter 5 Basic Properties of Light and Its Interactions with Matter ... 137

5.1 Quantized Electromagnetic Waves ... 137
5.1.1 General Form of 3D Quantized Electromagnetic Waves ... 144
5.1.2 Cavity Modes ... 145
5.1.3 Coherent States ... 147
5.2 Field Interactions with Atoms and Electrons ... 151
5.2.1 Jaynes–Cumming Model ... 151
5.2.2 Jaynes–Cumming Model: Example of Fock States ... 154
5.2.3 Jaynes–Cumming Model: Example of Coherent States ... 156

5.2.4 Jaynes–Cumming Model: Temperature Effects 156
5.3 Optical Correlations and Coherence 157
References 164

Chapter 6 Basic Properties of Lasers, Masers, and Spasers 165
6.1 Stimulated Emission 166
6.1.1 Deviations of the System from Thermal Equilibrium 168
6.2 Rate Equation Model of Laser Operations 169
6.3 Resonator Cavity 171
6.4 Maser 173
6.4.1 The Model 173
6.4.2 The Hamiltonian: Absence of Maser Fields 174
6.4.3 Density Matrix 176
6.4.4 Hamiltonian: Phenomenological Dissipative Terms 177
6.4.5 Development of the Solutions with Dissipative Effects 179
6.4.6 Density Matrix of the Maser 180
6.4.7 The General Atomic Passage Maser Processes 183
6.4.8 The Loss Terms in the Maser Processes 184
6.4.9 Statistical Properties of Maser Radiation 187
6.5 Spasers and Atom Lasers 188
References 189

Chapter 7 Semiconductor Junctions 191
7.1 Semiconductor Model 192
7.1.1 Thermal Occupancy 193
7.1.2 Extrinsic Semiconductors: *n*- and *p*- Type Materials 194
7.1.3 Positioning of the Chemical Potential 194
7.2 Semiconductor Junction Model 195
7.2.1 Electrostatics at the Junction 197
7.2.2 Application of a Potential across the Junction 199
7.2.3 Resulting Current versus Voltage Relationship of the Junction 201
References 202

Chapter 8 Rectifiers and Transistors 203
8.1 Rectifiers and Transistors 204
8.1.1 The Transition Region in a *p–n* Junction 205
8.1.2 The Transition Region Characteristics 206
8.2 Field Effect Transistor 208
8.2.1 *p–n–p* Transistor 209
8.2.2 Basic Transistor Circuit 210
8.2.3 Geometry of the Subregion of Net Positive Charge within the *n*-Material Layer 211
8.2.4 Connected Region of Charge-Neutral *n*-Material between the Source and Drain 212
8.2.5 Disconnected Region of Charge-Neutral *n*-Material between the Source and Drain 215
8.2.6 Conditions of Zero Drain Current 217

8.2.7 Switches and Amplifier Circuits ... 218
8.2.8 *n–p–n* Transistors ... 220
8.3 Bipolar Transistor ... 220
References ... 224

Chapter 9 Toward Single-Electron Transistors: Coulomb Blockade ... 225
9.1 A Single Island Device ... 227
9.2 Single-Electron Transistor ... 230
9.2.1 Electron Transitions between the Source and the Island ... 233
9.2.2 Electron Transitions between the Drain and the Island ... 235
9.2.3 Stability of *N* Net Uncompensated Electrons on the Island ... 236
9.3 Applications of Single-Electron Transistors ... 238
References ... 238

Chapter 10 Quantum Hall Effect ... 241
10.1 Two Quantum Hall Effects ... 242
10.1.1 Integer Quantum Hall Effect ... 242
10.1.2 Fractional Quantum Hall Effects ... 244
10.2 Classical Model of the Hall Effect ... 245
10.3 Theory of the Integer Quantum Hall Effect ... 247
10.3.1 Transverse Conductivity ... 249
10.3.1.1 Another Approach to the Conductivity ... 250
10.3.2 Shubnikov-de Hass Effect ... 252
10.4 Fractional Quantum Hall Effect ... 254
10.4.1 Free Electrons in High Fields ... 254
10.4.2 Wave Function for the Fractional Quantum Hall Effect ... 258
10.4.3 The Electron–Electron Interactions ... 260
10.4.4 Conductivity and Resistivity of the $\frac{1}{3}$ Fractional Hall State ... 261
10.4.5 Conclusions ... 262
10.5 Spin ... 263
10.6 Spin Hall Effect and the Spin Hall Effect of Light ... 264
References ... 265

Chapter 11 Resonance Properties ... 267
11.1 Metamaterial Responses and Simple Spatial Resonances ... 268
11.2 Standard Resonance Involving Quantum Wells ... 270
11.3 Fano Resonance Involving Quantum Wells ... 275
11.4 Topological Excitations ... 279
References ... 280

Chapter 12 Josephson Junction Properties and Basic Applications ... 281
12.1 Time-Dependent Ginzburg–Landau Free Energy ... 281
12.1.1 Schrodinger Gauge Symmetry ... 282
12.1.2 Ginzburg–Landau Form ... 283
12.1.2.1 An Example ... 284
12.2 Josephson Junction ... 285
12.3 Spatial Dependent Effects: Magnetic Fields ... 288

12.4 Josephson Junction in a Static Magnetic Field ... 289
12.5 Real-World versus Ideal Josephson Junction ... 290
12.6 Josephson Junctions of Finite Cross-Sectional Area ... 292
12.6.1 Two Junctions in Parallel ... 293
12.7 Type I and Type II Superconductors and Interfaces with Normal Metals ... 295
12.8 High-Temperature Superconductors ... 301
References ... 304

Chapter 13 Scaling and Renormalization ... 305

13.1 One-Dimensional Ferromagnet ... 307
13.2 Second-Order Phase Transitions ... 310
13.3 First-Order Phase Transitions ... 313
13.4 Scaling Theory ... 315
13.4.1 Examples of the Two-Dimensional Ising Model and Landau Theory ... 319
13.4.2 Natural Length Scales ... 321
13.5 Renormalization Group Approach ... 322
13.6 Application of the Renormalization Group to the Landau Formulation ... 332
Appendix ... 336
References ... 337

Index ... 339

Preface

This is an introductory text on topics of condensed matter physics with a focus towards device development in the nanosciences. It is meant to give a basic treatment of materials that have been formulated in the theory of condensed matter physics over the past 60–70 years. Consequently, a wide variety of subjects are included in electronics, optics, and acoustics, and to a certain extent, the topics chosen are essential in developing a theoretical background for working on nanoscience devices. All the materials are approached at a level accessible to advanced undergraduate students and are based on the simplest mathematical treatments. In this regard, to facilitate the student, the text is fashioned to be self-contained, with ample references to more advanced literature.

In the recent years, a focus of the physical sciences has been on the development of miniaturized devices based on advancing technologies in electronics, optics, and acoustics. Many of the techniques employed in these studies result from unusual processes for the transport of electrons, photons, and phonons available at small length scales and in systems of lower dimensionality. These processes may involve media either designed on small length scales in which the phase properties of electrons, photons, or phonon significantly determine the transport or, alternatively, on processes involving the highly correlated motion of single-particle excitations. In this regard, advances have been made in the applications of ballistic transport in place of the more traditional mechanisms of diffusive transport and in the use of highly correlated motions of interacting particles to manipulate single-particle flow within nano systems.

In ballistic transport, the excitations of a system move effectively with little or no scattering through a conducting medium, so that a small number of transmission and reflection processes dominate the motion in the system. On the other hand, in more traditional diffusive transport processes, excitations encounter a high degree of random collision within a disordered medium, and this greatly determines the nature of the transport. The difference between these two types of transport is a focus of our study, where it is shown to offer new directions in engineering applications of materials. Consequently, these differences are of essential importance in the study of nano devices.

In addition, as systems become small, the interaction of individual electrons through the coulomb force has potential to dominate the determination of the electron motion. This high correlation of electron motions gives rise to the coulomb blockade effect, which is important in the development of single electron transistors as well as in electronic formulations based on circuit assemblies from individual molecules. In this regard, the ability to manipulate single particles through nanocircuits is found to be an essential new feature offered in nano device technologies.

One important class of systems in device technology focused upon in our treatment are those displaying periodicity on meso- or macroscopic length scales. Examples of such materials are photonic crystals (in optical transport), phononic crystals (in acoustic transport), and electronic heterostructures (in electronic transport). Each of these involves periodic arrays of optical, acoustic, or electron media, which allow for a ballistic transport exhibiting the band structure filtering characteristics of electron motion in crystalline atomic materials. They all function due to the diffractive effects of the periodicity, which are engineered into the materials, offering transport responses in which certain frequencies of excitations are passed through the medium, while others are not. In addition, these frequency responses can be designed into the system by controlling the periodicity of the artificial media. The notions of band structure and its filtering applications form the basis of the ideas in the design of mesoscopic crystalline materials, and their applications as waveguides and resonators.

A second engineering technique that is considered involves the design of artificial materials known as meta-materials. These are developed as homogeneous media composed as arrays of mesoscopic resonant features, and are most often developed for optical or acoustic applications. In these materials, mesoscopic features exhibiting frequency resonances are embedded in a background medium, and the resulting artificial medium offers another important component of design technology. Such frequency resonances may be used in the design of homogeneous meta-material media exhibiting new refractive properties or as media exhibiting a continuous change of refractive index. Ultimately, these properties have led to applications in negative refractive index materials, cloaking devices, and in optical and plasmonic systems based on perfect lenses. Both artificial periodicity and the idea of frequency resonance have been applied as important aspects in the engineering designs of novel modern materials. Consequently, new types of macroscopic crystals and new types of homogeneous materials extend the ideas of condensed matter physics in alternative technological directions.

In line with the discussions of meso- and macroscopic periodicity, a treatment is given of quantum wells, quantum dots, and quantum heterostructures. These are, respectively, formed as slab inclusions, bounded inclusions, and layered arrays of media embedded into an otherwise uniform bulk electronic medium. Some of the basic properties of these electronic structures are presented within the context of the electron gas model that treats the conduction electrons as a gas of noninteracting quantum particles. In this regard, an important feature of these artificial structures is their ability to allow for the engineering of various electronic energy-level schemes. This can be done so that the interaction of light with the election transitions between the engineered levels has various important characteristics, leading to a variety of applications in laser and light display designs. Details of the applications of wells, dots, and heterostructure engineered materials are reviewed at an introductory level in the text, and references are given to more advanced topics for further study.

Returning to a consideration of atomic crystalline band structures, the electron orbital solutions of a simple model for the electronic band structure of graphene are presented. Graphene is a two-dimensional planar structure formed of carbon atoms that has become of great interest in the current research. The electron motion in graphene can be treated in the format of an elementary tight-binding model. An interesting feature of the model is that the conduction electrons exhibit a linear dispersion relation, so that some aspects of the electron motion mimic the behavior of light. Graphene also has unusual tensile properties that have made it of much recent discussion, and its structure can also be reformulated as nanotubules and buckyballs, which provide interesting applications in molecular electronics. All of these novel structures are currently of great significance to nanoscience studies.

Another topic in regard of diffusive transport is the mechanism of the metal–insulator transition known as the Anderson localization. At the Anderson transition, spatial disorder in the media causes the electron modes to become localized in space. The transition from metal to insulator is mediated by the degree of random disorder introduced into a system, and effects arising from the disorder have important dependences on the dimensionality of the medium being considered. Localization is an example of phase coherent modifications of diffusive transport and may be present in all systems with wavelike excitations. In this regard, it arises solely from phase coherence and is not a quantum effect. As shall be seen, localization is important component in the understanding of the quantum Hall effect, which is another phenomenon that has a great technological interest in the nanosciences.

With the onset of miniaturization in a system, the importance of coulomb interactions between the conduction electrons of the media can become important. Specifically, electron correlations arising from electron–electron interactions at some point begin to modulate the properties of the media. In macroscopic materials, an example of the importance of the coulomb interactions between electrons is the metal–insulator transition known as the Mott transition. This a transition in which the electrons become confined to finite spatial regions due to the coulomb interactions between electrons. (Note that the Mott transition is quite different from the Anderson transition, which arises from phase coherence.) An extreme limit of the Mott transition is the condensation of the interacting electrons of a Fermi liquid (i.e., an interacting fluid of electrons) into the form of a crystalline solid of electrons. This is known as a Wigner solid and represents an insulating state of the electrons. As noted earlier, similar types of effects are observed at nanoscales in the form of the so-called coulomb blockade. Here tunnel junctions created between an island and outside probe potentials are used to develop and manipulate charges on the island. The manipulation of the island charge is made by means of the coulomb interactions of the electrons on the island and their interactions with the potentials applied to the island. The resulting system operates at nanoscales as a type of single-electron transistor or turnstile controlling the motion of individual electrons through the system. This is an illustration of how interactions between individual electrons are becoming increasingly important and have applications on nanoscale dynamics.

In lower dimensional systems, interesting responses of nano systems have been observed. In the transport properties of two-dimensional electron gases, the quantum Hall effect has provided for a variety of important applications. Specifically, the quantum Hall effect involves the resistivity properties of the gas in the presence of an external magnetic field applied perpendicular to the plane of the gas. These effects include the integer quantum Hall effect, arising from the series of Landau levels characterizing the highly degenerate energy states of the electron orbitals of the two-dimensional electron gas. The energy states result from a magnetic field applied perpendicular to the plane of the gas and rely heavily on the independent (noninteracting) electron gas model. On the other hand, another Hall effect known as the fractional quantum Hall effect comes from the correlated electron motions introduced by the electron–electron interactions in the system. These interactions generate a further set of degenerate energy states. In this regard, the fractional quantum Hall effect presents a weak modification of the integer quantum Hall effect. A chapter on the consideration of these features and some of their applications is given as well as other discussions of interference effects in the two-dimensional electron gas in the absence of applied magnetic fields.

A brief introduction to some of the topics of quantum optics has been included in the text. This is an area of considerable importance in nanoscience and is of application in, for example, quantum computing and information science. In our discussions, an initial focus is on the basic ideas of optical coherence with a consideration of the nature of Fock and coherent states. A treatment of the interaction of light with matter is then presented, including a consideration of the Jaynes–Cumming model and its solutions. Some discussion is also given on the basic theory of lasers and the statistical properties of laser light that are shown to be closely related to the coherent states. The theory of masers and spasers are, in addition, developed in some detail. Finally, mention is made of optical analogies of electronic transport properties that can be found in certain types of optical materials.

For completeness, the text also presents a brief introduction to high- and low-temperature superconductivity, Josephson junctions, and the basic ideas of the field effect and bipolar transistors. These are standard systems of applications in technology and are topics necessarily included in an introduction for students. Some of the basic properties and applications of these subjects are presented along with references to more comprehensive treatments.

In addition, a final chapter is provided on critical phenomena, scaling, and the renormalization group. These topics contain important mathematical techniques for handling singularities found in materials undergoing phase transitions. They represent an interesting aspect of the physics of materials that has been developed in the last 60 years. A little understanding of these theories should be required of all students of condensed matter physics.

The focus of the presentation is theory and the development of the simplest basic models that relate the ideas at the basis of phenomena treated in current technological applications. It is endeavored to show how the great complexity of the many-body systems operating in all the phenomena discussed is simplified into a set of a small number of theoretical models. Each model then approximately treats the physics of the topic it is developed to describe. This, of course, requires various degrees of assumption and simplifying approximations. The approach is to use the simplest form of the theories that still relate reliable ideas of what is going on in the field. Where needed, a review or brief introduction is given to some of the more advanced theoretical techniques. This is done in the hope that the text is self-contained so that the reader need not constantly refer to outside books. In this regard, the text is meant for upper-level undergraduates and first-year graduate students.

Rancho Mirage, California, May, 2022.

1 Introduction

This text presents an introduction to some of the theoretical ideas that are of use in understanding the fundamental properties of nanoscale systems. The approach emphasizes theory and the development of elementary models of physical phenomena with a focus on generating a basic general understanding of the mechanisms at play in systems of current technological interest. In this regard, the preference of our approach is on analytical techniques applied to simplified models, rather than highly technical computer simulation studies applied to more detailed formulations of problems. Such analytical approaches are helpful with accounting for the broad physical principles at play in a physical phenomenon. This understanding, however, is often made at the sacrifice of reproducing the quantitative details of the physics that goes beyond the basic ideas of the fundamental operating principles.

The alternative to the analytical approach focuses on computer simulation methods that are commonly applied to perfect the design of efficient devices. Such computer simulation approaches certainly aid in developing the detailed quantitative properties and the nuances of particular many-body systems of particles and fields. They do not always offer insight into the underlying physics, and in this sense, simulations are of engineering and not of physical interest. In the simulation approach, the general understanding of a physical process is frequently lost in the mass of data generated by a computer. Consequently, the data is often of specific interest only to a particular system treated and does not always lead to an elegant development of the general physical mechanisms at play. In this book, the focus is on physics breadth rather than engineering depth, and on the simplicity of the development of physical ideas and not on the methodology available to finesse the quality of device operation. Consequently, the interested reader is then referred to the vast readily available literature on simulation methodology that exists outside this text.

In many-body physics, we are confronted with a microscopic system in which large numbers of degrees of freedom are involved in complex interactions. However, on the level of the macroscopic system composed from these microscopic elements, one often finds a highly correlated macroscopic dynamic. The problem of the physicist is to summarize the microscopic features of the system dynamics into a macroscopic model of dynamics that allows for the understanding of the physics of the macroscopic system, i.e., the properties of the macroscopic world must be generated from those of the microscopic world. In this sense, for example, an ocean wave is easier to understand in terms of fluid mechanics rather than in terms of the detailed motion of individual water molecules, or the physics of a superconductor device is best represented in terms of a thermodynamic model of the highly correlated macroscopic motion of its electrons than by a consideration of the detailed trajectories of its individual electrons. In this regard, the connections between the microscopic and macroscopic worlds are of fundamental importance in the development of technology. Such an evolution from the detailed microscopic dynamics to the understanding of a highly correlation macroscopic motion in a many-body system involves the formulation of simplified models exhibiting the gross properties of the material under consideration. This process is the goal of the text.

These general ideas are presented by focusing on a series of highly successful models that have accounted for the macroscopic physics of systems of importance to current technology. In the course of the study, the ideas introduced include a variety of engineering applications in semiconductor physics [1–3], the transport properties of lower dimensional systems [4–10], the treatment of diffusive and ballistic conductivity [5–10], hetero- and homojunctions [11], various resonant properties of materials and devices [5–10], the properties of high- and low-temperature superconducting materials and junctions [12–16], the formulation of metamaterials [17–20], and the properties and applications of systems operating on the principles of coulomb blockade [5–7]. In addition, some basic features of quantum optics [21–25] are introduced that are of interest in the manipulation of light,

DOI: 10.1201/9781003031987-1

and in fields such as quantum computing and information [20]. A discussion is also included on the nature of the photon statistics in laser light [21–25]. An introduction to semiconductor junctions and their basic devices is developed with a focus on understanding the principles of diode and transistor operation [1–10,26,27]. Superconducting devices are discussed, in general, with a characterization of their properties of phase coherence as applied to technology [2,12–16]. To enhance the students' background on the whole, a variety of the recently formulated basic treatments of phase transitions, the singularities associated with them, and the elementary ideas of renormalization group theory are also introduced [28–35]. The emphasis throughout the development is on the summary of a number of the high points in device modeling physics that have occurred over the last 60 years.

In this introductory chapter, a summary is given of some of the important physical phenomena considered in the text, with a qualitative review presented of the most useful models and results arising from their discussions. These have been the focus of current research in nanoscience, and their study facilitates a basic understanding of current devices and their technologies.

1.1 ELECTRICAL PROPERTIES

We begin with the study of electronic properties, starting from the earliest models developed for explaining conductivity [1–5]. In electronics, the first treatment of electron transport was the electron gas or Drude model of conductivity. Initially, it was based on a classical physics treatment, and later quantum mechanics was introduced into its formulation. The electron gas model is set to provide an understanding of both the DC and AC conductivity properties of metals and semiconductors. In the model, the conduction electrons are depicted as a uniform density gas of independent particles that do not interact with one another. This gas moves through a uniform viscous background of neutralizing positive charge so that the conductor formed of the gas and its background is charge neutral. The electrons of the gas, however, do interact with external electromagnetic fields and with the background through which they propagate. In this sense, the background is a scattering medium and is often represented as a viscous medium in which the electron motion is characterized by a velocity-dependent relaxation force. The resulting model of conductivity is usually first encountered in sophomore electronics classes where the mathematics of the motion of a single electron in the viscous medium is closely related to that of the terminal velocity problem of an object falling in the viscous atmosphere near the surface of the earth. It is highly successful as a basic semi-quantitative model of metal and semiconductor conductivity.

In more advanced treatments, quantum mechanics is introduced into the problem along with a detailed consideration of the electron scattering from impurities [1–5]. In this regard, often the neutralizing background of uniform positive charge density is replaced by a model of the periodic background array of positive ions remaining after the formation of the conduction electron gas. The new feature of the periodic background introduces an important scattering effect into the model, allowing us to distinguish between materials which are metals, semiconductors, and insulators. This distinction arises due to the phase carried by the quantum mechanical wavefunctions of the electron gas, causing them to Bragg scattering from the periodic array of positive ions.

The introduction of periodicity into the problem develops a frequency band structure in the electronic dispersion relations. (A schematic example of such a dispersion relation is presented in Figure 1.1a for a simplistic model of electrons in a periodic polymer chain.) In the band structure, the electron frequency modes are grouped into various stop and pass frequency bands of the system, and as a result, the properties of the material are fundamentally altered from the properties of electrons moving in a uniform background medium. Now only electrons with frequencies in the so-called pass bands can flow through the material, and electrons with frequencies outside the pass bands are in stop bands that do not allow their propagation through the material.

In the band structure scheme, the Pauli exclusion principle [1–3] enters as a fundamental restriction on how the electronic modes of the gas are occupied and is a basic determinator of whether the material is a metal, semiconductor, or insulator. Specifically, the Pauli principle limits the number

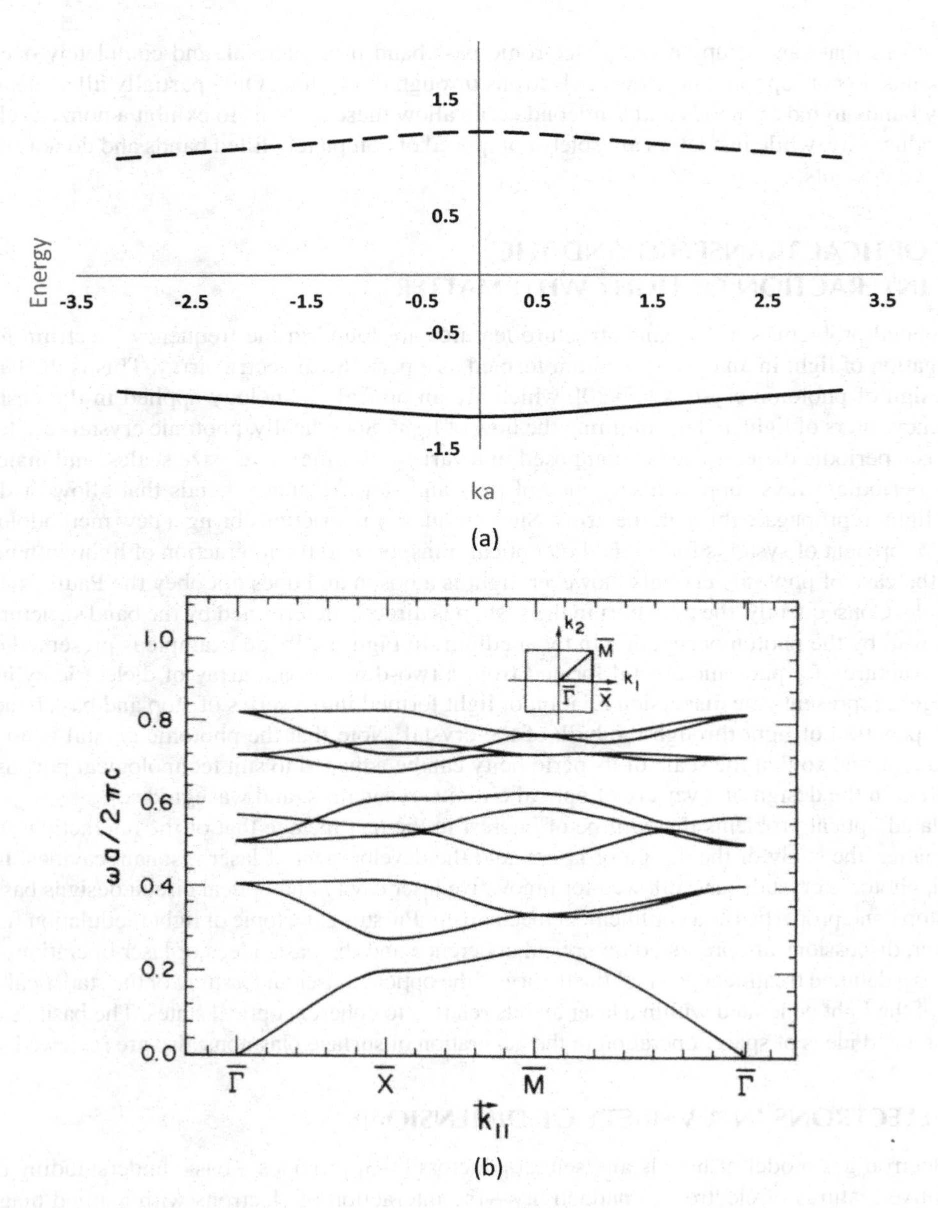

FIGURE 1.1 (a) Schematic dispersion relation for a simple electron band structure of a linear chain of atoms such as that studied in polymer physics. Specifically, the results are for a tight binding model of independent electrons hopping along an infinite chain of alternating *A*- and *B*-type atoms. The electrons move in two separate energy bands with energies given by $E = 1/2\left\{\varepsilon_A + \varepsilon_B \pm \sqrt{\left((\varepsilon_A - \varepsilon_B)^2\, 2 + 8t^2\left(1 + \cos fka\right)\right)}\right\}$, where ϵ_A (ϵ_B) is the site energy of an electron on an *A* (*B*) atom site, *t* is the nearest neighbor hopping energy between *A* and *B* atoms along the chain, *k* is the wavevector of the electron, and *a* is the nearest neighbor separation of the *A* (*B*) atoms along the chain. The plot is of E/ϵ_A versus *ka* for a chain in which $\epsilon_A = -\epsilon_B$ and $8\,t^2/(\epsilon_A^2) = 1$. The two energy bands, known as pass bands, are seen to be separated by an energy stop band region in which electrons do not propagate. The electrons in the chain only propagate in the dispersive energy pass bands. (b) Plot of the optical band structure of a two-dimensional photonic crystal defined on a square lattice. The plot is of frequency versus wavevector along several spatial axes, and the geometry of the wavevector space is indicated in the inset. Again, the dispersion exhibits pass bands in which the light propagates through the bulk of the crystal and stop bands in which the light does not propagate through the bulk of the crystal. (Reprinted with permission from Ref. [36], *Optical Society of America*. Note that both of these models will be treated in greater detail later in the text.)

of electrons that can occupy a given electronic pass band in a material, and completely occupied pass bands do not support a net flow of electrons through the system. Only partially filled electronic energy bands found in metals and semiconductors allow these systems to exhibit a nonzero electrical conductivity, while insulators are solely composed of completely filled bands and do not support electrical currents.

1.2 OPTICAL TRANSPORT AND THE INTERACTION OF LIGHT WITH MATTER

In a related problem, similar band structure features are found in the frequency spectrum for the propagation of light in an optical medium formed as a periodic dielectric array. This is the basis of the design of photonic crystals [17–20], which are an optical technology applied in the design of frequency filters of light and in confining the flow of light. Specifically, photonic crystals are formulated as a periodic dielectric array composed in a variety of dimensions, size scales, and materials. These periodic arrays support a sequence of pass and stop frequency bands that allow or do not allow light to propagate through the array. Such frequency restrictions bring a new methodology to the development of systems formulated on optical transport and the interaction of light with matter.

In the case of photonic crystals, however, light is a boson and does not obey the Pauli exclusion principle. Consequently, the transport in the system is directly determined by the band structure and not as well by the photon occupation in the medium. In Figure 1.1b, an example is presented of the band structure of a photonic crystal formed from a two-dimensional array of dielectric cylinders. The figure represents the dispersion relation of light formed into a series of stop and pass bands for the propagation of light through the bulk of the crystal. Note that the photonic crystal is an engineered material so that the scale of its periodicity can be adjusted to suit technological purposes. It is applied in the design of a variety of optical circuits, resonators, and waveguides.

Related optical problems that will be of interest in the text include that of the interaction of light with matter, the study of the design of lasers, and the development of laser resonant cavities. In this regard, photonic crystals have allowed for innovative laser cavity and optical circuit designs based on their stop band properties as a confinement mechanism. Pursuing the topic of light modulation further, however, discussions are presented on optical coherence and the basic ideas of laser operations. This includes a detailed treatment given of the theory of the optical maser and a study of the statistical properties of the light generated within a laser and its relation to coherent optical states. The basic features of the related ideas of spaser operation in the generation of surface plasmons also are reviewed.

1.3 ELECTRONS IN A VARIETY OF DIMENSIONS

The electron gas model of metals and semiconductors [1–3] provides a basic understanding of the qualitative features of electronic conductivity—the interaction of electrons with applied magnetic fields and with electromagnetic radiation. Specifically, the model is discussed in the context of conductivity problems involving one-, two-, and three-dimensional systems for both externally applied electric and magnetic fields [4–10] in the text. A variety of other transport properties also arise in the context of the independent motion of electrons in materials, including its applications in the study of quantum well systems and quantum dots, as well as in the treatment of the integer quantum Hall effect [37,38], tunnel junctions [5–11], and transistors [26,27]. Then, all the fundamental principles of these variety of electron effects are understood within the context of different aspects of the highly simplified electron gas model.

In addition, some discussions of Anderson localization [1,2,17–20] of the electron gas are also presented in the context of the metal–insulator transition arising from random disorder. This involves the transition of electron wavefunctions from being extended states developed throughout a random medium to being localized modes in which the wavefunctions are bounded and confined to a localized region in space. The transition between these two types of modes is found to depend

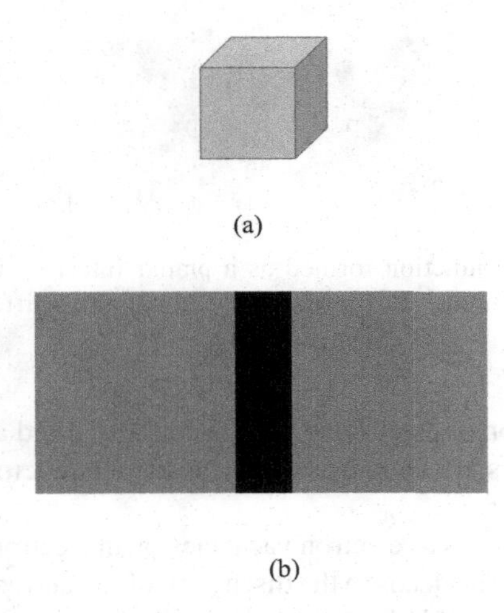

FIGURE 1.2 Schematic of: (a) A cubic quantum dot in a bulk background. (b) A quantum well formed as a layered medium composed as three infinite slabs of different materials. The two outer layers are semi-infinite in space occupancy, and the middle slab is an infinite slab of finite horizontal width.

on the degree of random disorder in the medium and is responsible for a metal–insulator transition driven by the degree of randomness in the system. The Anderson transition, which is an important consideration in systems of its own right, is also shown to be an important component of the theory explaining the quantum Hall effects.

Important applications of the electron gas model are found in the technology of quantum wells and dots [11]. (See Figure 1.2 for schematic examples of these features.) The structures involved are mesoscopic or macroscopic layers of different electronic materials in the case of quantum wells, and the focus is on the effect of the layers on the electron motion. The system shown in the figure represents one or more quantum barriers that are engineered in a material to modulate or trap the motion of electrons. An interest is in the technological applications of the electronic dispersion relations created within the layering. These prove helpful, for example, in the design of systems for lasers, switches, and the generation and detection of light.

On the other hand, quantum dots are mesoscopic or macroscopic bounded regions of one type of material introduced into another. The dot acts to confine electrons to a localized region of space. In this regard, a quantum dot is a bounded mesoscopic region of material that can trap an electron or electrons in a way similar to the binding of electrons by an atom. The resultant system can be made to act spectroscopically as a mesoscopic atom with various engineered optical transitions functioning as optical sources. Both quantum well and dot structures are widely applied in optical device technology.

1.4 SEMICONDUCTORS

Semiconductors are conductive materials that have totally filled pass bands and behave as insulators at zero temperature. At nonzero temperatures, however, some of their electrons are thermally promoted from the totally filled pass bands to higher energy unfilled pass bands. Consequently, the material exhibits a conductivity at nonzero temperatures and for small stop bands, which increases with increasing temperature. These types of pure semiconductors are known as intrinsic semiconductors in the absence of chemical impurities. Similarly, bound energy levels can be created in the stop bands with the addition of chemical impurities to a pure semiconduction. These are associated with the impurity ions and can contribute to the conductivity at nonzero temperatures through the ionization properties of the added impurities. This conductivity from impurity states occurs in

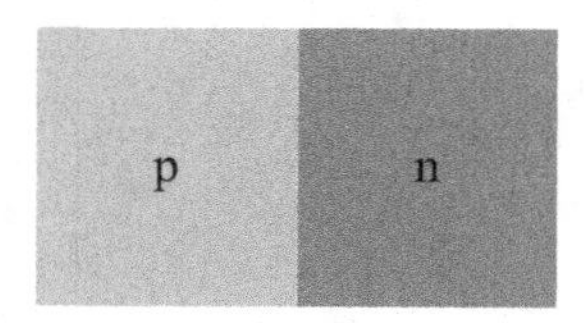

FIGURE 1.3 Schematic of a p–n junction formed as a planar interface between a semi-infinite p-type material and a semi-infinite n-type material. In this system, a potential difference is applied perpendicular to the interface.

addition to the thermal promotion of electrons between initially filled and unfilled pass bands. Such types of impurity semiconductors are known as extrinsic semiconductors, and may be classified into n-type and p-type semiconductors.

The ideas of positive charged holes as electron vacancies in an electron pass band are introduced for the treatment of semiconductors. This leads to the discussion of the current density of a material viewed as being the sum of both electron and hole currents [1–3]. In this picture, a negative charged electron that has been removed from an energy pass band leaves behind a positive charged hole as the electron transitions to another energy state of the band structure. (Note that under this operation, the system retains its charge neutrality.) Holes are also introduced into a semiconductor by introducing impurities into the material, i.e., the material can be doped with impurity atoms. Impurities introduced into the medium often bind electrons or holes to them, creating p-type or n-type semiconductors in which, respectively, the current is primarily due to holes or electrons. In pure or impure semiconductors, the conductivity is carried by negative (electrons), positive (holes), or both types of charge carriers.

The introduction of impurities into semiconductors gives rise to two important classes of semiconductors. Semiconductor materials in which the current flow is due predominately to electrons are known as n-type materials, whereas materials in which the current flow is due predominately to holes are known as p-type materials [1–4]. These types of adjustments in the nature of the charge carriers are made through the introduction of impurities into pure semiconducting materials, which become ionized impurity sites in the host semiconducting medium. This process is known as doping and represent a fundamental feature in the design of an important device known as a semiconductor junction. (See Figure 1.3 for a schematic of a p–n junction.)

In its basic form, the junction is an interface between a p-type and an n-type material. At the interface of the two different media, various electric potentials develop that, in the presence of a potential applied across the junction, are used to regulate the flow of electrons and holes across the junction. The modeling of the operation of such a junction [1–4] is essential to studying a variety of diode and transistor applications, and a discussion of such junction-based device applications is presented later. In particular, the basic operation of both the bipolar junction transistor and field effect transistor are a focus in the text.

1.5 THE LANDAUER APPROACH TO CONDUCTIVITY

Another approach to electron conductivity considered here is the Landauer approach. This is a model that is often used to discuss nanodevices in which the electron transport is of a ballistic rather than a diffusive nature. In ballistic transport, the electrons travel through a device with little or no scattering, and the dynamics is similar to that of the flow of electromagnetic waves through a lossless waveguide. This is quite different from electron propagation in a diffusive, highly random scattering medium in which phase coherent electron scattering is absent. While the Drude electron gas model focuses on the transport medium as a viscous or diffusive medium, the Landauer approach [5–10] treats the electrons as moving in a type of waveguide. In this regard, the Drude [1–4] and Landauer [5–10] approaches, respectively, treat the conductivity problem by starting from the oppositions' limits of diffusive and ballistic transport.

In the Landauer picture, impurities present in the system are characterized as tunneling barriers, which are represented by a reflection and a transmission coefficient. The scattering in the Landauer approach then retains phase coherence throughout the system, as the scatterers are few and their positions are well defined. However, in the limit of a system containing many scattering impurities, the net transmission through the waveguide tends to lose its scattering phase coherence. In this limit, the diffusive electron gas model is obtained. For many nanodevice applications, involving few impurities and short passages, the Landauer approach is more suitable than the diffusive picture of the Drude model. Systems in which the Landauer method applies are found to exhibit many types of interesting phase coherent effects.

A related phenomena to systems described by the Landauer method is that of universal conductance fluctuations. This involves a study of the conductance of one-dimensional conductors, which arises from scattering by randomly distributed impurities. In the limit of long samples, the conductance generated in a random sequence of samples can be shown to be distributed by a Poisson distribution. The ideas here are very similar to those employed in the study of the optical speckle found in the scattering of laser light from a random rough surface.

1.6 PHOTONIC CRYSTALS AND METAMATERIALS

Both of these ideas of ballistic and diffusive transport can also be generalized to optical systems. In this regard, optical media of the most recent importance have been those related to the development of the fields of photonic crystals and metamaterials. The photonic crystals are periodic arrays of dielectric media in one dimension (e.g., layered media), two dimensions (e.g., arrays of dielectric cylinders), or three dimensions (e.g., a stacking of dielectric cubes), exhibiting a band structure dispersion of light [17–20] for its ballistic transport. They provide an engineered series of pass and stop frequency bands in which light is allowed or not allowed to propagate through the bulk of the photonic crystal. Associated with these bands arise the designs of a variety of laser mirrors, optical waveguides, resonant cavities, optical switches, lasers, light filters, etc. While photonic crystals [17–20] modulate the properties of light with wavelengths of the order of the periodicity of the dielectric medium of the photonic crystal, metamaterials are engineered materials designed to appear as homogeneous media to the frequencies of light with which they are designed to interact. The metamaterials [17–20] rely on subwavelength optical resonance features in their formulation to create a medium with designed dielectric properties of permittivity and permeability. These features are engineered to manipulate the refractive properties as a function of frequency exhibited by the metamaterial. Metamaterials are used to extend the range of optical properties available for engineering applications. This has been helpful in developing materials characterized, for example, by negative index of refraction, applied in the formulation of perfect lenses, and entering in the design of cloak devices.

Such ideas of photonic crystals and metamaterials have also been developed in acoustics [17]. Here periodic arrays of acoustic materials form phononic crystals, which function on the principles of the diffraction of sound to create an acoustic band structure. On the other hand, phononic metamaterials are designed as subwavelength arrays of acoustic resonators, which compose special homogeneous materials exhibiting new refractive properties of sound.

1.7 QUANTUM OPTICS

In regard of device design and light manipulation, quantum optics has been of great recent interest. (See Figure 1.4 for a schematic of a quantum optics system, which will be considered in a later chapter of the text. It is composed as a two-level atom placed in an electromagnetic resonant cavity.) Some important applications in this field include schemes for quantum computation and information, laser design, the study of various entanglement properties, and in the development of encryption methods [17–20]. Consequently, an introduction is presented in the text of elementary principles

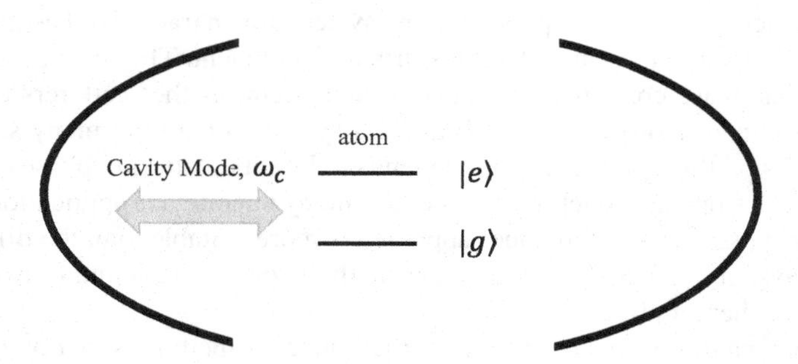

FIGURE 1.4 Schematic of the Jaynes–Cummings model of a two-energy-level atom within a resonant electromagnetic oscillator cavity. This system represents a common type of problem of interest in quantum optics. The object in its treatment is to study the interaction of the resonant electromagnetic modes of the cavity with the electronic transitions between the two energy levels of the model atom. Here $|g\rangle$ is the ground state of the atom, $|e\rangle$ is the excited state of the atom, and the cavity has a resonant electromagnetic mode at frequency ω_c.

of quantum optics along with the discussion of the interaction of light with an electron in a simple system of electron energy levels.

Such models of the interaction of electron energy levels with light have been an important recent focus in, for example, quantum computing and quantum information technology. In these schemes, the manipulation of the electron states is used as an information storage mechanism that can be configured from outside the electron system by the introduction of light pulses. The model has also found many applications in the study of lasers and the properties of laser light. In this regard, a discussion is presented of the general measurement of the statistical properties of light and in particular of the nature of the coherent light generated by a laser.

1.8 ANDERSON LOCALIZATION AND MOTT LOCALIZATION

Another important development of physics in the second part of the twentieth century is that of Anderson localization [1,2,17–20]. Anderson localization arises with the introduction of phase coherent scattering into the study of the transport properties of waves in random media. It predicts, with increasing disorder, a transition of the modes in the system from extended propagating modes to localized nonpropagating modes confined to a finite region of space. Specifically, it is shown that frequency bands of localized modal states are generated in three-dimensional randomly disordered media, and that these bands depend on the degree of disorder in the material. The localized modes have wavefunctions that are spatial confined to a bounded localized region of space and, consequently, do not participate in the zero-temperature transport of the material. At nonzero temperatures, however, the thermal fluctuations in the system yield a type of hopping conductivity between the localized states. As an interesting, important, result, it is found that all the modal states are localized at zero temperature in one- and two-dimensional randomly disordered systems. The spatial region over which the modes in these systems are localized can be very large, however, and this accounts for the conductivities that are found experimentally to be present in such lower dimensional systems in spite of the localization of the ideal modes in an infinite medium.

Anderson localization [2] arises from the wave nature of excitations in a random medium, and it is observed in electronic, phononic, and photonic systems [17–20]. In electronic systems, localization is associated with the Anderson transition, which is a metal–insulator transition associated with electron localized states. The Anderson transition may be observed, for example, in conductivity changes in a disordered system with changing pressure or the application of external magnetic fields.

This is different from the Mott metal–insulator transition [2], which originates in the Coulomb interactions between electrons and can be associated with structural changes in the lattice. The Coulomb interaction tends to spatially isolate and correlate the electrons with one another, and causes the electrons to crystalize into a Wigner solid in the extreme limit [2]. The Wigner solid that is formed is a condensed array of electrons, structured as a rigid lattice of electrons exhibiting insulator behavior.

Due to the strong Coulomb interactions between electrons, in general, a problem in electron systems is how to distinguish between Anderson and Mott localization [2]. This is less of a difficulty in optical systems where the photon–photon interactions from quantum electrodynamics are known to be very small. Consequently, optical systems have figured significantly in the development of Anderson localization.

1.9 QUANTUM HALL EFFECT

An important phenomenon that has been developed in the context of both the independent electron gas model and its Anderson localization treatment in disordered media is the integer quantum Hall effect [2,37,38]. Here a two-dimensional gas of electrons is considered in which the electrons again move as independent particles while interacting with an external magnetic field applied perpendicular to the plane of the gas. (See Figure 1.5 for a schematic of the Hall experiment.) In addition, the medium through which the electrons move is disordered, leading to the diffusive transport observed in the electron gas model in the absence of the applied magnetic field.

Considered as independent particles moving in weak disorder, the electron orbitals are the Landau solutions for a single electron moving in the plane perpendicular to the applied magnetic field. The electron solutions determined in this way are then occupied by the electrons of the gas according to the Pauli principle. Specifically, the Landau solutions occupy a sequence of highly degenerate energy states known as Landau levels, and these Landau levels are each composed of many degenerate electron modes. The generated Landau levels of the system are found to account for two important effects treated in the text. These are a diamagnetic response of the gas and the quantum Hall effect found in the conductivities of the system.

The Hall effect response of the quantum gas model is quite different from the Hall effect arising in the context of a theory based on classical mechanics and classical electrodynamics. When the dynamics of the Landau states are studied in terms of the resistivity tensor of the gas considered as a function of the applied magnetic field, the transverse Hall resistivity as a function of the applied magnetic field exhibits a series of stepped plateaus corresponding to the different occupied Landau levels of the gas. The resistivity plateaus originate in the Landau level quantum nature of the system along with the interplay of Anderson localized states that exist in the system in addition to and at energies outside of the Landau level solutions. These localized modes arise from random disorder in the system and its effect on the electron orbitals in the system.

Similarly, the diamagnetic response of the electron gas also exhibits anomalies associated with the filling of the Landau levels, as the intensity of the external magnetic field is changed. These properties again arise from the quantum nature of the solution and are not found in a classical mechanics treatment.

FIGURE 1.5 Schematic of the Hall geometry for the two-dimensional electron gas (in the plane of the page) with a perpendicularly applied magnetic field (into the plane of the page). A potential difference is applied between the right- and left-hand sides of the gas.

1.10 PHENOMENA RELATED TO THE HALL EFFECT

Some phenomena not involving the quantum Hall effect but related to the Hall effect have been of recent interest to electronic and optical device technology. These are the anomalous Hall effect, the spin Hall effect, and the spin Hall effect of light. A brief introduction to these important phenomena and their applications is presented in the text.

The anomalous Hall effect has a long history, being discover by Hall shortly after he found the classical Hall effect. In the anomalous Hall effect, it is noted that the transverse or Hall resistance, which is related to the applied magnetic field, can also be a function of the net magnetization of the conducting material. Here, an anisotropic spin-dependent scattering from the magnetization is observed in the conduction electrons of the system and is a basis for the dependence of the transverse resistance on the magnetization.

Another process closely related to the anomalous Hall effect is the spin Hall effect, which is concerned with a spin separation in the system. In certain materials, the spin Hall effect is observed in a strip of conducting material for a current flow along the length of the strip. Specifically, in the spin Hall effect, a spin separation occurs across the width of the strip separation in response to a current flow along its length. An optical analogy of the spin Hall effect also exists, which involves optical systems that can modulate the dynamics of light through mechanisms arising from the polarization of light. Some of these types of systems have involved metamaterials and are a basis of current interests in optics in which optical effects are developed to mimic effects from electron dynamics.

1.11 CORRELATED ELECTRON SYSTEMS

Moving from the idea of models based on the motion of independent electron modes into the area of models accounting for the collective or correlated motions of the electrons, the focus is on three important topics. The first phenomenon treated is that of superconductivity. Materials exhibiting this property are classified as low- and high-temperature superconducting materials. The low-temperature materials were first discovered at the beginning of the twentieth century, and high-temperature superconductors were encountered later near the end of the twentieth century. The physics of the mechanisms at operation in these two cases appear to be quite different. The next phenomenon of interest involving correlated motion is that of the fractional quantum Hall effect [2,37,38]. This involves a mechanism that is different from the integer quantum Hall effect, which is described by the independent particle model of electrons. It is a consequence of electron correlations due to their electron–electron Coulomb interactions and results in a variety of interesting physical effects. Finally, Coulomb blockade [5–10] as an important interaction in nanosystems designed to manipulate the motion of single electron devices is discussed. The Coulomb blockade effect is a direct result of the Coulomb interaction between electrons, as they are distributed within nanodevices. Specifically, it involves the interaction between net charges created within a nanocircuit and is a mechanism important in the development of single electron transistors.

1.11.1 SUPERCONDUCTIVITY

Superconductivity is one of the most important developments in the field of highly correlated motions of electrons in materials [12–16,39–42]. This state of matter is characterized by the Meisner effect, which is the exclusion of magnetic flux from the superconductor and by a perfect conductivity. At low critical temperatures (i.e., $T_c < 30°K$), superconductivity arises from a mechanism involving the interaction of the conduction electrons with phonons. This results in the creation of bound pairs of electrons known as Cooper pairs, which then condense into a low-energy state of the electronic system. Breaking a pair and removing it from the condensate is found to require a finite energy. On the other hand, the mechanism of superconductivity is different at higher critical temperatures (i.e., $30°K < T_c < 150°K$) than that in low-temperature materials. Its origins in high-temperature systems may be related to magnetic rather than phonon effects in the interaction of the electrons; however, the details of this are still considered as an open question. Nevertheless, both high- and low-temperature superconducting materials exhibit similar types of properties but on different temperature scales and with differing degrees of thermal fluctuation in the system properties.

Superconductivity arises from a highly correlated motion of the conduction electrons so that its theoretical treatment must be quite different from the independent particle models developed earlier. The approach to the study of superconductivity that is a focus for us is the Ginzburg–Landau theory [12]. In this model, the wavefunction of the system of electrons is characterized by a single function of position and phase, which represents the state of the correlated charge carriers in the model. It is used to express a hypothetical form for the free energy density of the superconductor as a function of position and a hypothetical form for the current density. In the Ginzburg–Landau theory, the free energy density of a superconductor is proposed using broad theoretical statements intended to represent its general physical properties as observed macroscopically. In this sense, it is known as a phenomenological theory as opposed to a theory, which is generated directly from first principles [12–16]. Though it is a model built only on observations at a macroscopic level, later the Ginzburg–Landau free energy form was validated for low-temperature materials with the development of the microscopic BCS theory of superconductivity. In this regard, the Ginzburg–Landau approach maintains its importance for technological applications and is the focus of our approach to superconductor technology. It is found to have a degree of validity for both low- and high-temperature materials.

The Ginzburg–Landau theory regards the electron motions in the superconductor condensate as spatially correlated to a considerable degree so that the transport properties are expressed in terms of a single macroscopic wavefunction of the system. The amplitude of the Ginzburg–Landau wavefunction is related to the density of correlated electrons (these later are shown in the BCS theory to be the density of bound electron Cooper pairs), while the phase enters into fixing the transport properties of the superconductor. The theory naturally develops into a discussion of two basic types of superconducting materials known as Type I and Type II superconductors [1,2,12–16,39,40], which arise from the relative importance of the surface and bulk-free energy of the systems. Here a fundamental distinction between these two types is the absence or presence of a mixed state of the material, i.e., a mixing of superconductor-normal metal regions.

A Type I material occurs as a homogeneous bulk superconducting medium. However, in the temperature–magnetic field phase diagram of a Type II superconductor, a mixed state of superconductor and normal material exists. This is known as a vortex state and involves the penetration of the regions of bulk superconductivity by tubules of normal nonsuperconducting, flux-bearing media. The phases are mixed in this sense, similar to those of an equilibrium mixture of ice and water, or a phase-separated metallic alloy. While low-temperature superconductors exhibit either Type I or Type II behaviors, it is found that the new high-temperature superconductors [39,40] only display Type II behaviors. In addition, for the elevated temperatures at which the high-temperature superconducting transition occurs, increased thermodynamic fluctuations in the system are present and make the Ginzburg–Landau theory more of an approximation.

1.12 JOSEPHSON JUNCTIONS

As an application of the Ginzburg–Landau formulation, the important topic of Josephson junction effects is studied. The Josephson junction is essentially a tunnel junction formed between two superconductors that are separated by a thin nonsuperconducting barrier, e.g., an insulating region. In this regard, the two superconductors of the Josephson junction are said to be connected by a weak link. (See Figure 1.6 for a schematic of a Josephson junction geometry [12–16].)

The transport of charge across the barrier of the junction exhibits a variety of interesting interference effects involving the phase difference between the superconductors on the two sides of the barrier. These include a DC Josephson effect in which a current flows through the barrier in the absence of an applied potential difference across the barrier and an AC Josephson effect involving

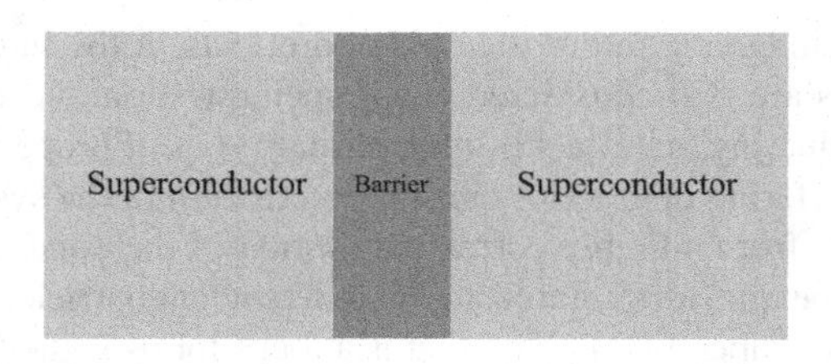

FIGURE 1.6 Schematic of a Josephson junction. The junction is composed as two semi-infinite superconductors that are separated by a thin infinite slab of nonsuperconducting barrier material. A potential difference can be applied perpendicular to the barrier in the operation of this system.

an alternating current through the barrier in the presence of an applied potential across the junction. In addition to these two important effects, there are a variety of other phase coherent effects associated with the junction geometry itself or the geometry of circuits into which multiple junctions are placed. All such properties have been found to be of technological and/or metrological importance, and form a basis for device designs.

Also, of interest related to some of these phase properties is the Aharonov–Bohm effect. This is not a property of superconductors but is a nanoscopic phenomenon involving applied magnetic fields to highly pure metals in the absence of superconductivity. Here, the important point of operation is that the material should be small and pure enough that electron scattering is not present in the system. Consequently, the electrons maintain their phase information during their motion. In these materials, due to the lack of impurity scattering, phase coherence is maintained in the single electron states, and the currents in some geometries (e.g., rings of material) can be made to exhibit phase interference properties. The relation between the superconducting phase coherence and the phase coherence in high-purity normal metals is discussed later.

1.13 FRACTIONAL QUANTUM HALL EFFECT

Another many-body property arising from the correlated motion of interacting electrons is the fractional quantum Hall effect which, like the integer quantum Hall effect [2,37,38,41,42], occurs in two-dimensional electron systems subject to an externally magnetic field. The effect is similar to the integer quantum Hall effect but comes from the further partitioning of the electron occupancy of the Landau levels [2,42] over that of the integer quantum Hall effect. Specifically, at high applied magnetic fields, the Landau levels of the integer quantum Hall effect are fragmented into a series of smaller sublevels about the original Landau levels. The fragmentation arises from the presence of strong electron–electron interactions in the system and results in energy sublevels occurring as satellites about the original Landau levels. From these energy satellites, additional modifications in the transverse resistivity considered as a function of the applied magnetic field are found in the form of subplateaus to the original plateaus observed in the integer quantum Hall effect. These sublevels are in addition to the original plateaus observed in the integer quantum Hall effect and have only been obtained in systems at high applied magnetic fields.

In summary, the additional energy levels introduced into the system by the electron correlations contribute to a new series of subplateaus in the study of the plot of the resistivity versus the applied magnetic field. It is generally found that the features of the fractional Hall effect are in addition to those of the integer Hall effect and require high applied fields for their observation, which is made primarily in the lower energy Landau levels of the system.

1.14 COULOMB BLOCKADE

A phenomenon based on electron–electron interactions, which is explained in terms of classical physics, is that known as Coulomb blockade [5–10,43]. In certain types of nanodevices, the spatial electron distribution can become important in such a way that the Coulomb interactions between individual electrons in the system affect the electrical transport properties exhibited. Examples are found in the development and operation of the so-called single electron transistor or in other types of so-called turnstile devices. In these types of devices, individual electrons are processed through the system by the modulation of the potentials applied across the device. The resulting electron processor [5–10,44,45] can be used to store or perform operations on information represented by the charged state of the system. A variety of circuit representations for the operation of single electron transistors and turnstiles are considered in the text. The importance of Coulomb blockade originates in the small size of the system, and in the degree to which electrons are isolated and concentrated in the device.

1.15 RESONANCE

Resonant frequency responses are another class of phenomena that have become of fundamental importance in the recent design of nanosystems and in the formulation of engineered materials. They are a basis for the formulation of nanodevices exhibiting specific engineered responses when subjected to external applied stimuli and have become particularly important in the development a new class of materials known as metamaterials. In this regard, metamaterials are homogeneous media exhibiting customized responses not otherwise available in natural materials.

Metamaterials [17–20] are engineered media that are composed as arrays of nano-resonators giving a resonant response at certain frequencies. The sizes of the individual resonator units and their separations are chosen to be small on the scale of the wavelengths of the radiation with which they are designed to interact. Consequently, the metamaterial appears to be a homogeneous medium on the scale of the radiation wavelength. However, the materials exhibit new refractive properties due to their resonant frequency response when used in optical or acoustical systems. In this regard, these types of materials enter in the design of many optical or acoustical engineered devices that are needed to display a required refractive response. They have figured in the design of negative refractive index materials, perfect lenses, and cloaking devices [46,47]. In particular, the development of optical materials with negative refractive index are significant, as such materials do not occur naturally.

The nature of resonances, as a general topic, is also important of its own right and will be studied in the context of a treatment of the topics of Fano resonances [5]. This allows for a characterization of the nature of the frequency response of a variety of interesting systems in nanoscience. Consequently, it is seen through this study that resonant responses can be classified into a variety of standard forms of frequency response.

1.16 SCALING AND RENORMALIZATION GROUP

Materials are found to exhibit a variety of phase transitions, and associated with these changes of state are important singularities in the material properties and media responses to external stimuli occurring at the phase transition. (As an example, see Figure 1.7 for a schematic of some of the critical properties of a ferromagnetic material [28–35,41]. Specifically, the figure shows as a function of temperature the cusp in the magnetization at its critical temperature and the divergence of the magnetic susceptibility at the critical temperature.) Consequently, understanding and classifying these singularities has become an important focus of developments in physics and material science during the last 70 years. In the final chapter of the text, a brief introduction is given to the development of the important fields of scaling theory and the theory of renormalization group transformations [28–35] that have been developed to study the singularities at phase transitions.

Both of these approaches allow for the understanding of many of the singularities associated with phase transitions. The scaling theory is obtained by realizing that the singularities of many critical phenomena are characterized by the forms of homogeneous functions. This allows for the development of a number of relationships between the singularities of different properties of the

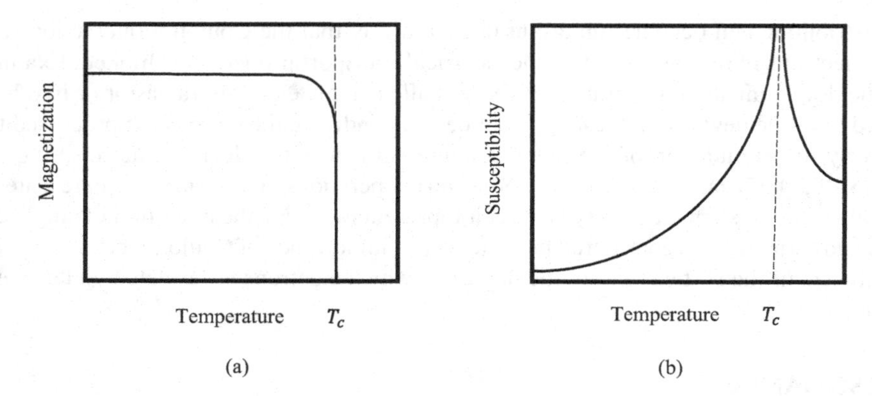

FIGURE 1.7 Schematic of some typical properties of a ferromagnet about its critical temperature T_c (at the dashed vertical line): (a) the magnetization versus temperature and (b) the susceptibility versus temperature. Specifically, the magnetization is found to drop to zero at the critical temperature, and the susceptibility is divergent at the critical temperature.

materials. On the other hand, the renormalization group is a set of transformations acting on the properties of the material, which leads to an understanding of the origins of many results from scaling theory.

The renormalization group theory is a study of the effects of changes in the length scales of systems and how they show up in changes of the thermal dynamic properties of the system. Contrary to its naming as a group, it is really not a study of the transformation groups of a thermodynamic system, because the transformations treated do not have inverses. In this regard, the renormalization group is technically a semigroup rather than a group of transformations. Nevertheless, the transformations studied in the renormalization group formulation are defined so as to generate change in the length scales of the systems under study. This is usually done by tracing out some of the degrees of freedom in the system partition function and considering the remaining partition function of untraced degrees of freedom. These changes of scales ultimately are used to allow for an explanation of the nature of the singularities exhibited by materials at their phase transformations. This aids us in characterizing the specific form of the singularities of the thermodynamic properties observed in a system at its phase transition, in terms of the fixed points of the renormalization group transformation.

Specifically, by studying changes of scale in systems exhibiting a phase transition, the properties of the singular behavior are determined at and in the neighborhood of the renormalization group transformation fixed points. Such treatments lead directly to a determination of the nature of the singularities exhibited as being represented in terms of power law singularities characterized by so-called critical exponents [28–35]. Since all the thermodynamics of a system is generated from the same singular free energy, the critical exponents of a given system are all found to be related to one another. In addition, they are found to depend on the dimension of the material, the range of the interactions in the material, and other general microscopic factors characterizing the material.

In summary, the topics of scaling and the renormalization group provide for the understanding of the power law nature of the singularities in the thermodynamic properties at a phase transition and their characterization in terms of a set of critical exponents. The relationship of the critical exponents to one another is also explained by the formulation, and various basic identities involving the critical exponents, in general, are given. In the text, the renormalization group theories are specifically developed and applied in the context of the two-dimensional Ising model and the Ginzburg–Landau-based free energy form. Both of these problems are important representative models in the theory of phase transitions, and offer basic historical examples of the ideas of scaling and renormalization group theory.

REFERENCES

1. Kittle, C. 1996. *Introduction to solid state physics, 7th Edition*, 393–410. New York: John Wiley & Sons, Inc.
2. Marder, M. P. 2000. *Condensed matter physics*. New York: John Wiley & Sons, Inc.
3. Animalu, A. O. E. 1977. *Intermediate quantum theory of crystalline solids*. Englewood Cliffs: Prentice-Hall.
4. Petty, M. C. 2007. *Molecular electronic: From principles to practice*. New York: John Wiley & Sons.
5. Ouisse, T. 2008. *Electron transport in nanostructures and mesoscopic devices: An introduction*. New York: Wiley.
6. Barnham, K. and D. Vvedensky 2001. *Low-dimensional semiconductor structures: Fundamentals and device applications*. Cambridge: Cambridge University Press.
7. Datta, S. 2005. *Quantum transport: Atom to transistor*. Cambridge: Cambridge University Press.
8. Landauer, R. 1957. Spatial variation of currents and fields due to localized scatterers in metallic conduction. *IBM Journal of Research and Development* 1: 223–231.
9. Landauer, R. 1990. Advanced technology and truth in advertising. *Physica A* 168: 75–87
10. Buttiker, M. 1989. Negative resistance fluctuations at resistance minima in narrow quantum Hall conductors. *Physical Review B* 38: 12724.
11. Harrison, P. and P. Harrison 2000. *Quantum wells, wires, and dots*. New York: John Wiley & Sons, Inc.
12. Fossheim, K. and A. Sudbo 2004. *Superconductivity: Physics and applications*, 141–198. New York: John Wiley & Sons, Inc.
13. Schrieffer, J. R. 1983. *Theory of superconductivity*. Boca Raton: CRC Press.
14. Barone, A. and G. Paterno 1982. *Physics and applications of the Josephson effect*. New York: John Wiley & Sons, Inc.
15. Wolf, E. L. 1985. *Principles of electron tunneling spectroscopy, 2nd Edition*. Oxford: Oxford University Press.
16. de Gennes, P. G. 1989. *Superconductivity of metals and alloys*. New York: Addison-Wesley.
17. McGurn, A. R. 2020. *Introduction to photonic and phononic crystals and metamaterials*. San Rafael: Morgan & Claypool Publishers.
18. McGurn, A. R. 2021. *Introduction to nonlinear optics of photonic crystals and metamaterials, 2nd Edition*. Bristol: IOP Publishing Ltd.
19. McGurn, A. R. 2015. *Nonlinear optics of photonic crystals and meta-materials*. San Rafael: Morgan & Claypool Publishers.
20. McGurn, A. R. 2018. *Nanophotonics*. Cham: Springer.
21. Gerry, C. and P. Knight 2005. *Introductory quantum optics*. Cambridge: Cambridge University Press.
22. Lambropoulos, P. and D. Petrosyan 2007. *Fundamentals of quantum optics and quantum information*. Berlin: Springer-Verlag.
23. Fox, M. 2006. *Quantum optics: An introduction*. Oxford: Oxford University Press.
24. Band, Y. B. 2006. *Light and matter: Electromagnetism, optics, spectroscopy and lasers*. Chichester: John Wiley & Sons, Ltd.
25. Walls, D.F. and G. J. Milburn 1995. *Quantum optics*. Berlin: Springer-Verlag.
26. Streetman, B. G. and S. Banerjee 2000. *Solid state electronic devices, 5th Edition*. New Jersey: Prentice Hall.
27. Ryndyk, D. A. 2016. *Theory of quantum transport at nanoscale: An introduction*. Cham: Springer.
28. Pilschke, M. and B. Bergersen 1989. *Equilibrium statistical physics*. New Jersey: Prentice Hall.
29. Pathria, R. K. 1996. *Statistical mechanics, 2nd Edition*. Oxford: Butterworth Heinemann.
30. McComb, W. D. 2004 *Renormalization methods: A guide for beginners*. Oxford: Clarendon Press.
31. Ma, S.-K. 1978. *Modern theory of critical phenomena*. London: The Benjamin/Cummings Publishing Company.
32. Fisher, M. E. 1983. Scaling, universality and renormalization group theory. In *Critical phenomena: Proceedings, Stellenbosch, South Africa 1982* Ed. F. J. W Hahne, 1–137. Berlin: Springer-Verlag.
33. Wilson, K. G. 1975. The renormalization group: Critical phenomena and the Kondo problem. *Reviews of Modern Physics* 47: 773–837.
34. Wilson, K. G. and J. Kogut 1974. The renormalization group and the ε expansion. *Physics Reports* 12: 75–200.
35. Chaikin, P. and T. C. Lubensky 1995. *Principles of condensed matter physics*. Cambridge: Cambridge University Press.
36. Maradudin, A. A. and A. R. McGurn 1993. Photonic band structure of a truncated two-dimensional, periodic dielectric medium. *Journal of the Optical Society of America B* 10: 302–313.

37. Prange, R. E. and S. M. Girvi 1987. *The quantum Hall effect*. New York: Springer.
38. Nazarov, Y. V. and Y. M. Blanter 2009. *Quantum transport: Introduction to nanoscience*. Cambridge: Cambridge University Press.
39. Saxena, A. K. 2012. *High-temperature superconductors, 2nd Edition*. Heidelberg: Springer.
40. Plakida, N. 2010. *High-temperature cuprate superconductors: Experiment, theory, and applications*. Heidelberg: Springer.
41. Getzlaff, M. 2008. *Fundamentals of magnetism*. Berlin: Springer.
42. Kroemer, H. 1994. *Quantum mechanics: For engineering, materials science, and applied physics*. New Jersey: Prentice Hall.
43. Stochman, M. I. 2013. Spaser, plasmonic amplification, and loss compensation. In *Active plasmonics and tuneable plasmonic metamaterials* Ed. A. V. Zayats and S. A. Maier. New York: John Wiley & Sons, Ltd.
44. Khursheed, R. C. and F. Z. Haque 2019. Review on single electron transistor (SET): Device in nanotechnology. *Austin Journal of Nanomedicine & Nanotechnology* 7: 1055–1066.
45. Goyal, S. and A. Tonk 2015. A review towards single electron transistor (SET). *International Journal of Advanced Research in Computer and Communication Engineering* 4: 36–39.
46. Pendry, J. B. 2000. Negative refraction makes a perfect lens. *Physical Review Letters* 85: 3966–3969.
47. Pendry, J. B., Schurig, D., Smith D. R., et al. Controlling electromagnetic fields. *Science* 312: 1780–1782.

2 Conductivity

A variety of simple theories have been developed to understand the conductivity properties of materials [1–3]. In this chapter, the focus is on theories related to Fermi gas models in which the electric current arises from the motion of charged particles through a viscous medium. The Fermi models are essentially restatements of the classical mechanical Drude model of conduction (put forth in 1900) in terms of quantum mechanics [1,2]. With the introduction of the ideas of quantum theory, they provide a semiquantitative understanding of the thermal and transport properties of many metals and semiconductors, acting as highly successful physical descriptions of the behaviors of these materials.

In the following, a treatment is given of a Fermi gas of particles with charge q where $q > 0$ or $q < 0$ and in which the particles obey the Pauli exclusion principle [1,2]. While the particles interact with external electric and magnetic fields and the viscous medium that transports them, in most cases they do not interact with one another. The transport medium and the motion of the charge particles through it then constitutes a model for the motion of an electrical current in metals or semiconductors. In this context, discussions are given of the conductivity of such a system in the absence or presence of magnetic interactions, the diamagnetic and paramagnetic properties of the system, and the Anderson localization of the gas [1,2].

Extending the model to include interactions between the particles forming the gas introduces correlations in the motions of the charged particles [2–10]. These correlations show up in the Stoner theory of ferromagnetism [2,10], which provides a qualitative understanding of magnetic ordering in metals [10] such as Ni, Co, and Fe; in the Mott transition [2], which is a metal–insulator transition driven by repulsive Coulomb interactions between the charge carriers; in the formation of a Wigner solid [2], where at very low carrier densities the charge carriers crystalize into a lattice; and in the theory of superconductivity [3–9].

In the given presentation, first the Drude model will be reviewed, and then its generalization to quantum systems is discussed. The magnetic properties will next be studied, followed by a treatment of Anderson and Mott localization, and the final topic is the Ginzburg–Landau theory of superconductivity.

2.1 BASIC IDEAS OF CONDUCTIVITY

The basic property that characterizes the flow of charge in a system is the current density. This is given for an isotropic medium by the general form [1,2,11–17]:

$$\vec{j} = qn\vec{v}. \tag{2.1}$$

Here n is the carrier density of the system, which in three-dimensional gases is defined as the number of carriers of charge q per volume, $\vec{v}$ is the average drift velocity of each carrier in the viscous medium, and $\vec{j}$ is the charge per second flowing through a unit area perpendicular to $\vec{j}$. In the case of a two-dimensional medium, however, n becomes the carrier density given as the number of carriers of charge q per area, $\vec{v}$ is again the average drift velocity in the viscous medium, and $\vec{j}$ becomes the charge per second through a unit length perpendicular to $\vec{j}$. Finally, in a similar extension to one-dimensional media, n is the carrier density defined as the number of carriers of charge q per length, $\vec{v}$ is the average drift velocity, and $\vec{j}$ is the charge per second in the direction of $\vec{j}$ past a point. (See Figure 2.1 for a schematic of the geometries of each of these media.) The study of the electrical conductivity in each of these types of systems is then reduced to developing theories for

DOI: 10.1201/9781003031987-2

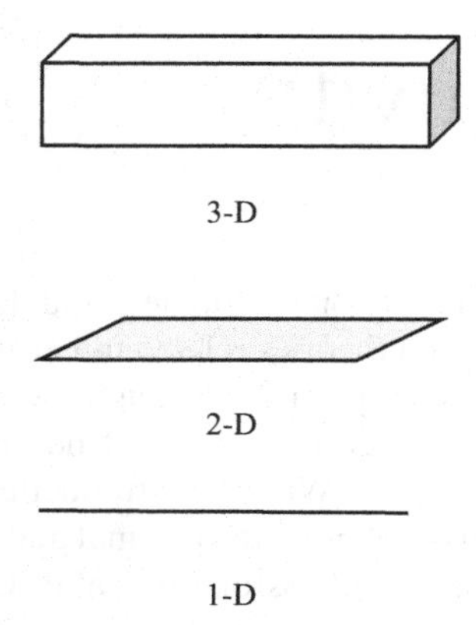

FIGURE 2.1 Schematic drawing of conductors in three, two, and one dimensions. In three dimensions, the conductor is a parallelepiped. In two dimensions, the conductor is a rectangle; and the conductor is a line in one dimension. A potential difference is applied between the left and right edges of the conductors, causing a current to flow.

the determination of their corresponding n and $\vec{v}$, and the nature of the conductivity is found to be highly dependent on the dimension of the system treated.

In this regard, the earliest attempt at a study of the current density is the Drude theory, which is based on classical mechanics. In the Drude approach the charged particle q is considered as moving in a viscous medium with dynamics described by the equation [1,2]

$$m\left[\ddot{x} + \dot{x}/\tau + \omega_0^2 x\right] = qE. \tag{2.2}$$

Here, the externally applied electric field E and the motion of the charge are along the x-direction, τ has units of time and sets the viscous drag of the medium, ω_0 is the oscillator frequency of a harmonically bound carrier, and m is the effective mass of the charge carrier. Equation (2.2) is recognized as the damped harmonic oscillator of classical mechanics, and it reduces to the classical mechanics equation of a free carrier moving through a viscous medium for $\omega_0^2 = 0$.

Later the case of Eq. (2.2) in which $\omega_0 \neq 0$ will be used to model the response to externally applied electric fields of electrons bound to atoms [1,2]. This is important in developing theories of the dielectric response of crystals. Currently, however, the $\omega_0 = 0$ case is used to study the response of free electrons traveling in a viscous medium to an applied field; this last $\omega_0 = 0$ case is recognized as a simple form of the terminal velocity problem treated in elementary mechanics.

For a time-independent applied electric field the particles of the Fermi gas rapidly approach a drift velocity determined by the condition that $\dot{v} = 0$ and is obtained from the $\omega_0 = 0$ form of Eq. (2.2) as

$$v = \dot{x} = \frac{q\tau}{m}E = \frac{q}{|q|}\mu E. \tag{2.3}$$

where $\mu = |q\tau/m|$ is defined as the mobility of the charge carrier. This is the basic result from the Drude treatment, and applying Eq. (2.3) in Eq. (2.1) for a three-dimensional medium yields Ohm's law relationship [1,2,12–14]

$$I_3 = Aj_3 = \left(\frac{q^2 n_3 \tau}{m}\frac{A}{L}\right)(LE) = \frac{1}{R_3}V \tag{2.4}$$

where I_3 is the current through a rod of length L and cross-sectional area A (See the schematic in Figure 2.2 for the geometry of the current flow.), j_3 is the three-dimensional current density, V is the potential difference between the ends of the rod that are separated by L, n_3 is the three-dimensional carrier density, and R_3 is the resistance between the ends separated by L. Similarly, in two dimensions the current is given by

$$I_2 = L_T j_2 = \left(\frac{q^2 n_2 \tau}{m}\frac{L_T}{L}\right)(LE) = \frac{1}{R_2}V \tag{2.5}$$

where I_2 is the current through a rod of length L and cross-sectional length L_T (See the schematic in Figure 2.2), j_2 is the two-dimensional current density, V is the potential difference between the ends that are separated by L, n_2 is the two-dimensional carrier density, and R_2 is the resistance between the ends of the two-dimensional rod that are separated by L. For a one-dimensional medium then

$$I_1 = j_1 = \left(\frac{q^2 n_1 \tau}{m}\frac{1}{L}\right)(LE) = \frac{1}{R_1}V \tag{2.6}$$

where I_1 is the current through a rod of length L (See the schematic in Figure 2.2.), j_1 is the one-dimensional current density, V is the potential difference between the ends that are separated by L, n_1 is the one-dimensional carrier density, and R_1 is the resistance between the ends separated by L.

From Eqs. (2.4) to (2.5), it then follows that the conductivity, defined as $\vec{j}_i = \sigma_i \vec{E}$ for $i = 1,2,3$ dimensions, is given as

$$\sigma_i = \frac{q^2 n_i \tau}{m} = \frac{1}{\rho_i} \tag{2.7a}$$

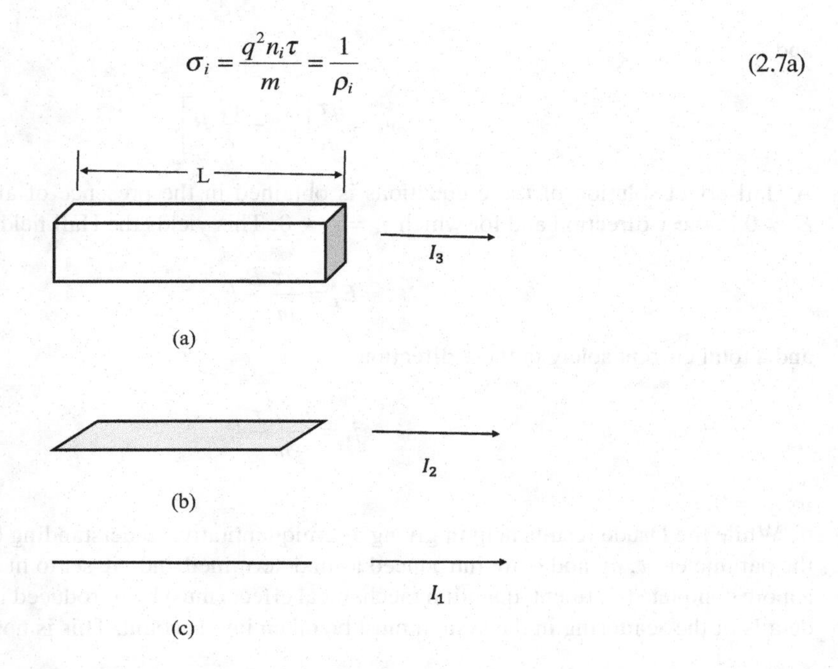

FIGURE 2.2 Schematic drawing of conductors in (a) three, (b) two, and (c) one dimensions. In three dimensions, the conductor is a parallelepiped; in two dimensions, the conductor is a rectangle; and the conductor is a line in one dimension. A potential difference (0-V) is applied between the left and right edges of the conductors to, respectively, cause currents I_3, I_2, and I_1 to flow through the media. The horizontal length of the objects is L.

where ρ_i is the related resistivity of the material being considered. Since the conductivity and resistivity are properties of the medium and not the geometry of the rods, the resistance of the rods in Figure 2.2 is dependent on their geometric parameters and, consequently, scale with the dimensionality of the system. In particular, the resistance is found to be of the general form

$$R_d \propto \frac{L}{L_T^{d-1}} \tag{2.7b}$$

where d is the dimensionality of the system, L is the length of the rod across which the potential is applied, and L_T is a characteristic cross-sectional length of the area through which the current passes. In the case of Eq. (2.7) applied to the three-dimensional system, use has been made of the relationship $A \propto L_T^2$.

The Drude approach generalizes to a consideration of the conductivity in the presence of an externally applied magnetic field. In this case, Eq. (2.2) for $\omega_0 = 0$ becomes [1,2,11,13–17]

$$m\left[\ddot{\vec{r}} + \dot{\vec{r}}/_{\tau}\right] = q\left[\vec{E} + \frac{1}{c}\dot{\vec{r}} \times \vec{B}\right]. \tag{2.8}$$

where $\vec{B}$ is an external field, and in a steady state solution $\ddot{\vec{r}} = 0$ with the drift velocity $\vec{v} = \dot{\vec{r}}$. For a two-dimensional gas in the $x - y$ plane subject to a constant magnetic induction along the z-axis, it then follows that

$$v_x = \frac{q\tau}{m}\left[E_x + \frac{v_y}{c}B\right] \tag{2.9a}$$

and

$$v_y = \frac{q\tau}{m}\left[E_y - \frac{v_x}{c}B\right]. \tag{2.9b}$$

A Hall effect solution of these equations is obtained in the presence of an applied electric field $E_x \neq 0$ in the x-direction and for which $j_y = v_y = 0$. This yields the Hall field in the y-direction

$$E_y = \frac{q\tau}{m}\frac{B}{c}E_x \tag{2.10a}$$

and a total current solely in the x-direction

$$j_x = \frac{n_2 q^2 \tau}{m}E_x, \tag{2.10b}$$

for $j_y = 0$.

While the Drude results help in giving a semiquantitative understanding of conducting systems, the parameters τ, m, and n are introduced as undetermined and are set to fit experimental data. For a more complete treatment, quantum mechanical effects must be introduced into the system, and the details of the scattering in the system must be taken into account. This is now addressed.

2.2 QUANTUM EFFECTS

With the introduction of quantum mechanics into the considerations, the charged particle states in the Fermi gas become planewaves of the form [1,2]

$$\psi_k(\vec{r}) \propto e^{i\vec{k}\cdot\vec{r}}, \tag{2.11a}$$

each having momenta $\vec{p} = \hbar\vec{k}$. (Note: For an additional simplicity, in the following, the electrons will be considered as spinless particles, i.e., in the so-called spinless fermion model. Later, the modifications due to spin will be treated in Sections 2.3.4 and 2.4 within the context of magnetic properties arising from electron spin.) Confining these particles in a cubic box of side L, subject to periodic boundary conditions between opposite faces of the box, the wavevectors of the particles in a three-dimensional gas are found to be of the form

$$k_x = \frac{2\pi}{L} n_x, k_y = \frac{2\pi}{L} n_y, k_z = \frac{2\pi}{L} n_z \tag{2.11b}$$

where n_x, n_y, and n_z are integers. In three dimensions the ground state of the Fermi gas is then a sphere in wavevector space centered about $\vec{k} = 0$. All of the states in the sphere are completely occupied by the particles of the gas. Similarly, considerations can be made for two- and one-dimensional gases of Fermi particles, respectively, confining particles in a square of side L in two dimensions or to a line segment of length L in one dimension. In this way, the ground state of the gas in wavevector space is a completely filled circle centered about the origin in two dimensions, while the ground state is a completely filled line segment centered about the origin in wavevector space in one dimension.

In the presence of a uniform electric field E applied in the x-direction, the Hamiltonian of a charge in the Fermi gas is given in the Heisenberg formulation of quantum mechanics by [1,2,18]

$$H = \frac{p^2}{2m} - qEx. \tag{2.12}$$

In this representation, the operator equations of motion in the y- and z-directions are

$$\dot{p}_y = \frac{1}{i\hbar}\left[p_y, H\right] = \dot{p}_z = \frac{1}{i\hbar}\left[p_z, H\right] = 0 \tag{2.13a}$$

so that these components of momenta are constant in time, but in the x-direction

$$i\hbar\dot{p}_x = \left[p_x, H\right] = i\hbar qE \tag{2.13b}$$

with momentum and wavevector solutions are given by

$$p_x(t) = p_x(0) + qEt, \tag{2.14a}$$

$$k_x(t) = k_x(0) + \frac{qE}{\hbar}t. \tag{2.14b}$$

Consequently, the x- components of the wavevectors in Eq. (2.11b) evolve in time according to Eq. (2.14b), while the other components remain unchanged, if existent.

Given the solutions in Eqs. (2.13) and (2.14), a quantum mechanical treatment of the electrical conductivity can be approached. To do this, first the number of wavevector solution states in the gas needs to be determined. This requires a counting of the total number of states occupied by the electrons in the system. After this, a consideration of the change in wavevector of each of the states arising from the field interactions is applied to a determination of the conductivity of the system.

In the absence of a field the free-particle dispersion relation of the gas has the form [1,2]

$$\varepsilon = \frac{\hbar^2 k^2}{2m}. \tag{2.15}$$

This provides a mapping from wavevector space onto the energies of the particles and can be used to link the density of solution states in wavevector space to the density of energy solutions for the gas.

From Eq. (2.11b), the density of states in wavevector space of the three-dimensional gas of spinless particles is $D_3(\vec{k}) = \frac{L^3}{(2\pi)^3}$, so that given $D_3(\vec{k})$ the number of solutions of the gas that exist in a volume d^3k of wavevector space is determined as $dN = D_3(\vec{k})d^3k$. (Note: If the two states of spin of the charge carriers were considered, the density of states would change to $D_3^s(\vec{k}) = 2\frac{L^3}{(2\pi)^3}$). The density of states $D_3(\vec{k})$ is readily converted to a statement regarding the density of energy states. Using the dispersion relation in Eq. (2.15), the density of states in energy of the gas is then obtained as [1,2,12,14]

$$D_3(\epsilon) = \frac{L^3}{2\pi^2}\sqrt{\frac{2m^3}{\hbar^6}}\sqrt{\epsilon} = \frac{L^3}{\sqrt{2}\pi^2}\frac{m^{3/2}}{\hbar^3}\sqrt{\epsilon}, \tag{2.16a}$$

and when the two degenerate spin states are considered

$$D_3^s(\epsilon) = 2D_3(\epsilon) = 2\frac{L^3}{\sqrt{2}\pi^2}\frac{m^{3/2}}{\hbar^3}\sqrt{\epsilon}. \tag{2.16b}$$

In the ground state of the spinless gas of N wavevector eigenstates, all of the wavevectors of the eigenstates are in a sphere of radius k_F centered about the origin and defined by

$$N = \int_0^{k_F} 4\pi k^2 dk D_3(\vec{k}) = \int_0^{\epsilon_F} d\epsilon D_3(\varepsilon) \tag{2.17}$$

where $\epsilon_F = \frac{\hbar^2 k_F^2}{2m}$ is the maximum energy of the system. In these expressions, the maximum wave number k_F is termed the Fermi wave number, and the largest energy in the gas ϵ_F is known as the Fermi energy.

Similarly, from Eq. (2.11b) the density of states in wavevector space of the spinless two-dimensional gas is $D_2(\vec{k}) = \frac{L^2}{(2\pi)^2}$, and using the dispersion relation in Eq. (2.15) the density of states in energy of the gas is [12]

$$D_2(\epsilon) = \frac{L^2}{2\pi}\frac{m}{\hbar^2}. \tag{2.18}$$

(Again note that for a gas with spin one-half the density of states with spin $D_2^s(\vec{k}) = 2D_2(\vec{k})$ and $D_2^s(\epsilon) = 2D_2(\epsilon)$.) The density of states in wavevector space of the one-dimensional gas is $D_1(\vec{k}) = \frac{L}{2\pi}$, and from the dispersion relation in Eq. (2.15) the density of states in energy of the spinless gas is given by [12]

$$D_1(\epsilon) = \frac{L}{\sqrt{2}\pi}\frac{\sqrt{m}}{\hbar}\frac{1}{\sqrt{\epsilon}}. \tag{2.19}$$

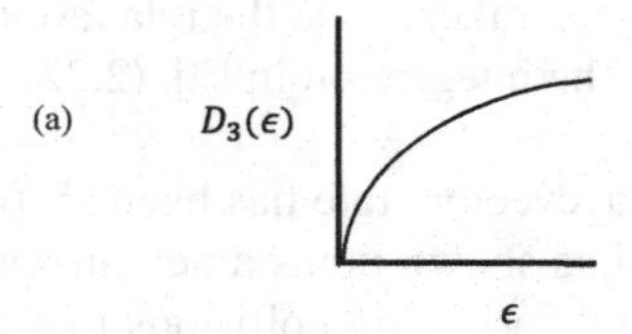

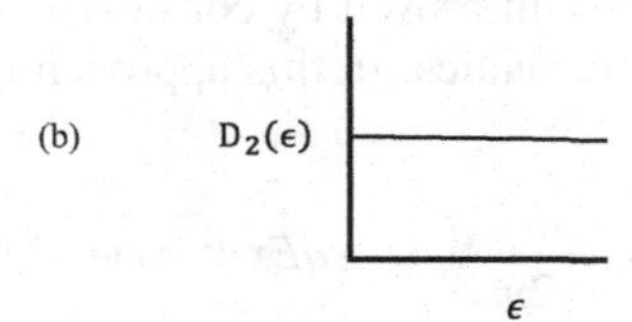

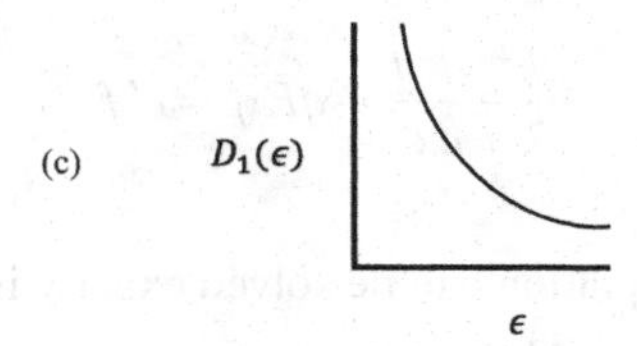

FIGURE 2.3 Schematic representation of the density of electron energy levels $D_d(\epsilon)$ versus energy ϵ in (a) three dimensions for $D_3(\epsilon) \propto \epsilon^{(1/2)}$, (b) two dimensions for $D_2(\epsilon) \propto \epsilon^0$, and (c) one dimension for $D_1(\epsilon) \propto \epsilon^{(-1/2)}$.

(Again for a gas with spin one-half the density changes to $D_1^s\left(\vec{k}\right) = 2D_1\left(\vec{k}\right)$ and $D_1^s(\epsilon) = 2D_1(\epsilon)$.) For these considerations, the ground state of the two-dimensional system is a circle of radius k_F and that of the one-dimensional system is a segment of length k_F. Both of the wavevector space geometries of the spinless gases are related to the number of modes of the system, N, through similar relationships to those in Eq. (2.17), which also define energy maxima ϵ_F. In general dimension, it is then found that

$$D_d(\epsilon) \propto \varepsilon^{(d-2)/2} \text{ and } D_d^s(\epsilon) \propto \varepsilon^{(d-2)/2} \tag{2.20}$$

where $d = 1,2,3$ is the dimension of the system, and a plot of $D_d(\epsilon)$ versus ϵ is shown in Figure 2.3. Next, a treatment of the current density is given both in the absence and presence of an applied electric field. The conductivity of the system is obtained based on considerations as to how the wavevector solutions change with the introduction of an externally applied electric field.

Using the definition of the current density in Eq. (2.1), the $x-$ component of the current density in the absence of an applied electric field is given for the spinless gas by [1,2,12–14]

$$j_x = q\frac{1}{L^d}\left(\frac{L}{2\pi}\right)^d \int d^d k \frac{\hbar k_x}{m} = 0 \tag{2.21}$$

where the integration is over the sphere, circle, or line of states in wavevector space. This expression is generalized to treat the presence of an applied electric field by introducing the changes in wavevector introduced by the field in the integrand. In this way, the presence of an applied electric field in the x-direction obtained from Eq. (2.14b) changes Eq. (2.21) to yield a nonzero-current density of the form

$$j_x = q\frac{1}{L^d}\left(\frac{L}{2\pi}\right)^d \int d^d k \left[\frac{\hbar k_x}{m} + \frac{qE\tau}{m}\right] = \frac{n_d q^2 \tau}{m} E \tag{2.22}$$

where the time in Eq. (2.14b) have been taken to be the relaxation time, τ, of the particle scattering in the viscous medium. In addition, the integration in Eq. (2.22) is over the same wavevector states as in Eq. (2.21).

Note, in this regard, that each wavevector state has been shifted in wavevector space through a displacement of $qE\tau$, and the shift has shown up as a net current in the x-direction. In Eq. (2.22), τ represents a typical time between disorienting collisions that relax the system back into a zero-current state. In the steady state of the system the wavevectors of each of the particles are shifted by a uniform amount, which is determined by the relaxation time in the viscous medium.

The above treatment can be further improved by considering the quantum particles within the Schrodinger approach to quantum mechanics. In this approach the single particle states are given by [12–18]

$$-\frac{\hbar^2}{2m}\nabla^2\psi - qEx\psi = \epsilon\psi \tag{2.23a}$$

and for a wavefunction of the form $\psi(\vec{r}) \propto e^{i(k_y y + k_z z)} f(x)$, Eq. (2.23a) becomes [18]

$$-\frac{\hbar^2}{2m}\frac{d^2 f}{dx^2} - qExf = \epsilon' f \tag{2.23b}$$

where $\epsilon' = \epsilon - \frac{\hbar^2}{2m}\left(k_y^2 + k_z^2\right)$. This equation can be solved exactly in terms of Airy functions, but it is helpful to first consider its treatment in the context of first-order perturbation theory.

In a first-order perturbation theory in the potential $-qEx$, the lowest order of $f(x) \cong e^{ik_x x}$ and the formal first correction to the energy ϵ' is $-qE\langle x\rangle$, where $\langle x\rangle$ is the expectation in the lowest order correction to the normalized wave function. The first-order corrected energy in the system is then written in the form

$$\begin{aligned}\epsilon' &= \epsilon - \frac{\hbar^2}{2m}\left(k_y^2 + k_z^2\right) + \frac{\hbar^2}{2m}\left[k_x - \frac{m}{\hbar^2 k_x} qE\langle x\rangle\right]^2 \\ &= \epsilon - \frac{\hbar^2}{2m}\left(k_y^2 + k_z^2\right) + \frac{\hbar^2}{2m}s^2\end{aligned}, \tag{2.24a}$$

where

$$s = k_x - \frac{m}{\hbar^2 k_x} qE\langle x\rangle. \tag{2.24b}$$

Taking

$$\langle x\rangle = \frac{p_x}{m}t = \frac{\hbar k_x}{m}t \tag{2.24c}$$

in Eq. (2.24b) then gives

$$s = k_x - \frac{m}{\hbar^2 k_x} qE\langle x\rangle = k_x - \frac{qE}{\hbar}t \tag{2.25}$$

For constant k_y, k_z, s and constant total energy ϵ, it then follows from Eq. (2.25) that

$$k_x = s + \frac{qE}{\hbar}t. \tag{2.26}$$

This is in agreement with Eq. (2.14), which is the operator statement for the momentum evolution in the Heisenberg treatment of the problem. The results in Eqs. (2.14) and (2.26) are weak field statements in which the applied electric fields are a perturbation on the other interactions of the gas particles in the system, e.g., the ion backgrounds in metals and semiconductors, etc. This is generally the case in most problems of interest, but there are important exceptions that require a full treatment of Eq. (2.23b).

In the more general case, a full solution of Eq. (2.23) is required. Returning to a consideration of this equation, Eq. (2.23b) can be rewritten in the form [18]

$$\frac{d^2 f}{dx^2} - (\alpha + \beta x) f = 0 \tag{2.27}$$

where $\alpha = -\frac{2m}{\hbar^2}\epsilon'$ and $\beta = -\frac{2m}{\hbar^2} qE$. Making a change of position variables to redefine the $x-$ variable as

$$\tilde{x} = \frac{\alpha + \beta x}{\beta^{2/3}}, \tag{2.28}$$

Equation (2.27) takes the form of the Airy equation [18]

$$\frac{d^2 f}{d\tilde{x}^2} - \tilde{x} f = 0. \tag{2.29}$$

This has the general solution written in terms of the Airy functions $Ai(\tilde{x})$ and $Bi(\tilde{x})$ given by

$$f(\tilde{x}) = aAi(\tilde{x}) + bBi(\tilde{x}) \tag{2.30}$$

where a and b are set by the boundary conditions of the problem, and for $q > 0$ or $q < 0$

$$\tilde{x} = -\left[\frac{2m}{\hbar^2}\right]^{\frac{1}{3}} \left[\frac{\epsilon'}{(qE)^{2/3}} + \frac{qE}{(qE)^{2/3}} x\right]. \tag{2.31}$$

2.3 MAGNETIC FIELD EFFECTS

An interesting model of a conducting system is that of a two-dimensional quantum mechanical Fermi gas subject to an externally applied magnetic field [1,2,12–19]. For this system, the charge carriers are treated as confined to movement in the $x-y$ plane while interacting with an applied magnetic induction B perpendicular to the conduction plane and parallel to the z-direction. In addition, the charges are subject to a harmonic confining potential along the y-direction in the $x-y$ plane. This limits the current flow in the plane to be along the $x-$ direction and provides for an easily solvable example of the properties of a finite strip. It also provides a basis for discussions of the quantum Hall effect and of thin films deposited on a substrate.

The Hamiltonian equation of motion of a single particle of charge q confined about the x-axis by a harmonic potential and interacting with such a magnetic field is given by

$$\left[\frac{\left(p_x - \frac{q}{c}A_x\right)^2 + \left(p_y - \frac{q}{c}A_y\right)^2}{2m} + \frac{1}{2} m\omega_0^2 y^2\right] \psi(x,y) = \epsilon \psi(x,y), \tag{2.32}$$

where the first term in the brackets represents the interaction with the magnetic induction, and the second term is the harmonic confining potential with a strength characterized by the oscillator frequency ω_0. Once the vector potential of the magnetic field is determined from the standard relationship $\vec{B} = \nabla \times \vec{A}$ and the boundary conditions are set, the solutions of Eq. (2.32) are readily obtained analytically in terms of standard functions.

A convenient choice of vector potential for the conduction problem that has been posed is to take the form

$$\vec{A} = -By\hat{i}. \tag{2.33}$$

While other choices for $\vec{A}$ can be made, Eq. (2.33) provides a representation of the uniform magnetic induction along the z-direction following from the relationship $\vec{B} = \nabla \times \vec{A} = B\hat{k}$ and, in addition, leads to a simple form for the Hamiltonian of the problem being treated. Introducing Eq. (2.33) into (2.32), it then follows that [1,2,12–19]

$$\left[\frac{\left(p_x + \frac{q}{c}By \right)^2 + \left(p_y \right)^2}{2m} + \frac{1}{2}m\omega_0^2 y^2 \right] \psi(x,y) = \epsilon\psi(x,y), \tag{2.34}$$

and upon assuming a wavefunction of the form

$$\psi(x,y) = \frac{1}{\sqrt{L}} e^{ikx} f(y) \tag{2.35}$$

where $L \to \infty$ is the length of the Fermi gas along the x-direction, the Schrodinger equation in Eq. (2.34) is reduced to a one-dimensional problem specified by

$$\left[\frac{\left(\hbar k + \frac{q}{c}By \right)^2 + p_y^2}{2m} + \frac{1}{2}m\omega_0^2 y^2 \right] f(y) = \epsilon f(y). \tag{2.36}$$

In addition, if periodic boundary conditions are applied along the x-direction in order to consider conduction problems, it follows that

$$k = \frac{2\pi}{L} n_x \tag{2.37}$$

where n_x is an integer. Equations (2.35)–(2.37) then specify the complete formulation of the problem of a charge q propagating through the system.

First consider the case of Eq. (2.36) for which $B = 0$ so that the problem is reduced to treating only the harmonic confinement in the system. In this limit the Schrodinger equation becomes

$$\left[\frac{p_y^{\,2}}{2m} + \frac{1}{2}m\omega_0^2 y^2 + \frac{\hbar^2 k^2}{2m} \right] f(y) = \epsilon f(y), \tag{2.38}$$

and the system simplifies to that of a one-dimensional harmonic oscillator along the y-axis and a planewave propagation along the x-direction. The solutions of Eq. (2.38) are then given by

$$f_n(y) = H_n\left(\sqrt{\frac{m\omega_0}{\hbar}}y\right)e^{-\frac{1}{2}\frac{m\omega_0}{\hbar}y^2} \tag{2.39a}$$

$$\epsilon_{n,k} = \frac{\hbar^2k^2}{2m} + \left(n + \frac{1}{2}\right)\hbar\omega_0 \tag{2.39b}$$

where $\{H_n(y')\}$ are the normalized Hermite polynomials. Here, the harmonic motion is found to be centered about the x-axis and independent of the planewave motion along the x-direction.

In this limit the group velocity of the charged particle is obtained from [1,2,12–19]

$$v_g = \frac{1}{\hbar}\frac{\partial\epsilon_{n,k}}{\partial k} = \frac{\hbar k}{m} = \frac{p_x}{m}, \tag{2.40}$$

giving the velocity of charge and energy transport in the system. It represents the particle motion as the combination of a constant momentum translation along the x-direction and a harmonic confinement in the y-direction. In this regard, the degree of the confinement of the charges about the x-axis is increased with increasing oscillator frequency, ω_0, but otherwise does not affect the group velocity.

Returning to Eq. (2.36) in the $B \neq 0$ limit, the dynamical properties of the equation are most effectively revealed by rewriting it into the form

$$\left[\frac{p_y^{\;2}}{2m} + \frac{m\omega_c^2}{2}(y - y_k)^2 + \frac{m\omega_0^2}{2}y^2\right]f(y) = \epsilon f(y). \tag{2.41}$$

Here $\omega_c = \frac{|q|B}{mc}$ is the cyclotron frequency for the circular orbit of the charged particle in a magnetic induction B, and $y_k = -\frac{\hbar c}{qB}k$ is a shift of the wavefunction solution along the y-axis, which is seen to depend on both the particle charge and wavenumber. (Note that oppositely charged particles are shifted in opposite directions along the y-axis as are oppositely directed particles.) After some algebra, Eq. (2.41) takes the form [1,2,12–19]

$$\left[\frac{p_y^{\;2}}{2m} + \frac{m(\omega_0^2 + \omega_c^2)}{2}\left(y - \frac{\omega_c^2}{\omega_0^2 + \omega_c^2}y_k\right)^2 + \frac{1}{2}\frac{\hbar^2k^2}{m}\frac{\omega_0^2}{\omega_0^2 + \omega_c^2}\right]f(y) = \epsilon f(y) \tag{2.42}$$

from which it follows that both the confining frequency and the cyclotron frequency enter into both the oscillator properties and the spatial shift of the wavefunctions of the charged particles. In addition, it is seen that in the limit that $\omega_c = 0$ Eq. (2.42) reduces to Eq. (2.38).

Equation (2.42) is now of the general form of Eq. (2.38) and can be solved in the same manner. In this way, the functions $f(y)$ are now given by

$$f_n(y) = H_n\left(\sqrt{\frac{m(\omega_c^2 + \omega_0^2)^{1/2}}{\hbar}}\left(y - \frac{\omega_c^2}{\omega_c^2 + \omega_0^2}y_k\right)\right)e^{-\frac{1}{2}\frac{m\sqrt{\omega_c^2 + \omega_0^2}}{\hbar}\left(y - \frac{\omega_c^2}{\omega_c^2 + \omega_0^2}y_k\right)^2} \tag{2.43a}$$

$$\epsilon_{n,k} = \frac{\hbar^2k^2}{2m}\frac{\omega_0^2}{\omega_0^2 + \omega_c^2} + \left(n + \frac{1}{2}\right)\hbar\sqrt{\omega_0^2 + \omega_c^2} \tag{2.43b}$$

where $\{H_n(y')\}$ are the normalized Hermite polynomials. In this limit the group velocity of the charged particle is obtained from

$$v_g = \frac{1}{\hbar}\frac{\partial \epsilon_{n,k}}{\partial k} = \frac{\hbar k}{m}\frac{\omega_0^2}{\omega_0^2 + \omega_c^2} = \frac{p_x}{m}\frac{\omega_0^2}{\omega_0^2 + \omega_c^2}, \tag{2.44}$$

representing the particle motion as the combination of a constant momentum translation along the x-direction and a harmonic confinement in the y-direction. Now, however, the group velocity depends on both ω_0 and ω_c, which parameterize the confinement in the y-direction. In addition, the harmonic wavefunction describing the particle motion in the y-direction is shifted above or below the x-axis, depending on the sign of the charge q and the direction of the wavevector for the x-motion.

Note that as $\omega_c \to 0$ Eqs. (2.43) and (2.44), respectively, reduce to Eqs. (2.39) and (2.40). Similarly, in the limit that $\omega_0 \to 0$ the problem reduces to that of a Fermi gas of charges only interacting with the externally applied magnetic induction. It is interesting to note that the group velocity of the particle is zero in this last limit.

2.3.1 Classical Treatment of a Two-Dimensional Gas in a Harmonic Confining Potential

For a comparison with the quantum solution just given, consider the solution of the related model in classical mechanics [1,2,10–19]. In particular, the effects of the harmonic oscillator confining potential will be examined in the classical solution of the Hall effect problem for motion in a viscous medium. For this solution, the harmonic confining potential is again taken in the y-direction, the applied magnetic induction is constant and uniform and in the z-direction, and there is an applied electric field in the x-direction.

The equation of motion of the charge q in the classical mechanics treatment of this configuration is then given by

$$m\left[\dot{\vec{v}} + \vec{v}/\tau\right] = q\left[\vec{E} + \frac{1}{c}\vec{v}\times\vec{B}\right] - m\omega_0^2 y\hat{j} \tag{2.45}$$

where $\vec{E} = E\hat{i}$, $\vec{B} = B\hat{k}$. In the steady state, $\dot{\vec{v}} = 0$ and the solutions for the constant drift velocities in the system are

$$v_x = \frac{q\tau}{m}\left(E_x + \frac{v_y}{c}B\right) \tag{2.46a}$$

$$v_y = \frac{q\tau}{m}\left(E_y - \frac{v_x}{c}B\right) - \omega_0^2\tau y \tag{2.46b}$$

and

$$v_z = \frac{q\tau}{m}E_z \tag{2.46c}$$

The Hall effect solution that is of interest to us is obtained from setting $E_z = v_y = 0$ in Eq. (2.46).

In this solution, there is a current density flowing in the x-direction given by

$$j_x = \frac{nq^2\tau}{m}E_x \tag{2.47a}$$

and the $y-$ component of the electric field is

$$E_y = \frac{q\tau}{m}\frac{B}{c}E_x + \frac{m\omega_0^2}{q}y. \tag{2.47b}$$

It is again found that the $x-$ component of current density is unchanged from that in the system without the harmonic confining potential, but now the Hall field in Eq. (2.47b) is found to depend on the $y-$ coordinate.

2.3.2 Orbital and Landau Diamagnetism: Magnetism of a Two-Dimensional Fermi Gas

An important property of the Fermi gas is its response to a uniform applied magnetic field [10]. Specifically, what is the net magnetic moment displayed by the gas as a response to an applied magnetic field? There are actually two components to the gas's magnetic response. The first is the magnetic response of the Landau levels and the Landau diamagnetism arising from the orbital motion of the charges in the presence of a field, and the second is Pauli paramagnetism. Pauli paramagnetism arises from the intrinsic spin carried by the charged particle, and the electron spin has not yet been discussed. In this section the properties of Landau levels and Landau diamagnetism are treated, and this is followed in the next section by a treatment of Pauli paramagnetism. For the present discussions, the charged particles will continue being considered as spinless. It is then a simple exercise to modify the presentation to include spin.

For the consideration of Landau levels and Landau diamagnetism in a two-dimensional Fermi gas of spinless particles, it is convenient to treat the model of a conducting strip of length $L \to \infty$ in the x-direction and a width W in the y-direction subject to a magnetic induction B applied to the strip in the z-direction [10]. (See Figure 2.4a for a schematic of this geometry.) The gas is now confined in the y-direction by the width of the strip and not by the harmonic confinement considered in the previous section. Consequently, in the absence of the harmonic confinement (i.e. for $\omega_0 = 0$) the Hamiltonian of the gas is from Eq. (2.41) given by [1,2,10–19]

$$\left[\frac{p_y^{\,2}}{2m} + \frac{m\omega_c^2}{2}(y - y_k)^2\right] f(y) = \epsilon f(y), \tag{2.48}$$

where again $\omega_c = \dfrac{|q| B}{mc}$ is the cyclotron frequency for the circular orbit of the charged particle in a magnetic induction B, and $y_k = -\dfrac{\hbar c}{qB} k$ is the shift of the wavefunction solution along the y-axis. The confinement of the particles to the finite width of the strip now enters the problem as part of the boundary conditions.

In the $\omega_0 = 0$ limit the solutions of the eigenvalue problem in Eq. (2.48) are of the form [1,2,10–19]

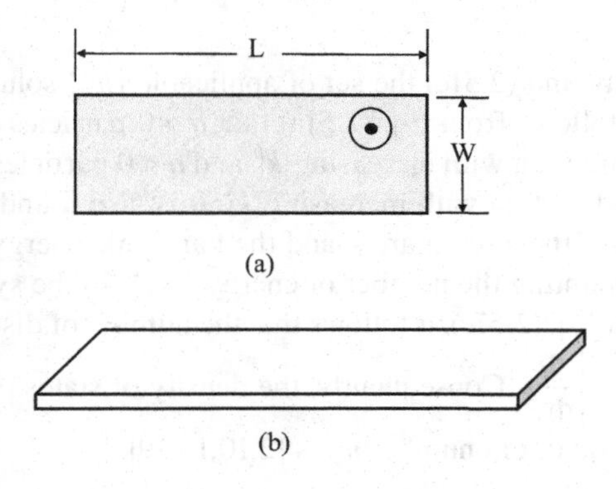

FIGURE 2.4 Schematic for the flow of current in a strip of media. In (a) is a planar rectangle of media with the x-axis horizontal and the y-axis vertical, and the magnetic field along the z-axis, and (b) the same rectangle, given a small height in the z-direct.

$$\psi_{n,k}(x,y) = \frac{1}{\sqrt{L}} e^{ikx} f_n(y) = \frac{1}{\sqrt{L}} e^{ikx} H_n\left(\sqrt{\frac{m|\omega_c|}{\hbar}}(y - y_k)\right) e^{-\frac{1}{2}\frac{m|\omega_c|}{\hbar}(y-y_k)^2} \tag{2.49a}$$

$$\epsilon_n = \left(n + \frac{1}{2}\right)\hbar|\omega_c| \tag{2.49b}$$

where $\{H_n(y')\}$ are the normalized Hermite polynomials. Again applying periodic boundary conditions in the x-direction yields the conditions on k,

$$k = \frac{2\pi}{L} n_x \tag{2.50}$$

where n_x is an integer. Another boundary condition arises in the context of the parameter

$$y_k = -\frac{\hbar c}{qB} k, \tag{2.51a}$$

which must be restricted to be inside the strip. Consequently, it follows that $0 < y_k < W$.
Applying this condition for $q > 0$, it is found from Eq. (2.50) that n_x must satisfy

$$0 > n_x > -\frac{m|\omega_c|LW}{2\pi\hbar} = -\frac{\Phi}{\Phi_q} \tag{2.51b}$$

where $\Phi = BLW$ is the magnetic flux through the strip, and $\Phi_q = \left|\frac{2\pi\hbar c}{q}\right|$ is a basic quantum unit of magnetic flux. Similarly for $q < 0$ the condition that $0 < y_k < W$ requires

$$0 > -n_x > -\frac{m|\omega_c|LW}{2\pi\hbar} = -\frac{\Phi}{\Phi_q} \tag{2.51c}$$

or

$$0 < n_x < \frac{m|\omega_c|LW}{2\pi\hbar} = \frac{\Phi}{\Phi_q}. \tag{2.51d}$$

Note that in both Eqs. (2.51b) and (2.51c) the set of applicable $\{n_x\}$ solutions are dependent on the sign of q. In this regard, it follows from Eq. (2.51a) that $q > 0$ particles are states of $k < 0$ that are displaced upward in the y-direction with increasing $|k|$, and $q < 0$ particles are states of $k > 0$ that are displaced upwards in the y-direction with increasing $|k|$. It is also found that the relevant quantum numbers labeling the states of the system are k and the harmonic energy level quantum number n, and both of these apply in counting the number of energy levels in the system.

From either Eq. (2.51b) or Eq. (2.51c), it follows that the number of distinct k solutions for a given energy level n is $\frac{m|\omega_c|}{2\pi\hbar} LW = \left|\frac{\Phi}{\Phi_q}\right|$. Consequently, the density of states per level n (i.e., the number of states in each level of the quantum number n.) is [2,10,12,19]

$$D(n) = \frac{m|\omega_c|}{2\pi\hbar} LW = \left|\frac{\Phi}{\Phi_q}\right|, \tag{2.52}$$

where it should be remembered that this is for spinless particles. This agrees with the earlier result in Eq. (2.18) that the density of energy states in the two-dimensional gas confined in an area A is independent of the energy and given by

$$D_2(\epsilon) = A\frac{m}{2\pi\hbar^2} \tag{2.53a}$$

Specifically, from Eq. (2.53a) it follows that the number of states in an energy interval $\hbar\omega_c$ is expressed as

$$\tilde{N}(n) = D_2(\epsilon)\hbar\omega_c = A\frac{m}{2\pi\hbar^2}\hbar|\omega_c| = \left|\frac{\Phi}{\Phi_q}\right|. \tag{2.53b}$$

so that each quantum level n contains the same number of energy states. This is an important relationship that will be used later to determine the ground state energy of the Fermi gas in an external magnetic field. First, however, consider the ground state energy of the gas in the absence of a field.

At $T = 0$ in the absence of an applied magnetic field, the total number of particles, N, in the Fermi gas is related to the Fermi energy (i.e., the highest occupied energy level.) of the gas of spinless particles by

$$N = \int_0^{\epsilon_F} D_2(\epsilon)d\epsilon = A\frac{m}{2\pi\hbar^2}\epsilon_F, \tag{2.54}$$

where ϵ_F denotes the Fermi energy. From Eq. (2.54) it is seen that

$$\epsilon_F = n\frac{2\pi\hbar^2}{m} \tag{2.55}$$

relates the Fermi energy of the $B = 0$ gas to the density of particles $n = \frac{N}{A}$ in the gas. In addition, the total energy per particle of the particles in the gas is obtained from $D_2(\epsilon)$, as

$$\epsilon_T(T = 0, B = 0) = \frac{1}{N}\int_0^{\epsilon_F} \epsilon D_2(\epsilon)d\epsilon = \frac{\epsilon_F}{2}. \tag{2.56}$$

This again relates the total energy per particle to the density of particles in the gas. Next consider the changes of the energy per particle of the gas upon the application of an external magnetic field to the system and how this compares with that of the $B = 0$ gas.

Now for $T = 0$ and $B \neq 0$, the counting of states in the system involves the harmonic oscillator quantum numbers n rather than a direct counting of wavenumbers [10]. In particular, at $T = 0$, the N particles of the gas occupy the lowest energy oscillator states. Let us say that the N particles of the gas are spread among the lowest r oscillator states of the systems. In this case the oscillator levels $n = 0, 1, 2, \ldots, r-1$ will be totally occupied, and the $n = r$ oscillator state will be the last perhaps partially occupied level. From Eq. (2.53b) for $n = 0, 1, 2, \ldots, r-1$

$$\tilde{N}(n) = A\frac{m}{2\pi\hbar^2}\hbar|\omega_c| = \left|\frac{\Phi}{\Phi_q}\right| = N_L \tag{2.57a}$$

is the number of states in each of the totally occupied levels and for the remaining partially occupied level [1,2,10,12–19]

$$\tilde{N}(r) = \alpha A\frac{m}{2\pi\hbar^2}\hbar|\omega_c| = \alpha N_L, \tag{2.57b}$$

where $0 \le \alpha \le 1$. Consequently, the total number of particles in the system is then, summing all of these states,

$$N = (r+\alpha)N_L = sN_L = s\left|\frac{\Phi}{\Phi_q}\right|, \tag{2.57c}$$

where $r \le s \le r+1$ for $s = r+\alpha$.

From the definition of the flux $\Phi = BA$ and the carrier density in two dimensions $n = \frac{N}{A}$, it then follows from Eq. (2.57c) that

$$s = n\frac{|\Phi_q|}{B} = n\frac{1}{B}\frac{hc}{|q|} = \frac{B_0}{B}, \tag{2.58a}$$

where $B_0 = n|\Phi_q|$ and

$$\alpha = s - r \tag{2.58b}$$

Since each filled oscillator level contains N_L charges, and the energies of the levels are given by Eq. (2.49b), it follows that the total ground state energy per particle of the gas is given by [1,2,10,12–19]

$$\epsilon_T = \left[\sum_{n=0}^{r-1}\left(n+\frac{1}{2}\right)\hbar|\omega_c| + \alpha\left(r+\frac{1}{2}\right)\hbar|\omega_c|\right]\frac{N_L}{N}. \tag{2.59}$$

Applying the summation formula $\sum_{l=1}^{n} l = \frac{1}{2}n(n+1)$ and following some further arithmetic, this expression is simplified. From this it is found that [1,2,10]

$$\epsilon_T = \left[\frac{r^2}{2} + \left(r+\frac{1}{2}\right)\alpha\right]\frac{\hbar|\omega_c|}{s} = \frac{\epsilon_F}{2} + \frac{\hbar|\omega_c|}{2s}\alpha(1-\alpha) \tag{2.60}$$

where ϵ_F is defined in Eq. (2.55) as the Fermi energy of the $B = 0$ gas, and α is the fractional occupancy of the highest r oscillator level accepting ground state particles.

In terms of the result in Eq. (2.60), the ratio of the total energy in the presence and absence of the applied field B becomes [1,2,10]

$$\frac{\epsilon_T(B)}{\epsilon_T(0)} = 1 + \left(\frac{B}{B_0}\right)^2 \alpha(1-\alpha). \tag{2.61}$$

In addition, from Eq. (2.60) the magnetization per particle response of the gas to the applied field B is expressed in terms of the total energy and obtained from

$$M = -\frac{\partial \epsilon_T}{\partial B}. \tag{2.62a}$$

For infinitesimal changes in B, r is a constant in Eq. (2.58b) so that [10]

$$\frac{\partial \alpha}{\partial B} = \frac{\partial s}{\partial B} = -\frac{B_0}{B^2} \tag{2.62b}$$

and, applying this in Eq. (2.62), the magnetization of the gas with its highest energy particles occupying the r oscillator level, the magnetization per particle is found to be

$$M = \frac{\epsilon_T(0)}{B_0}\left[1 - 2\alpha - \frac{2\alpha(1-\alpha)}{r+\alpha}\right]. \tag{2.63a}$$

In the limit that there are many oscillator levels so that $r \gg 1$, the last term in the bracket in Eq. (2.63a) drops out, and it is seen that [10]

$$M \to \frac{\epsilon_T(0)}{B_0}[1 - 2\alpha]. \tag{2.63b}$$

It is interesting to note that the application of an external magnetic field to the Fermi gas changes the excitation spectrum of the gas from a continuum set of energy levels given by $\epsilon = \frac{\hbar^2}{2m}k^2$ to a discrete set of harmonic oscillator levels of the form $\epsilon = \left(n + \frac{1}{2}\right)\hbar\omega_c$. In this transformation of the particle dispersion relations between the two systems, particles with a range of k having different energies $\epsilon = \frac{\hbar^2}{2m}k^2$ are regrouped into states, each of which is characterized by the energy of a single oscillator quantum number n. Consequently, many particles of the gas with a range of energies are transformed to have the same energy that is characterized by the harmonic oscillator quantum number n. The discrete set of oscillator energy states, in the presence of the applied field, shows up in the new physical properties of the system, provided that the temperature of the system satisfies the condition, $k_BT \ll \hbar\omega_c$. Specifically, in this limit of the temperature the discrete nature of the energy levels is not washed out by thermal effects.

In the opposite limit, in which $k_BT \gg \hbar\omega_c$, the thermal fluctuations dominate over the discrete nature of the Landau levels in the gas. This is the region of temperatures in which the gas exhibits the effects known as Landau diamagnetism. The properties of the gas under a Landau diamagnetic response requires a study of the thermal occupancy of the Landau levels involving an application of the Grand Canonical Partition Function, and for this the reader is referred to Refs. [1] and [2]. Here it is only stated that in this limit the magnetization per electron is related to the applied magnetic induction B by [1,2,10]

$$M = \chi_{LD}B \tag{2.64a}$$

where the Landau diamagnetic susceptibility of the two-dimensional gas is given by

$$\chi_{LD} = -\frac{q^2}{12\pi mc^2}. \tag{2.64b}$$

2.3.3 Magnetic Properties and Landau Diamagnetism of a Slab of Fermi Gas

The earlier considerations of Landau diamagnetism in a two-dimensional Fermi gas of spinless particles can be extended to treat a slab of gas of finite thickness. In this model, the conducting strip of length $L \to \infty$ in the x-direction and width W in the y-direction is now given a small finite thickness H in the z-direction. The resulting slab is again subject to a magnetic induction B applied in the z-direction. (See Figure 2.4b for a schematic.) Under these conditions, the Hamiltonian of the gas is modified from Eq. (2.48) to include a $z-$ component of momentum, making it of the form [1,2,10,12–19].

$$\left[\frac{p_y^2}{2m} + \frac{p_z^2}{2m} + \frac{m\omega_c^2}{2}(y - y_k)^2\right]f(y) = \epsilon f(y), \tag{2.65}$$

where again $\omega_c = \frac{|q|B}{mc}$ is the cyclotron frequency of the particle of charge q, and $y_k = -\frac{\hbar c}{qB}k$ is the shift of the wavefunction solution along the y-axis. The confinement of the particles to the slab again enters through the boundary conditions that are set to limit the current of particles to be in the x-direction.

The solutions of the eigenvalue problem in Eq. (2.65) are of the form [10]

$$\psi_{n,k}(x,y,z) = \frac{1}{\sqrt{L}}e^{ikx}f_n(y,z) = \frac{1}{\sqrt{L}}e^{ikx}H_n\left(\sqrt{\frac{m|\omega_c|}{\hbar}}(y-y_k)\right)e^{-\frac{1}{2}\frac{m|\omega_c|}{\hbar}(y-y_k)^2}\sqrt{\frac{2}{H}}sin(k_z z), \quad (2.66a)$$

$$\epsilon_n = \frac{\hbar^2 k_z^2}{2m} + \left(n + \frac{1}{2}\right)\hbar|\omega_c| \quad (2.66b)$$

where $\{H_n(y')\}$ are the normalized Hermite polynomials, and the particles are now confined to the slab in the z-direction by the condition $0 \le z \le H$. Again, applying periodic boundary conditions in the x-direction yields the condition on k for conductive solutions in the x-direction so that

$$k = \frac{2\pi}{L}n_x \quad (2.67a)$$

where n_x is an integer. In addition, the need for new confining boundary conditions in the z-direction implies that

$$k_z = \frac{\pi}{H}n_z, \quad (2.67b)$$

where n_z is a positive integer. As a final condition, it remains from our earlier considerations that in the y-direction the boundary condition on the parameter

$$y_k = -\frac{\hbar c}{qB}k \quad (2.68)$$

is the condition that $0 < y_k < W$, restricting y_k to remain within the slab. With these requirements the form of the particle wavefunctions is then completely specified [10].

For a consideration of the eigenvalue energies of the particles, it follows from introducing Eq. (2.67) into Eq. (2.66b) that the energy dispersion relation is written as [1,2,10,12–19]

$$\epsilon_n = \frac{\hbar^2}{2m}\left(\frac{\pi}{H}\right)^2 n_z^2 + \left(n + \frac{1}{2}\right)\hbar|\omega_c|. \quad (2.69)$$

From the first term on the right-hand side it is found that as H becomes small, the first term in Eq. (2.69) becomes very large so that the energy spectrum is composed of a series of $\left(n + \frac{1}{2}\right)\hbar|\omega_c|$ oscillator spectra, each of which is separated by large energy intervals of the form $\frac{\hbar^2}{2m}\left(\frac{\pi}{H}\right)^2 n_z^2$. In the limit that $H = 0$, only one oscillator spectrum remains.

2.3.4 Pauli Paramagnetism

In this section the second type of magnetic interaction possible in the Fermi gas is considered. This is the Pauli paramagnetism, which arises from the interaction of the applied magnetic field with the magnetic moments associated with the spins of the charged particles [10]. For the first time in this chapter, then, the charged particles are now considered to have an intrinsic spin with an associated magnetic moment. In particular, the treatment given is for a Fermi gas composed of spin one-half particles.

Let us specify to considerations of a Fermi gas of electrons in an externally applied magnetic induction B. For each of the electrons in the gas, the magnetic moment is related to the electron spin, $\vec{s}$, by [1,2,10]

$$\vec{\mu} = -\frac{ge}{2mc}\vec{s} = -\frac{g}{2}\frac{e\hbar}{2mc}\vec{\sigma} = -\frac{g}{2}\mu_B\vec{\sigma} \tag{2.70}$$

where $g \approx 2$ is the gyromagnetic factor for the electron, $e > 0$ is the unit of electric charge, $\vec{s} = \frac{\hbar}{2}\vec{\sigma}$ where $\vec{\sigma}$ are the Pauli matrices, and $\mu_B = \frac{e\hbar}{2mc}$ is the Bohr magneton. The energy of interaction U of the electron spin with the applied magnetic field is given by $U = -\vec{\mu}\cdot\vec{B}$, so that, ignoring the magnetic contribution of the orbital motion, the orbital energy of an electron is of the form

$$\begin{aligned} \epsilon &= \frac{\hbar^2}{2m}k^2 - \mu_B B, \text{ where } \vec{B} \text{ and } \vec{\mu} \text{ are parallel} \\ &= \frac{\hbar^2}{2m}k^2 + \mu_B B, \text{ where } \vec{B} \text{ and } \vec{\mu} \text{ are anti-parallel} \end{aligned} \tag{2.71}$$

for $\vec{k} = (k_x, k_y)$ in a two-dimensional gas or $\vec{k} = (k_x, k_y, k_z)$ in a three-dimensional gas. The model in Eq. (2.71) is essentially that of a gas of uncharged particles that, however, interact with the external field through their magnetic moments. In this regard, the introduction of spin into the general problem acts only to split the energies of the states of the spinless system by $\pm\mu_B B$.

Consider the case of the two-dimensional gas of electrons in the $x-y$ plane at $k_B T \ll \epsilon_F$ with an applied magnetic field in the z-direction. The energies of the electrons in the system are given in Eq. (2.71) so that for a two-dimensional gas it follows from Eq. (2.18) that the number of spin-down electrons (with magnetic moments parallel to the magnetic field) in the gas is

$$N_+ = \int_{-\mu_B B}^{\epsilon_F} d\epsilon D_2(\epsilon + \mu_B B) = \frac{A}{2\pi}\frac{m}{\hbar^2}(\epsilon_F + \mu_B B), \tag{2.72a}$$

while the number of spin-up electrons (with magnetic moments antiparallel to the magnetic field) in the same gas is [10]

$$N_- = \int_{\mu_B B}^{\epsilon_F} d\epsilon D_2(\epsilon - \mu_B B) = \frac{A}{2\pi}\frac{m}{\hbar^2}(\epsilon_F - \mu_B B). \tag{2.72b}$$

Here A is the area of the gas, and $D_2(\epsilon)$ is the density of energy states for a spinless gas of particles that is used to separately describe the spin-up and spin-down states of the gas of particles with spins. It is important to note in Eq. (2.72) that the Fermi energies of both the spin-up and spin-down particles are the same in the ground state of the gas.

In terms of Eq. (2.72), the magnetization per particle of the Fermi gas of $N = N_+ + N_-$ electrons for $k_B T \ll \epsilon_F$ is [10]

$$M = \frac{\mu_B(N_+ - N_-)}{N} = 2\mu_B^2\frac{A}{2\pi}\frac{m}{\hbar^2}\frac{B}{N} = \frac{2\mu_B^2 D_2(\epsilon_F)}{N}B, \tag{2.73}$$

and a net magnetization of the system is generated parallel to the applied magnetic induction. In addition, the Fermi energy of the gas can be related to the total particle density of the gas from Eq. (2.72) and is given in two dimensions by [10]

$$\epsilon_F = \frac{\pi}{A}\frac{\hbar^2}{m}(N_+ + N_-) = \frac{\pi}{A}\frac{\hbar^2}{m}N = \pi\frac{\hbar^2}{m}n = \frac{N}{2D_2(\epsilon_F)} \tag{2.74}$$

where N is the total number of electrons in the gas. It is seen from Eq. (2.74) that the Fermi energy from the spin interaction of the $B \neq 0$ system is unchanged from that of the $B = 0$ system.

Note that the orbital effects of the applied magnetic field were ignored in Eq. (2.71). They could have been included but at low fields would have introduced only a small correction. This comes from the fact that the spins and orbital coordinates are separate from one another, and separately interact with the applied fields. The essential physics of Pauli paramagnetism comes from the energy splitting of the energy levels of the zero-spin particles by $\pm\mu_B B$ upon the application of the field B into the system with the introduction of the spin to the particles.

In conclusion, for the two-dimensional electron gas with spins the orbital contribution to the magnetization per particle at low temperatures, $k_B T \ll \hbar\omega_c \ll \epsilon_F$, is [1,2,10]

$$M = \frac{\pi\hbar^2}{2m}n\frac{2e}{nch}(1-2\alpha) = \frac{\hbar e}{2mc}(1-2\alpha), \tag{2.75}$$

and the Paul paramagnetic spin contribution is given by Eq. (2.73). However, for the limit that $k_B T \gg \hbar\omega_c$, the magnetization per particle from the orbital motion is given from the Landau diamagnetism result in Eq. (2.64), which for the electron gas takes the form

$$M = \chi_{LD} B, \tag{2.76a}$$

where the Landau diamagnetic susceptibility is given by

$$\chi_{LD} = -\frac{e^2}{12\pi mc^2}, \tag{2.76b}$$

and the Pauli spin contribution to the magnetization per particle is given from Eq. (2.73). In both the limits, to leading order in the model, the Landau and Pauli contributions to the magnetization just add to give the leading order magnetic response of the Fermi electron gas.

2.4 STONER THEORY OF PERMANENT MAGNETISM OF METALS

To this point, the properties of a Fermi gas of spin-up and spin-down particles have been described by theories in which the particles do not interact with one another but, otherwise, are subject to interaction with external applied fields [1,2,10]. These theories are for the most part successful in describing basic transport properties associated with electrical currents and magnetic properties that focus on the single particle responses to applied magnetic fields. However, if one allows for interactions between the particles of the gas themselves, it is possible to arrive at theories for much more complex phenomena involving correlations in the particle motions due to their particle–particle interaction. In particular, in this section, such consideration will be made in discussions of permanent ferromagnetism in the gas of particles carrying spin states. The effects of particle–particle interactions as addressed here are important in approaching an understanding of technologically important ferromagnetic materials such as iron, cobalt, and nickel. For the Fermi gas in the ferromagnetic state, there is an ordered state of the spins in the gas in which one spin state (i.e., spin up or spin down) is numerically preferred in the system over the other. In transitioning from the state in which no preferred spin state exists in the gas to the ferromagnetic state, the gas is said to have undergone a phase transition that involves developing a highly collective state of many body motions [1,2,10]. It shall be seen that this transition is driven solely by the interactions between the particles in the gas.

In this regard, a simple treatment of the particle interactions is that made within the context of what is generally termed as a mean field theory. This offers a semiquantitative understanding of the origin of the ferromagnetic state, and in its simplest form provides the basis of the Stoner theory of ferromagnetism in metallic materials exhibiting permanent magnetic ordering. Specifically, in the Stoner theory the particles of the system are assumed to interact with an average effective field (i.e., a mean field) created from their interactions with the other particles in the gas. Upon inclusion of this effective field it is found that under certain conditions the system prefers the occupancy of one spin state of the gas particles over that of the other spin state.

To see how this works, consider a gas of spin-up and spin-down particles, which in the absence of an applied external field both have dispersion relations of the form

$$\epsilon_{\uparrow} = \epsilon_{\downarrow} = \frac{\hbar^2}{2m} k^2. \tag{2.77a}$$

For a three-dimensional gas the density of spin-up modes is given by $D_3(\epsilon_{\uparrow})$, and the density of spin-down modes is given by $D_3(\epsilon_{\downarrow})$ where $D_3(\epsilon)$ is defined in Eq. (2.16) in terms of spinless particles, i.e., $D_3(\epsilon)$ is the density of states for spinless particles. (See Figure 2.5 for a schematic for the gas of spin-up and spin-down particles.) It follows from these considerations that at zero temperature and in the absence of particle–particle interactions, the number of up and down spins in the gas are given by [10]

$$N_{\uparrow} = \int_0^{\epsilon_F} d\epsilon_{\uparrow} D_3(\epsilon_{\uparrow}) = N_{\downarrow} = \int_0^{\epsilon_F} d\epsilon_{\downarrow} D_3(\epsilon_{\downarrow}). \tag{2.77b}$$

These results reflect the energy degeneracy of the particles with respect to spin and the fact that the Fermi energy is the same for both spin-up and spin-down particles. Specifically, the gas has no

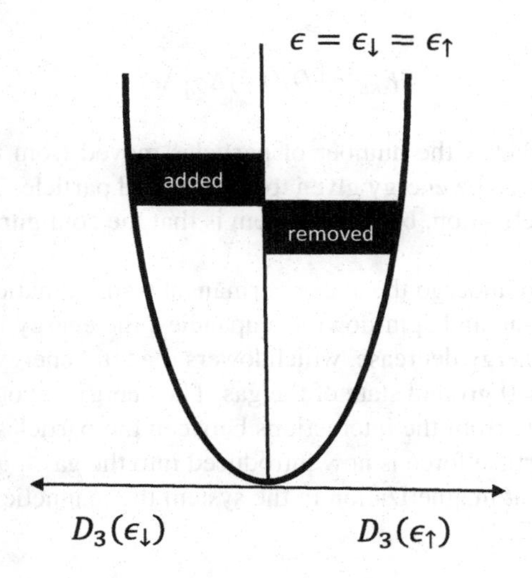

FIGURE 2.5 Density of energy states $D_3(\epsilon_{\uparrow})$ and $D_3(\epsilon_{\downarrow})$ for, respectively, up- and down-spin electrons versus energy ϵ denoted either for the up spins $\epsilon_{\uparrow}$ or for the down spins $\epsilon_{\downarrow}$. The shaded regions represent the states subtracted from $D_3(\epsilon_{\uparrow})$ and added to $D_3(\epsilon_{\downarrow})$, as discussed in the text for the treatment of ferromagnetic ordering. In the plot, the density of states are plotted on the horizontal axis, and the electron energies are plotted on the vertical axis.

preference in the occupancy of one spin state over the other, so there is no net magnetic moment developed in the gas. From Eq. (2.77b), it also is seen that the total number of particles in the gas is

$$N = N_{\uparrow} + N_{\downarrow}. \tag{2.77c}$$

Next, as a comparison, the effects of particle–particle interactions will be introduced in the context of the ideas of mean field theory. From these discussions it will be found that a net difference between the number of spin-up and spin-down requires intraparticle forces.

First consider the system in Eq. (2.77) at $T = 0$ for the case in which an energy band of states $\Delta\epsilon$ is taken from the top of the spin-up particle states and moved to the top of the spin-down states, then the spin-up number of particles is decreased by

$$\Delta N_{\uparrow} = D_3(\epsilon_F)\Delta\epsilon, \tag{2.78a}$$

and the spin-down number of particles is increased by

$$\Delta N_{\downarrow} = D_3(\epsilon_F)\Delta\epsilon, \tag{2.78b}$$

where the Fermi energy is the same for both the up- and down-spin particles and $\Delta\epsilon \ll \epsilon_F$ at the beginning of the process. (Note that the particle spins as well as their energies must be changed to move particles from the states of spin-up to the spin-down states.) For this process, the net number of particles in the gas is unchanged, and only the numbers of up- and down-spin particles have been adjusted.

As a consequence of this shuffling of the particle occupancy, in the final configuration of the gas there are

$$\Delta N_{\downarrow} + \Delta N_{\uparrow} = 2D_3(\epsilon_F)\Delta\epsilon \tag{2.79}$$

net more particles in the spin-down states of the gas than in the spin-up states. This assures a net nonzero magnetization of the gas. In addition, the increase in the kinetic energy of the gas from the $T = 0$ ground state is [10]

$$E_{KE} = \left[D_3(\epsilon_F)\Delta\epsilon\right]\Delta\epsilon, \tag{2.80}$$

where the term in the bracket is the number of particles moved from the up spin particles to be down-spin particles, and $\Delta\epsilon$ is the energy given to each moved particle. The resulting configuration of the gas gives a net magnetization, but the problem is that the configuration is no longer a ground state of the gas.

In order for the system to undergo the above permanent transformation to a state with an imbalance of the number of spin-up and spin-down occupancies, the energy increase in Eq. (2.80) must be counterbalanced by an energy decrease, which lowers the total energy of the imbalanced system so as to make it the new $T = 0$ ground state of the gas. The energy to counteract the increase in Eq. (2.80) is now shown to come from the interactions between the particles of the gas. For these considerations, the particle–particle force is now introduced into the gas.

In the case of a permanent magnetization in the system the magnetic moment per particle, m, is given from Eq. (2.78) by [1,2,10]

$$m = \frac{M}{N} = \mu_B\left(n_{\downarrow} - n_{\uparrow}\right) = \frac{1}{N}\mu_B 2D_3(\epsilon_F)\Delta\epsilon \tag{2.81}$$

where $n_{\uparrow}$ ($n_{\downarrow}$) are the fraction of spin-up (spin-down) particles in the gas, μ_B is Bohr magneton, and we have assumed that the fermions are negatively charged. In the case that $M \neq 0$ in the ground state of the gas, it is assumed that the magnetization is caused by a net effective field arising from the interactions between the particles. To leading order, the effective field can reasonably be represented as being related to the permanent magnetization by [10]

$$B_{eff} = \lambda M, \tag{2.82}$$

where λ parameterizes the strength of the effective field. This characterization gives a zero effective field in the absence of a permanent magnetization of the gas but is nonzero when a permanent magnetization is present in the gas.

For a small magnetization ΔM in the gas, its energy of interaction in the presence of the effective field is given by

$$\Delta E_{\text{int}} = -B_{\text{eff}}\Delta M = -\lambda M \Delta M = \Delta\left[-\frac{1}{2}\lambda M^2\right]. \tag{2.83}$$

Consequently, for a general field

$$E_{\text{int}} = -\frac{1}{2}\lambda M^2 = -\frac{1}{2}\lambda\left[N\mu_B\left(n_\downarrow - n_\uparrow\right)\right]^2 = -U\left[D_3(\epsilon_F)\Delta\epsilon\right]^2 \tag{2.84}$$

where $U = 2\lambda\mu_B^2$, and from Eqs. (2.80) to (2.84) the total change in energy of the gas upon shifting its configuration is [10]

$$E_{Total} = D_3(\epsilon_F)\Delta\epsilon^2 - UD_3(\epsilon_F)^2\Delta\epsilon^2 = D_3(\epsilon_F)\Delta\epsilon^2\left[1 - UD_3(\epsilon_F)\right]. \tag{2.85}$$

The condition for the stability of the shifted configuration is then $E_{\text{Total}} = 0$, requiring that

$$UD_3(\epsilon_F) = 1. \tag{2.86}$$

A phase transition to the ferromagnetic state then occurs in the gas under the condition in Eq. (2.86), which directly relates the transition to the density of states at the Fermi energy and the effective field parameter λ. In this regard, it is interesting to note that the ferromagnetic metals iron, nickel, and cobalt have particularly high density of energy states at their Fermi energies, and this is surely a factor in presence of their magnetic transitions.

Another interaction of importance that can be easily treated in the ferromagnetic gas is that from an applied magnetic field. Upon the application of an externally applied B to the gas of interacting particles, an additional term is added to Eq. (2.85), which represents the additional energy in the system arising from the interaction of the gas magnetization M with B. Specifically, the energy change in the gas from this interaction is represented by

$$E_B = -MB. \tag{2.87}$$

Adding this into Eq. (2.85) it follows that total additional energy in the system is now given by

$$E_{\text{Total}} = D_3(\epsilon_F)(\Delta\epsilon)^2\left[1 - UD_3(\epsilon_F)\right] - MB = \frac{M^2}{4\mu_B^2 D_3(\epsilon_F)}\left[1 - UD_3(\epsilon_F)\right] - MB, \tag{2.88}$$

where Eq. (2.81) has been used to write $D_3(\epsilon_F)(\Delta\epsilon)^2 = \dfrac{M^2}{4\mu_B^2 D_3(\epsilon_F)}$. In Eq. (2.88), the first term represents the energy change in the gas from the differential occupancy in the spin-up and spin-down particles of the gas when a magnetization M exists in the gas, and the second term is the energy from the interaction of the field with the magnetization. The properties of the gas of interacting particles in the presence of an applied magnetic field now follow directly from the study of Eq. (2.88).

The energy in Eq. (2.88), which shows how the gas depends on its net magnetization, provides for the determination of the nature of the equilibrium state of the gas. In this regard, the equilibrium state of the gas is obtained by minimizing the energy form in Eq. (2.88) with respect to the gas magnetization M. Proceeding in this way from the conditions of minimization, it follows that [10]

$$\frac{\partial E_{\text{Total}}}{\partial M} = 0 = \frac{M}{2\mu_B^2 D_3(\epsilon_F)}\left[1 - UD_3(\epsilon_F)\right] - B, \tag{2.89}$$

so that solving Eq. (2.89) for the magnetization in terms of the applied field it is found that

$$M = \frac{2\mu_B^2 D_3(\epsilon_F)}{1 - UD_3(\epsilon_F)} B. \tag{2.90}$$

From Eq. (2.90) it is seen that in the $0 \le UD_3(\epsilon_F) \le 1$ region of physically interesting solutions the gas is not permanently ordered into the ferromagnetic state, and the nonzero magnetization is linearly related to the applied field through a paramagnetic relationship [10].

From Eq. (2.90), the susceptibility of the gas of interacting particles is then given by

$$\chi = \frac{\partial M}{\partial B} = \frac{2\mu_B^2 D_3(\epsilon_F)}{1 - UD_3(\epsilon_F)}. \tag{2.91}$$

Here for $U=0$ Eq. (2.91) reduces to the susceptibility of the noninteracting gas of particles. In the case that $UD_3(\epsilon_F) = 1$, however, the susceptibility becomes divergent, and this divergence signals the instability of the system to the permanently magnetized gas. Between the $U = 0$ limit and the divergence at $UD_3(\epsilon_F) = 1$, the magnetization and paramagnetic susceptibility response of the gas are enhanced from that of the gas of noninteracting particles.

2.5 LOCALIZATION PROPERTIES OF THE FERMI GAS MODEL

In the previous section, aside from a brief study of the Stoner model of ferromagnetism, the discussions have focused on elementary treatments of conductivity and simple magnetic responses of the conduction particles in the context of the Fermi gas model. For the most part, the particles of the gases did not interact with one another, and the scattering properties of the particles with the viscous medium ignored any types of phase-coherent processes. In particular, upon interacting with the viscous medium, the particles did not carry phase information that could be directly related to the phase processes impressed upon them during the scattering. As a result, the particles failed to retain the phase changes characterizing their total transport through the system. The discussions will now address the consequences of including memory effects into the system, along with a presentation of treatments of additional phenomena arising from the details of the particle–particle interactions.

In this section some discussions will be given of Mott and Anderson localization [20–30]. These two types of localization arise, respectively, from the inclusion of particle–particle interactions [20–24] and phase-coherent scatterings into the Fermi gas model [22–24]. Mott localization involves materials formed as crystalline arrays of atoms and molecules with strong electron–electron interactions, and is concerned with the transformation of these materials between conducting and nonconducting states of matter. The transformation between the two states of conduction arises from the repulsive interactions that are introduced between the particles in the Fermi gas of the conduction electrons, transforming it into a Fermi liquid of strongly correlated electrons. In the absence of particle–particle interactions, the Fermi gas of conduction electrons is a conducting system, but a point of interaction is reached in which the Fermi liquid becomes nonconductive as the particle–particle interaction is turned on and increased in intensity. This is then a metal–insulator transition, and is found to account for a variety of metal–insulator phase transitions observed in materials.

Anderson localization, on the other hand, is a conductor to nonconductor transition that arises due to disorder in the system [23,24]. It comes from phase-coherent impurity scattering in the system that confines the particles of the Fermi gas to be trapped in certain localized region of space. Consequently, the particles cannot travel throughout the material, and there is an absence of conduction. This type of localization is similar to the trapping of water in a geographic topography. In Michigan, for example, there are many lakes that limit the flow of their waters through the confining contours of the land so that there is no flow between lakes. In a quantum mechanical potential landscape, electrons can similarly become trapped within a potential landscape. In quantum systems, however, the phase properties of the particle motion enter into the localization process so as to enhance the ability of the landscape to localize the particles over that found in classical mechanical systems. In particular, it will be shown that Anderson localization always occurs in one- and two-dimensional disordered systems, and these types of systems are insulators.

As temperature is introduced into an Anderson localized media, it is found that thermal hopping of charged particles between two different regions of localization begins to occur [23,24]. Under these conditions, the transmission of charged particles through the material is assisted by means of the thermal fluctuations in the system. This assisted hoping shows up in a temperature-dependent conductivity based on the so-called Mott variable-range hopping mechanism. Characteristic exponential temperature forms for the electrical conductivity are observed in materials supporting these processes, and the details of the temperature dependence are sensitive to the geometric characteristics of the localized wavefunctions of the charge carriers.

In the following, first a presentation on Anderson localization and the ideas of the Mott variable-range hopping mechanism are given. This is followed by discussion of the Mott transition and the related ideas of a Wigner solid [20,21] composed as a crystalized array of electrons. It is good to begin the study of Anderson localization by treating a simple example involving the localization of a one-dimensional Fermi gas, which is scattered by a random array of delta-function potentials.

Consider a one-dimensional Fermi gas propagating along the x-axis in the absence of particle–particle interactions [25]. The gas of noninteracting particles, however, interacts with a randomly disordered scattering media characterized as an array of random delta-function scattering potentials that are fixed in space. It will be shown that the wavefunctions of the gas particles in this gas each become localized to a finite region of the one-dimensional space by the random medium in which the particles propagate. This represents a radical transition from the case in the absence of the delta-function scatterings in which the particle wavefunctions are not localized but extend throughout the entire x-axis.

To begin with the treatment of the scattering problem, first consider the description of the solutions of a gas particle propagating in the region $x_a \le x \le x_b$ of the x-axis with a motion described by the Schrodinger equation [25]

$$-\frac{\hbar^2}{2m}\frac{d^2\psi}{dx^2} + u\psi = \epsilon\psi, \tag{2.92}$$

where u is a constant over the interval $x_a \le x \le x_b$. The general solution of Eq. (2.92) in the interval has the form $\psi(x) = Ae^{ikx} + Be^{-ikx}$ corresponding to the energy $\epsilon = \frac{\hbar^2 k^2}{2m} + u$, and this can be written in a matrix form for the wavefunction and its $x-$ derivative as

$$\begin{vmatrix} \psi(x) \\ \psi'(x) \end{vmatrix} = \begin{vmatrix} e^{ikx} & e^{-ikx} \\ ike^{ikx} & -ike^{-ikx} \end{vmatrix} \begin{vmatrix} A \\ B \end{vmatrix} = M(x) \begin{vmatrix} A \\ B \end{vmatrix}, \tag{2.93}$$

where $k=\sqrt{\frac{2m}{\hbar^2}(\epsilon-u)}$. The matrix formulation in Eq. (2.93) can now be used to develop the general form of the wavefunction solutions in the random delta-function scattering medium.

In this development, the delta-function potentials in the medium are represented as step functions defined over an infinitesimal region of space. Specifically, the delta-function potential at the position x' is defined over the interval $x'-b\le x\le x'+b$ in the limit that $b\to 0$ as a constant potential having intensity $u=\frac{v}{2b}$. Consequently,

$$u(x)=v\delta(x-x')=\begin{cases} v\frac{1}{2b}, & x'-b\le x\le x'+b \\ 0, & \text{otherwise}\end{cases} \tag{2.94}$$

Aside from the delta-function interactions, the gas will be assumed to otherwise propagate in a region in which $u=0$ in Eq. (2.92).

Under these provisions, the Fermi gas will be treated as propagating along the x-axis and being elastically scattered from a countably infinite number of delta-function potentials of the form

$$u(x)=\sum_n v_n\delta(x-x_n), \tag{2.95}$$

where n runs over the integers, and $x_n<x_{n+1}$ and v_n are random. The object in the following is to generate the solution for the gas in the region $x_n\le x\le x_{n+1}$ from that in the region $x_{n-1}\le x\le x_n$. Proceeding in this way from left to right along the x-axis generates the solution of the particle motion in the random medium.

Consider, then, the representation of the gas in the interval $x_n\le x\le x_{n+1}$ between the scattering potentials at x_n and x_{n+1}. In this interval, the gas encounters the potential

$$u(x)=\begin{cases} v_n\frac{1}{2b}, & x_n\le x\le x_n+b \\ 0, & x_n+b\le x\le x_{n+1}-b \\ v_{n+1}\frac{1}{2b}, & x_{n+1}-b\le x\le x_{n+1}\end{cases} \tag{2.96}$$

Using the form of the solution in Eq. (2.93), it follows that in the interval $x_n\le x\le x_n+b$ the solution is represented by

$$\begin{vmatrix}\psi_{n-}(x)\\ \psi'_{n-}(x)\end{vmatrix}=\begin{vmatrix}e^{iq_nx} & e^{-iq_nx}\\ iq_ne^{iq_nx} & -iq_ne^{-iq_nx}\end{vmatrix}\begin{vmatrix}A_{n-}\\ B_{n-}\end{vmatrix}=M_{n-}(x)\begin{vmatrix}A_{n-}\\ B_{n-}\end{vmatrix} \tag{2.97a}$$

where $q_n=\sqrt{\frac{2m}{\hbar^2}\left(\epsilon-\frac{v_n}{2b}\right)}$, in the interval $x_n+b\le x\le x_{n+1}-b$ the solution is represented by

$$\begin{vmatrix}\psi_{n}(x)\\ \psi'_{n}(x)\end{vmatrix}=\begin{vmatrix}e^{ikx} & e^{-ikx}\\ ike^{ikx} & -ike^{-ikx}\end{vmatrix}\begin{vmatrix}A_{n}\\ B_{n}\end{vmatrix}=M_{n}(x)\begin{vmatrix}A_{n}\\ B_{n}\end{vmatrix} \tag{2.97b}$$

where $k=\sqrt{\frac{2m}{\hbar^2}\epsilon}$, and in the interval $x_{n+1}-b\le x\le x_{n+1}$ the solution is represented by

$$\begin{vmatrix} \psi_{n+}(x) \\ \psi'_{n+}(x) \end{vmatrix} = \begin{vmatrix} e^{iq_{n+1}x} & e^{-iq_{n+1}x} \\ iq_{n+1}e^{iq_{n+1}x} & -iq_{n+1}e^{-iq_{n+1}x} \end{vmatrix} \begin{vmatrix} A_{n+} \\ B_{n+} \end{vmatrix} = M_{n+}(x) \begin{vmatrix} A_{n+} \\ B_{n+} \end{vmatrix} \tag{2.97c}$$

where $q_{n+1} = \sqrt{\frac{2m}{\hbar^2}\left(\epsilon - \frac{v_{n+1}}{2b}\right)}$. Next, the ψ 's and their $x-$ derivatives are related to one another at the interfaces in Eq. (2.96) where the constant scattering potentials change in value. This is done using the conditions that the ψ's and their $x-$ derivatives are everywhere continuous.

In this regard, from Eq. (2.97a) it follows that

$$\begin{vmatrix} A_{n-} \\ B_{n-} \end{vmatrix} = M_{n-}^{-1}(x_n) \begin{vmatrix} \psi_{n-}(x_n) \\ \psi'_{n-}(x_n) \end{vmatrix} \tag{2.98a}$$

so that from Eq. (2.97a) and the continuity of wavefunction and its derivative with respect to x it is found that

$$\begin{vmatrix} \psi_n(x_n+b) \\ \psi'_n(x_n+b) \end{vmatrix} = \begin{vmatrix} \psi_{n-}(x_n+b) \\ \psi'_{n-}(x_n+b) \end{vmatrix} = M_{n-}(x_n+b) \begin{vmatrix} A_{n-} \\ B_{n-} \end{vmatrix} = M_{n-}(x_n+b)M_{n-}^{-1}(x_n) \begin{vmatrix} \psi_{n-}(x_n) \\ \psi'_{n-}(x_n) \end{vmatrix} \tag{2.98b}$$

Similarly, from Eq. (2.97b), it follows that

$$\begin{vmatrix} A_n \\ B_n \end{vmatrix} = M_n^{-1}(x_n+b) \begin{vmatrix} \psi_n(x_n+b) \\ \psi'_n(x_n+b) \end{vmatrix} \tag{2.99a}$$

so that from Eq. (2.97b) and the continuity of the wavefunction and its derivative that

$$\begin{vmatrix} \psi_{n+}(x_{n+1}-b) \\ \psi'_{n+}(x_{n+1}-b) \end{vmatrix} = \begin{vmatrix} \psi_n(x_{n+1}-b) \\ \psi'_n(x_{n+1}-b) \end{vmatrix} = M_n(x_{n+1}-b)M_n^{-1}(x_n+b) \begin{vmatrix} \psi_n(x_n+b) \\ \psi'_n(x_n+b) \end{vmatrix}, \tag{2.99b}$$

and again we have from Eq. (2.97c)

$$\begin{vmatrix} A_{n+} \\ B_{n+} \end{vmatrix} = M_{n+}^{-1}(x_n-b) \begin{vmatrix} \psi_{n+}(x_{n+1}-b) \\ \psi'_{n+}(x_{n+1}-b) \end{vmatrix} \tag{2.100a}$$

and

$$\begin{vmatrix} \psi_{n+1-}(x_{n+1}) \\ \psi'_{n+1-}(x_{n+1}) \end{vmatrix} = \begin{vmatrix} \psi_{n+}(x_{n+1}) \\ \psi'_{n+}(x_{n+1}) \end{vmatrix} = M_{n+}(x_{n+1})M_{n+}^{-1}(x_{n+1}-b) \begin{vmatrix} \psi_n(x_{n+1}-b) \\ \psi'_n(x_{n+1}-b) \end{vmatrix}. \tag{2.100b}$$

In this way, successively applying Eqs. (2.98b), (2.99b), and (2.100b)

$$\begin{vmatrix} \psi_{n+1-}(x_{n+1}) \\ \psi'_{n+1-}(x_{n+1}) \end{vmatrix} = K(n+1) \begin{vmatrix} \psi_{n-}(x_n) \\ \psi'_{n-}(x_n) \end{vmatrix} \tag{2.101a}$$

where

$$K(n+1) = M_{n+}(x_{n+1})M_{n+}^{-1}(x_{n+1}-b)M_n(x_{n+1}-b)M_n^{-1}(x_n+b)M_{n-}(x_n+b)M_{n-}^{-1}(x_n) \quad (2.101b)$$

and the two by two matrix $K(n+1)$ hops the wavefunction and its derivative over the interval between x_n and x_{n+1}. The operators $\{K(i)\}$ then generate the entire solution along the x-axis.

In order to evaluate $K(n+1)$ the products of each of the terms $M_{n-}(x_n+b)M_{n-}^{-1}(x_n)$, $M_n(x_{n+}-b)M_n^{-1}(x_n+b)$, $M_{n+}(x_{n+1})M_{n+}^{-1}(x_{n+1}-b)$ must be evaluated. These three terms all have the general form

$$\begin{aligned} M(x)M^{-1}(x') &= \begin{vmatrix} e^{irx} & e^{-irx} \\ ire^{irx} & -ire^{-irx} \end{vmatrix} \frac{-1}{2ir} \begin{vmatrix} -ire^{irx'} & -e^{-irx'} \\ -ire^{irx'} & e^{-irx'} \end{vmatrix} \\ &= \begin{vmatrix} \cos r(x-x') & \frac{1}{r}\sin r(x-x') \\ -r\sin r(x-x') & \cos r(x-x') \end{vmatrix} \end{aligned} \quad (2.102)$$

and can be obtained in terms of the expression on the far right-hand side of Eq. (2.102).

The results generated in Eqs. (2.96)–(2.102) are now used to obtain a set of recursion relations from which the wavefunctions of the particles are determined over the x-axis. To this end, consider Eq. (2.101a) over the interval x_n and x_{n+1} and the matching expression over the interval between x_{n-1} and x_n, which is given by

$$\begin{vmatrix} \psi_{n-}(x_n) \\ \psi'_{n-}(x_n) \end{vmatrix} = K(n) \begin{vmatrix} \psi_{n-1-}(x_{n-1}) \\ \psi'_{n-1-}(x_{n-1}) \end{vmatrix}. \quad (2.103)$$

It then follows from Eq. (2.101a) that

$$\psi_{n+1-}(x_{n+1}) = K_{11}(n+1)\psi_{n-}(x_n) + K_{12}(n+1)\psi'_{n-}(x_n) \quad (2.104a)$$

and from Eq. (2.103) that

$$\psi_{n-}(x_n) = K_{11}(n)\psi_{n-1-}(x_{n-1}) + K_{12}(n)\psi'_{n-1-}(x_{n-1}) \quad (2.104b)$$

$$\psi'_{n-}(x_n) = K_{21}(n)\psi_{n-1-}(x_{n-1}) + K_{22}(n)\psi'_{n-1-}(x_{n-1}) \quad (2.104c)$$

Rewriting Eq. (2.104b) gives

$$\psi'_{n-1-}(x_{n-1}) = \frac{\psi_{n-}(x_n) - K_{11}(n)\psi_{n-1-}(x_{n-1})}{K_{12}(n)}, \quad (2.105a)$$

and applying this in Eq. (2.104c) yields

$$\psi'_{n-}(x_n) = K_{21}(n)\psi_{n-1-}(x_{n-1}) + K_{22}(n)\frac{\psi_{n-}(x_n) - K_{11}(n)\psi_{n-1-}(x_{n-1})}{K_{12}(n)}. \quad (2.105b)$$

Applying Eq. (2.105b) in Eq. (2.104a) and using the fact that the determinant of $K(n)$ is one, it then follows that

$$\psi_{n+1-}(x_{n+1}) = \left[K_{11}(n+1) + \frac{K_{12}(n+1)}{K_{12}(n)} K_{22}(n) \right] \psi_{n-}(x_n) - \frac{K_{12}(n+1)}{K_{12}(n)} \psi_{n-1-}(x_{n-1}). \quad (2.106)$$

(Note that in evaluating Eq. (2.106) the determinant of a product of matrices is the product of the determinants.) The result in Eq. (2.106) is a recursion relation giving $\psi_{n+1-}(x_{n+1})$ in terms of $\psi_{n-}(x_n)$ and $\psi_{n-1-}(x_{n-1})$. In this way, in conjunction with the relation in Eq. (2.93), the solution is defined all along the x-axis. In addition, for the treatment of the model of delta-function interactions discussions of the $b \to 0$ limit must be considered.

To evaluate $K(n+1)$ and $K(n)$ in the limit that $b \to 0$, note that considering the factors entering into the product in Eq. (2.101b) for $b \to 0$ gives

$$M_{n-}(x_n + b) M_{n-}^{-1}(x_n) \to \begin{vmatrix} 1 & 0 \\ \frac{m}{\hbar^2} v_n & 1 \end{vmatrix}, \quad (2.107a)$$

$$M_{n+}(x_{n+1}) M_{n+}^{-1}(x_{n+1} - b) \to \begin{vmatrix} 1 & 0 \\ \frac{m}{\hbar^2} v_{n+1} & 1 \end{vmatrix}, \quad (2.107b)$$

and

$$M_n(x_{n+} - b) M_n^{-1}(x_n + b) \to \begin{vmatrix} \cos k(x_{n+1} - x_n) & \frac{1}{k} \sin k(x_{n+1} - x_n) \\ -k \sin k(x_{n+1} - x_n) & \cos k(x_{n+1} - x_n) \end{vmatrix}. \quad (2.107c)$$

From Eq. (2.101b), it is seen that the product of these three factors yields $K(n+1)$ and from this $K(n)$. The forms in Eqs. (2.107) and (2.101b) are then applied directly in Eq. (2.106) to obtain a difference equation recursion relation for the general particle wavefunction solutions. An example will now be given of the use of this method to determine the solutions of a periodic array of identical delta-function potentials in the case that $x_{n+1} - x_n = a$ and $v_{n+1} = v_n = \ldots = v$.

2.5.1 Periodic Potential

Consider a one-dimensional medium formed by placing identical delta-function potentials for which $v_n = v$ periodically along the x-axis with the nearest-neighbor separations a. This is a model for electrons propagating within a one-dimensional crystal, which is known in elementary condensed matter physics as the Kronig–Penning Model [1,2,25]. Its solutions illustrate some of the basic properties of electronic modes within crystalline materials.

The first thing to note about the problem is that due to the periodic nature of the system $K(n+1) = K(n)$. This equality is just the statement that the potentials are the same at the ends of each of the cells of length a as they are repeated along the x-axis. As a consequence, Eq. (2.106) becomes

$$\psi_{n+1-}(x_{n+1}) = \left[K_{11}(n+1) + K_{22}(n) \right] \psi_{n-}(x_n) - \psi_{n-1-}(x_{n-1}), \quad (2.108a)$$

and from Eq. (2.107) it then follows that in the $b \to 0$ limit

$$K_{11}(n+1) + K_{22}(n) = 2\left[\cos ka + \frac{m}{\hbar^2} \frac{v}{k} \sin ka \right], \quad (2.108b)$$

which is again seen to be independent of n.

The solution of Eq. (2.108) is now obtained by assuming that it has the form of a planewave given by $\psi_{n-} = \psi e^{iqna}$. (It is important to note that a planewave solution of this form is not localized in space but rather extends throughout all space, having the same amplitude ψ through the lattice.) Substituting this wavefunction expression into Eq. (2.108a) gives

$$\cos qa = \cos ka + \frac{m}{\hbar^2}\frac{v}{k}\sin ka = \cos\sqrt{\frac{2m}{\hbar^2}\epsilon}\,a + \frac{m}{\hbar^2}\frac{v}{\sqrt{\frac{2m}{\hbar^2}\epsilon}}\sin\sqrt{\frac{2m}{\hbar^2}\epsilon}\,a \tag{2.109a}$$

as the condition for the solutions to exist. The relation in Eq. (2.109) is a transcendental equation for the particle energy, ϵ, expressed as a function of the particle wavenumber q and is readily solved numerically. A condition for the solution of Eq. (2.109) in terms of planewave states is that

$$\left|\cos\sqrt{\frac{2m}{\hbar^2}\epsilon}\,a + \frac{m}{\hbar^2}\frac{v}{\sqrt{\frac{2m}{\hbar^2}\epsilon}}\sin\sqrt{\frac{2m}{\hbar^2}\epsilon}\,a\right| \le 1; \tag{2.109b}$$

otherwise, if Eq. (2.109b) is not satisfied, Eq. (2.109a) has only complex solutions for q, which do not lead to bounded wavefunctions. Consequence, there are bands of energies (energy pass bands) in which solutions of the form of planewaves exist and bands of energies in which planewave solution do not exist (energy stop bands).

It is seen from Eq. (2.109) that the equation is periodic in q, and the unique planewave solutions of the equation exist over the interval $-\frac{\pi}{a} \le q \le \frac{\pi}{a}$. Outside this interval, there are only duplicates of the solutions within the interval. In addition, from the periodic nature of $\cos\sqrt{\frac{2m}{\hbar^2}}\epsilon a$ and $\sin\sqrt{\frac{2m}{\hbar^2}}\epsilon a$ treated as a function of energy, it also follows that there are many planewave solutions for each q within the interval $-\frac{\pi}{a} \le q \le \frac{\pi}{a}$. These solutions combine to form the energy pass bands of the system.

For the Fermi gas in the periodic scattering media, it is seen that at pass band energies the wavefunctions extend over the entire x-axis, being delocalized in space. In the stop bands planewave solutions do not exist, as the wavefunctions are not bounded. The unbounded wavefunctions are not complete physical states of the system.

Now let us return to consider the gas in the disordered scattering delta-functions of the disordered scattering medium. In this system the bounded wavefunction modes of the gas are no longer extended states but are localized to be contained within a bounded spatial region of the gas. To begin with, a discussion of how to quantify the localized nature of the wavefunctions of the gas will be given.

2.5.2 Modes in the Anderson Localized Gas

The particle wavefunctions of a Fermi gas in a random one-dimensional medium are composed as a set of localized modes [22–25]. The best way to understand the nature of these modes as they are determined by the disordered medium is to return to a study of Eq. (2.106), which recursively generates the wavefunctions in terms of the parameters characterizing the medium. From such a study, a means of classifying the geometry of the localized modes will be developed. In particular, in the following a well characterized example of a scattering problem is treated from which the

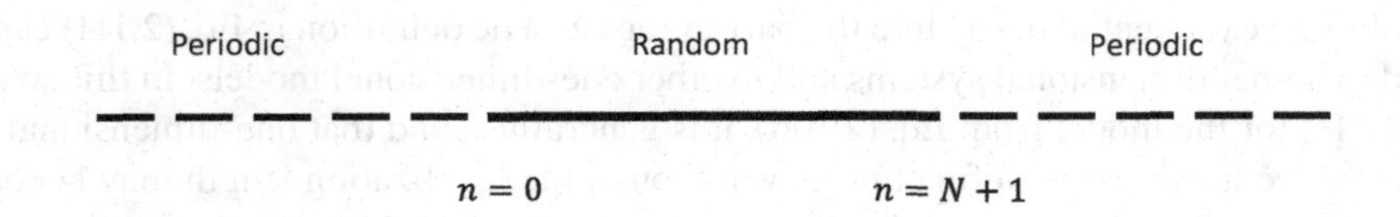

FIGURE 2.6 A one-dimensional scattering system. It is composed of two semi-infinite regions of periodic media between which is found a region of random disorder media.

geometric properties of the localized wavefunctions are determined in terms of those of the disordered medium. The method used will be seen to readily generalize to other types of disordered systems.

Specifically, consider a model consisting of a periodic medium into which a random disordered segment of scattering medium is embedded to develop a characterization of localized modes. In this model a periodic lattice is defined on the x-axis as the set $\{x_n\}$ where $x_n = na$ and n is an integer, and delta-function scattering potentials are located on each site x_n of the lattice. (See Figure 2.6 for a schematic of this one-dimensional geometry.) In the regions $n \leq 0$ and $N+1 \leq n$, take $v_n = v$ as the amplitudes of the delta-function, but otherwise treat these amplitudes v_n as a random variable. This model is readily solved by means of Eq. (2.106) for the scattering of modes originating from the periodic media and scattering from the barrier of random disordered sites.

Now study the scattering problem of a planewave of unit amplitude that is incident from left infinity in the periodic medium onto the left of the embedded barrier consisting of the random array of scattering potentials. In terms of the incident wave, the amplitude t_N of the transmitted wave into the periodic region to the right of the barrier of random disorder and the amplitude r_N of the reflected wave back into the region to the left of the barrier of random disorder are determined from an application of Eq. (2.106). The transmission coefficient of the transmitted wave is then given in terms of the transmission amplitude by

$$T_N = |t_N|^2 . \tag{2.110}$$

Unlike the case of the extend wavefunctions of the delocalized system treated earlier, it is expected that as $N \to \infty$ the transmission coefficient $T_N \to 0$ for wavefunctions localized in a finite region of the x-axis. This follows as at zero temperature the particles of the system of localized modes will be trapped to the bound region of space in which the modal wavefunctions are localized.

A measure of the extent of the localization of the modes in the disorder medium can now be provided in terms of the rate at which T_N approaches zero with increasing N. This rate is characterized by the localization length ℓ, which is defined as [22–25]

$$\frac{1}{\ell} = -\lim_{N\to\infty} \left\langle \frac{\ln T_N}{N} \right\rangle . \tag{2.111}$$

where the notation $\langle\ \rangle$ indicates an average over the statistical disorder of the random scattering potential. In this regard, the potentials v_n of the random medium are considered to be characterized by some statistical distribution, e. g., Gaussian, Poisson, uniform distributions. The limit in Eq. (2.111) will depend on the nature of this characterization, and ℓ should provide an order of magnitude estimate of the typical spatial length involved in the confining geometry of the localized wavefunction.

In this regard, note that if $T_N \propto e^{-\frac{Na}{L}}$, then from Eq. (2.111) it follows that the localization length is given by $\ell = \frac{L}{a}$. The exponential form for T_N agrees with our understanding, for example, of the one-dimensional barrier transmission problem in which the wavefunction within a one-dimensional barrier exhibits an exponential decay into the barrier region. The definition in Eq. (2.111) can also be generalized to higher dimensional systems and to other one-dimensional models. In this way, evaluating Eq. (2.111) for the model from Eq. (2.106), it is generally found that one-dimensional random media always have localized wavefunctions, even though the localization length may become very large. More discussion of this point of the dependence of the localization on system dimensionality will be given later.

In regard of the dimension of the system, for now, however, it is interesting to note that the discussions of the one-dimensional system can be generalized to treat a particular type of three-dimensional conducting slab modeled in the context of a Fermi gas [31]. Specifically, consider a Fermi gas confined between two parallel planes [31] in the form of an infinite slab of thickness d. (Here the parallel planes are separated by d in the z-direction.) Disorder is then introduced into the system by ruling a one-dimensional random rough grating on one of the slab surfaces, and the disorder is chosen to be weak such that the amplitude of the roughness perpendicular to the average confining planes is much less than the now mean slab thickness d. (The grating is ruled so as to destroy the translational symmetry of the rough surface in the x-direction.) For arbitrary weak surface disorder, it can be shown that the localization length for motion in the x-direction is $l \propto \exp\left(\frac{d}{b}\right)^3$, where b is a parameter that depends on the Fermi energy and the statistical parameters characterizing the surface roughness [31]. It is seen that, as the thickness of the slab increases, the localization length of the electrons rapidly increases. In addition, as the disorder in the system is reduced, the localization length also increases. As a general rule, it is found that the localization length is found to increase as the disorder in the slab decreases in significance. Consequently, considered as an infinite slab with surface roughness, the model always results in an insulator, but the localization length can be made so large that if the slab is cut into a finite piece that piece will display a nonzero conductance.

As a final note on the one-dimensional treatment, in Eq. (2.111) the average over the site potentials is made for the logarithm of the transmission coefficient. The reason for this is that the average of the logarithm maintains the phase information in the scattering processes leading to T_N. This information is carried by the excitations, as they travel throughout the disordered medium and contributes to the net ability of the particles to travel through the random medium. It needs to be preserved for each of the scattering potential configurations of the system to correctly characterize the nature of the wavefunctions in the system. We shall next demonstrate how the phase information in the scattering process from the random medium enters into determining the localization properties of the medium.

2.5.3 Phase Coherence in Localization

Consider the scattering of a quantum mechanical particle entering a two- or three-dimensional disordered medium composed of point scattering potentials, i.e., delta-functions [26]. (See Figure 2.7 for an illustration of one set of scattering processes in such a medium.) Of particular interest to the discussion of localization are scattering processes in which a particle wave enters the random medium with an incident wavevector $\vec{k}$ and interacts elastically with the medium through a series of scattering events that cause the particle to exit the random medium in a final state of wavevector $-\vec{k}$. The scattering of $\vec{k}$ to $-\vec{k}$ is referred to as backscattering or a retroreflection, as it reverses the direction of the incident wave. These processes are particularly important to the development of the localization properties in a random medium.

In Figure 2.7, one such process is illustrated in which a wave enters a medium with wavevector $\vec{k}$. In this process the wave undergoes n scatterings involving the sequential momentum exchanges $\vec{g}_1, \vec{g}_2, \ldots, \vec{g}_n$ with the random medium. After exchanging a total momentum $\vec{g}_1 + \vec{g}_2 + \vec{g}_3 + \ldots + \vec{g}_n$ with the medium, the wave emerges from the medium traveling with a wavevector $-\vec{k}$. The amplitude for this particular process is denoted A, and is composed of both a magnitude and phase. Note,

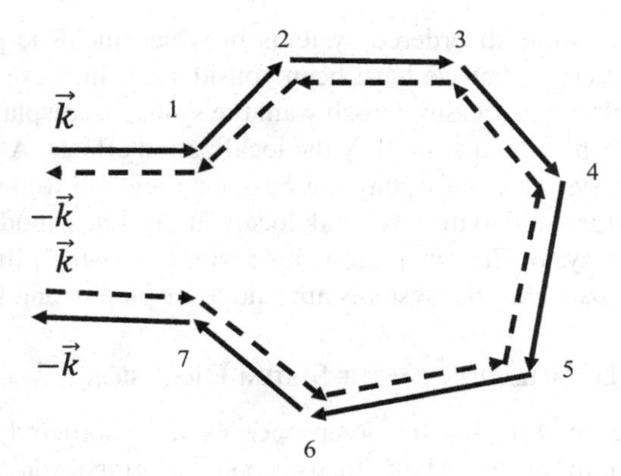

FIGURE 2.7 Schematic of two scattering paths in a random medium. The incident waves have wavevectors $\vec{k}$, and the final waves exiting the medium have wavevectors $-\vec{k}$. The two paths scatter from the medium in an opposite order of scattering. In the example given here, there are seven encounters, each with momentum transfers $\vec{g}_i$ for $i=1, 2, \ldots, 7$.

however, that another scattering process exists for the wave interacting with the random medium in which the wave enters the medium with wavevector $\vec{k}$ and scatters in the reverse sequence $\vec{g}_n,\ldots,\vec{g}_2,\vec{g}_1$ of scattering events. In this process the same momentum is transferred to the random medium (i.e., $\vec{g}_n+\ldots+\vec{g}_3+\vec{g}_2+\vec{g}_1=\vec{g}_1+\vec{g}_2+\vec{g}_3+\ldots+\vec{g}_n$), and the wave also emerges from the random medium with final wavevector $-\vec{k}$. The amplitude for this process is the same as that of the first process considered, i.e., it has the same amplitude and phase. (Note in regard to these two processes for the scattering of $\vec{k}$ to $-\vec{k}$ that $\vec{k}_{i+1}-\vec{k}_i=\left(-\vec{k}_i\right)-\left(-\vec{k}_{i+1}\right)$ so that the first process involving the transfers $\vec{g}_1,\vec{g}_2,\ldots,\vec{g}_n$ is just the reverse process of that involving the transfers $\vec{g}_n,\ldots,\vec{g}_2,\vec{g}_1$.)

Consequently, the probability of these two events occurring in a quantum system is obtained from the sum of the amplitudes of the two events so that their combined probability is proportional to [26]

$$|A+A|^2=4|A|^2. \tag{2.112}$$

The factor of four in Eq. (2.112) comes from the fact that in quantum mechanics the mechanics of the system is studied in terms of the probability amplitude of the events, and the probability of an event is related to the modulus squared of the probability amplitude. In classical mechanics, the probability of two separate events is the sum of the probabilities of each event of the independent events. The result in Eq. (2.112) is therefore twice of what would be found treating the particles classically.

As a side point, it is interesting to note here that the ideas discussed in the earlier context of backscattering are closely related to those involved in the Young's double slit experiment in optics. Both the phenomena rely heavily on the passage of the relative phase information of the excitations carried through the medium. In both enhanced backscattering and wave interference the nature of the paths provides for the little difference between the phases of the participating waves. This results in a phase enhancement contribution to the total scattering. A similar phase-related phenomenon to both of these phase-coherent processes is again observed in Bragg reflection, which is the basis of X-ray diffraction experiments.

The conclusion of these discussions, then, is that in quantum systems there is an enhancement for backscattering events that arises due to the phase-coherent processes occurring in the system. This enhancement of the backscattering shows up as an opposition to the motion of a particle through the random medium. Ultimately, it gives rise to Anderson localization.

In this regard, there are some disordered systems in which inelastic processes are present in addition to the elastic scattering that we have been considering. In these systems, there can be a competition between the elastic processes, which want the system to display Anderson localization, and dissipative processes, which tend to mollify the localization effects. A backscattering enhancement is observed for such systems, but it may not be strong enough to dominate and localize the excitations. These systems are said to display weak localization. Their modes are localized in space but eventually dissipate away. In the other limit, for systems in which the backscattering effects dominate and localize the particles, the systems are said to display strong localization.

2.5.4 Dependence of Localization on the System Dimension

In this section the dependence of the localization properties of a disordered system on its dimensionality are discussed [1,2,27]. In this regard, the focus is on presenting a conjecture that the wavefunctions of the modes in disordered one- or two-dimensional systems are always localized, while the modes in disordered three-dimensional systems may or may not be localized. In addition, the existence of localized modes in three dimensions is found to depend on the strength of the disorder, while all modes are localized independent of the strength of the disorder in one- and two-dimensional systems. The conjecture is made on the basis of scaling theory arguments, which have been successful in developing an understanding of systems that undergo phase transitions. In the present cases, the transition is a metal–insulator transition that is driven by the resistance of the sample.

Early on, near the beginning of this chapter, the dimensional dependence of the resistance of an object was discussed for the cases in which a system is weakly disordered and exhibits electrical transport properties characterized by a conductivity or resistivity. The disorder in these systems was such that, if any of the wavefunctions in the material were localized, the length scales of the regions of the confinement of the wavefunctions were much larger than the dimensions of the sample being considered. In such finite-sized samples, there is a transport of current characterized by a resistance, even though an infinite sample of the same material would have no current flow. A study of these type of systems lead to the relation $R_d \propto \frac{L}{L_T^{d-1}}$ given in Eq. (2.7b), where d is the dimension of the system, L is the length over which a potential is applied, and L_T is a cross-sectional length of the sample [27].

In this section we first consider some of the scaling properties of weakly disorder systems of Eq. (2.7b) in terms of the lengths characterizing the resistance of a sample. To facilitate the discussions, it is assumed that $L_T = L$ so that Eq. (2.7b) simplifies to become

$$R_d = \rho_d L^{2-d}, \tag{2.113}$$

where from Eq. (2.7a) $\rho_d = \frac{m}{q^2 n_d \tau}$ and d is the dimension of the system, i.e., $d = 1$ for a line, $d = 2$ for a square, and $d = 3$ for a cube of material. A useful first-order differential equation for studying the resistance as a function of sample length L then follows from Eq. (2.113) and is given by [2,27]

$$\frac{dR_d}{dL} = \rho_d (2-d) L^{1-d} = (2-d)\frac{R_d}{L}. \tag{2.114a}$$

This can alternatively be rewritten into another form as

$$\frac{L}{R_d}\frac{dR_d}{dL} = \frac{d \ln R_d}{d \ln L} = (2-d). \tag{2.114b}$$

Consequently, from Eq. (2.114), it is found that $\frac{d \ln R_d}{d \ln L} = (2-d) \geq 0$ and $\frac{dR_d}{dL} = (2-d)\frac{R_d}{L} \geq 0$ for $d = 1, 2$ so that both $\ln R_d$ and R_d tend to increase with increasing L in samples with these dimensionalities. By similar reasoning, however, for samples with $d = 3$, both $\ln R_d$ and R_d decrease with increasing L. Next, we look at the dependence of both $\ln R_d$ and R_d on L in the opposite limit of large disorder in which R_L is so large that the sample wavefunctions are becoming localized, and their localization length is much less that the characteristic length of the sample.

For a sample in which the localization length $\ell \ll L$, the wavefunctions are not extended throughout the sample in modified planewaves states but are confined to a finite region of space with wavefunctions asymptotically characterized by the exponential form $\psi(\vec{r}) \propto e^{-\frac{|\vec{r}|}{\ell}}$. As a consequence of the localized nature of the wavefunctions at zero temperature, the resistance of a sample of length $L \gg \ell$ should be of the form [2,27]

$$R_d = A_d e^{C_d \frac{L}{L_0}}, \tag{2.115}$$

where A_d, C_d, $L_0 > 0$ are parameters characterizing the exponential increase of the sample resistance with increase in its characteristic length L. As per our earlier discussion of samples with small R_d, computing the derivatives in Eq. (2.114) for the sample described by Eq. (2.115) gives

$$\frac{dR_d}{dL} = \frac{C_d}{L_0} R_d > 0 \tag{2.116a}$$

and

$$\frac{d \ln R_d}{d \ln L} = \ln R_d . \tag{2.116b}$$

Note that here, now, in the limit of large R_d, it is found that $\ln R_d \gg 1$.

Note that the derivatives in Eqs. (2.114b) and (2.116b) share a common form that can be expressed as [2,27]

$$\frac{d \ln R_d}{d \ln L} = \beta_d (R_d) \tag{2.117}$$

where it is found that $\beta_d(R_d) = (2-d)$ for small R_d, and it is found that $\beta_d(R_d) = \ln R_d$ for large R_d. The conjecture that is introduced at this point is that Eq. (2.117) is valid for all R_d, and that $\beta_d(R_d)$ is an increasing function of R_d. In this view the function $\beta_d(R_d)$ interpolates between the small R_d limit of $\beta_d(R_d) = (2-d)$ and the large R_d limit where $\beta_d(R_d) = \ln R_d$. It should be noted in these discussions, however, that assuming $\beta_d(R_d)$ only depends on R_d for general R_d is a major assumption placed on the form of the derivative in Eq. (2.117).

A result of these assumptions on $\beta_d(R_d)$ is that for all R_d

$$\frac{dR_d}{dL} = \frac{R_d}{L} \beta_d (R_d), \tag{2.118}$$

and, due to the monotonic increasing nature of $\beta_d(R_d)$, it follows that for general one- and two-dimensional systems $\beta_d(R_d) > 0$ so that $\frac{dR_L}{dL} > 0$, and the resistance increases with increasing L. Consequently, for these dimensionalities, an infinite sample at zero temperature is always an insulator, composed of localized states. In three-dimensional systems, however, $\beta_d(R_d) < 0$ for small R_d and $\beta_d(R_d) > 0$ for large R_d. This means three-dimensional samples in which $\beta_d(R_d) < 0$ approach metallic behavior as $L \to \infty$ because $\frac{dR_d}{dL} < 0$, and three-dimensional samples with $\beta_d(R_d) > 0$ approach insulating behavior as $L \to \infty$ because $\frac{dR_d}{dL} > 0$. In this regard, for the three-dimensional system the point at which $\beta_d(R_d) = 0$, which separates these two regions of opposite signs of $\beta_d(R_d)$, is a point of unstable equilibrium at the boundary separating the two regions of different scaling behaviors.

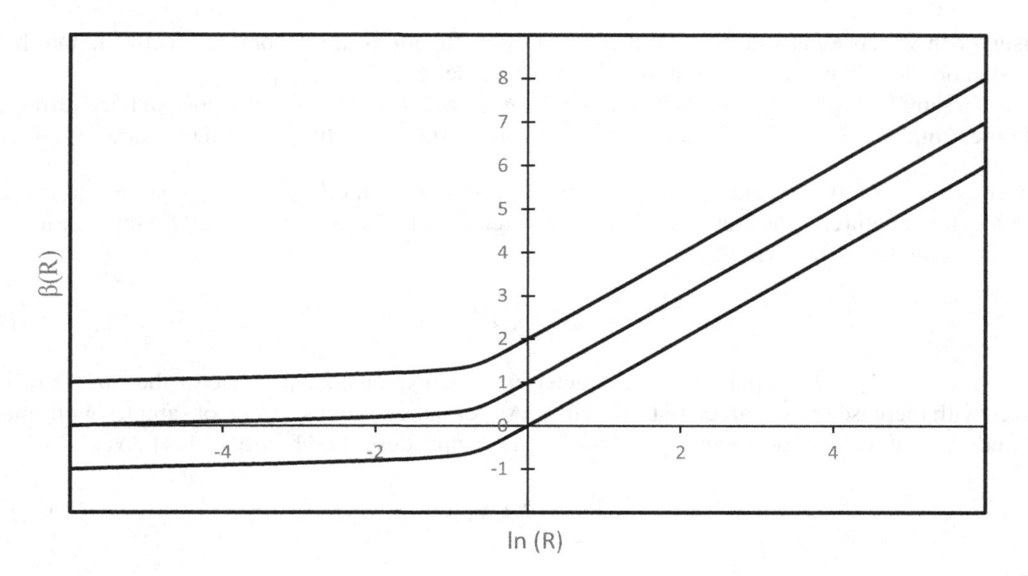

FIGURE 2.8 Schematic illustrating the proposed behavior of $\beta(R)=\beta_d(R_d)$ versus $\ln(R)=\ln(R_d)$. In the plot are three solid lines representing schematic results for the $d=$one-, two-, and three-dimensional systems. These are given, respectively, from top to bottom in the figure.

In Figure 2.8, a schematic drawing of a plot of $\beta_d(R_d)$ versus $\ln(R_d)$ is shown. This was given in Ref. 27 as a proposed dependence of $\beta_d(R_d)$ on $\ln(R_d)$ and was meant as a qualitative indication of behavior. To this end note that at the left side of the drawing the correct small resistance limits of $\beta_d(R_d)$ in one-, two-, and three-dimensional systems are shown. The right side of the plot displays the divergence of $\beta_d(R_d)$, as the system becomes localized in the limit of high resistance. This dependence comes from the exponential form of the localized wavefunctions in the localized medium. Between these two limiting forms, they are connected by a monotonically increased function of R_d. In addition, there is an unstable fixed point at $\ln R_d = 0$ in the three-dimensional system.

The scaling conjectures just discussed, regarding localization and their effects on the resistance of disordered media, have been investigated numerically in certain random models of transport properties where they have been verified for a number of particular systems [2,27–31]. The interested reader is referred to the literature for discussions of these results.

2.5.5 Hopping and Variable-Range Hopping Conductivity

Next, we shall extend the earlier considerations that have focused on zero-temperature systems to treat the conductivity of Anderson localized materials at nonzero temperatures [23,24,29–32]. Due to the thermal fluctuations in these systems, the localized charged particles of the medium now hop between different localized modes. These hopping transitions, which are absent in the zero-temperature materials, are assisted by energy exchanges with thermal fluctuations. In addition, the mobile particles at nonzero temperatures can interact with externally applied electric fields so as to propagate a current through the random medium. The dynamics of such currents are basic types of diffusion process, which are described by theories that have been developed to understand general diffusive transport behaviors.

The hopping processes between different localized modes in the disordered medium are thermally activated with relative probabilities dependent on the Boltzmann factor $e^{-\frac{W}{k_BT}}$, where W is an activation energy for the hopping between the initial and final states. In addition, the hopping in the medium also occurs over varying distances$R$ having relative probabilities $e^{-2\frac{R}{\ell}}$, where ℓ is the localization length characterizing the diameter of the wavefunctions of the modes in the system. In this regard, the longer the distance traveled during a hop the less probable is the hop. Combining these factors characterizing the hopping behavior, the transition rate, r, for the hopping in the systems takes the form [2,23,24,32]

$$r = \nu_{th} e^{-\left\{2\frac{R}{\ell}+\frac{W}{k_B T}\right\}} \tag{2.119}$$

where ν_{th} is a frequency characterizing the rate of hopping attempts. The rate of hopping in Eq. (2.119) is then seen to depend significantly on variations in the variables R, W, and T. To study the conductivity, it remains to introduce into Eq. (2.119) the effects of the application of an applied electric field.

In the presence of an applied field E along the x-axis, the hopping of a particle of charge q through a length $\pm R$ in the x-direction represents an energy change of the particle by $\pm qRE$. Consequently, the difference in the hopping rates, $r_\pm$, in these two opposite directions enters in terms of the particle energy change during the hop. This energy change modifies the activation energy W of the particle as it hops, so that from Eq. (2.119)

$$r_\pm = \nu_{th} e^{-\left\{2\frac{R}{\ell}+\frac{W \pm qRE}{k_B T}\right\}}. \tag{2.120}$$

In the case that $|qRE| \ll k_B T$, the difference in the $r_\pm$ hopping rates is then given by

$$r_- - r_+ = 2\nu_{th} e^{-\left\{2\frac{R}{\ell}+\frac{W}{k_B T}\right\}} \sinh\frac{qRE}{k_B T} = 2\nu_{th} e^{-\left\{2\frac{R}{\ell}+\frac{W}{k_B T}\right\}} \frac{qRE}{k_B T}. \tag{2.121}$$

The result in Eq. (2.121), therefore, indicates the anisotropy in the medium caused by the presence of the applied electric field. This shows up in a current flow through the medium, which will now be determined.

The current density in the system along the x-direction is obtained from Eq. (2.121) and its definition in terms of the charge of the carriers, the carrier density per volume, and the carrier drift velocity as

$$j = \sigma E = e\left[2n(E_F)k_B T\right]\left|R(r_- - r_+)\right|. \tag{2.122}$$

Here $e = |q| > 0$ is the unit of electrical charge, $2n(E_F)$ is the density of states per volume of both spin-up and spin-down carriers, $2n(E_F)k_B T$ is the number of up- and down-spin electrons within a thermal fluctuation of the Fermi energy, and $R(r_- - r_+)$ is the velocity of carriers as they hop through the medium. Combining Eqs. (2.121) and (2.122), it follows that

$$\sigma = 4e^2 R^2 n(E_F) \nu_{th} e^{-\left\{2\frac{R}{\ell}+\frac{W}{k_B T}\right\}} \tag{2.123}$$

gives the temperature-dependent conductivity in the localized medium where $\sigma \to 0$ as $T \to 0$.

Two case of the hopping are of interest: In the first case the nearest-neighbor separation of sites in the system, R_0, is such that $R_0 \gg \ell$, and a nearest-neighbor hop must be well outside of the particle localization length. Under this condition, due to the $e^{-2\frac{R}{\ell}}$ factor in Eq. (2.120), nearest-neighbor hopping is dominant in the medium, i.e., hopping longer than R_0 is highly improbable. Consequently, the hopping final states in the media are confined to the volumes $\frac{4\pi}{3}R_0^3$ around the position of the initial state of the hopping particle. Within this volume there must be at least one state to receive the hop.

In this regard, $2n(E_F)W$ are the number of modes per volume that can be hoped into in a system with an activation energy W, and the number of modes contained in $\frac{4\pi}{3}R_0^3$ are $\frac{4\pi}{3}R_0^3 2n(E_F)W$. For at least one receiving mode to be present in $\frac{4\pi}{3}R_0^3$ about the jumping site, it follows that $\frac{4\pi}{3}R_0^3 2n(E_F)W = 1$. This condition requires that the activation energy must be of the form

$$W = \frac{1}{\frac{4\pi}{3}R_0^3 2n(E_F)}. \tag{2.124a}$$

for the nearest-neighbor hopping to be the mechanism of conduction.

From Eqs. (2.123) to (2.124a), it then follows that the medium has a resultant conductivity given by

$$\sigma = 4e^2R^2n(E_F)\nu_{th}e^{-\left\{2\frac{R_0}{\ell}+\frac{1}{k_BT}\frac{1}{\frac{4\pi}{3}R_0^3 2n(E_F)}\right\}}. \tag{2.124b}$$

In this limit it is seen from Eq. (2.124b) that the conductivity depends on temperature with the general form

$$\sigma \propto e^{-\frac{A}{k_BT}}. \tag{2.125}$$

The second case is that of the so-called variable-range hopping conductivity. In this limit $R_0 \sim \ell$ or $< \ell$ so that the range of hopping $R > \ell$, and again there must be at least one mode to hop to in the volume $\frac{4\pi}{3}R^3$. This requires that for there to be one state to hop into in the sphere of volume $\frac{4\pi}{3}R^3$ about the hopping site, the activation energy must satisfy $\frac{4\pi}{3}R^3 2n(E_F)W = 1$. Consequently, the activation energy needed for this condition to be met is that [2,23,24,32]

$$W = \frac{1}{\frac{4\pi}{3}R^3 2n(E_F)} \tag{2.126}$$

in order to characterize the variable-range hopping transitions.

Unlike the nearest-neighbor hopping case, the range of the hopping for variable-range hopping is not fixed but changes with the temperature of the medium. For these types of materials, it is now important to determine how the hop range changes with the temperature. In this regard, the sphere of volume $\frac{4\pi}{3}R^3$ contains many points that can statistically occur as a final point of the particle hop. An average hoping distance $\bar{R}$ for jumping from the center of the sphere to another point in the sphere can be estimated by the form

$$\bar{R} = \frac{\int_0^R r^3 dr}{\int_0^R r^2 dr} = \frac{3}{4}R. \tag{2.127}$$

Here the average or typical hopping distance $\bar{R}$ is obtained, assuming a uniform distribution of hopping over the sphere and dividing by a uniform distribution over the largest circular section of the sphere. The range and conductivity are next treated and determined as functions of the temperature.

Applying Eqs. (2.127) and (2.126) in Eq. (2.123) then gives the conductivity

$$\sigma = 4e^2 R^2 n(E_F) \nu_{th} e^{-\left\{2\frac{\bar{R}}{\ell} + \frac{W}{k_B T}\right\}}, \tag{2.128}$$

where the hoping rate of the system in the absence of the field is characterized by the factor $\nu_{th} e^{-\left\{2\frac{\bar{R}}{\ell} + \frac{W}{k_B T}\right\}}$. Maximizing the hopping rate determines the typical range of the hopping and from Eq. (2.128) the conductivity. The condition for the maximum hopping rate occurs when the argument inside { } is minimized with respect to R so that

$$0 = \frac{d}{dR}\left\{2\frac{\bar{R}}{\ell} + \frac{W}{k_B T}\right\} = \frac{d}{dR}\left[2\frac{1}{\ell}\frac{3}{4}R + \frac{1}{k_B T}\frac{3}{4\pi R^3 2n(E_F)}\right] = \frac{3}{2}\frac{1}{\ell} - \frac{1}{k_B T}\frac{9}{4\pi R^4 2n(E_F)} \tag{2.129}$$

and

$$R = \frac{3^{1/4} \ell^{1/4}}{\left[2\pi k_B T 2n(E_F)\right]^{1/4}}, \tag{2.130}$$

and from Eq. (2.127) the average hoping distance $\bar{R} = \frac{3}{4}R$. Applying Eqs. (2.126) and (2.130) in Eq. (2.128), it follows that the conductivity of a three-dimensional localized medium has the general temperature-dependent form

$$\sigma \propto e^{-\frac{B}{T^{1/4}}} \tag{2.131}$$

where $B > 0$. To conclude, Eqs. (2.127), (2.130), and (2.131) provide the forms of the temperature dependence for the average hopping range and conductivity in three-dimensional media.

The results for the conductivity, under the assumption that the charged carriers do not interact with one another, can similarly be determined for two- and one- dimensional materials. In this way, it is found that

$$\sigma \propto e^{-\frac{B}{T^{1/(d+1)}}} \tag{2.132}$$

where d is the dimensionality of the system. Additional considerations have also been made by Efros and Shklovskii to include the particle–particle interactions between the charge carriers [32]. They find that, with the inclusion of particle–particle interactions, the conductivity is

$$\sigma \propto e^{-\frac{B}{T^{1/2}}} \tag{2.133}$$

in all the dimensions.

The results in this section treated considerations of temperature-dependent hopping effects in localized systems. These arise from the introduction of thermal fluctuations to the systems and are firmly based on the underlying localization foundations of zero-temperature localization. Going back to considerations of the zero-temperature system, some of the early results on zero-temperature Anderson localization have offered semiquantitative understanding of the origins of localization in this limit. While highly sophisticated techniques have been developed to study localization, these early considerations are still very useful in understanding the mechanisms and conditions on a system to exhibit the transition from a metallic to a nonconducting state with the changing degree of disorder. Consequently, these early developments in localization are now discussed.

2.5.6 Ioffe–Regel Criterion and Minimum Metallic Conductivity

Early on in the study of localization, some semiquantitative notions of the nature of the localized state were put forth. These include the Ioffe–Regel criterion and the idea of the minimum metallic conductivity [2,23,24,29–32]. Both relate to the nature of the zero-temperature localized system in the vicinity of its transition from a metallic conducting state to a localized state exhibiting zero conductivity. In addition, both provide useful insights into the development of the Anderson localization of a disordered system as a function of the degree of the disordering. In the following discussions, these two ideas regarding the nature of the localized state are considered for systems at zero-temperature and in a variety of dimensions.

The Ioffe–Regel criterion considers the relationship of the mean free path l_m of an excitation between scattering events within a randomly disordered medium to the wavevector $\vec{k}$ of the excited mode. In particular, a useful parameter to characterize the nature of the scattering in the medium is obtained by examining the magnitude of kl_m where $k = |\vec{k}|$. If the scattering of the mode by the medium is very weak, then l_m must be very large, such that $kl_m \gg 1$. In this limit, there are many wavelengths of the mode between the separate scattering sites, and the dispersion relation and lifetime of the mode are little affected by its scattering in the medium. However, as the means free path shortens, kl_m decreases, and the effect of the disorder on the modal excitations increases. The Ioffe–Regel criterion suggests that if $kl_m \le 1$, the wavelength must be greater than the mean free path so that a well-defined mode characterized by a renormalized wavevector $\vec{k}$ should not be possible under these conditions. In particular, it is suggested that the condition $kl_m \sim 1$ is necessary and sufficient for the modal excitation to be localized by the disorder.

Related to the idea of the Ioffe–Regel criterion is that of the minimum metallic conductivity. This starts out with a consideration of the Drude result for the conductivity, i.e.,

$$\sigma = \frac{ne^2\tau}{m} = \frac{ne^2 l_m}{mv_F} = \frac{ne^2 l_m}{\hbar k_F} = \frac{ne^2 k_F l_m}{\hbar k_F^2} \tag{2.134}$$

where n is the carrier density, m is the effective mass of the carriers, and τ is the average time between scattering interactions. Here v_F is the speed of a carrier at the Fermi surface and $\tau = \frac{l_m}{v_F}$ gives a rough average value of the typical time between successive collisions with the disorder. In addition, use is made in Eq. (2.134) of the relationship $mv_F = \hbar k_F$ for the Fermi momentum.

In general, the carrier density

$$n = \frac{N}{L^d} = c_d k_F^d \tag{2.135}$$

where $d = 1, 2, 3$ is the spatial dimension of the system; N is the number of carriers; L is the length of the line, square, or cube of material in these respective dimensions; $0.034 < C_d < 0.32$ is a constant that is fixed for a given dimension; and k_F is the Fermi wavenumber. From the far right-hand side of Eqs. (2.134) and (2.135), it follows that

$$\sigma = \frac{c_d k_F^d e^2 k_F l_m}{\hbar k_F^2} = \frac{c_d k_F^{d-2} e^2 k_F l_m}{\hbar}. \tag{2.136}$$

Since from the Ioffe–Regel criterion, the condition for the metal–insulator transition is that $k_F l_m \sim 1$, it follow that the minimum metallic conductivity possible in the system is

$$\sigma_{\min} \approx \frac{c_d k_F^{d-2} e^2}{\hbar} \propto \frac{e^2}{h} k_F^{d-2}. \tag{2.137}$$

Additionally, if it is considered that $l_m \sim a$, where a is the nearest neighbor, interatomic distance provides a lower bound on the mean free path, then

$$\sigma_{\min} \propto \frac{e^2}{h} a^{2-d} \tag{2.138}$$

offers an estimate based on a. In the case of two-dimension systems it is interesting to note that Eqs. (2.137) and (2.138) reduce to $\sigma_{\min} \propto \frac{e^2}{h}$.

2.6 MOTT TRANSITION

The Mott transition is another type of metal–insulator transition [2,23,24,32]. Unlike the Anderson transition, which arises due to the phase-coherent scattering of the charged carriers from disorder in the material, the Mott transition arises from the particle–particle scattering interactions between the charge carriers. In this regard, the Mott transition can occur in either ordered or disordered materials. Anderson transitions only are found in disordered media, and the further introduction of particle–particle interactions in their study act to modify our earlier considerations regarding the localization length and its dependence on the disorder.

To understand the Mott transition, consider a system of silver atoms, accounting for all of the electron–electron interactions in the system. A silver wire is conducting, but at some separation, the system must become an insulator if the separations of the silver atoms in the wire are uniformly increased. The reason for this is that the atoms are separated by microscopic distances and behave as in a metal in the silver wire, while the array of isolated silver atoms behaves as an insulator in the system formed as an array of macroscopically separated silver atoms. In this regard, to move an electron through the array of macroscopically separated silver atoms, it must tunnel through space to occupy a higher energy level within another silver atom. Between the metal and insulator states of the atoms, there must be an atomic separation at which the system changes between a metal and an insulator.

An example of the Mott transition occurs in a crystal of silicon with concentration density n_{Si} in which a concentration density of phosphorous n_P impurities is introduced. At some critical concentration of phosphorous n_P^{Cr} it is found that the system undergoes a metal–insulator transition. A simple theory of the nature of the metal–insulator transition is obtained by studying the permittivity of the impurity material as a function of the phosphorous impurity concentration. This is done by considering the Clausius–Mossotti equation relating to the molecular polarizability of the phosphorous to the dielectric constant of the impurity medium.

The Clausius–Mossotti equation relates the dielectric constant of the silicon–phosphorus medium ε to an effective molecular polarizability of the molecules in the system by [2]

$$\gamma_{\text{mol}} = \frac{3}{4\pi(n_P + n_{Si})}\left(\frac{\varepsilon - 1}{\varepsilon + 2}\right). \tag{2.139}$$

Here, the average molecular polarizability is defined as $\gamma_{\text{mol}} = \frac{n_{Si}\gamma_{Si} + n_P\gamma_P}{n_P + n_{Si}}$, where γ_{Si} and γ_P are the molecular polarizabilities of silicon and phosphorous, respectively. Solving Eq. (2.139), it is then found that

$$\varepsilon = \frac{3 + 8\pi(n_P + n_{Si})\gamma_{\text{mol}}}{3 - 4\pi(n_P + n_{Si})\gamma_{\text{mol}}}. \tag{2.140}$$

From Eq. (2.140), it is seen that for $n_P = 0$ the dielectric constant ϵ is that of silicon, and for $n_{Si} = 0$ the dielectric constant ϵ is that of phosphorous. Between these two points of concentration, the system is a mixture. The essential point in the considerations in Eqs. (2.139) and (2.140), however, is that the dielectric constant becomes divergent in the case that the denominator in Eq. (2.140) is zero. This infinity indicates that at this point the system is unstable, giving an infinite response to an infinitesimal change in the electric field. As a consequence, this point is critical for the metal–insulator transition so that

$$n_P^{Cr} = \frac{3}{4\pi}\frac{1}{\gamma_{\text{mol}}} - n_{Si}. \tag{2.141}$$

These considerations are fairly crude, and usually a more sophisticated theory is required to obtain accurate theoretical results. From experiment it is found that at the density of silicone the metal–insulator transition takes place at a phosphorous impurity density of $n_P^{Cr} = 3.74 \times 10^{18}\,\text{cm}^{-1}$.

2.7 WIGNER CRYSTAL

As a final consideration of the Fermi gas of interacting charged carriers, a discussion is given of very low carrier density materials. Specifically, an important point regarding these materials is that in their extremely low density limit the carriers are found to condense into a crystalline structure [2,20,21]. It is important to note, in addition, that the crystal is assumed to be part of an electrically neutral medium for a treatment of these materials so that the charges on the lattice are embedded within a uniform homogeneous medium of an opposite charged background material termed as "jellium". This type of crystal is known as a Wigner crystal, and these types of materials can be composed as three-dimensional systems in which the charges crystalize to form a body-centered cubic lattice and as two-dimensional systems in which the charges condense into a triangle lattice. In one dimension the condensate is a one-dimensional crystal composed as an equally spaced linear array of carriers.

In the dilute limit in which the intra-carrier separation $r_s \to \infty$, the carriers crystalize into a body-centered cubic crystalline lattice with the energy per carrier given by

$$\frac{E}{N} = -\frac{1.792}{r_s} + \frac{2.66}{r_s^{3/2}} + \frac{b}{r_s^2} \tag{2.142}$$

where $r_s = \dfrac{r_0}{a_0}$, $r_0 = \left(\dfrac{3}{4\pi}\rho\right)^{1/3}$ for density ρ, $a_0 = \dfrac{\hbar^2}{me^2}$, m is the carrier mass, and the energy is in Rydberg $\dfrac{e^2}{2a_0}$. Quantum Monte Carlo simulations have indicated that crystallization first occurs at $r_s = 106$ for three-dimensional systems and $r_s = 31$ for two-dimensional systems.

2.8 SUPERCONDUCTIVITY

In this section, some discussion is given of the theory of low-temperature superconductivity with a focus on the Ginzburg–Landau theory of the superconducting transition [2–9]. The superconducting state of matter is very important in technological considerations and has been studied for many years because of these applications. Specifically, in superconducting materials the characterizing properties are that, below a critical temperature T_c, the materials display perfect conductivity and an exclusion of magnetic flux from their bulk. This last property of flux exclusion is known as the Meissner effect [2–9]. Both of these features of the superconducting state come from the condensation of the charge carriers of the material into a highly correlated state of motion, which has sometimes been referred to as a macroscopic quantum state of the system. In the following, the formulation of the properties of superconductors is discussed in the context of the model of a Fermi gas of charged carrier that interacts with one another through attractive interactions and the ultimate realization of the carrier properties, as described in the Ginzburg–Landau theory [3].

In models of superconductivity the particles of the Fermi gas include particle–particle interactions that are attractive. This is unlike the previous discussions of the Fermi gas, which in some cases included repulsive particle–particle interactions, and the attractive interactions now introduced into the Fermi gas lead to a totally different state of matter peculiar to quantum mechanical systems of fermions. In this regard, in the presence of attractive interactions the fermion ground state of the systems (i.e., the sphere of smallest wavevector states centered about the origin at $\vec{k} = 0$) become unstable and transitions to a new ground state of a highly correlated gas of bound pairs of carriers known as Cooper pairs [2–9]. Consequently, superconductivity can occur in systems of either positive or negative charge carriers, but the examples that are commonly found in solid state physics predominantly involve electrons as charge carriers.

In the case of low-temperature superconductivity, the attractive force between the electrons arises from the interaction of the electrons with a background of positive ions. The particulars of the interaction involve electron interactions mediated by the lattice of ions so that an electron traveling through the lattice distorts the positions of the positive ions by means of its Coulomb interaction with them. This distortion causes the positive ions to screen the negative charge of the electron, and the screened electron creates a Coulomb field felt by another electron within the material. The net effect is a resultant attractive force between the two electrons in the system.

The basic properties of low-temperature superconductors are well described by the Ginzburg–Landau theory, which was developed before the microscopic BCS (Bardeen, Cooper, Schreifer) theory of superconductivity, and in many engineering applications it is more useful than the full microscopic treatment. For this reason, our focus here will be on the Ginzburg–Landau theory and its many important technological applications. For the details of the microscopic theory developed by BCS, the reader is referred to the literature.

The Ginzburg–Landau theory involves making certain reasonable thermodynamic assumptions regarding the form of the free energy of the superconductive state [2–9]. These assumptions and the Ginzburg–Landau theory were essentially justified later by the BCS theory. In this sense, it is a type of theory known as a phenomenological theory, as it is not based on microscopic considerations of the system but is designed at a macroscopic level to explain the macroscopic observations of the system properties.

To understand the origins of the Ginzburg–Landau theory, first consider the uncorrelated system of noninteracting charged carriers with charge q in an external applied field. Specifically, it will now be shown how the Schrodinger equation of the uncorrelated Fermi gas can be obtained from a variational problem made on a functional integral. The particles in this Fermi gas are described by a free-particle Schrodinger equation of the form [2–9]

$$\frac{1}{2m}\sum_{i=1}^{N}\left(\vec{\mathrm{p}}_i - \frac{q}{c}\vec{A}_i\right)^2 \psi(\vec{r}_1,\vec{r}_2,..,\vec{r}_N) = E\psi(\vec{r}_1,\vec{r}_2,..,\vec{r}_N) \tag{2.143}$$

where each of the N particles of the system moves independently of the others while interacting with an external field represented by $\vec{A}_i$. As a theoretical technique one method of arriving at the Schrodinger equation in Eq. (2.143) is by considering a variational problem based on extremizing the functional form

$$\int\left[\prod_{i=1}^{N} d^3r_i\right]\left\{\frac{1}{2m}\sum_{i=1}^{N}\left|\left(\vec{p}_i - \frac{q}{c}\vec{A}_i\right)\psi(\vec{r}_1,\vec{r}_2,..,\vec{r}_N)\right|^2 - E\left|\psi(\vec{r}_1,..,\vec{r}_N)\right|^2\right\} \tag{2.144}$$

with respect to the independent variables $\psi(\vec{r}_1,\vec{r}_2,..,\vec{r}_N)$ and $\psi^*(\vec{r}_1,\vec{r}_2,..,\vec{r}_N)$. (Note that $\psi(\vec{r}_1,\vec{r}_2,..,\vec{r}_N)$ and $\psi^*(\vec{r}_1,\vec{r}_2,..,\vec{r}_N)$ can be varied independently of one another, as they are complex functions having both real and imaginary parts.) In this consideration of Eq. (2.144), the integrals are over all space with the probability density of the particles $n(\vec{r}_1, \vec{r}_2, ..., \vec{r}_N) \propto |\psi(\vec{r}_1, \vec{r}_2, ..., \vec{r}_N)|^2$, and both of the variations of Eq. (2.144) with respect to ψ or ψ^* lead to the result in Eq. (2.143). These variational considerations are the basis of the variational method of approximation in quantum mechanics, which, for example, is useful in generating the Hartree–Fock approximation. As shall be seen next, the ideas of the variational method are also a root in considerations of the Ginzburg–Landau theory and are helpful in developing a theory of the density of Cooper pairs in the superconducting system.

In the superconductor problem, on the other hand, the charge carriers are highly correlated in space. As a consequence, we will consider a type of wavefunction defined in space to be of the form $\psi(\vec{r})$ so that the density of carriers (i.e., the Cooper pairs) in the correlated phase of the superconductor is given by $n(\vec{r}) = |\psi(\vec{r})|^2$. In particular, due to the high correlation of the state, $n(\vec{r})$ depends only on one space variable. In the superconductor, the free energy per volume at temperatures near the critical temperature T_c is further postulated to be of the simple form [3]

$$F_s(\vec{r}) = F_N + \frac{1}{2m_s}\left|\left(-i\hbar\nabla - \frac{q}{c}\vec{A}(\vec{r})\right)\psi(\vec{r})\right|^2 + \alpha|\psi(\vec{r})|^2 + \frac{\beta}{2}|\psi(\vec{r})|^4 + \frac{B^2}{8\pi} \tag{2.145}$$

where F_N is the free energy per volume of the metal in the absence of superconductivity (Note: For the purposes here this is just a constant.), m_s is the effective mass of the carriers, q is the charge of the carriers, $\alpha = \alpha_0 \frac{T - T_c}{T_c}$, $\alpha_0 > 0$ is a constant, T is the temperature, T_c is the critical temperature for the metal–superconductor transition of the materials, $\beta > 0$ is a constant, and $\vec{A}$ with $\vec{B} = \nabla \times \vec{A}$ describes an externally applied magnetic induction. The change in the free energy with the condensation of Cooper pairs in the system is then represented by the second, third, and fourth terms on the right in Eq. (2.145). As shall be seen later, above the critical temperature these terms are zero, and below the critical temperature they become nonzero.

Subtracting the constant value of F_N from the superconductor free energy per volume and integrating over the spatial extent of the medium being studied gives [3]

$$\Delta\mathcal{F} = \int d^3r\left[F_s(r) - F_N\right] = \int d^3r\left[\frac{1}{2m_s}\left|\left(-i\hbar\nabla - \frac{q}{c}\vec{A}(\vec{r})\right)\psi(\vec{r})\right|^2 + \alpha|\psi(\vec{r})|^2 + \frac{\beta}{2}|\psi(\vec{r})|^4 + \frac{B^2}{8\pi}\right] \tag{2.146}$$

where $\Delta\mathcal{F}$ is the difference in free energy between the superconductor and normal phases of the material. An equation for $\psi(\vec{r})$ is now obtained by extremizing Eq. (2.146). Extremizing Eq. (2.146) with respect to the three independent variables ψ, ψ^*, and A, it follows that

$$\begin{aligned}&\int d^3r\left\{\frac{1}{2m_s}\left(-i\hbar\nabla - \frac{q}{c}\vec{A}\right)^2\psi + \alpha\psi + \beta|\psi|^2\psi\right\}\delta\psi^* + \text{c.c.}\\ &+\int d^3r\left\{\frac{-q}{2m_s c}\left[-i\hbar\psi^*\nabla\psi + i\hbar\psi\nabla\psi^*\right] + \frac{q^2}{m_s c^2}\vec{A}|\psi(\vec{r})|^2 + \frac{1}{4\pi}\nabla\times\vec{B}\right\}\cdot\delta\vec{A} = 0\end{aligned} \tag{2.147}$$

where in obtaining the $\delta\vec{A}$ term in Eq. (2.147) the identity $\nabla\cdot\left(\vec{a}\times\vec{b}\right) = \vec{b}\cdot\nabla\times\vec{a} + \vec{a}\nabla\times\vec{b}$ was used, and *c.c.* denotes the complex conjugate of the first term.

From a consideration of the coefficients in Eq. (2.147), it is found that the first two terms on the left-hand side both give [3]

$$\frac{1}{2m_s}\left(-i\hbar\nabla - \frac{q}{c}\vec{A}\right)^2\psi + \alpha\psi + \beta|\psi|^2\psi = 0, \tag{2.148a}$$

and the third term on the left implies that

$$\frac{-q\hbar}{2m_s c}\left[-i\psi^*\nabla\psi+i\hbar\psi\nabla\psi^*\right]+\frac{q^2}{m_s c^2}\vec{A}+\frac{1}{4\pi}\nabla\times\vec{B}=0 \tag{2.148b}$$

From Eq. (2.148b), it is then found that

$$\nabla\times\vec{B}=\frac{4\pi}{c}\left\{\frac{q}{2m_s}\left[-i\hbar\psi^*\nabla\psi+i\hbar\psi\nabla\psi^*\right]-\frac{q^2}{m_s c}\vec{A}|\psi|^2\right\}=\frac{4\pi}{c}\vec{j}_s \tag{2.149}$$

where $\vec{j}_s$ is the current density in the superconductor. The two equations that describe the superconductor in the presence of an external magnetic field are Eq. (2.148a), which is of the form of a nonlinear Schrodinger equation and a modified version of Amperes law given in Eq. (2.149). In this regard, the right-hand side of Eq. (2.149) represents the current flow through the material.

As an example of Eqs. (2.148a) and (2.149) consider a homogeneous bulk medium in the absence of an externally applied magnetic field. From Eq. (148a), it follows that [3]

$$\alpha\psi+\beta|\psi|^2\psi=\left(\alpha+\beta|\psi|^2\right)\psi=0. \tag{2.150}$$

Consequently, for the case that $T\geq T_c$, where $\alpha,\beta>0$, it is found that $\psi=0$ so that $\Delta\mathcal{F}=0$, and there is no superconductivity. In the case, however, where $T<T_c$, with $\alpha<0$ and $\beta>0$, it follows that

$$|\psi|^2=-\frac{\alpha}{\beta}=-\frac{\alpha_0}{\beta}\frac{T-T_c}{T_c}>0 \tag{2.151}$$

and the material exhibits superconductivity. In the superconductor limit Eq. (2.146) becomes

$$\Delta\mathcal{F}=-\frac{1}{2}\frac{\alpha^2}{\beta}V=-\frac{1}{2}\frac{\alpha_0^2}{\beta}\frac{(T-T_c)^2}{T_c^2}V, \tag{2.152a}$$

where V is the volume of the material, and the free energy is lowered from that of the normal metallic state. In the presence of a magnetic induction B applied exterior to the superconductor Eq. (2.152a) becomes [3]

$$\Delta\mathcal{F}=\left[-\frac{1}{2}\frac{\alpha^2}{\beta}+\frac{B^2}{8\pi}\right]V=\left[-\frac{1}{2}\frac{\alpha_0^2}{\beta}\frac{(T-T_c)^2}{T_c^2}+\frac{B^2}{8\pi}\right]V, \tag{2.152b}$$

and it is seen that there is a critical field $B_c=H_c$ defined by $\Delta\mathcal{F}=0$. If fields greater than this critical field are applied to the superconductor, it transforms into a normal metal and is completely penetrated by the flux. This is an example of a Type I superconductor behavior. Later in another chapter discussions will be made of Type I and Type II superconductors, and the properties of Type I and Type II superconductors are treated in greater detail. For now, we go on to address further general superconductor properties.

2.8.1 Planar Interface between a Superconductor and Normal Metal

The next type of problem of interest is the study of a superconductor–normal metal interface. Specifically, consider the case of an infinite superconductor with a field greater than the critical field applied to it in the region $x\leq 0$, and in the region $x>0$ there is no field. The portion with the applied field will revert to the normal metallic state, while the portion without field will remain a superconductor [2,3].

In order to develop some simplifications in the notation, we first consider the one-dimensional form of Eq. (2.148) along the x-direction but retain the three-dimensional form of Eq. (2.149). These equations are described for a wavefunction that is only dependent on x, by Eq. (2.148a) written as

$$\frac{1}{2m_s}\left(-i\hbar\frac{\mathrm{d}}{\mathrm{dx}}-\frac{q}{c}A_x\right)^2\psi+\frac{1}{2m_s}\frac{q^2}{c^2}\left(A_y^2(x)+A_z^2(x)\right)\psi+\alpha\psi+\beta|\psi|^2\psi=0 \quad (2.153a)$$

and the superconducting current density from Eq. (2.149) again given by

$$\vec{j}_s(x)=\frac{q}{2m_s}\left(-i\hbar\psi^*\nabla\psi+i\hbar\psi\nabla\psi^*\right)-\frac{q^2}{m_s c}|\psi|^2\vec{A} \quad (2.153b)$$

From the earlier discussion of the bulk it is seen that $\psi=0$ above the critical temperature, and, consequently, Eq. (2.153) is only considered for the case that $T<T_c$. In addition, it should be remembered that, for fields above the critical field, the wavefunction is again $\psi=0$, yielding a normal metallic state solution.

An approximate solution of the interface problem is obtained as follows: In obtaining an approximate solution for $\psi(x)$ in the superconductor, we first assume $\vec{A}=0$ in the superconductor and solve Eq. (2.153a) for ψ. An approximation for $\vec{j}_s(x)$ is then obtained from Eq. (2.153b) by assuming that the ψ in Eq. (2.153b) is the constant value of ψ within the bulk of the superconductor. These two considerations then give the leading order dependence of the wavefunction and the current.

In the temperature region where $T<T_c$ and in the absence of an external field, Eq. (2.153a) is rewritten into the form [2,3]

$$-\xi^2\frac{d^2\varphi}{dx^2}-\varphi+|\varphi|^2\varphi=0 \quad (2.154)$$

where $\xi^2=\frac{\hbar^2}{2m_s|\alpha|}$ and $\varphi=\psi\sqrt{\frac{\beta}{|\alpha|}}$. Note that in making this rewrite, φ has been normalized so that $\lim_{x\to\infty}\varphi(x)\to 1$. This notation is now adopted to treat the superconductor–normal metal interface.

For a consideration of our system, the region $x\le 0$ is normal metal, and the region $x>0$ is a homogeneous superconductor described by Eqs. (2.153) and (2.154). A solution of Eq. (2.154) that satisfies the boundary conditions (requiring a vanishing superconductor carrier density in the normal metal and an approach to the bulk superconductor carrier density far from the surface) has the form

$$\begin{aligned}\varphi(x)&=\tanh\frac{x}{\sqrt{2}\xi},x\ge 0\\ &=0,\ \text{otherwise.}\end{aligned} \quad (2.155)$$

Here ξ is termed the coherence length and represents the typical length over which the superconductor wavefunction changes from the bulk to the nonsuperconducting phases. In this sense, it is a measure of the rigidity of the wavefunction.

Note that Eq. (2.155) is obtained under the assumption that the magnetic field does not penetrate into the superconductor. This is only a reasonable assumption in so-called Type I superconductors and more will be said of this later.

A first approximation for the small penetration of the magnetic field into the superconductor is obtained from Eq. (2.153b) by assuming that the superconductor wavefunction at the interface is the same as that in the bulk superconductor, i.e., the bulk wavefunction given from Eq. (2.151) by

$\psi_0 = \sqrt{-\dfrac{\alpha}{\beta}}$ is a constant in the entire region of superconductor. Since the wavefunction is then a constant, it follows from Eq. (2.153b) that the current in the superconductor is [2,3]

$$\vec{j}_s(x) = -\frac{q^2}{m_s c}|\psi_0|^2 \vec{A} \tag{2.155}$$

and taking the curl of both sides of this equation gives

$$\vec{B} = -\frac{m_s c}{q^2|\psi_0|^2}\nabla\times\vec{j}_s(x) = -\frac{c^2 m_s}{4\pi q^2|\psi_0|^2}\nabla\times\nabla\times\vec{B} \tag{2.156}$$

From the identity $\nabla\times\nabla\times\vec{C} = \nabla\nabla\cdot\vec{C} - \nabla^2\vec{C}$, it is then found that

$$\nabla^2\vec{B} = \vec{B}/\lambda_L^2 \tag{2.157}$$

where $\lambda_L^2 = \dfrac{c^2 m_s}{4\pi q^2|\psi_0|^2}$. The result in Eq. (2.157) is the famous London equation for superconductivity, and λ_L is known as the London penetration depth, which is a measure of the penetration of the magnetic flux into the superconductor.

Note alternatively that taking the curl of Eq. (2.155) twice and applying Amperes law to the result gives [2,3]

$$\nabla^2\vec{j}_s = \vec{j}_s/\lambda_L^2. \tag{2.158}$$

Solutions for Eqs. (2.157) and (2.158) pertaining to the metal–superconductor interface with $\vec{A}(x) = B_s x\hat{j}$ give

$$\vec{B}(x) = \vec{B}_s e^{-x/\lambda_L} \tag{2.159a}$$

and

$$\vec{j}_s(x) = \vec{j}_{s0} e^{-x/\lambda_L} \tag{2.159b}$$

where $\vec{B}_s = B_s\hat{k}$ is the intensity of the magnetic induction at the surface and $\vec{j}_{s0} = j_{s0}\hat{j}$. In both cases, the field and the current are found to decay to zero within the bulk of the superconductor.

The solution of the problem is given from Eqs. (2.155) to (2.158), provided that $\xi \gg \lambda_L$. This, as will be discussed in a later chapter, is the condition for a Type I superconductor.

2.8.2 Flux Quantization

Another interesting property of superconductors is that of flux quantization. This arises in consideration of the following problem. A planar ring of superconducting material carries a superconducting current $\vec{j}_s$ that is given from Eq. (2.149) by [2,3]

$$\vec{j}_s = \frac{e}{m_s}\left[i\psi^*\hbar\nabla\psi - i\hbar\psi\nabla\psi^*\right] - \frac{4e^2}{m_s c}|\psi|^2\vec{A} \tag{2.160}$$

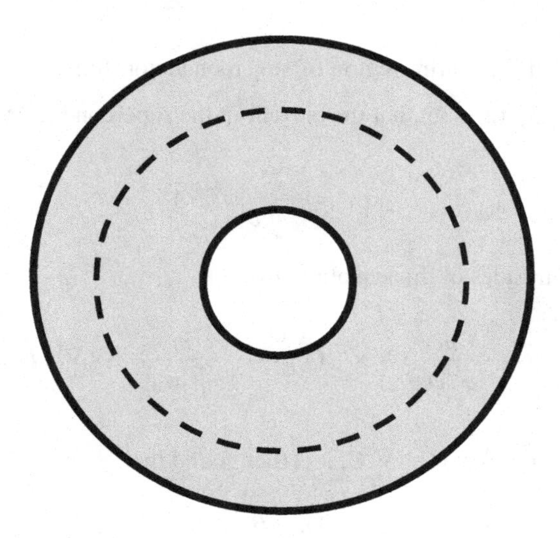

FIGURE 2.9 Schematic of a superconducting ring for the discussion of flux quantization. The dashed line is a path deep within the superconducting bulk of the ring. The current density is zero along the dashed line, and a magnetic induction perpendicular to the page is within the inner radius of the ring.

where here we have taken $q = -2e$ for Cooper pairs in an electron system. (See Figure 2.9 for a schematic of the system.) The object of the problem is to relate the magnetic flux through the ring to the phase of the superconducting wavefunction.

To obtain this relationship, consider the superconducting wavefunction in the ring to be of the form

$$\psi(\vec{r}) = \psi_0 e^{i\theta(\vec{r})}, \tag{2.161}$$

i.e., of a constant-amplitude and a position-dependent phase. A consequence of this is that the pair density is uniform over the superconducting ring, but a current can exist within parts the ring. Substituting into Eq. (2.160) gives a superconducting current density of the form

$$\vec{j}_s = -\frac{2e\hbar}{m_s}|\psi_0|^2 \nabla\theta - \frac{4e^2}{m_s c}|\psi_0|^2 \vec{A}. \tag{2.162}$$

At a radius r which is deep within the superconducting bulk of the ring the current $\vec{j}_s = 0$ so that integrating around the circumference of r within the depth of the ring and applying Stoke's theorem gives [2,3]

$$\oint \nabla\theta \cdot dl = 2\pi n = \frac{2e}{c\hbar}\oint \vec{A} \cdot dl = \frac{2e}{c\hbar}\int B \cdot dS = \frac{2e}{c\hbar}\Phi \tag{2.163}$$

where Φ is the magnetic flux through the ring, and n is an integer. It then follows that the flux through the ring is quantized so that

$$\Phi = n\Phi_{2e}, \tag{2.164}$$

where $\Phi_{2e} = \frac{\pi\hbar c}{e}$.

2.9 AHRONOV–BOHM EFFECT IN NORMAL METALS

Another type of flux quantization effect occurs in mesoscopic systems, which are normal metals. For this effect to be present, the scattering of the electrons in the Fermi gas of conduction electrons must be minimal so that the media does not appear as a viscous media. In this limit the conduction electrons move in the material as though they were traveling in a waveguide (ballistic transport) and not a wire with a resistivity. The individual electrons are therefore described by the free-particle Schrodinger equation [33]

$$\frac{1}{2m}\left(-i\hbar\nabla+\frac{e}{c}\vec{A}\right)^2\psi=\epsilon\psi \tag{2.165}$$

with solutions of the form

$$\psi(\vec{r})\propto e^{i\left(\vec{k}\cdot\vec{r}-\frac{e}{c\hbar}\int_{\vec{r}_0}^{\vec{r}}d\vec{r}\cdot\vec{A}\right)}. \tag{2.166}$$

In particular, consider applying the solution in Eq. (2.166) to the problem of an electron entering at one side of a conductive ring, traveling through either branch of the ring to the other side of the ring, and exiting the ring at the side opposite where it entered. (See Figure 2.10 for a schematic of the ring geometry.) Furthermore, for the discussions presented here, a magnetic induction is assumed to be confined to the center of the ring, i.e., the inner radius in the plane of the ring in Figure 2.10. Such a system, for example, could be realized by generating a field in a solenoid passing through the center of the ring.

For this system, the probability amplitude for an electron entering A at $\vec{r}_A$ and traveling in the upper semicircle to the exit B at $\vec{r}_B$ is of the form [2,3,33]

$$T_u=t_u e^{i\frac{e}{c\hbar}\int_{\vec{r}_A}^{\vec{r}_B}d\vec{r}\cdot\vec{A}}=t_u e^{i\theta_u}. \tag{2.167a}$$

Similarly, the probability amplitude for entering A at $\vec{r}_A$ and traveling in the lower semicircle to the exit B at $\vec{r}_B$ is of the form

$$T_l=t_l e^{i\frac{e}{c\hbar}\int_{\vec{r}_A}^{\vec{r}_B}d\vec{r}\cdot\vec{A}}=t_l e^{i\theta_l}. \tag{2.167b}$$

The probability of both of these processes is then given by

$$P\propto\left|t_u e^{i\theta_u}+t_l e^{i\theta_l}\right|^2=|t_u|^2+|t_l|^2+t_u t_l^* e^{i(\theta_u-\theta_l)}+t_l t_u^* e^{i(\theta_l-\theta_u)}. \tag{2.168a}$$

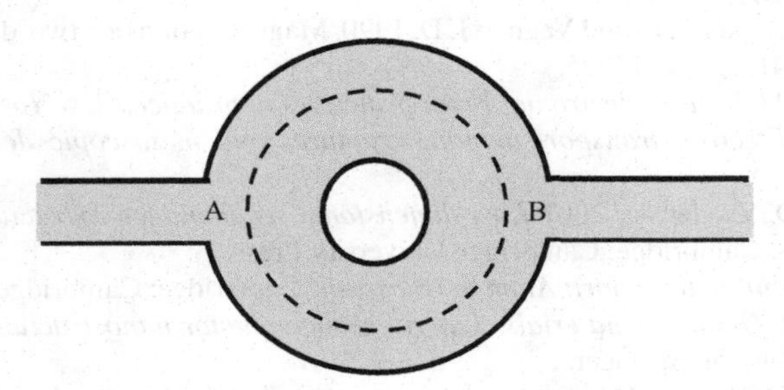

FIGURE 2.10 Schematic for the Aharonov–Bohm effect. The dashed line is an integration path between A and B within the bulk metal of the ring. A magnetic induction, within the region of the inner radius of the ring, is perpendicular to the plane of the page.

Now assume the symmetry of the two semicircles of the upper and lower paths so that for going from A to B $t_u = t_l = t$, and, consequently,

$$P \propto 2|t|^2 \{1 + \cos[(\theta_u - \theta_l)]\}. \tag{2.168b}$$

It then follows that

$$(\theta_u - \theta_l) = \frac{e}{c\hbar}\oint d\vec{r} \cdot \vec{A} = \frac{e}{c\hbar}\Phi = \pi\frac{\Phi}{\Phi_0} \tag{2.169}$$

where in Eq. (2.169) the path integral is now over the complete circumference within the ring, Φ is the solenoid flux, and $\Phi_0 = \frac{\pi\hbar c}{e}$.

It is seen from Eqs. (2.168) to (2.169) that P is a periodic function of the flux Φ through the ring. Similar to the flux quantization in the superconducting ring, the normal metal has a flux quantization effect that comes from the phase of the wavefunction. In the normal metal the phase is maintained during the electron propagation, as the ring is smaller than the typical distance between electron scatterings. In the superconductor, the wavefunction is macroscopic and retains its phase properties independent of scattering sites. In one system the coherence arises from the size of the paths treated while in the other the coherence is a property of the macroscopic wave function.

The type of phase interference discussed here for normal metals on a small scale is of particular interest in the discussions of a second method of studying conductivity and transport properties known as the Landauer approach. Such types of phase-coherent effects developed in the Landauer approach are found of great importance in nano-circuits. This is now addressed in Chapter 3.

REFERENCES

1. Kittle, C. 1996. *Introduction to solid state physics, 7th Edition*, 393–410. New York: John Wiley & Sons, Inc.
2. Marder, M. P. 2000. *Condensed matter physics*. New York: John Wiley & Sons, Inc.
3. Fossheim, K. and A. Sudbo 2004. *Superconductivity: Physics and applications*, 141–198. New York: John Wiley & Sons, Inc.
4. Schrieffer, J. R. 1983. *Theory of superconductivity*. Boca Raton: CRC Press.
5. Barone, A. and G. Paterno 1982. *Physics and applications of the Josephson effect*. New York: John Wiley & Sons, Inc.
6. Wolf, E. L. 1985. *Principles of electron tunneling spectroscopy, 2nd Edition*. Oxford: Oxford University Press.
7. de Gennes, P. G. 1989. *Superconductivity of metals and alloys*. New York: Addison-Wesley.
8. Saxena, A. K. 2012. *High-temperature Superconductors, 2nd Edition*. Heidelberg: Springer.
9. Plakida, N. 2010. *High-temperature cuprate superconductors: Experiment, theory, and applications*. Heidelberg: Springer.
10. Jauregui, K., Marchenko, V. I., and Vagner, I. D. 1990. Magnetization of a two-dimensional electron gas. *Physical Review* B41, 12922–12925.
11. Petty, M. C. 2007. *Molecular electronic: From principles to practice*. New York: John Wiley & Sons.
12. Ouisse, T. 2008. *Electron transport in nanostructures and mesoscopic devices: An introduction*. New York: Wiley.
13. Barnham, K. and D. Vvedensky 2001. *Low-dimensional semiconductor structures: Fundamentals and device applications*. Cambridge: Cambridge University Press.
14. Datta, S. 2005. *Quantum transport: Atom to transistor*. Cambridge: Cambridge University Press.
15. Heitmann, D. 2010. *Quantum materials: Lateral semiconductor nanostructures, hybrid systems, and nanocrystals*. Heidelberg: Springer.
16. Ryndyk, D. A. 2016. *Theory of quantum transport at nanoscale: An introduction*. Cham: Springer.
17. Streetman, B. G. and S. Banerjee 2000. *Solid state electronic devices, 5th Edition*. New Jersey: Prentice Hall.
18. Harrison, P. and P. Harrison 2000. *Quantum wells, wires, and dots*. New York: John Wiley & Sons, Inc.

19. Prange, R. E. and S. M. Girvi 1987. *The quantum Hall effect*. New York: Springer.
20. Wigner, E. 1934. On the interaction of electrons in metals. *Physical Review* 46: 1002–1011.
21. Jeno, S. 2010. *Fundamentals of the physics of solids*. Vol. 3. Normal, broken-symmetry, and correlated systems. Berlin: Springer Science & Business Media.
22. Lee, P. A. and T. V. Ramakrishnan 1985. Disordered electron systems. *Reviews of Modern Physics* 57: 287–333.
23. Mott, N. F. and E. A. Davis 2012. *Electronic processes in non-crystaline materials*. Oxford: Oxford University Press.
24. Mott, N. F. and W. D. Twose 1961. The theory of impurity conductivity. *Advances in Physics* 10: 107–155.
25. Kirkman, P. D. and J. B. Pendry 1984. The statistics of one-dimensional resistances. *Journal of Physics C: Solid State Physics* 17: 4327–4344.
26. Bergmann, G. 1984. Weak localization in thin films a time-of-flight experiment with conduction electrons. *Physics Reports* 107: 1–58.
27. Abrahams, E. Anderson, P. W., Licciardello, D. C., et al. 1979. Scaling theory of localization: Absence of quantum diffusion in two dimensions. *Physical Review Letters* 42: 673–677.
28. Roy, S. B. 2019. *Mott insulators: Physics and applications*. Bristol: IOP Publishing.
29. McGurn, A. R. 2021. *Introduction to nonlinear optics of photonic crystals and metamaterials, 2nd Edition*. Bristol: IOP Publishing Ltd.
30. McGurn, A. R. 2018. *Nanophotonics*. Cham: Springer.
31. McGurn, A. R. and A. A. Maradudin 1984. Localization of electrons in thin films with rough surfaces, *Physical Review B* 30: 3136–3146.
32. Mott, N. F. 1969. Conductivity in non-crystalline materials. *Philosophical Magazine* 19: 835–852.
33. Aharonov, Y. and Bohm, D. 1959. Significance of electromagnetic potentials in quantum theory. *Physical Review* 115: 485–491.

3 Conductivity
Another View

An alternative approach to the study of conductivity is presented [1–8] in this chapter. This is an approach due to Landauer, which treats a conducting wire as a type of waveguide for the electronic wavefunctions [1–3]. In the Landauer approach [1–3], the electrons move in a nondissipative waveguide into which impurity scattering can ultimately be introduced, whereas in the previous chapter, the electrons were considered to propagate through a viscous, dissipative medium. As before, the electrons are treated in the electron gas model in the absence of electron–electron interactions. In this regard, the Landauer formulation allows for the determination of the conductance of a material in the absence of viscosity and other loss effects, with a later systematic introduction of these effects. In addition, the Landauer formulation can be applied to the study of the transport properties of other types of excitations in electronic, photonic, and phononic materials and is not just limited to electronic systems.

In the new formulation, the quantum mechanical nature of the electrons is introduced from the very beginning, and the phase properties of the electrons are retained in the course of the calculations. The previous formulation in Chapter 2 tended to ignore or average out phase coherent effects. As shall be seen in many nanoscience systems, one cannot ignore or average out the phase coherent effects so that the Landauer formulation [4–8], which retains these effects, becomes important in the study of such systems. For example, it shall be found that nano- and mesoscopic systems exhibit a quantum of electrical conductance, and the conductance of general materials is characterized in units of this conductance. The quantum of conductance will be seen to be a direct consequence of the wave nature of the excitations propagated through the systems [1–4].

In the following, the Landauer formulation is introduced along with a discussion of the quantum of electrical conductance. This is followed by a treatment showing how the Landauer formulation reduces to that discussed in the previous chapter, as the macroscopic limit is attained. This will prove the equivalence of these two methods of studying conductance in macroscopic and mesoscopic systems.

3.1 THE LANDAUER FORMULATION

The Landauer formulation treats the motion of charge as a waveguide problem [1–4]. In this section, the example of a two-dimensional waveguide is studied in the context of the Landauer formulation. (See Figure 3.1a for a schematic of the waveguide.) The waveguide is of a rectangular shape of length L and width W. To the left of the waveguide, the waveguide contacts with a reservoir of electrons at an electric potential $V \geq 0$ and to the right of the waveguide the waveguide contacts with a reservoir of electrons at an electric potential 0. The noninteracting particles (i.e., no electron–electron interactions are present) within the waveguide have a free particle dispersion relation of the form [4]

$$\epsilon\left(k_x, n_y\right) = \frac{\hbar^2 k_x^2}{2m} + \frac{\hbar^2}{2m}\left(\frac{\pi n_y}{W}\right)^2 = \frac{\hbar^2 k_x^2}{2m} + \epsilon_0\left(n_y\right), \tag{3.1}$$

where $k_x = \frac{2\pi n_x}{L}$ for n_x an integer, $\epsilon_0\left(n_y\right) = \frac{\hbar^2}{2m}\left(\frac{\pi n_y}{W}\right)^2$, and $n_y = 1, 2, 3,\ldots$ Here, the first term in Eq. (3.1) comes from the $x-$ motion, and the second term is from the $y-$ motion, and it is assumed

DOI: 10.1201/9781003031987-3

that the electron wavefunctions satisfy boundary conditions requiring them to be zero at the $y = 0$ and $y = W$ edges of the waveguide channel. As in the treatment in Chapter 2, the particles do not interact with one another.

For the moment consider only electrons in the single quantum level n_y, so that from Eq. (3.1) the x – component of kinetic energy is given by [4]

$$\epsilon_r = \epsilon\left(k_x, n_y\right) - \epsilon_0\left(n_y\right) = \frac{\hbar^2 k_x^2}{2m}. \tag{3.2}$$

Following the discussions in Chapter 2, the density of modes as a function of ϵ_r is given by the one-dimensional form

$$D_1^s\left(\epsilon_r\right) = 2D_1\left(\epsilon_r\right) = \frac{L}{\pi\hbar}\sqrt{\frac{2m}{\epsilon_r}} \tag{3.3}$$

where $D_1\left(\epsilon_r\right) = \frac{L}{\sqrt{2}\pi}\frac{\sqrt{m}}{\hbar}\frac{1}{\sqrt{\epsilon_r}}$ is the number of modes per energy in the spinless gas, and $D_1^s\left(\epsilon_r\right)$ is the number of modes per energy in the spin one-half gas. In our discussions in this chapter, consequently, we now consider a gas of spin one-half electrons in which the energy of the gas does not depend on the electron spin.

To begin the study of this system, first consider the model at zero temperature for the case of two identical reservoirs on each side of the waveguide, i.e., the case in which $V = 0$. The electric potential is zero in the reservoir to the right of the waveguide, and the modes in the reservoir are occupied to the Fermi energy ϵ_F. At equilibrium the electric potential of the reservoir to the left of the waveguide also is at zero electric potential, with its reservoir modes occupied to the Fermi energy ϵ_F. (See Figure 3.1b for a schematic.) In this configuration, no current flows through the waveguide.

Next consider changing the configuration of the system by applying an electric potential $V > 0$ to the reservoir on the left. Now each electron of charge $-e$ in the left reservoir is shifted by the potential energy $U = -eV$ so that the Fermi energy of the left reservoir is $\epsilon_F' = \epsilon_F + U = \epsilon_F - eV$, while the reservoir to the right is left at zero potential. (See Figure 3.1c for a schematic.) In this case, the system is no longer in equilibrium, and electrons flow from the right to the left reservoir, generating a net current I in the waveguide [1–4].

The zero-temperature current flowing through the waveguide between the reservoirs is given for $|eV| \ll \epsilon_F$ by [4–8]

$$I = e\int_{\epsilon_F - eV}^{\epsilon_F} \frac{D_1^s\left(\epsilon_r\right)}{2L} v\left(\epsilon_r\right) d\epsilon_r \tag{3.4}$$

where $e > 0$ is the unit of electric charge, $\frac{D_1^s\left(\epsilon_r\right)}{2L}$ is the density of energy states per length considering only modes traveling from the right to the left, and

$$v\left(\epsilon_r\right) = \frac{\hbar k_x}{m} = \hbar\sqrt{\frac{2\epsilon_r}{m\hbar^2}} \tag{3.5}$$

is the magnitude of the velocity of an electron traveling in the waveguide. Consequently, from Eqs. (3.3) to (3.5)

$$\frac{D_1^s\left(\epsilon_r\right)}{2L} v\left(\epsilon_r\right) = \frac{2}{h}, \tag{3.6}$$

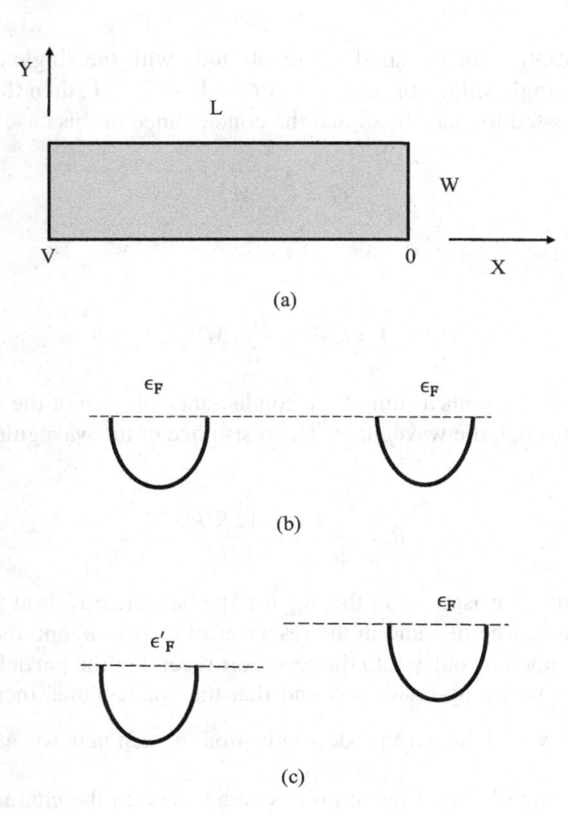

FIGURE 3.1 Schematic drawings of: (a) The rectangular waveguide of length L and width W. On the left the waveguide contacts a reservoir of electrons at an electric potential $V \geq 0$, and to the right the waveguide contacts a reservoir of electrons at an electric potential 0. (b) Comparison of the parabolic dispersion of the electrons in the left and right reservoirs. In the case of zero potential difference between the reservoirs, the Fermi levels ϵ_F of the reservoirs (dashed line) are the same; and no current flows between the reservoirs. (c) Comparison of the parabolic dispersion of the electrons in the left and right reservoirs. In the case of a nonzero potential difference between the reservoirs, the Fermi levels ϵ_F of the reservoirs (dashed lines) are different; and a current flows between the reservoirs.

so that it follows from Eq. (3.4) that

$$I = e^2 \frac{2}{h} V. \tag{3.7}$$

Equation (3.7) is of the form

$$V = IR_s = \frac{I}{G}, \tag{3.8}$$

where R_s is the effective resistance, and G is the conductance given by [4]

$$G = \frac{2e^2}{h}. \tag{3.9}$$

Note that it is important to emphasize that the factor of 2 in the conductance arises from the spin degeneracy of the electron dispersion relation in Eq. (3.9) [6,8]. This degeneracy is lifted, for example, in the presence of an applied magnetic field.

Note that G in Eq. (3.9) is for the single channel mode with the single quantum number n_y. If n_y is not restricted to a single value but ranges over $n_y = 1, 2, \ldots, M$, then the above calculations in Eqs. (3.4)–(3.9) are repeated for each n_y so that the conductance in this case becomes [4]

$$G = \frac{2e^2}{h}M, \tag{3.10a}$$

and from Eqs. (3.7) to (3.8)

$$I = GV = \frac{2e^2}{h}MV. \tag{3.10b}$$

The result in Eq. (3.10b) represents a sum of the conductance of each of the M channels, giving the resultant total current through the waveguide. The resistance of the waveguide passage is then from Eq. (3.10a) given by [4]

$$R_s = \frac{h}{2e^2 M} = \frac{12.9\ k\Omega}{M}, \tag{3.11}$$

and is known as the contact resistance. In this regard, the particle travels at the same constant electric potential in both the waveguide and in the reservoir of its origin, and the electric potential it is subject to only changes upon its entry into the reservoir receiving the particle.

From Eq. (3.10) the interesting result is found that the conductance increases by the quantum unit of $\frac{2e^2}{h}$ with the entry into the waveguide conduction of each new wavefunction state of n_y. (In this sense, $G_0 = \frac{2e^2}{h}$ is regarded as a quantum of conductance in the characterization of nanosystems.) This entry of a new state into the conduction can be made by changing the value of the Fermi energy, accomplished, for example, by changing the electric potential of the reservoir. An illustration of this is indicated in Figure 3.2, which shows the two quantum levels $n_y = n$ and $n_y = n+1$ with the Fermi level positioned between the two levels. By raising the Fermi level above that shown in the figure, conduction by means of the $n_y = n+1$ level can be introduced into the waveguide.

It is also important to note that in Eq. (3.11) the resistance is independent of the length of the waveguided. This is different from the Ohmic resistance of a viscous medium such as that which was discussed in Chapter 2. There it was shown that the resistance of a length of viscous material was proportional to its length. The reason for this difference is that unlike the viscous medium [4–7] there is no scattering in the waveguide. Consequently, the electrons in the waveguide are transported by modes that propagate undisturbed through the waveguide channel. In this regard,

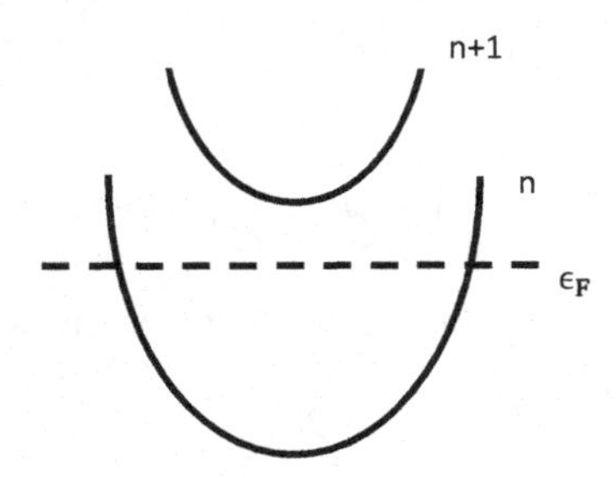

FIGURE 3.2 A schematic showing the two parabolic quantum energy levels $n_y=n$ and $n_y=n+1$ with the Fermi level ϵ_F (dashed line) positioned between the two levels. By raising the Fermi level above that shown in the figure, conduction by means of the $n_y=n+1$ level can be introduced into the waveguide.

we shall now consider how the waveguide model needs to be modified in order to arrive at a model of a material exhibiting an Ohmic length-dependent resistance.

3.2 SCATTERING WITHIN THE WAVEGUIDE: OHMIC LIMIT

Consider now that a scattering site is introduced into the channel of the waveguide. (This is illustrated schematically in Figure 3.3.) The scattering site or barrier is placed within the waveguide so that a fraction of the wave traveling in the waveguide is transmitted through the barrier with a transmission coefficient $0 \leq T \leq 1$ and a fraction of the wave is reflected back from the barrier with a reflection coefficient $0 \leq R \leq 1$. In this regard, the transmission and reflection coefficients satisfy

$$T + R = 1, \tag{3.12}$$

so that all of the wave is either transmitted or reflected by the barrier. As an additional simplification, it will also be assumed that the incident, reflected, and transmitted modes only involve one state of fixed quantum number n_y for each of the n_y modes introduced into the waveguide in the initial considerations. This last point is essentially the assumption that the scattering is elastic at the barrier.

In the waveguide transport process, there are three currents due to the introduction of the barrier. These are the current I introduced into the waveguide and sent toward the barrier, and the transmitted current [4]

$$I_T = \frac{2e^2}{h} MTV \tag{3.13a}$$

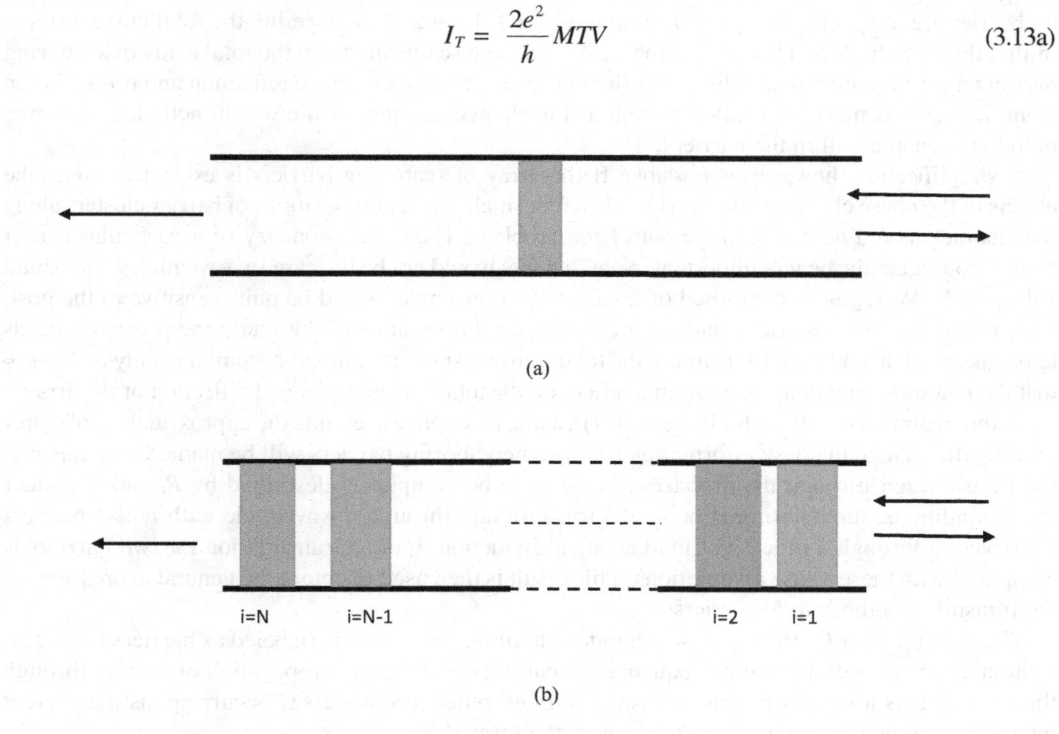

FIGURE 3.3 Schematic of a waveguide containing many scattering barriers along its length. The barriers are randomly placed but on average of a uniform density over the length of the waveguide. The barriers are labeled consecutively $i = 1, 2, 3, \ldots, N$ along the length of the waveguide channel, so that the reflection and transmission coefficients of the ith barrier are, respectively, R_i and T_i with $R_i + T_i = 1$. An electron is incident from the right (left directed arrow), and it is reflected (right directed arrow on the right) and transmitted (left directed arrow) on the left. A waveguide with a single barrier is illustrated in (a), and the full system of N barriers is illustrated in (b).

and the reflected current

$$I_R = \frac{2e^2}{h} MRV = \frac{2e^2}{h} M(1-T)V \tag{3.13b}$$

of the barrier. Note that here V is the potential difference between the two reservoirs at the opposite ends of the waveguide. Since the currents are conserved, it follows that

$$I = I_R + I_T, \tag{3.13c}$$

and the conductance of the waveguide changes from that in Eq. (3.10) for the waveguide in the absence of the barrier to [4]

$$G = \frac{2e^2}{h} MT, \tag{3.14}$$

i.e., it is reduced by a factor of the transmission coefficient through the barrier.

Now consider a waveguide containing many scattering barriers along the length of the waveguide channel that are randomly placed but on average of a uniform density over the length of the waveguide. (See Figure 3.3 for a schematic.) The barriers are labeled consecutively $i = 1, 2, 3, \ldots, N$ along the length of the waveguide channel so that the reflection and transmission coefficients of the ith barrier are, respectively, R_i and T_i with $R_i + T_i = 1$. In order to determine the total current transmitted through the $N \gg 1$ barriers of the waveguide, the scattering from the total array of scattering barriers must be summed up. This is a difficult process, because both the reflection and transmission from the barriers must be handled as well as the change in phase of the wavefunction, as it moves both between and within the barriers.

A simplification, however, is possible. If the array of scattering barriers is extremely large, the effects of the phase changes at the various different small, finite combinations of barrier clusters along the channel should tend to average out of the problem. The exact geometry of a particular cluster would, consequently, be less important. Note that this would not be the case in waveguides with small values of N. Waveguides composed of three sites, for example, would be quite sensitive to the positions of the scattering barriers, and the detailed phase information of the scattering processes needs to be taken into account in determining the total transmission through the system. It is only as $N \to \infty$ that the phase information has a minimal effect on the total transmission and reflection of the array.

In the treatment to follow, for large $N \gg 1$ in a macroscopic waveguide the approximation of disregarding the change in phase information between neighboring barriers will be made. Consequently, the transmission through the ith barrier is taken to be completely described by R_i and T_i. Under these conditions, the determination of the transmission through a waveguide with $N \gg 1$ barriers is arrived at through a process of mathematical induction. First, a transmission for two barriers is computed with these phase assumptions. This result is then used to obtain the general expression for the transmission through N barriers.

The transmission $T_{1,2}$ through a waveguide containing two barriers (labeled as barriers 1 and 2) is written as a sum over an infinite sequence of scatting events in the propagation of current through the system. It is a sum of distinct transmission and reflection processes occurring, as the current interacts with the two barriers and is given by the form [4,8]

$$T_{1,2} = T_1T_2 + T_1R_2R_1T_2 + T_1R_2R_1R_2R_1T_2 + \ldots = \frac{T_1T_2}{(1-R_1R_2)}. \tag{3.15a}$$

Here, the first term in the sum is an event in which the current is first transmitted through barrier one and then transmitted through barrier two. The second term of the sum involves first a transmission through barrier one, followed by reflection from barrier two, and followed by a reflection from barrier one and a final transmission through barrier two. The high-order terms in the sum involve a transmission through barrier one, followed by higher multiples of reflections between the two barriers, until there is a final transmission through barrier two. On the far right of the equation is the total sum of all of such scattering events. In a similar fashion, the reflection coefficient from the barrier of two sites is given by

$$R_{1,2} = R_1 + T_1 R_2 T_1 + T_1 R_2 R_1 R_2 T_1 + \ldots = R_1 + T_1 R_2 \frac{1}{1 - R_1 R_2} T_1, \tag{3.15b}$$

so that $R_{1,2} + T_{1,2} = 1$.

Equation (3.15a) can readily be shown to obey the identity [4,8]

$$\frac{1 - T_{1,2}}{T_{1,2}} = \frac{1 - T_1}{T_1} + \frac{1 - T_2}{T_2}. \tag{3.16}$$

Consequently, if $T_{1,j}$ denotes the total transmission through the consecutive barriers from 1 to j where $T_{1,1} = T_1$ and if it is assumed that

$$\frac{1 - T_{1,N-1}}{T_{1,N-1}} = \frac{1 - T_{1,N-2}}{T_{1,N-2}} + \frac{1 - T_{N-1}}{T_{N-1}}, \tag{3.17}$$

then it follows from Eqs. (3.15) to (3.17) that

$$\frac{1 - T_{1,N}}{T_{1,N}} = \frac{1 - T_{1,N-1}}{T_{1,N-1}} + \frac{1 - T_N}{T_N}. \tag{3.18}$$

Repeatedly applying the recursive identity in Eq. (3.18), it follows that

$$\frac{1 - T_{1,N}}{T_{1,N}} = \sum_{i=1}^{N} \frac{1 - T_i}{T_i}, \tag{3.19}$$

where $T_{1,N}$ is the total transmission through the consecutive barriers $1, 2, \ldots, N$. In this process, the step going from Eqs. (3.17) to (3.18) comes by lumping together the transmission $T_{1,N-1}$ from the first $N-1$ barriers as that from a single effective barrier T_{eff}. The resulting T_{eff} is then treated as the first barrier in Eq. (3.15a), while T_N is treated as the second barrier. In this way $T_{1,N}$ is generated from Eq. (3.15a), and the result yields Eq. (3.18) and subsequently Eq. (3.19).

Taking $T_i = T$ for all of the barriers along the waveguide, it is found from Eq. (3.19) that [4,6,8]

$$\frac{1 - T_{1,N}}{T_{1,N}} = N \frac{1 - T}{T}, \tag{3.20}$$

and solving Eq. (3.20) for $T_{1,N}$ gives, in terms of T for an array of N scattering sites,

$$T_{1,N} = \frac{T}{N(1 - T) + T}. \tag{3.21}$$

If the length of the waveguide array of N sites is L, it follows then from Eq. (3.21) that

$$T_{1,N} = \frac{TL}{N(1 - T)L + TL} = \frac{l}{L + l}, \tag{3.22}$$

where $l = \frac{L}{N}\frac{T}{1-T} = \frac{1}{n}\frac{T}{1-T} = \frac{1}{n}\frac{T}{R}$ for $n = \frac{N}{L}$, the constant density of scattering barriers per length.

In the limit $T \approx 1$ of weak scattering [4]

$$n(1-T)l = nRl = T \approx 1, \tag{3.23}$$

so that l is a measure of the length the particles travel before they are completely reflected by the disorder in the waveguide, i.e., it is the mean free path of the particles propagating in the waveguide.

Up until now the dependence on the waveguide length L of the transmission coefficient has been computed for an array of $N \gg 1$ barriers for a single n_y mode. The total transmission down the waveguide channel, however, is obtained by summing over all of the n_y modes that can carry a current in the guide channel. In this regard, in order to generalize the result for the current flow in the system to include multiple n_y modes, an expression for the number M of conducting n_y modes is also needed. As in the discussion in Eqs. (3.13)–(3.14), it then follows that for the current I introduced into the waveguide and sent toward the barrier array the transmitted current is given by [4]

$$I_T = \frac{2e^2}{h} M T_{1,N} V, \tag{3.24a}$$

and the reflected current is given by

$$I_R = \frac{2e^2}{h} M R_{1,N} V = \frac{2e^2}{h} M\left(1 - T_{1,N}\right)V. \tag{3.24b}$$

Again, here V is the potential difference between the two reservoirs at the opposite ends of the waveguide. The conductance of the waveguide then follows as [4]

$$G = \frac{2e^2}{h} M T_{1,N}. \tag{3.25}$$

where the value of M is obtained by considering the waveguide boundary conditions along the $y-$ axis. Along this axis the separation of the $k_y > 0$ components of the modal wavevectors is $\frac{\pi}{W}$, so that $M = \frac{k_F}{\pi/W} = \frac{k_F W}{\pi}$ is the number of k_y solutions consistent with the boundary conditions where $0 \le k_y \le k_F$ and from Eq. (3.1) $\epsilon_F\left(k_x = 0, n_{yF}\right) = \frac{\hbar^2 k_F^2}{2m}$ with $k_F = \frac{\pi n_{yF}}{W}$. This provides a relationship of M to the width W of the conductive strip.

Applying this and the expression for the transmission coefficient from Eq. (3.22) in Eq. (3.25) gives

$$G = \frac{2e^2}{h} M \frac{l}{L+l} = \frac{2e^2}{h}\frac{k_F W}{\pi}\frac{l}{L+l} \tag{3.26a}$$

and the resistance

$$R_s = \frac{1}{G} = \frac{h}{2e^2}\frac{\pi}{k_F l}\frac{1}{W}(l+L). \tag{3.26b}$$

Here, the first term in Eq. (3.26b) gives the contact resistance of the waveguide, and the second term, which is proportional to the length of the waveguide, is the Ohmic resistance of the waveguide. In the limit of $L \gg l$ Eq. (3.26) becomes [4,6,8]

$$G = \frac{2e^2}{h}\frac{k_F l}{\pi}\frac{W}{L} \propto \frac{W}{L} \tag{3.27a}$$

and

$$R_s = \frac{h}{2e^2}\frac{\pi}{k_F l}\frac{L}{W} \propto \frac{L}{W}. \tag{3.27b}$$

These two expressions display the correct dependence on L and W in the Ohmic limit of the two-dimensional waveguide. Due to the approximations made during the discussions, however, Eqs. (3.25) and (3.26) only represent approximations for the constant of proportionality.

3.3 LANDAUER–BUTTIKER FORMULA

A general formulation for the waveguide conduction problem has been developed, which is used to treat models of waveguide scattering and networks of branching waveguides. This generalization of the Landauer treatment is known as the Landauer–Buttiker formula [1–4,6,8], and it expresses the current flow in a system composed as a network of ballistic transport waveguides containing reflecting and transmitting features. In the new formulation, the currents in the system are expressed in terms of the chemical potentials of the reservoirs, which are jointed by the waveguide network, and the various network reflection and transmission coefficients. (Remember that in the zero-temperature systems considered here the chemical potentials are the Fermi energies.) In addition, for the single waveguide problems treated in the previous section, the new formulation reduces to the Landauer formulation used there. In the following discussions, the Landauer–Buttiker formulation is introduced and studied for two-dimensional electron gas models of greater complexity than those considered in the earlier sections.

The Landauer–Buttiker formulation considers a system of K waveguides jointed to K reservoirs (See Figure 3.4a for a schematic.), giving the electron current in each waveguide as [4,6,8]

$$I_i = \frac{2e}{h}\left[(N_i - R_{ii})\mu_i - \sum_{j \neq i} T_{ij}\mu_j\right], \tag{3.28}$$

where I_i is the current flow in the ith waveguide connecting to the ith reservoir (Note that the current flow is opposite the flow of electrons, and the electrons flow from a region of high chemical potential to one of low chemical potential.), N_i are the number of different quantum states of transverse quantum number n_y carrying current longitudinally along the length of the waveguide channel, and μ_j is the chemical potential of the jth reservoir attached to the waveguide carrying current I_j away from the jth reservoir. The reflection coefficient of the ith waveguide channel is

$$R_{ii} = \sum_{n,m} R_{ii}(n,m), \tag{3.29a}$$

where $R_{ii}(n,m)$ is the reflection coefficient in the ith channel characterizing the reflection of the mth transverse quantized channel mode into the nth transverse quantized mode, and, similarly,

$$T_{ij} = \sum_{n,m} T_{ij}(n,m) \tag{3.29b}$$

where $T_{ij}(n,m)$ is the transmission coefficient into the ith channel of current in the jth waveguide channel so that the mth transverse quantized mode in j is transmitted into the nth transverse quantized mode in i. The sums in Eq. (3.29) are over all of the transverse quantized modes carrying current.

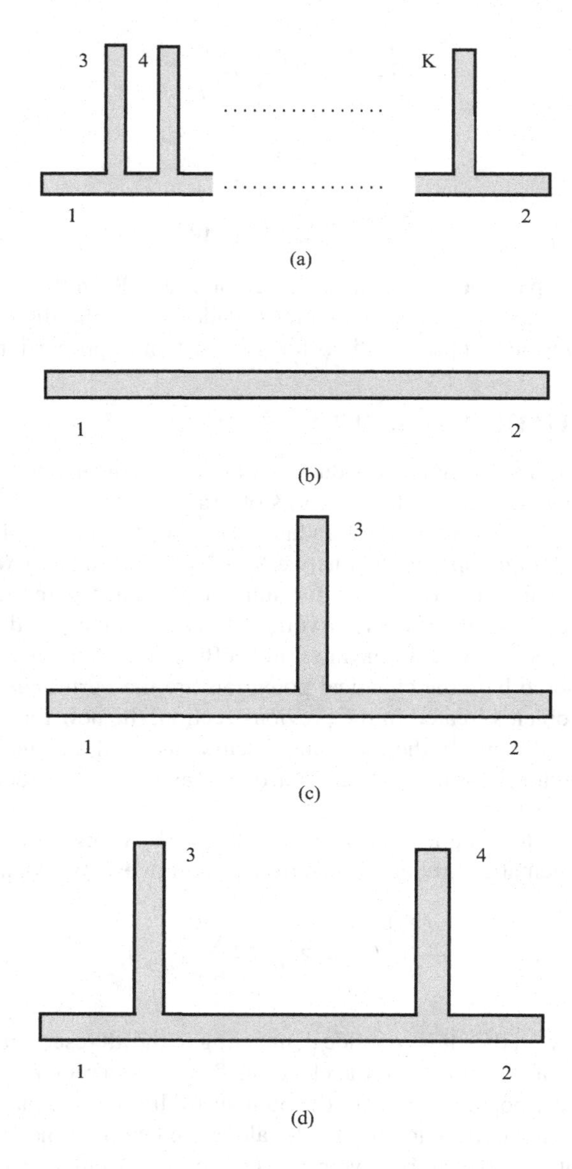

FIGURE 3.4 Schematics illustrating the Landauer–Buttiker formulation: (a) a system of K waveguides joined to K reservoirs, (b) a single waveguide between two reservoirs attached at 1 and 2, (c) a system of three waveguides jointed to three reservoirs, and (d) a system of four waveguides joined to four reservoirs.

Again, as in our earlier discussions, the factor of 2 in Eq. (3.28) comes from a consideration of a system in which the dispersion relation for the electrons is independent of the electron spin.

The statement in Eq. (3.28) is that the currents flowing in the waveguides of the network are the results of various reflections and transmission processes among the different waveguide modes of the network. In particular, if current is only carried in one transverse quantized mode of the waveguide, Eq. (3.28) reduces to [4,6,8]

$$I_i = \frac{2e}{h}\left[(1 - R_{ii})\mu_i - \sum_{j \neq i} T_{ij}\mu_j\right], \tag{3.30}$$

where the sums in Eq. (3.29) are over the single transverse quantized current-carrying mode.

As an example of the formulation, the earlier conductance result for a single waveguide is now reconsidered. (See Figure 3.4b for a schematic of the single waveguide problem.) Here the potential is $V \geq 0$ at reservoir 1, and the potential is zero at reservoir 2. Applying Eq. (3.28) to the system in the figure for the case of N current-carrying transverse guided modes gives

$$I_1 = \frac{2e}{h}\left[(N-R)\mu_1 - T\mu_2\right] = -I \tag{3.31a}$$

$$I_2 = \frac{2e}{h}\left[(N-R)\mu_2 - T\mu_1\right] = I, \tag{3.31b}$$

where account is taken that $I_1 = -I_2$, so that the positive current in the channel flows from reservoir 1 to reservoir 2, $T_{12} = T_{21} = T$, $R_{11} = R_{22} = R$, and $\mu_2 > \mu_1$. In addition, another important relationship is

$$N = R + T. \tag{3.31c}$$

This is a statement of the current conservation, indicating that if N different transverse carrier modes are sent into the waveguide, R of them are reflected, and T are transmitted by the wave guide.

Subtracting Eq. (3.31a) from Eq. (3.31b) and using $N - R = T$, it follows that

$$I = \frac{2e}{h}(\mu_2 - \mu_1)T. \tag{3.32}$$

so that the conductance is then obtained from the relationship [4,6,8]

$$G = \frac{eI}{(\mu_2 - \mu_1)} = \frac{2e^2}{h}T. \tag{3.33}$$

This expression for the conductance is found to agree with the earlier result in Eq. (3.14) after some adjustment of the notation. In particular, in comparing Eqs. (3.33) and (3.14) it should be noted that the transmissions T are defined differently in the two formulae. In Eq. (3.33) $T = \sum_{n,m}(n,m)$, and scattering between different modes $n \neq m$ is allowed. In Eq. (3.14) $T = T(n,n)$ so that only the scattering of the carrier mode n into itself is considered, and the M in Eq. (3.14) accounts for the $M = N$, different n modes carrying currents. With these two notational adjustments, Eqs. (3.33) and (3.14) are the same.

A more complex example is that Figure 3.4c involves three waveguides between three reservoirs. The reservoirs are all made of the same materials, and are connected by a network of three waveguides that are able to pass currents between each other and the reservoirs. The configuration of Figure 3.4c is an example of a network that can operate as a type of voltage probe.

For the three-probe operation, a current I is passed between reservoirs 1 and 2, while no current flows in or out of reservoir 3. The interest in the network is to measure the potential of the reservoir 3, related by μ_3, and correlate it to the conductance between reservoirs 1 and 2.

The waveguide and reservoir arrangement in Figure 3.4c is described by general equations, which from Eq. (3.28) are of the form [4,6,8]

$$I_1 = \frac{2e}{h}\left[(N_1 - R_{11})\mu_1 - T_{12}\mu_2 - T_{13}\mu_3\right] \tag{3.34a}$$

$$I_2 = \frac{2e}{h}\left[(N_2 - R_{22})\mu_2 - T_{21}\mu_1 - T_{23}\mu_3\right] \tag{3.34b}$$

$$I_3 = \frac{2e}{h}\left[(N_3 - R_{33})\mu_3 - T_{31}\mu_1 - T_{32}\mu_2\right] \tag{3.34c}$$

In addition, an important relationship of present usefulness is obtained by considering Eq. (3.34) for the case in which $\mu_1 = \mu_2 = \mu_3 = \mu$. In this limit of the system, the currents $I_1 = I_2 = I_3 = 0$. Consequently, under these conditions, it follows from Eq. (3.34) that

$$N_i - R_{ii} = \sum_{j \neq i} T_{ij}. \tag{3.35}$$

A solution for μ_3 is obtained from Eq. (3.34) under the consideration that $I_3 = 0$. It follows from Eq. (3.34c) and an application of the relation in Eq. (3.35) that

$$\mu_3 = \frac{T_{31}\mu_1 + T_{32}\mu_2}{T_{31} + T_{32}}, \tag{3.36}$$

and further if $T_{31} = T_{32} = T$, then

$$\mu_3 = \frac{\mu_1 + \mu_2}{2}. \tag{3.37}$$

For the case that $\mu_1 < \mu_2$, the current I out of reservoir 1 and into reservoir 2 is from Eqs. (3.34a) and (3.35) then given by

$$I_1 = -I = \frac{2e}{h}\left[(T_{12} + T_{13})\mu_1 - T_{12}\mu_2 - T_{13}\mu_3\right] = \frac{2e}{h}\left[T_{12} + \frac{1}{2}T_{13}\right](\mu_1 - \mu_2), \tag{3.38}$$

where $\mu_1 - \mu_2 < 0$ is the chemical potential difference between reservoir 1 and 2. As a point of interest, note here that for $T_{13} = T_{23} = 0$ the result in Eqs. (3.32) and (3.33) is reproduced.

As an interesting particular case of the system in Eq. (3.38), consider that $T_{ij} = T_{ji}$, $R_{11} = 0$, $N_i = 1$, and $T_{12} + T_{13} = 1$. In addition, upon taking $\mu_2 = 0$ as the zero of energy, it follows from Eq. (3.38) that [4,8]

$$I_1 = -I = \frac{2e}{h}\left[1 - \frac{1}{2}T_{13}\right]\mu_1 = -\frac{2e^2}{h}\left[1 - \frac{1}{2}T_{13}\right]eV, \tag{3.39a}$$

where $\mu_1 - \mu_2 = \mu_1 = -eV$. Consequently, the conductance in the presence of the probe is

$$G = \frac{2e^2}{h}\left[1 - \frac{1}{2}T_{13}\right]. \tag{3.39b}$$

Note that the conductance of the three-waveguide probe in Eq. (3.39b) is found to be less than the $G_0 = \frac{2e^2}{h}$ conductance of the single waveguide in the absence of the probe waveguide. In this regard, the arrangement of $R_{11} = 0$ as part of the conditions for the discussions leading to Eq. (3.39) was made in order to provide a comparison of the three-probe model with that of the single ballistic waveguide between two reservoirs.

Another useful consideration is that of a network of four waveguides between four reservoirs. (This is schematically shown in Figure 3.4d.) The arrangement in Figure 3.4 is described by equations from Eq. (3.28) and the four waveguide generalization of Eq. (3.35) yielding the form [4,6,8]

$$I_1 = \frac{2e}{h}\left[(T_{12}+T_{13}+T_{14})\mu_1 - T_{12}\mu_2 - T_{13}\mu_3 - T_{14}\mu_4\right] \tag{3.40a}$$

$$I_2 = \frac{2e}{h}\left[(T_{21}+T_{23}+T_{24})\mu_2 - T_{21}\mu_1 - T_{23}\mu_3 - T_{24}\mu_4\right] \tag{3.40b}$$

$$I_3 = \frac{2e}{h}\left[(T_{31}+T_{32}+T_{34})\mu_3 - T_{31}\mu_1 - T_{32}\mu_2 - T_{34}\mu_4\right] \tag{3.40c}$$

$$I_4 = \frac{2e}{h}\left[(T_{41}+T_{42}+T_{43})\mu_4 - T_{41}\mu_1 - T_{42}\mu_2 - T_{43}\mu_3\right]. \tag{3.40d}$$

As a further example of a waveguide network, these equations are now solved for the case of a system in which $I_1 = -I_2 = -I$ and $I_3 = I_4 = 0$. This is a generalization of the three-probe system to a considerations of four probes. The object in the following study of the network is again to relate the current I flowing in waveguides 1 and 2 to the difference in the chemical potentials $\mu_3 - \mu_4$ between the two probe waveguides. Ultimately, a consideration of the model will display a number of new ballistic transport properties.

A simplification of Eq. (3.40) is made by defining the zero of energy such that $\mu_2 = 0$. By conservation of current, the number of equations in Eq. (3.40) can then be reduced to the following three [6]

$$I_1 = -I = \frac{2e}{h}\left[(T_{12}+T_{13}+T_{14})\mu_1 - T_{13}\mu_3 - T_{14}\mu_4\right] \tag{3.41a}$$

$$I_3 = 0 = \frac{2e}{h}\left[(T_{31}+T_{32}+T_{34})\mu_3 - T_{31}\mu_1 - T_{34}\mu_4\right] \tag{3.41b}$$

$$I_4 = 0 = \frac{2e}{h}\left[(T_{41}+T_{42}+T_{43})\mu_4 - T_{41}\mu_1 - T_{43}\mu_3\right]. \tag{3.41c}$$

These are solved for μ_1, μ_3, μ_4 as functions of I.

Applying standard methods, the inverse of the matrix in Eq. (3.41) is computed, and it is found in the case of networks with reciprocity (i.e., $T_{ij} = T_{ji}$) that [6,8]

$$\mu_3 - \mu_4 = -\left[\frac{2e}{h}\right]^2 \frac{I}{Det}\left[T_{13}T_{24} - T_{14}T_{23}\right] = -e\Delta V_{34} \tag{3.42}$$

where Det is the determinant of the system of equations, and ΔV_{34} is the difference in electric potential between reservoirs 3 and 4. The resistance defined as $R_{s12,34} = \frac{\Delta V_{34}}{I}$ is then given by

$$R_{s12,34} = \frac{1}{e}\left[\frac{2e}{h}\right]^2 \frac{T_{13}T_{24} - T_{14}T_{23}}{Det}. \tag{3.43}$$

As an important point, it is noted that the network has an interesting property that it is possible for the resistance to be negative when $T_{13}T_{24} < T_{14}T_{23}$. This property of the resistance has been observed in a number of experiments, which are now briefly reviewed [5].

In recent experiments, the four-probe resistance measurement has been made on a number of systems formed as waveguides joined together into the configuration of a cross. In Figure 3.5, two experimental systems are shown with waveguide terminal labeled 1, 2, 3, and 4. The resistance for the geometry in Figure 3.5a [5] is given in terms of the form [4–6,8]

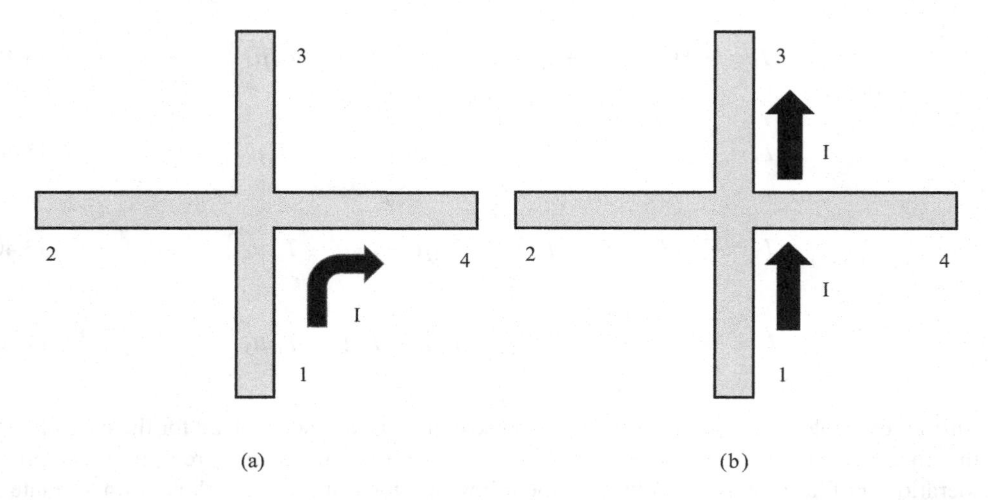

FIGURE 3.5 Schematic of two experimental systems are shown with waveguide terminal labeled 1, 2, 3, and 4. A potential is applied across terminals 2 and 3 in (a), and a current is measured flowing from 1 to 4. A potential is applied across terminals 2 and 4 in (b), and a current is measured flowing from 1 to 3.

$$R_{s14,23} = \frac{1}{e}\left[\frac{e}{h}\right]^2 \frac{T_{21}T_{34} - T_{24}T_{31}}{Det}, \tag{3.44}$$

where in Eq. (3.44) the order of the indices on the transmission coefficients matters, because the experiments involved the application of a symmetry-breaking external magnetic field applied perpendicular to the plane of the circuit. (In this regard, also note the absence of the factor of 2 arising from the spin degeneracy of the electron dispersion.) Specifically, the presence of the magnetic field makes for the possibility that $T_{ij} \neq T_{ji}$. In the experiment, a current flows from 1 to 4, and the potential is measured across 2 and 3. It is found that initially in zero field the resistance $R_{s12,34}$ was negative, but as the field was increased $R_{s12,34}$ decreased in magnitude, eventually becoming positive. The measured results were symmetric for antiparallel configurations of the applied fields.

For the system in Figure 3.5b, the current flows from 1 to 3, and the voltage is measured between 2 and 4. This configuration is meant to simulate the measurement [4–6,8] of a Hall effect (see Chapter 9). The resistance of the system is given by [5]

$$R_{s13,42} = \frac{h}{2e^2}\frac{T_{41}T_{23} - T_{43}T_{21}}{(T_{21}+T_{41})\left[(T_{21}+T_{31})^2 + (T_{41}+T_{31})^2\right]} \tag{3.45}$$

where, again, because of an applied magnetic field perpendicular to the plane of the circuit the order of the subscripts is important. For the system the resistance in Eq. (3.45) is a Hall resistance that is zero at zero applied field but otherwise an antisymmetric function of the field. The resistance then ranges between positive and negative values between antiparallel field orientations.

For the details of both of these experimental systems, the reader is referred to the original literature and to [5].

3.4 UNIVERSAL CONDUCTANCE FLUCTUATIONS

An interesting statistical problem in regard to the transport properties of a single waveguide between two different reservoirs occurs in the discussion of universal conductance fluctuations [4,5]. If we consider identical scattering impurities that are randomly distributed along the waveguide channel with a constant density per length n_s, there are many different configurations of channel impurities consistent with the constant n_s. For a great length of waveguide, it has been shown in Section 3.2

that the transport in the system is described on average by Ohms law. This characterization ignores the phase fluctuations in the carrier scattering, which are assumed to average out in the macroscopic system. On shorter mesoscopic scales, however, an exact account of the phase changes in the system must be made in order to arrive at the correct conductivity [5]. In these systems different impurity configurations with the same n_s may have significantly different conductance. Consequently, for fixed n_s the conductance that is measured fluctuates about its mean value so that the mean conductance becomes a less useful description of the conductance. The fluctuations about the mean conductance are known as universal conductance fluctuations and as shall be seen later are very similar to the speckle that occurs in the transmission of laser light through a randomly disordered optical medium. The universal conductance fluctuations are now discussed.

From the earlier discussion of the single waveguide channel in the presence of scattering impurities, it was shown in Eq. (3.33) that the conductance of the channel is of the form [4,5,9,10]

$$G = \frac{2e^2}{h}T. \tag{3.46}$$

For the most general case, in which scattering is allowed between the various dispersive bands of the waveguide modes, it follows from Eq. (3.29) that [4,5,9,10]

$$T = \sum_{n,m} T(n,m), \tag{3.47a}$$

where $T(n,m)$ is the transmission coefficient of a channel mode in the mth dispersive band to be scattered into the nth channel mode. Similarly,

$$R = \sum_{n,m} R(n,m), \tag{3.47b}$$

where $R(n,m)$ is the reflection coefficient for the mth channel mode to be reflected into the nth channel mode. In this notation, it follows that

$$N = T + R \tag{3.48}$$

where N is the number of different bands of dispersive channel modes contributing to the conductance.

In Eqs. (3.46)–(3.48), the most general Landauer–Buttiker formulation is adopted. It includes the participations of multiple dispersive conductive bands in the conductance and allows for multiple impurities in the channel. In this regard, it should be remembered that discussions prior to those of the Landauer–Buttiker formulation did not include scattering between the different dispersive bands of the waveguide modes. This most general Landauer–Buttiker formulation for the waveguide transport problem is now used to study the statistical nature of the mean waveguide current and its variance about the mean. This relationship defines the universal conductance fluctuations of the waveguide transport [9,10].

The statistical measure of the conductance fluctuations of the carrier transport is measured by the statistical variance of the conductance defined as [4,5,8–10]

$$var(g) = \sqrt{\langle g^2 \rangle - \langle g \rangle^2}, \tag{3.49}$$

where $g \equiv \frac{G}{2e^2/h} = T = N - R$. Here $\langle\ \rangle$ represents an average overall random configurations of the system subject to a constant impurity density n_s. Consequently, from Eq. (3.49) and the definition of g the variance of the conductance in the waveguide becomes [9,10]

$$var(g)=\sqrt{\langle R^2\rangle-\langle R\rangle^2}, \tag{3.50a}$$

and, for uncorrelated $R(n,m)$ it follows that

$$var(g)\approx Nvar(R(n,m)). \tag{3.50b}$$

In order to evaluate the variance, it remains to discuss R and $R(n,m)$ in terms of the amplitude for propagation of the carrier wavefunctions in the disorder medium. By following these paths from the injection of a carrier into the medium to its final state as a reflected mode, R is obtained from quantum mechanics. The details of this discussion are similar to the treatment of backscattering enhancement in the study of weak localization given in Figure 2.7 of the previous chapter.

The reflection coefficient for carrier propagation in the scattering medium is written in terms of the propagation amplitude of the carriers along the various different reflective paths through the disordered medium. In particular,

$$R(n,m)=\left|\sum_i A_{nm}(i)\right|^2, \tag{3.51a}$$

where $A_{nm}(i)$ is the wavefunction amplitude for the contribution of the ith path to the $R(n,m)$ component of reflection, i.e., incoming mode m is reflected into final state mode n [9,10]. Note that in the quantum mechanical system the different amplitudes add, and the modulus square of the sum gives the reflective component.

Averaging Eq. (3.51) over the various random configurations of the random medium consistent with the constant density n_s, it follows that [4,5]

$$\langle R(n,m)\rangle=\left\langle\left|\sum_i A_{nm}(i)\right|^2\right\rangle=\sum_i\left\langle\left|A_{nm}(i)\right|^2\right\rangle, \tag{3.51b}$$

where it is assumed that the phases between the different $A_{nm}(i)$ are random. Under the same assumption

$$\langle R^2(n,m)\rangle=\left\langle\sum_{ii'jj'}A_{nm}(i)A^*_{nm}(i')A_{nm}(j)A^*_{nm}(j')\right\rangle=2\sum_i\left\langle\left|A_{nm}(i)\right|^2\right\rangle\sum_j\left\langle\left|A_{nm}(j)\right|^2\right\rangle \tag{3.52}$$

so that from Eq. (3.51a) it follows that

$$R^2(n,m)=2R(n,m)^2 \tag{3.53}$$

From Eqs. (3.50), (3.51) and (3.53), the variance of the refection coefficient is obtained as

$$var(R(n,m))=\sqrt{\langle R^2(n,m)\rangle-\langle R(n,m)\rangle^2}=\langle R(n,m)\rangle. \tag{3.54}$$

Configurational averaging of Eq. (3.48) for mesoscopic systems including many impurities

$$N=\langle T\rangle+\langle R\rangle\approx\langle R\rangle, \tag{3.55}$$

where the far-right-hand approximation assumes that after a long-enough interaction with conductive scattering of the waveguide, most carriers are back reflected by the medium. (See Refs. [4–10] for a rigorous proof of this.) On average, using Eq. (3.47b), it is, consequently, reasonable to approximate

$$\langle R(n,m)\rangle \approx \frac{1}{N} \tag{3.56a}$$

(i.e., the incident wave is equally scattered into the reflected wave channels), and from Eq. (3.50) it then follows that

$$var(g) \approx 1. \tag{3.56b}$$

The variance of the conductance $var(G)$ is found to be of order $\frac{2e^2}{h}$. This is a statement of the nature of universal conductance fluctuations in mesoscopic systems, and it has been observed in the noise encountered in various measurements made on systems of this type. As has been seen in the earlier discussions, many mesoscopic networks display average conductance of the order $\frac{2e^2}{h}$, which is the same as the conductance fluctuations about the average. The average conductance and their fluctuations are computed from a statistical ensemble of random systems, all of which exhibit an average density of scattering sites n_s.

A closely related phenomenon in optics known as speckle involves the statistical characterization of the intensity pattern of bright and dark patches for coherent light transmitted through a random scattering medium. The intensity distribution of speckle patterns generally obey Poisson intensity distributions [4,5] where the shape of the Poisson distribution is set by the average intensity of the pattern. This results in similar properties being present in the patterns of bright and dark patches of light, and the conductance fluctuations in the realization of random disordered conducting systems.

3.5 NONZERO TEMPERATURE

The above discussions were made for zero-temperature systems. This is often not a bad approximation for waveguide networks, because the Fermi energies ϵ_F of the reservoirs involved in the problems are much greater than the energy $k_B T_s$ of the thermal fluctuations of the system. (Here, T_s is the temperature of the network and reservoirs.) The primary corrections to the zero-temperature results come from the inclusion of the Fermi occupation in the equations and the integration over the energy modes of the reservoirs. In the following, the treatment of nonzero-temperature systems is discussed by considering the illustrative example of a waveguide between two reservoirs. This model was initially treated in Eqs. (3.5) and (3.6), and later in the discussions of waveguide scattering.

The Fermi occupation of an energy level of the quantum mechanical system is given by [4,5,8]

$$f_0(\epsilon) = \frac{1}{1+\exp\left[\frac{\epsilon-\mu}{k_B T_s}\right]}. \tag{3.57a}$$

This represents the probability that a mode of energy ϵ is occupied for a system at temperature T_s and chemical potential μ. In the limit of zero temperature, Eq. (3.57a) reduces to the form of a Heaviside function

$$\begin{aligned} f_0(\epsilon) &= 1 \quad \text{if } \epsilon < \mu \\ &= 0 \quad \text{otherwise} \end{aligned}, \tag{3.57b}$$

where the chemical potential μ reduces to the Fermi energy ϵ_F. In this limit the states below the Fermi energy are all occupied with probability one, and those above the Fermi energy ϵ_F have zero probability of occupation. At nonzero temperatures T_s, the occupancy in Eq. (3.57b) is primarily perturbed for energies within the range $\varepsilon_F - k_B T_s < \epsilon < \epsilon_F + k_B T_s$, where for most applications $k_B T_s \ll \epsilon_F$.

Consider the ballistic waveguide first treated in Eqs. (3.4)–(3.6) and described schematically in Figure 3.1. From Eq. (3.4) for a ballistic waveguide in the absence of impurities

$$I = e\int_{\epsilon_F - eV}^{\epsilon_F} \frac{D_1^s(\epsilon_r)}{2L} v(\epsilon_r)d\epsilon_r = e\int_{\epsilon_F - eV}^{\epsilon_F} \frac{2}{h} d\epsilon_r = \frac{2e^2}{h}V, \tag{3.58}$$

where Eqs. (3.5) and (3.6) have been applied. The modification of this equation, taking into account Eq. (3.57a) for a nonzero temperature, is then given by [4,5,8]

$$\begin{aligned} I &= e\int \frac{2}{h}\left[f_2(\epsilon_r)(1 - f_1(\epsilon_r)) - f_1(\epsilon_r)(1 - f_2(\epsilon_r))\right]d\epsilon_r \\ &= e\int \frac{2}{h}\left[f_2(\epsilon_r) - f_1(\epsilon_r)\right]d\epsilon_r \end{aligned}, \tag{3.59}$$

where

$$f_1(\epsilon) = \frac{1}{1 + \exp\left[\dfrac{\epsilon - \mu_1}{k_B T_s}\right]} \tag{3.60a}$$

and

$$f_2(\epsilon) = \frac{1}{1 + \exp\left[\dfrac{\epsilon - \mu_2}{k_B T_s}\right]} \tag{3.60b}$$

are, respectively, the Fermi distributions for the left and right reservoirs and the current through the waveguide flows from the left to the right reservoir, respectively.

Note: In Eq. (3.59), $f_2(\epsilon)(1 - f_1(\epsilon))$ is the probability that there is an electron of energy ϵ that can leave reservoir 2 and be received into an empty state of energy ϵ in reservoir 1. Similarly, $f_1(\epsilon)(1 - f_2(\epsilon))$ is the probability that there is an electron of energy ϵ that can leave reservoir 1 and be received into an empty state of energy ϵ in reservoir 2. The total current in the waveguide is that from the sum of these two transport processes.

In addition, if there are scattering sites in the waveguide, described by an energy dependent scattering transmission coefficient $T(\epsilon)$, Eq. (3.59) then becomes [4,5,8]

$$I = e\int \frac{2}{h}\left[f_2(\epsilon_r) - f_1(\epsilon_r)\right]T(\varepsilon_r)d\epsilon_r. \tag{3.61}$$

This is the system described schematically in Figure 3.3a, which is essentially the same as that discussed in Eq. (3.59) but now including a transmitting barrier in the waveguide. Note, in this expression two changes have been introduced into the treatment. In the earlier zero-temperature discussions the transmission coefficient was taken to be a constant, independent of the energy, but in Eq. (3.61) it is allowed to be dependent on the electron energy. In addition, in both Eqs. (3.59) and (3.61) the thermal mode occupancy of the two reservoirs are accounted for by the factor of $f_2(\epsilon_r) - f_1(\epsilon_r)$, rather than in the integration limits. For applications, some simplifications are now introduced in Eq. (3.61).

In the case that $\mu_1 \approx \mu_2$, $f_2(\epsilon)$ can be expanded about $f_1(\epsilon)$ in a Taylor series in $\mu_1 - \mu_2 = -eV$ so that

$$f_2(\epsilon_r) \approx f_1(\epsilon_r) + (\mu_1 - \mu_2)\frac{df_1(\epsilon_r)}{d\epsilon_r} = f_1(\epsilon_r) - eV\frac{df_1(\epsilon_r)}{d\epsilon_r}. \tag{3.62a}$$

Consequently, in Eqs. (3.59) and (3.61)

$$f_2(\epsilon_r) - f_1(\epsilon_r) \approx -eV\frac{df_2(\epsilon_r)}{d\varepsilon_r}. \tag{3.62b}$$

And applying this in Eq. (3.61) it follows that

$$I = \frac{2e^2}{h}V\int\left[-\frac{df_2(\epsilon_r)}{d\varepsilon_r}\right]T(\varepsilon_r)d\epsilon_r, \tag{3.63}$$

where V is the electric potential of reservoir 1, and reservoir 2 is at zero electric potential. The derivative in the integrand is highly peaked about the chemical potential of reservoir 2 so that the integral in Eq. (3.63) yields quickly to a numerical evaluation.

As an example of Eq. (3.63), at zero temperature $f_2(\epsilon_r)$ becomes the Heaviside function for which

$$-\frac{df_2(\epsilon_r)}{d\varepsilon_r} = \delta(\epsilon_r - \mu_2). \tag{3.64}$$

Applying Eq. (3.64) in Eq. (3.63), the earlier results in Eqs. (3.13) and (3.14) are reproduced. Specifically,

$$I = \frac{2e^2}{h}MTV \tag{3.65a}$$

from which it follows that the conductance of the system is [4,5,6,8]

$$G = \frac{2e^2}{h}MT \tag{3.65b}$$

where T in Eqs. (3.65) and (3.66) is the constant transmission coefficient of the waveguide, and M is the subband degeneracy. To obtain Eq. (3.58) in this limit, it is only needed to take $T = 1$.

REFERENCES

1. Landauer, R. 1957. Spatial variation of currents and fields due to localized scatterers in metallic conduction. *IBM Journal of Research and Development* 1: 223–231.
2. Landauer, R. 1990. Advanced technology and truth in advertising. *Physica A* 168: 75–87
3. Buttiker, M. 1989. Negative resistance fluctuations at resistance minima in narrow quantum Hall conductors. *Physical Review B* 38: 12724.
4. Ouisse, T. 2008. *Electron transport in nanostructures and mesoscopic devices: An introduction.* New York: Wiley.
5. Barnham, K. and D. Vvedensky 2001. *Low-dimensional semiconductor structures: Fundamentals and device applications.* Cambridge: Cambridge University Press.
6. Datta, S. 2005. *Quantum transport: Atom to transistor.* Cambridge: Cambridge University Press.
7. Nazarov, Y. V. and Y. M. Blanter 2009. *Quantum transport: Introduction to nanoscience.* Cambridge: Cambridge University Press.
8. Ryndyk, D. A. 2016. *Theory of quantum transport at nanoscale: An introduction.* Cham: Springer Series in Solid State Physics.
9. Lee, P. A. 1986. Universal conductance fluctuations in disordered metals. *Physica A* 140A, 169–174.
10. Joners, P. A. E. 2000. Quantum transport in disordered magnetoresistive systems. Thesis, University of Groningen.

4 Properties of Periodic Media

In this chapter, the properties of periodic media [1–12] are introduced within the context of their application to studies in nanoscience. Periodic media represent periodic arrangements of matter in space, and are one of the important classes of materials that are studied in condensed matter physics and engineering technologies. Examples of periodic media exist on a full variety of length scales. On the microscopic or atomic scale, many naturally occurring materials are formed as crystals in which atoms and molecules are regularly situated on a periodic lattice [1–3]. In such types of crystalline materials, the periodicity plays an important role in determining the general nature of the conductivity displayed by the material. In addition to these materials, an important class of media are artificial or engineered materials that are designed as periodic arrays of mesoscopic or macroscopic features regularly positioned on a periodic lattice [4–12]. In these materials the periodicity again shows up in the general transport properties of the medium. The difference between the various periodic systems is the length scales and frequencies of the excitations or radiations on which the periodic media act in setting the transport properties and the type of excitation or radiation that is transported. In this way, periodic media have been configured to modulate the transport of electrons, electromagnetic waves, or acoustic waves [1–12].

An important periodicity in condensed matter physics is the periodicity of atoms and molecules arranged in a crystalline medium [1,2]. The electrical conduction properties of metals, insulators, and semiconductors result from these periodic arrangements. Specifically, the electron frequencies in crystalline media separate into frequency pass bands in which electrons are transported by the medium and frequency stop bands in which electrons do not transport through the medium. In the crystal a given frequency of electron is either in a pass band or a stop band, depending on the periodicity of the material, and the pass and stop bands are finite continuum frequency intervals covering the entire frequency spectrum from zero to infinity. See Figure 4.1 for a schematic of such a system of pass and stop bands of a crystalline medium [1,2].

The atomic and molecular crystalline materials are metals, insulators, or semiconductors, depending on how the electrons occupy the pass and stop bands in the material [1–3]. At zero temperature the lowest frequency levels are occupied in the material in a manner that is consistent with the Pauli exclusion principle. If this leaves a partially occupied pass band in the system (see Figure 4.1a), then the energy of the electrons in the material can be infinitesimally increased through the application of an external electric field. The promotion of the electrons in this way allows them to flow through the material, generating a net electrical current. For this case, the electrons easily flow through the material that then behaves as a metal.

If, on the other hand, the highest frequency pass band in the system is fully occupied (see Figure 4.1b and c), then the energy of the electrons in the material cannot be infinitesimally increased through the application of an external electric field [1–3]. A finite amount of energy must be added to the electrons to promote an electron through the upper stop band and into the next higher frequency pass band. In this case, the material behaves as an insulator or a semiconductor. For a semiconducting material, the upper stop band widths are generally of order 1 eV or less so that thermal disorder allows them to support a current at room temperatures [1,2]. (See Figure 4.1b for the schematic of a semiconductor.) Insulators, however, have upper stop band widths greater than 1 eV and are stable against thermal excitations of electrons. (See Figure 4.1c for the schematic of an insulator.)

The above statements of the band structure properties of metals, insulators, and semiconductors are made in the context of a model of electrons in which there are no electron–electron interactions [1–3]. These conditions are later modified in more sophisticated treatments in which electron–electron interactions and other many-body effects are included [3].

DOI: 10.1201/9781003031987-4

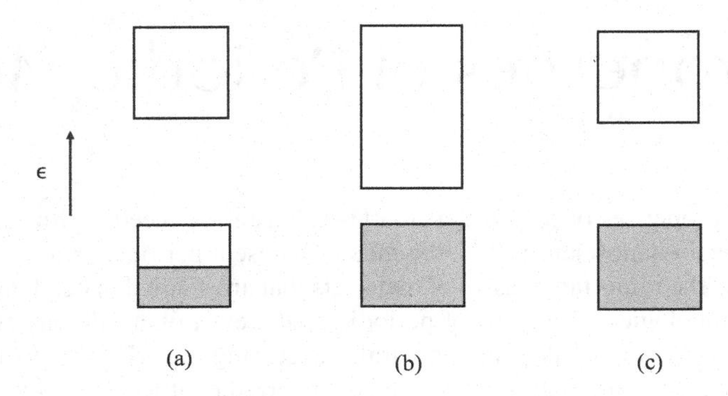

FIGURE 4.1 A highly schematic two pass band representation of the electron filling in: (a) a metal with a partially filled lower pass band, (b) a semiconductor with a totally filled lower pass band and a small gap (stop band) between the two pass bands, and (c) an insulator with a totally filled lower pass band and a large stop band gap between the two pass bands. In each figure, there are two pass bands represented by an upper and lower rectangle that are separated by a vertical space of stop band. The energy of the electrons, ϵ, occupying the bands is on the vertical scale (see arrow on the left of the figure), with increasing energies toward the top and lower energies toward the bottom. The gray shading indicates the electron energy levels that are filled at zero temperature, and the Fermi energy at zero temperature is indicated by a black horizontal line at the top of the gray shading of occupied electron levels.

When dielectric materials are arrayed in an artificial crystalline arrangement, they form what are known as photonic crystals. These exhibit interesting optical properties that find applications in optics and optical technologies. In free space light exhibits a continuum frequency spectrum, but in the presence of a periodic dielectric medium the frequency spectrum of light is partitioned into a sequence of pass and stop band frequency intervals. This is a similar partition to that of the pass and stop band structure of electronic systems. However, light obeys Bose–Einstein statistics in the optical systems, rather than the Fermi–Dirac statistics displayed in electronic systems [1,2]. As in the case of electron systems, the system of stop and pass bands limits the transport properties of the medium. Light at stop band frequencies will not propagate into the bulk of a photonic crystal that only allows pass band frequencies of light to propagate through the bulk of the crystal [4–7]. In this way, photonic crystals can be used to filter and confine light [4–7]. The sequence of pass and stop bands are found to be dependent on the dielectric composition and the size of the features from which the photonic crystal is composed.

Similarly, when acoustic materials are arrayed in an artificial crystalline arrangement, a phononic crystal is formed [4]. Again, due to the periodicity, the acoustic frequency spectrum is broken into a sequence of pass and stop bands. These exhibit interesting acoustic filtering and confinement properties for applications in acoustic technologies.

The modal wavefunctions in each of the periodic media for electron, electromagnetic, or acoustic excitations also have specific general functional forms [1–9]. In homogeneous, isotropic materials, the modes of the system are planewaves [4–7]. This is due to the complete invariance of the medium under translations, i.e., under any translation in space the medium looks the same before and after the translation [13]. It is also a statement of the conservation of momentum for the particle translating in space. In periodic media the general form of the wavefunction is a planewave times a function with the same spatial periodicity as that of the periodic medium. The periodic function arises as a statement that the energy density of the excitation must have the same periodicity as that of the medium in which the wave propagates. The planewave part of the wavefunction is a statement of the conservation of crystal momentum where the crystal momentum is different than that of the momentum of the mode. In this regard, the excitation modes in the periodic medium are all labeled as distinct states of crystal momentum.

Upon the introduction of impurities into either homogeneous isotropic media or periodic media, the two different momenta enter into descriptions of the resulting scattering processes in each of the respective media [1–4]. In homogeneous isotropic media, the scattering process proceeds with a conservation of momentum. In a periodic medium, the scattering proceeds by a conservation process involving crystal momentum.

In the following, first a discussion is given of the tight-binding model of electrons and its application to the electronics of graphene. Graphene has recently been of great interest in a variety of technological applications [14–16]. For example, it is found to exhibit unusual electronic properties normally associated with relativistic systems as well as displaying important mechanical properties for applications. Following this, a treatment of some of the elementary electronics and optics of quantum wells, quantum dots, and quantum heterostructures are discussed. These have been of current technological interest for many applications [4–7,17]. In particular, quantum heterostructures exhibit useful electronic band structures on mesoscopic and macroscopic scales. Finally, the properties of photonic and phononic crystals are presented with an introduction to basic technologies associated with them [4–7,17,18–20].

4.1 TIGHT-BINDING MODEL

Let us consider one of the first type of models that was used to study the motion of conduction electrons in a metal. This is the tight-binding model that treats the conduction electrons as fermions moving through a background of positive ions formed in a periodic lattice [1–9]. In the context of this picture the electrons are considered as localized on the ions, and the motion of the conduction electrons proceeds by their hoping from ion to ion as they progress through the material. Consequently, the ions are located at the sites of a periodic lattice, and the occupied site becomes an atom of the system when an electron occupies an ion, but it remains an unoccupied ion when there is no electron on the site.

For a simple treatment, in the following it is assumed that the conduction electron occupancy at an ion site is in one of two different states. In the first state denoted by $|0\rangle$, the ion does not have a conduction electron residing on it, and the occupied atom has a conduction electron residing on it in the second state denoted by $|1\rangle$. These are the only two occupancy states possible in the model. More complex models, however, can be introduced in a straightforward way by increasing the occupancy number of conduction electrons possible per ion site. In these systems, to satisfy the Pauli exclusion principle, additional quantum numbers are needed so that the conduction electrons on a given ion site are distinguishable from one another. For developing the topic of tight-binding models, here only models with one conduction electron occupancy per ion site are considered [2].

In studying the dynamics of the electron hopping, it is helpful to introduce a set of operators that describe the state of occupancy of the conduction electrons on an ion site [2,4–7]. Once the properties of these are correctly defined, their operator dynamics and consequently that of the electrons in the system are given by quantum mechanical considerations. Specifically, the dynamics is obtained by a standard application of the rules of quantum mechanics in the Heisenberg representation. This representation develops the operator equations of motion in terms of their commutation relations with the Hamiltonian.

In this regard, consider introducing a set of two different occupation operators a, a^+ that are set up to transform the atomic and ionic states at a given site into one another, i.e., they are designed to either remove or place a conduction electron on an ionic site of the lattice. In particular, the two operators are defined by the transformations [2,13]

$$a^+|0\rangle = |1\rangle \tag{4.1}$$

where a^+ turns the ionic state $|0\rangle$ into an atomic state $|1\rangle$, and

$$a|1\rangle = |0\rangle \tag{4.2}$$

where a turns the atomic state $|1\rangle$ into an ionic state $|0\rangle$. It should be noted that both of these operators can ultimately be expressed in terms of a 2×2 matrix in the space of $|1\rangle$, $|0\rangle$ states.

In the first Eq. (4.1), the a^+ operator places an electron onto an ionic site, turning it into an atomic site. Similarly, in the second Eq. (4.2), the a operator removes an electron from an atomic site, making it into an ionic site. Since two conduction electrons cannot be placed on the same site, it follows that [2,13]

$$a^+|1\rangle = 0 = \left(a^+\right)^2|0\rangle, \tag{4.3}$$

and, similarly, a conduction electron cannot be removed from an ionic site so that

$$a|0\rangle = 0 = a^2|1\rangle. \tag{4.4}$$

Consequently, Eqs. (4.1)–(4.4) completely define the 2×2 matrices representing a and a^+ in the space of $|1\rangle$, $|0\rangle$ states, and these matrices along with their operators are found to be adjoint to one another.

Following from Eqs. (4.1) to (4.4), some useful operator relations of a and a^+ are summarized by the identities [2,13]

$$a^+a + aa^+ = 1 \tag{4.5}$$

and

$$\left(a^+\right)^2 = a^2 = 0. \tag{4.6}$$

These are the basic operator identities that summarize the relationships needed to treat the occupancy transformations and electron dynamics of a single atomic–ionic site. In systems with many ionic sites on a lattice, some modification of these is required in order to describe the dynamics.

To generalize the set of operations a, a^+ to treat many atomic and ionic sites requires a little care. A generalization can be made by a study of the application of the operator identities in Eqs. (4.5) and (4.6). To see this, consider the set of occupancy states $\left\{|1\rangle_n, |0\rangle_n\right\}$ where now the subscripts $n = 1,2,3,\ldots,N$ label the N different possible atomic–ionic sites in the system, and 1 and 0 provide the atom–ion occupancy identification of the site. (Note: In the earlier considerations only one lattice site $|0\rangle$ or $|1\rangle$ was treated so that no subscript site index n was needed on the states.) In the new representation, a state of the N sites in which all the sites are atomic sites is [2,13]

$$|1\rangle_1|1\rangle_2|1\rangle_3\ldots|1\rangle_N, \tag{4.7a}$$

and a state in which all the sites are ionic is

$$|0\rangle_1|0\rangle_2|0\rangle_3\ldots|0\rangle_N. \tag{4.7b}$$

Between these limiting states are a variety of mixtures of atomic and ionic sites of the system indicated by a sequence of 0 's or 1 's.

The set of operators a, a^+ now are generalized to depend on the site labels n so that they become the set a_n, a_n^+ for each of the subscripts $n = 1,2,3,\ldots,N$. In this notation, for example [2,13]

$$a_n|1\rangle_n = |0\rangle_n \tag{4.8a}$$

and similarly

$$a_n^+|0\rangle_n = |1\rangle_n. \tag{4.8b}$$

(Note: The operators a_n, a_n^+ only operate on the states of the nth lattice site.) In addition, in this extended operator notation, Eq. (4.5) then becomes

$$a_n^+ a_m + a_m a_n^+ = \delta_{n,m} \tag{4.9}$$

while Eq. (4.6) generalizes to [2,13]

$$a_n^+ a_m^+ + a_m^+ a_n^+ = a_n a_m + a_m a_n = 0. \tag{4.10}$$

It is important to note that the sums in Eqs. (4.9) and (4.10) are zero in the case $n \neq m$. This is an additional assumption placed on the operators in generalizing Eqs. (4.5) and (4.6) to many-body systems. As shall be seen in the following, it is a condition that allows for the separation of the electron dynamics into distinct modes. Such a separation is consistent with experiments and with intuition based on classical systems.

As an illustration of the many-body generalization and the treatment of its dynamics, a study is now given of the tight-binding model solutions for a one-dimensional chain of atoms with a single conduction electron per each ion–atom lattice site.

4.1.1 A One-Dimensional Model of a Chain of Atoms

Consider a one-dimensional model of conduction electrons hopping along a chain of $2N+1 \to \infty$ equally spaced atomic–ionic sites. The sites of the chain are positioned along the x-axis such that the location of the nth atomic–ionic site of the chain is at $x_n = n\tilde{a}$, where n is an integer. In the chain dynamics, only hopping between the nearest-neighbor sites of the chain is allowed [1,2].

The simplest tight-binding model arranged for the nearest-neighbor hopping is given by the Hamiltonian [2]

$$H = \sum_{n=-N}^{N} \left[t\left(a_{n+1}^+ a_n + a_{n-1}^+ a_n\right) + u a_n^+ a_n \right]. \tag{4.11}$$

In this Hamiltonian, the terms multiplying t represent hopping between the sites along the chain, taking an electron from an atomic site n, and placing it on neighboring ionic sites $n+1$ or $n-1$. In the process an energy related to t is shuttled about the chain so that t relates to the kinetic energy of electron motion, as the electron travels from site to site. The second term contributes an occupancy energy u to the system. Specifically, the energy of an electron on the nth site is

$$u a_n^+ a_n |1\rangle_n = u|1\rangle_n, \tag{4.12a}$$

but in the absence from the site of an occupying electron

$$u a_n^+ a_n |0\rangle_n = 0. \tag{4.12b}$$

From both Eqs. (4.12a) and (4.12b), it is seen that the number operator [2]

$$n_m = a_m^+ a_m \tag{4.13}$$

measures the number of conduction electrons on the mth atomic–ionic site in the lattice. For the case that $u < 0$, the terms in u each represent an energy of attraction between an electron occupying the site and its positive ion on the chain. Consequently, the model in Eq. (4.11) is the most basic quantum mechanical system composed of a kinetic and a potential energy for electrons traveling along the chain.

Due to the translational symmetry of the Hamiltonian on the infinite lattice, it is helpful to develop the set a_n, a_n^+ in terms of a Fourier series in the lattice. Specifically, these operators are represented by the forms [2]

$$a_n = \frac{1}{\sqrt{2N}} \sum_k a_k e^{ikx_n} \tag{4.14a}$$

and

$$a_n^+ = \frac{1}{\sqrt{2N}} \sum_k a_k^+ e^{-ikx_n}, \tag{4.14b}$$

where applying periodic boundary conditions between the ends of the chain requires that

$$k = \frac{n\pi}{N\tilde{a}} \tag{4.14c}$$

for $n = 0, \pm 1, \pm 2, \ldots, \pm N$ for $N \to \infty$. Conversely, inverting the Fourier series defines the k-space creation and destruction operators

$$a_k = \frac{1}{\sqrt{2N}} \sum_n a_n e^{-ikx_n} \tag{4.15a}$$

and

$$a_k^+ = \frac{1}{\sqrt{2N}} \sum_n a_n^+ e^{ikx_n}, \tag{4.15b}$$

where the sums are over the lattice sites.

Applying Eq. (4.14) in Eqs. (4.9) and (4.10), it then follows that

$$a_k^+ a_{k'} + a_{k'} a_k^+ = \delta_{k,k'} \tag{4.16}$$

and similarly

$$a_k^+ a_{k'}^+ + a_{k'}^+ a_k^+ = a_k a_{k'} + a_{k'} a_k = 0. \tag{4.17}$$

The similarity of the k-space relations in Eqs. (4.16) and (4.17) with those of Eqs. (4.9) and (4.10) in direct space depends on the definition of the $n \neq m$ relations of occupation operators in the direct space relations of Eqs. (4.9) and (4.10). It is now seen that these allow for the Fourier transformed modes of Eqs. (4.16) and (4.17) to represent the occupation operators of the wavevector modes of the Hamiltonian in Eq. (4.11).

An application of a_k^+, a_k to the set of states generalized from Eq. (4.7) to describe the chain of $2N+1 \to \infty$ sites shows that they affect transformations between states of wavevector k. For example, applying a_k^+ to the generalization of Eq. (4.7b) given by [2,13]

$$|0\rangle_{-N} \ldots |0\rangle_{-1} |0\rangle_0 |0\rangle_1 \ldots |0\rangle_N, \tag{4.18a}$$

it is found that

$$a_k^+ |0\rangle_{-N} \ldots |0\rangle_{-1} |0\rangle_0 |0\rangle_1 \ldots |0\rangle_N = \frac{1}{\sqrt{2N}} \sum_{n=-N}^{N} e^{ikx_n} a_n^+ |0\rangle_{-N} \ldots |0\rangle_{-1} |0\rangle_0 |0\rangle_1 \ldots |0\rangle_N \tag{4.18b}$$

creates a linear combination of single atomic states, each weighted by the factor $\frac{1}{\sqrt{N}}e^{ikx_n}$ related to the lattice position of the created atomic state. The linear combination in Eq. (4.18b) is then regarded as a state of wavevector k and is denoted as $|k\rangle$. Consequently,

$$a_k^+|0\rangle_{-N}\ldots|0\rangle_{-1}|0\rangle_0|0\rangle_1\ldots|0\rangle_N = |k\rangle \tag{4.19}$$

creates an excitation of wavevector k in the chain of ions. Similarly, applying a_k to the same generalization of Eq. (4.7b) gives zero as

$$a_k|0\rangle_{-N}\ldots|0\rangle_{-1}|0\rangle_0|0\rangle_1\ldots|0\rangle_N = \frac{1}{\sqrt{N}}\sum_{n=1}^{N} e^{-ikx_n}a_n|0\rangle_{-N}\ldots|0\rangle_{-1}|0\rangle_0|0\rangle_1\ldots|0\rangle_N = 0, \tag{4.20}$$

but from Eq. (4.16) it follows that

$$\begin{aligned} a_k|k\rangle &= a_k a_k^+|0\rangle_{-N}\ldots|0\rangle_{-1}|0\rangle_0|0\rangle_1\ldots|0\rangle_N \\ &= \left(-a_k^+a_k+1\right)|0\rangle_{-N}\ldots|0\rangle_{-1}|0\rangle_0|0\rangle_1\ldots|0\rangle_N = |0\rangle_{-N}\ldots|0\rangle_{-1}|0\rangle_0|0\rangle_1\ldots|0\rangle_N \end{aligned} \tag{4.21}$$

removes the state $|k\rangle$ from the chain. To summarize, operators a_k^+ and a_k transform between states in which a mode of wavevector k is absent and a mode of wave vector k is present, in particular, $a_k^+|0\rangle_{-N}\ldots|0\rangle_0|0\rangle_1\ldots|0\rangle_N = |k\rangle$ and $a_k|k\rangle = |0\rangle_{-N}\ldots|0\rangle_0|0\rangle_1\ldots|0\rangle_N$. The a_n, a_n^+ and a_k^+, a_k sets of operators are seen to have analogous properties in their respective spaces of operation.

Similarly, it is readily shown that for $k \neq k'$

$$a_k^+a_{k'}^+|0\rangle_{-N}\ldots|0\rangle_0|0\rangle_1\ldots|0_N = |k\rangle|k'\rangle = |k,k'\rangle, \tag{4.22}$$

where $|k,k'\rangle$ is a state containing two different wavevector modes: one of wavevector k and one of wavevector k'. In this regard, it also follows that [2,13]

$$n_k = a_k^+a_k \tag{4.23}$$

is the number operator in k-space. When acting on a state in k-space, it gives the number of modes of wavevector k present on the chain.

In summary, the occupation properties of the operators a_k^+, a_k on the k-space states are analogous to those of a_n^+, a_n in the direct lattice of the chain. They include creating, destroying, and counting of the states of k consistent with Fermi–Dirac statistics. These properties will now be used to solve for the modes and energies of the Hamiltonian in Eq. (4.11).

Applying Eq. (4.14) in Eq. (4.11) for the Hamiltonian defined over $2N+1\to\infty$ sites, it follows after a little Fourier series manipulation that [1,2,13]

$$H = \sum_k [2t\ \cos\ k\tilde{a}+u]a_k^+a_k = \sum_k [2t\ \cos\ k\tilde{a}+u]n_k. \tag{4.24a}$$

In this format, the energy of the system is found to depend on the number of individual wavevector modes in the system. This is a great simplification of the study of the systems, as the Hamiltonians (4.24a) can now be solved as a simple eigenvalue problem. In fact, the Hamiltonian is diagonal in the wavevectors $|k\rangle$ with the corresponding eigenvalues [1,2]

$$\epsilon_k = [2t\cos k\tilde{a}+u], \tag{4.24b}$$

where $k = \frac{n\pi}{N\tilde{a}}$ for $n = 0, \pm 1, \pm 2, \ldots, \pm N$ and $N \to \infty$.

An interesting additional consideration is to treat the dynamics of the set of operators a_k, a_k^+ using standard quantum mechanics considerations. In this regard, the dynamics of any operator of the system are obtained from its Heisenberg equations of motion. In the Heisenberg formulation, the time dependence of a general operator denoted A is given by [1,2,13]

$$\frac{dA}{dt} = \frac{i}{\hbar}[H, A] + \frac{\partial A}{\partial t}, \tag{4.25}$$

where the second term on the right comes from an explicit time dependence of the operator, whereas the first term does not treat the explicit time dependence of A. Applying Eq. (4.25) to a_k^+, it follows that

$$\frac{da_k^+}{dt} = \frac{i}{\hbar}\epsilon_k a_k^+ \tag{4.26a}$$

and, similarly, for a_k

$$\frac{da_k}{dt} = -\frac{i}{\hbar}\epsilon_k a_k. \tag{4.26b}$$

(Note: In this development neither of these operators depends explicitly on the time.)

The solutions of the equations in Eq. (4.26) then give [2,13]

$$a_k^+(t) = a_k^+(0)e^{\frac{i}{\hbar}\epsilon_k t} \tag{4.27a}$$

and

$$a_k(t) = a_k(0)e^{-\frac{i}{\hbar}\epsilon_k t}. \tag{4.27a}$$

Note that the chain has $2N+1 \to \infty$ distinct modes with energies given in Eq. (4.24b) in the preceding treatment. The modes are occupied applying Fermi–Dirac statistic, and this occupation is a consequence of the anticommutation properties of the a_n, a_n^+ operators in Eqs. (4.9) and (4.10) and the anticommutation properties of the a_k^+, a_k operators in Eqs. (4.16) and (4.17).

The model just discussed treating the electrons as spinless particles. A simple modification of the theory allows for the handling of electrons, which are spin-up or spin-down particles. In this modified model, the wavevector states are generalized to the states $|k\uparrow\rangle$ and $|k\downarrow\rangle$, which are of wavevector k and spin-up or -down, respectively. The operators creating spin-up and spin-down modes of k are $a_{k\uparrow}^+$, $a_{k\downarrow}^+$, and the operators destroying states of spin-up and spin-down modes of k are the $a_{k\uparrow}$, $a_{k\downarrow}$. Similarly, the localized modes in the space of ionic–atomic lattice sites are $|0\uparrow_n\rangle$, $|0\downarrow_n\rangle$ and $|1\uparrow_n\rangle$, $|1\downarrow_n\rangle$ for spin-up and -down, and $a_{n\uparrow}^+$, $a_{n\downarrow}^+$ and $a_{n\uparrow}$, $a_{n\downarrow}$ are related operators.

In terms of these modifications, the Hamiltonian in Eq. (4.24) becomes [2,13]

$$H = \sum_{k\sigma}[2t \cos k\tilde{a} + u]a_{k\sigma}^+ a_{k\sigma} = \sum_{k\sigma}[2t \cos k\tilde{a} + u]n_{k\sigma}, \tag{4.28}$$

where $\sigma = \uparrow$ or $\downarrow$, and the sum in k-space remains over the same range of k. As with Eq. (4.24), the Hamiltonian is diagonal in the wavevectors $|k\sigma\rangle$ with corresponding eigenvalues

$$\epsilon_{k\sigma} = [2t \cos k\tilde{a} + u], \tag{4.29}$$

where $k = \frac{n\pi}{N\tilde{a}}$ for $n = 0, \pm1, \pm2, \ldots, \pm N$ and $N \to \infty$. In this regard, the eigen energies are independent of spin.

Note that if in Eq. (4.28) there is one conduction electron per lattice site, at $T = 0$ the lowest energy states of the $t < 0$ model fill the band of states between $-\frac{\pi}{2} \leq k \leq \frac{\pi}{2}$. In this case the system behaves as a metal, because the electrons in the system can be pushed by an applied field to occupy higher energy states. However, if there are two conduction electron per lattice site (i.e., both spin-up and -down states), the lowest energy states fill the band of states between $-\pi \leq k \leq \pi$ at $T = 0$. In this case all of the wavevector and spin states are filled. No other energy states are available for the electrons to be pushed into so that the system must model an insulator.

Next, the tight-binding model in two dimensions is investigated. The focus is on the solution of the model within the context of its particularly interesting application as a model of graphene.

4.1.2 A Two-Dimensional Tight-Binding Model of Graphene

In this section, a tight-binding model that is applicable to the study of graphene is discussed [14]. Graphene is an interesting material for studies in nanoscience because of some of the unusual dispersive properties of its conduction electrons [1,2,14–16]. It is formed as a planar sheet of carbon atoms composed through sp^2 bonding and exhibiting the conductive properties of a semiconductor. The dispersive properties of the conduction electrons, however, cause it to display features that simulate those of relativistic systems. In fact, the dynamics of the electrons can be expressed in a type of Dirac equation formulation [9–12,15,16]. Being two dimensional, graphene is of greater complexity than the one-dimensional model treated in the previous section, but it is not so complex as a fully three-dimensional material. Consequently, it offers a useful introduction to the theoretical development of the dynamics of tight-binding models. The properties of graphene are developed applying the descriptive tools common to all higher dimensional electron dynamics, but the applications are in a more transparent format than those in higher dimensional materials.

The crystal structure of graphene is shown in Figure 4.2. Graphene has a crystal lattice that is composed of two sublattice lattices of carbon atoms [14]. One sublattice is denoted as the A lattice

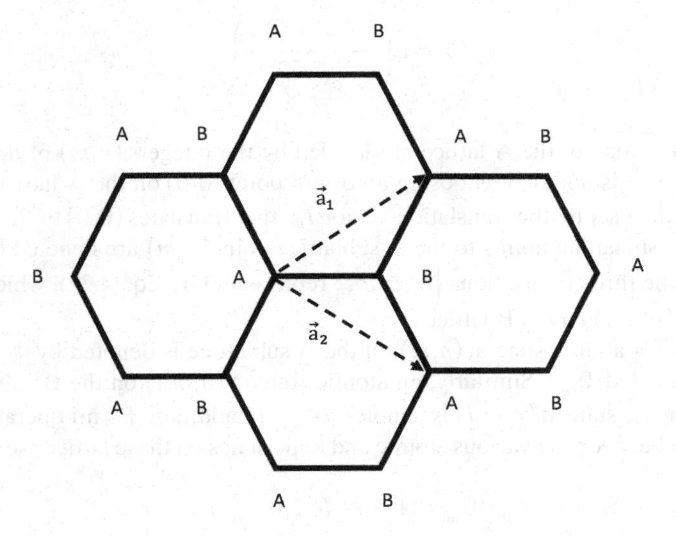

FIGURE 4.2 A schematic of a portion of the graphene lattice composed of two shifted lattices of A and B carbon atoms. In the figure, the carbon atoms are located at the lattice vertices, which are labeled A or B, and the two vectors $\vec{a}_1$ and $\vec{a}_2$ are shown on the figure. The lattices of A and B atoms are identical but shifted relative to one another in space.

with the atoms on this lattice denoted as A atoms, while the other sublattice is denoted as the B lattice with the atoms on it denoted as B atoms. The two sublattices are shifted relative to one another so that the A and B atoms see a different arrangement of their surrounding neighbors.

The A lattice has a translational symmetry such that it is translated into itself by any so-called translation vector of the form [9–12]

$$\vec{T}_{nm} = n\vec{a}_1 + m\vec{a}_2 \tag{4.30}$$

where n and m are integers, and

$$\vec{a}_1 = \frac{\tilde{a}}{2}\left(3\hat{i} + \sqrt{3}\hat{j}\right) \tag{4.31a}$$

$$\vec{a}_2 = \frac{\tilde{a}}{2}\left(3\hat{i} - \sqrt{3}\hat{j}\right). \tag{4.31b}$$

In this notation, a translation of the lattice points of the A lattice through a translation of $\vec{T}_{nm}$ sends the lattice into itself. The B lattice is obtained from the A lattice by translating each point of the A lattice through the vector translation $\vec{b}_1 = \tilde{a}\hat{i}$ so that the $\vec{T}_{nm}$ are also translation vectors of the B lattice. In this regard, the nearest-neighbor lattice points of a point on the A lattice are B lattice points, and conversely the nearest-neighbor lattice points of a point on the B lattice are A lattice points.

Specifically, given a point on the A lattice, its nearest-neighbor B lattice points are shifted from it by the vectors [9–12]

$$\vec{b}_1 = \tilde{a}\left(\hat{i} + 0\hat{j}\right), \tag{4.32a}$$

$$\vec{b}_2 = \tilde{a}\left(-\frac{1}{2}\hat{i} + \frac{\sqrt{3}}{2}\hat{j}\right), \tag{4.32b}$$

$$\vec{b}_3 = \tilde{a}\left(-\frac{1}{2}\hat{i} - \frac{\sqrt{3}}{2}\hat{j}\right). \tag{4.32c}$$

In this scheme, the points of the A lattice are labeled by the integers (n,m) of the translation vector in Eq. (4.30). For this notation, choosing an origin point $(0,0)$ on the A lattice, the point (n,m) on the A lattice is then set by the translation vector $\vec{T}_{nm}$ that translates $(0,0)$ to the point (n,m). The nearest-neighbor B sublattice points to the A sublattice point (n,m) are denoted by (n,m,i), where $i = 1,\ 2,\ 3$ denotes the three translations $\left\{\vec{b}_1, \vec{b}_2, \vec{b}_3\right\}$ represented by Eq. (4.32), which translate (n,m) on the A lattice to (n,m,i) on the B lattice.

In this notation, an atomic state at (n,m) on the A sublattice is denoted by $|1\rangle_{nm}$, while an ionic state at (n,m) is denoted $|0\rangle_{nm}$. Similarly, an atomic state at (n,m,i) on the B sublattice is denoted by $|1\rangle_{nmi}$, while an ionic state at (n,m,i) is denoted $|0\rangle_{nmi}$. In addition, Fermi operators a_{nm}^{+}, b_{nmi}^{+} and a_{nm}, b_{nmi} transform between the various atomic and ionic states on these lattices so that, for example,

$$a_{nm}^{+}|0\rangle_{nm} = |1\rangle_{nm}, \quad b_{nmi}^{+}|0\rangle_{nmi} = |1\rangle_{nmi} \tag{4.33a}$$

$$a_{nm}|1\rangle_{nm} = |0\rangle_{nm}, \quad b_{nmi}|1\rangle_{nmi} = |0\rangle_{nmi}. \tag{4.33b}$$

Here a_{nm}^{+} and a_{nm} operate only on the A lattice site (n,m), and b_{nmi}^{+} and b_{nmi} operate only on the B lattice site (n,m,i).

The earlier considerations in Eqs. (4.9) and (4.10) of the fermion properties of the conduction electrons now generalize to the creation and destruction operators of the conduction electrons of graphene. This generalization follows in a straightforward way so that in the present notation the following statements regarding the operators on both the A and B lattices hold [9–12,14]

$$a^+_{nm}a_{n'm'} + a_{n'm'}a^+_{nm} = \delta_{nm,n'm'}, \tag{4.34a}$$

$$a^+_{nm}a^+_{n'm'} + a^+_{n'm'}a^+_{nm} = a_{nm}a_{n'm'} + a_{n'm'}a_{nm} = 0, \tag{4.34b}$$

$$b^+_{nmi}b_{n'm'i'} + b_{n'm'i'}b^+_{nmi} = \delta_{nmi,n'm'i'}, \tag{4.34c}$$

$$b^+_{nmi}b^+_{n'm'i'} + b^+_{n'm'i'}b^+_{nmi} = b_{nmi}b_{n'm'i'} + b_{n'm'i'}b_{nmi} = 0, \tag{4.34d}$$

$$a^+_{nm}b_{n'm'i'} + b_{n'm'i'}a^+_{nm} = 0, \tag{4.34e}$$

$$a^+_{nm}b^+_{n'm'i'} + b^+_{n'm'i'}a^+_{nm} = a_{nm}b_{n'm'i'} + b_{n'm'i'}a_{nm} = 0, \tag{4.34f}$$

$$b^+_{nmi}a_{n'm'} + a_{n'm'}b^+_{nmi} = 0. \tag{4.34c}$$

In addition, as per the study of the one-dimensional tight-binding model, the operators a^+_{nm}, b^+_{nmi} and a_{nm}, b_{nmi} can be re-expressed in k-space by Fourier series, which ultimately is used to obtain the modal solutions of the tight-binding model of graphene.

To this end, the Fermi operators a^+_{nm}, b^+_{nmi} and a_{nm}, b_{nmi} are expressed in terms of Fourier series as [9–12,14]

$$a^+_{nm} = \frac{1}{2N}\sum_{\vec{k}} e^{-i\vec{k}\cdot\vec{r}_{nm}}\, a^+_{\vec{k}} \tag{4.35a}$$

$$b^+_{nmi} = \frac{1}{2N}\sum_{\vec{k}} e^{-i\vec{k}\cdot\vec{r}_{nm}}\, e^{-i\vec{k}\cdot\vec{b}_i}\, b^+_{\vec{k}} \tag{4.35b}$$

$$a_{nm} = \frac{1}{2N}\sum_{\vec{k}} e^{i\vec{k}\cdot\vec{r}_{nm}}\, a_{\vec{k}} \tag{4.35c}$$

$$b_{nmi} = \frac{1}{2N}\sum_{\vec{k}} e^{i\vec{k}\cdot\vec{r}_{nm}}\, e^{i\vec{k}\cdot\vec{b}_i}\, b_{\vec{k}} \tag{4.35d}$$

and, conversely,

$$a^+_{\vec{k}} = \frac{1}{2N}\sum_{n.m} e^{i\vec{k}\cdot\vec{r}_{nm}}\, a^+_{nm} \tag{4.36a}$$

$$b^+_{\vec{k}} = \frac{1}{2N}\sum_{n,m,i} e^{i\vec{k}\cdot\vec{r}_{nm}}\, e^{i\vec{k}\cdot\vec{b}_i}\, b^+_{nmi} \tag{4.36b}$$

$$a_{\vec{k}} = \frac{1}{2N}\sum_{n,m} e^{-i\vec{k}\cdot\vec{r}_{nm}}\, a_{nm} \tag{4.36c}$$

$$b_{\vec{k}} = \frac{1}{2N}\sum_{n,m,i} e^{-i\vec{k}\cdot\vec{r}_{nm}} e^{-i\vec{k}\cdot\vec{b}_i} b_{n,m,i} \tag{4.36d}$$

where

$$\vec{r}_{nm} = \vec{T}_{nm} \tag{4.37}$$

are now the position vectors of the lattice sites on the A lattice, and in summing over i in Eq. (4.36) care is made not to multiply count the B lattice sites. These sums are similar to those of the one-dimensional system in Eq. (4.14), but are modified to treat both the A and B lattices and the peculiar characteristics of higher dimensional systems. In this regard, the A and B lattices each contain $(2N+1)^2 \to \infty$ lattice sites so that there is a total of $2(2N+1)^2 \to \infty$ sites in the crystal, and this accounts for the factor of $2N$ entering these expressions. A further important point to note is that care must be taken in defining and performing the k-space sums, as they now occur over a lattice defined by two linearly independent vectors $\vec{a}_1, \vec{a}_2$ that are not orthogonal to one another. These k-space wavevectors and summations are now discussed.

Given the translation vectors in Eq. (4.31), it is readily shown that the vectors $\vec{g}_1$, $\vec{g}_2$ defined by

$$\vec{g}_1 = \frac{1}{\sqrt{3}\tilde{a}}\left(\frac{1}{\sqrt{3}}\hat{i} + \hat{j}\right) \tag{4.38a}$$

$$\vec{g}_2 = \frac{1}{\sqrt{3}\tilde{a}}\left(\frac{1}{\sqrt{3}}\hat{i} - \hat{j}\right) \tag{4.38b}$$

have orthogonality properties with the vectors $\vec{a}_1$, $\vec{a}_2$, which are expressed as follows

$$\vec{g}_1 \cdot \vec{a}_1 = 1, \quad \vec{g}_1 \cdot \vec{a}_2 = 0, \tag{4.39a}$$

$$\vec{g}_2 \cdot \vec{a}_1 = 0, \quad \vec{g}_2 \cdot \vec{a}_2 = 1. \tag{4.39b}$$

Presently, it shall be seen that the orthogonality properties of vectors $\vec{g}_1$, $\vec{g}_2$ in Eq. (4.39) are very important in developing the set of wavevectors $\vec{k}$ entering into the Fourier series sums in Eq. (4.35) and (4.36). In this regard, in terms of the vectors $\vec{g}_1$, $\vec{g}_2$, the $(2N+1)^2 \to \infty$ wavevectors in the sums in Eqs. (4.35) and (4.36) can be written in the form [9–12,14]

$$\vec{k} = \frac{\pi l}{N}\vec{g}_1 + \frac{\pi p}{N}\vec{g}_2, \tag{4.40a}$$

where $l, p = 0, \pm 1, \pm 2, \ldots, \pm N$. In this way, the set in Eq. (4.40a) form the points in k-space for the sums in Eqs. (4.35) and (4.36).

To see that this is a correct expression for the wavevectors in the Fourier series, consider the weights in the Fourier series in Eq. (4.35). These are given in the context of a variety of different notations by

$$e^{i\vec{k}\cdot\vec{r}_{nm}} = e^{i\vec{k}\cdot\vec{T}_{nm}} = e^{i\left(\frac{\pi l}{N}\vec{g}_1 + \frac{\pi p}{N}\vec{g}_2\right)\cdot(n\vec{a}_1 + m\vec{a}_2)} = e^{i\frac{\pi}{N}(ln+pm)}. \tag{4.40b}$$

Here the weight $e^{i\vec{k}\cdot\vec{r}_{nm}}$ in Eq. (4.40b) has been rewritten into the form $e^{i\frac{\pi}{N}(ln+pm)}$ so that the sums over $\vec{k}$ in Eq. (4.35) are now re-expressed in the more traditional Fourier series notation as sums over the integers l, p. In this rewrite, the l, p indices each run over the same range of integers as in the one-dimensional system in Eq. (4.14). In addition, the sums over k-space now become sums over these integers, so that in the Eq. (4.35)

$$\sum_{\vec{k}} = \sum_{l,p}. \tag{4.40c}$$

With these identifications, a standard Fourier series involving sums over integers is retrieved from the sums over $\vec{k}$'s or $\vec{r}_{nm}$'s. On the contrary, in the one-dimensional case in Eq. (4.14), only one integer $n = 0, \pm1, \pm2, \ldots, \pm N$ was summed, now in the two-dimensions case two integers $l, p = 0, \pm1, \pm2, \ldots, \pm N$ are summed. Consequently, the Fourier analysis in two dimensions comes from the applications of two one-dimensional Fourier analyses: one over the $\vec{a}_1$- direction and the other over the $\vec{a}_2$- direction.

As an example, consider the general function defined by $f_{nm} = f(\vec{r}_{nm})$ over the lattice generated by $\vec{T}_{nm}$ in Eq. (4.30). From Fourier theory, the function is represented by the planewave series

$$f_{nm} = f(\vec{r}_{nm}) = \frac{1}{2N} \sum_{l,p} \tilde{f}_{n,p} e^{i\frac{\pi}{N}(ln+pm)} \tag{4.41a}$$

where $l, p = 0, \pm1, \pm2, \ldots, \pm N$ and $N \to \infty$. In terms of the results in Eq. (4.40b), however, the series can also be represented by the sum

$$f_{nm} = f(\vec{r}_{nm}) = \frac{1}{2N} \sum_{l,p} \tilde{f}_{n,p} e^{i\left(\frac{\pi l}{N}\vec{g}_1 + \frac{\pi p}{N}\vec{g}_2\right)\cdot(n\vec{a}_1 + m\vec{a}_2)} = \frac{1}{2N} \sum_{\vec{k}} \tilde{f}_{\vec{k}} e^{i\vec{k}\cdot\vec{r}_{nm}} \tag{4.41b}$$

for $\vec{k}$ defined in Eq. (4.40a) with $l, p = 0, \pm1, \pm2, \ldots, \pm N$ and $N \to \infty$. Whereas the exponential weighting in Eq. (4.41a) involves scalars, the exponential weighting in Eq. (4.41b) is written in terms of the scalar product of vectors.

Consequently, the wavevectors in Eq. (4.40a) represent a direct generalization of the expression for the wavevectors of the one-dimensional system given in Eq. (4.14c) to the treatment of a two-dimensional lattice where now account needs to be taken that $\vec{a}_1$, $\vec{a}_2$ are linearly independent but not necessarily orthogonal vectors. A little explanation, however, remains necessary regarding the geometry of the region of $\vec{k}$ over which the sum is made.

In two dimensions, the wavevectors for the sums in Eq. (4.35) range within the parallelogram region in k-space illustrated in Figure 4.3a. The parallelogram contains the points defined in Eq. (4.40a), which are then the points summed over. While the parallelogram gives a mathematically correct result for the sums in Eq. (4.35), it is not an esthetically useful region to sum over from the point of view of a physicist. Specifically, physicists generally prefer to illustrate properties of the system in a region that is symmetric about the origin in k-space. An important property of the points in k-space, however, allows for the distortion of the region of summation of Eq. (4.35) into one with symmetry about the origin in k-space. This distortion generates an identical result for the sums of Eq. (4.35) but evaluates the sums over a different closed region of k-space than that shown in Figure 4.3a. This distortion and its resulting summation are now discussed.

Another important property of the sums in Eq. (4.35) is that if a vector of the form [9–12,14]

$$\vec{G} = 2\pi(m_1\vec{g}_1 + m_2\vec{g}_2) \tag{4.42}$$

where m_1, m_2 are integers is added to the $\vec{k}$ wavevector in Eq. (4.40a), it is found that the weights in the Fourier series remain unchanged, i.e.,

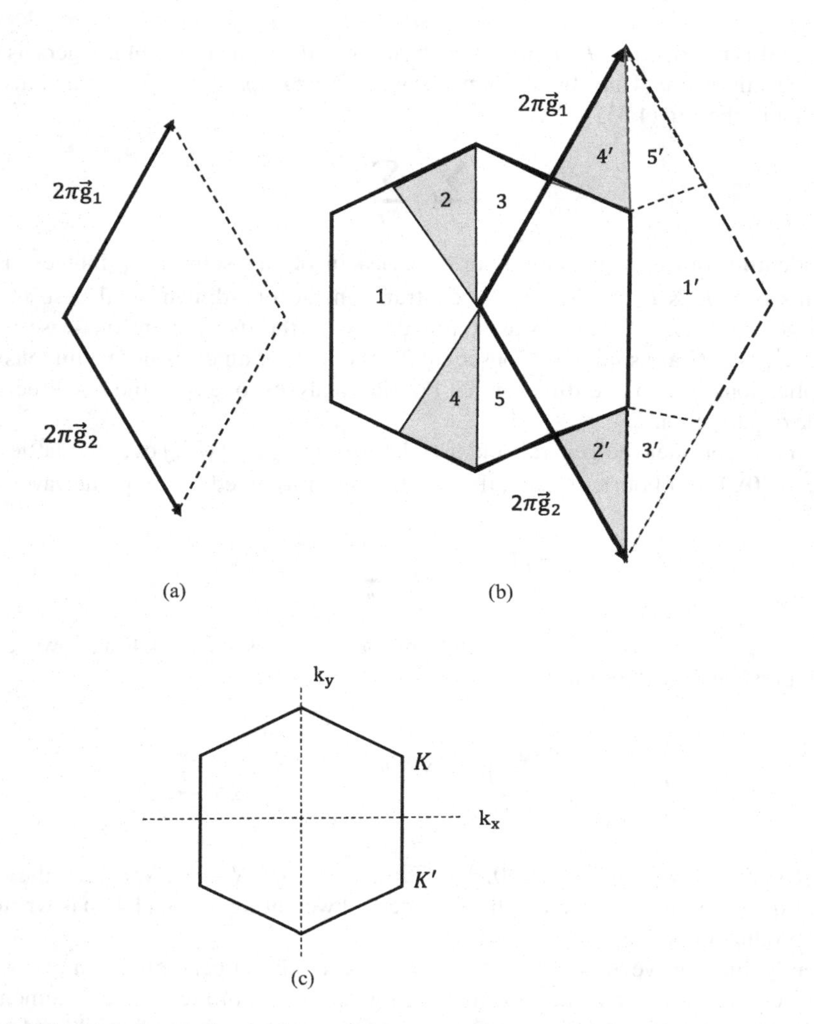

FIGURE 4.3 Schematic figures for the k-space sums. In (a) the two vectors $2\pi\vec{g}_1$ and $2\pi\vec{g}_2$ are shown (solid arrows) along with the k-space parallelogram (dashed lines) they define. The parallelogram defines the region over which k-sums are made. In (b), to understand the summation of $\vec{k}$ over a unit cell, the region of summation in k-space is broken into areas labeled 1´, 2´, 3´, 4´, and 5´ and some adjacent k-space parts 1, 2, 3, 4, and 5. The relation between these regions are discussed in the text, and an indication is given of how the parallelogram cell of k-summation can be distorted into an equivalent sum over the hexagram region of k-space. In (c) a hexagram unit cell in k-space is shown, and reference is made to the points K and K'.

$$e^{i\vec{k}\cdot\vec{r}_{nm}} = e^{i\left(\vec{k}+\vec{G}\right)\cdot\vec{r}_{nm}} = e^{i\vec{k}\cdot\vec{r}_{nm}}e^{i2\pi(m_1n+m_2m)}. \tag{4.43}$$

Consequently, adding $\vec{G}$ to a wavevector in k-space takes you to a new k-space point $\vec{k}+\vec{G}$ having the same weight in the Fourier series. A consequence of this is that the region of the k-space sum can be distorted from its parallelogram shape.

The distorted region in k-space, which is of interest to physicists, is of the form of a hexagon. This is seen in Figure 4.3b, which shows the parallelogram of Figure 4.3a conjoined with the hexagonal region of interest in k-space. In this figure, the points in region 2 are obtained from the points in region 2′ by translating the points in 2′ by $-2\pi\vec{g}_2$, and the points in region 3 are, similarly, obtained from the points in region 3′ by translating the points in 3′ by $-2\pi\vec{g}_2$. The points in region 4 are obtained from the points in region 4′ by translating the points in 4′ by $-\overrightarrow{2\pi g_1}$, and the points in region 5 are, similarly, obtained from the points in region 5′ by translating the points in 5′ by $-2\pi\vec{g}_1$. In addition, the points in region 1 are obtained from the points in region 1′ by translating the points in 1′ by $-2\pi(\vec{g}_1 + \vec{g}_2)$. Consequently, the new region of summation is a hexagon, and the sum in Eq. (4.31) is invariant under the change in the regions of the summation from that of the parallelogram to that of the hexagon [9–12,14].

4.1.2.1 The Tight-Binding Hamiltonian of Graphene

The very simplest model of graphene is obtained by studying a tight-binding Hamiltonian of the form [1,2,9–12,14]

$$H = -t\sum_{n,m,j}\left(b^{+}_{nmj}a_{nm} + a^{+}_{nm}b_{nmj}\right) + u\sum_{n,m}a^{+}_{nm}a_{nm} + u\sum_{n,m,i}b^{+}_{nmi}b_{nmi}. \tag{4.44}$$

Here the first term on the right (with j running over the three B lattice sites that are nearest neighbors of the nm site on the A lattice) is the kinetic energy associated with the hopping of the electrons from site to site, and the last two terms are the energies associated with the bonding interactions of the conduction electrons with their ionic sites. (Note: In the third term on the right n,m,i, run only once through the sites of the B lattice.) The conduction electrons travel over both the A and B lattices, as well as between the two. Substituting Eq. (4.35) into Eq. (4.44), it follows that re-expressed in k-space the Hamiltonian in Eq. (4.44) takes the form [9–12,14]

$$\begin{aligned} H &= -t\sum_{\vec{k}}\left(\Delta_{\vec{k}}b^{+}_{\vec{k}}a_{\vec{k}} + \Delta^{*}_{\vec{k}}a^{+}_{\vec{k}}b_{\vec{k}}\right) + u\sum_{\vec{k}}\left(a^{+}_{\vec{k}}a_{\vec{k}} + b^{+}_{\vec{k}}b_{\vec{k}}\right) \\ &= -t\sum_{\vec{k}}\left(b^{+}_{\vec{k}},\vec{a}^{+}_{\vec{k}}\right)\begin{vmatrix} -\frac{u}{t} & \Delta_{\vec{k}} \\ \Delta^{*}_{\vec{k}} & -\frac{u}{t}\end{vmatrix}\begin{vmatrix} b_{\vec{k}} \\ a_{\vec{k}}\end{vmatrix} \end{aligned} \tag{4.45a}$$

where

$$\Delta^{*}_{\vec{k}} = \sum_{i} e^{i\vec{k}\cdot\vec{b}_i} = e^{ik_x\tilde{a}} + e^{-i\frac{1}{2}k_x\tilde{a}}\,2\cos\frac{\sqrt{3}}{2}k_y\tilde{a}. \tag{4.45b}$$

For the evaluation of Eq. (4.45) the sums over $\vec{k}$ range throughout the hexagonal region discussed earlier in Figure 4.3b. A simplification of the form of the Hamiltonian in Eq. (4.45) can be made, and a discussion is given of its properties and its relation to the Dirac equations. These are now presented in the following.

Consider the case of Eq. (4.45) in which $t > 0$ and $u = 0$ as the most basic model of planar graphene. In this limit, the Hamiltonian is readily solved and provides a rough estimate of the properties of the graphene conduction electrons for a parameterization of $t \approx 2.8\ eV$. In particular, a simplification of Eq. (4.45) can be made by expressing it in terms of the eigenvalues and eigenvectors of the matrix in Eq. (4.45a). This involves the application of an appropriated unitary transformation.

In this regard, the matrix on the far-right-hand side of Eq. (4.45a) is put into diagonal form through the application of a unitary transformation generated by the unitary matrix [9–12,14]

$$U(\vec{k}) = \frac{1}{\sqrt{2}}\begin{vmatrix} \frac{\Delta_{\vec{k}}}{|\Delta_{\vec{k}}|} & -\frac{\Delta_{\vec{k}}}{|\Delta_{\vec{k}}|} \\ 1 & 1\end{vmatrix} \tag{4.46a}$$

so that

$$U^{+}\left(\vec{k}\right)\begin{vmatrix} 0 & \Delta_{\vec{k}} \\ \Delta_{\vec{k}}^{*} & 0 \end{vmatrix}U\left(\vec{k}\right)=\begin{vmatrix} \left|\Delta_{\vec{k}}\right| & 0 \\ 0 & -\left|\Delta_{\vec{k}}\right| \end{vmatrix}. \tag{4.46b}$$

Consequently, from Eq. (4.45a) and the unitary transformation, it follows that

$$\begin{aligned} H &= -t\sum_{\vec{k}}\left(b_{\vec{k}}^{+},\vec{a}_{\vec{k}}^{+}\right)\begin{vmatrix} 0 & \Delta_{\vec{k}} \\ \Delta_{\vec{k}}^{*} & 0 \end{vmatrix}\begin{vmatrix} b_{\vec{k}} \\ a_{\vec{k}} \end{vmatrix} \\ &= -t\sum_{\vec{k}}\left(b_{\vec{k}}^{+},\vec{a}_{\vec{k}}^{+}\right)U\left(\vec{k}\right)U^{+}\left(\vec{k}\right)\begin{vmatrix} 0 & \Delta_{\vec{k}} \\ \Delta_{\vec{k}}^{*} & 0 \end{vmatrix}U\left(\vec{k}\right)U^{+}\left(\vec{k}\right)\begin{vmatrix} b_{\vec{k}} \\ a_{\vec{k}} \end{vmatrix} \\ &= -t\sum_{\vec{k}}\left(b_{\vec{k}}^{+},\vec{a}_{\vec{k}}^{+}\right)U\left(\vec{k}\right)\begin{vmatrix} \left|\Delta_{\vec{k}}\right| & 0 \\ 0 & -\left|\Delta_{\vec{k}}\right| \end{vmatrix}U^{+}\left(\vec{k}\right)\begin{vmatrix} b_{\vec{k}} \\ a_{\vec{k}} \end{vmatrix}. \end{aligned} \tag{4.46c}$$

Looking at the terms to the right and left of the matrix in Eq. (4.46c), it follows that a set of new operators are defined in the system by the unitary relations

$$\begin{vmatrix} c_{\vec{k}} \\ d_{\vec{k}} \end{vmatrix}=U^{+}\left(\vec{k}\right)\begin{vmatrix} b_{\vec{k}} \\ a_{\vec{k}} \end{vmatrix}=\frac{1}{\sqrt{2}}\begin{vmatrix} \dfrac{\Delta_{\vec{k}}^{*}}{\left|\Delta_{\vec{k}}\right|}b_{\vec{k}}+a_{\vec{k}} \\ -\dfrac{\Delta_{\vec{k}}^{*}}{\left|\Delta_{\vec{k}}\right|}b_{\vec{k}}+a_{\vec{k}} \end{vmatrix}, \tag{4.47a}$$

and

$$\left(c_{\vec{k}}^{+},\vec{d}_{\vec{k}}^{+}\right)=\left(b_{\vec{k}}^{+},\vec{a}_{\vec{k}}^{+}\right)U\left(\vec{k}\right)=\frac{1}{\sqrt{2}}\left(\frac{\Delta_{\vec{k}}}{\left|\Delta_{\vec{k}}\right|}b_{\vec{k}}^{+}+\vec{a}_{\vec{k}}^{+},-\frac{\Delta_{\vec{k}}}{\left|\Delta_{\vec{k}}\right|}b_{\vec{k}}^{+}+\vec{a}_{\vec{k}}^{+}\right). \tag{4.47b}$$

Collecting together all the transformations in Eqs. (4.46) and (4.47), the Hamiltonian of Eq. (4.45) is rewritten into a new form. In its new format, the Hamiltonian then becomes [9–12,14]

$$H=-t\sum_{\vec{k}}\left|\Delta_{\vec{k}}\right|\left(c_{\vec{k}}^{+}c_{\vec{k}}-d_{\vec{k}}^{+}d_{\vec{k}}\right). \tag{4.48}$$

Now, $c_{\vec{k}}^{+}$, $c_{\vec{k}}$, $d_{\vec{k}}^{+}$, $d_{\vec{k}}$ are shown to be new creation and destruction operators that generate the modal solutions for the conduction electrons of graphene, and their energies are obtained directly from the Hamiltonian in Eq. (4.48).

It is interesting to note that the operators $c_{\vec{k}}^{+}$, $c_{\vec{k}}$, $d_{\vec{k}}^{+}$, $d_{\vec{k}}$ were obtained by a sequence of unitary transformations made on the lattice site occupation operators a_{nm}^{+}, a_{nm}, b_{nmi}^{+}, b_{nmi}. The first transformation was the Fourier transform to k-space, and this was later followed by a change of basis that diagonalized the Hamiltonian in Eq. (4.45). Under this sequence of transformations, it can be shown that the $c_{\vec{k}}^{+}$, $c_{\vec{k}}$, $d_{\vec{k}}^{+}$, $d_{\vec{k}}$ obey the standard fermion anticommutation relations given by [9–12,14]

$$c_{\vec{k}}^{+}c_{\vec{k}'}+c_{\vec{k}'}c_{\vec{k}}^{+}=\delta_{\vec{k},\vec{k}'}, \tag{4.49a}$$

$$c_{\vec{k}}^{+}c_{\vec{k}'}^{+} + c_{\vec{k}'}^{+}c_{\vec{k}}^{+} = c_{\vec{k}}c_{\vec{k}'} + c_{\vec{k}'}c_{\vec{k}} = 0, \tag{4.49b}$$

$$d_{\vec{k}}^{+}d_{\vec{k}'} + d_{\vec{k}'}d_{\vec{k}}^{+} = \delta_{\vec{k},\vec{k}'}, \tag{4.49c}$$

$$d_{\vec{k}}^{+}d_{\vec{k}'}^{+} + d_{\vec{k}'}^{+}d_{\vec{k}}^{+} = d_{\vec{k}}d_{\vec{k}'} + d_{\vec{k}'}d_{\vec{k}} = 0, \tag{4.49d}$$

and in addition

$$d_{\vec{k}}^{+}c_{\vec{k}'}^{+} + c_{\vec{k}'}^{+}d_{\vec{k}}^{+} = d_{\vec{k}}c_{\vec{k}'} + c_{\vec{k}'}d_{\vec{k}} = 0, \tag{4.49e}$$

$$c_{\vec{k}}^{+}d_{\vec{k}'} + d_{\vec{k}'}c_{\vec{k}}^{+} = d_{\vec{k}}^{+}c_{\vec{k}'} + c_{\vec{k}'}d_{\vec{k}}^{+} = 0. \tag{4.49f}$$

These relations follow directly from the site commutation properties in Eq. (4.34), and as a consequence the operators $c_{\vec{k}}^{+}$, $c_{\vec{k}}$, $d_{\vec{k}}^{+}, d_{\vec{k}}$ create and remove or destroy sets of fermion excitations from the system. These fermion excitations happen to be the eigenmodes of the transformed Hamiltonian in Eq. (4.48).

In this regard, the operators $c_{\vec{k}}^{+}$, $c_{\vec{k}}$ act only on the eigenmodes of the Hamiltonian $|1\rangle_{c\vec{k}}$, $|0\rangle_{c\vec{k}}$ and the $d_{\vec{k}}^{+}, d_{\vec{k}}$ act only on the eigenmodes of the Hamiltonian $|1\rangle_{d\vec{k}}$, $|0\rangle_{d\vec{k}}$ so that

$$c_{\vec{k}}|1\rangle_{c\vec{k}} = |0\rangle_{c\vec{k}} \tag{4.50a}$$

$$c_{\vec{k}}^{+}|0\rangle_{c\vec{k}} = |1\rangle_{c\vec{k}} \tag{4.50b}$$

$$d_{\vec{k}}|1\rangle_{d\vec{k}} = |0\rangle_{d\vec{k}} \tag{4.50c}$$

$$d_{\vec{k}}^{+}|0\rangle_{d\vec{k}} = |1\rangle_{d\vec{k}}, \tag{4.50d}$$

and all other actions of the operators $c_{\vec{k}}^{+}$, $c_{\vec{k}}$, on the $|1\rangle_{c\vec{k}}$, $|0\rangle_{c\vec{k}}$, or the $d_{\vec{k}}^{+}$, $d_{\vec{k}}$, on the $|1\rangle_{d\vec{k}}$, $|0\rangle_{d\vec{k}}$, give zero. Similarly, the number operators formed from these operators are

$$n_{c\vec{k}} = c_{\vec{k}}^{+}c_{\vec{k}} \tag{4.51a}$$

and

$$n_{d\vec{k}} = d_{\vec{k}}^{+}d_{\vec{k}}, \tag{4.51b}$$

with the properties that $n_{c\vec{k}}|1\rangle_{c\vec{k}} = |1\rangle_{c\vec{k}}$, $n_{c\vec{k}}|0\rangle_{c\vec{k}} = 0$ and that $n_{d\vec{k}}|1\rangle_{c\vec{k}} = |1\rangle_{d\vec{k}}$, $n_{d\vec{k}}|0\rangle_{c\vec{k}} = 0$. In the context of the number operators, Eq. (4.48) reduces to [9–12,14]

$$H = -t\sum_{\vec{k}}|\Delta_{\vec{k}}|\left(n_{c\vec{k}} - n_{d\vec{k}}\right), \tag{4.52}$$

which counts the energy within the various $n_{c\vec{k}}$ and $n_{d\vec{k}}$ modes existing in the system.

4.1.2.2 Dispersive Properties of the Excitations in Graphene

The dispersion relation of the particle excitations in graphene is obtained from Eqs. (4.45) to (4.52), where it is found that the individual conduction electrons in the model have energies $\epsilon_{\vec{k}}$ of the form [9–12,14]

$$\epsilon_{\vec{k}} = \pm t \left| \Delta_{\vec{k}} \right|. \tag{4.53}$$

The dispersion relations are composed of two different energy bands defined over the hexagonal region of k-space of Figure 4.3b and c discussed earlier. Both of the excitation spectra in Eq. (4.53) are centered about zero energy with the energy extreme located at $\vec{k} = 0$, where $\epsilon_0 = \pm 3t$. In addition, the closest approach of the two bands is located at the points labeled K and K' in Figure 4.3c. The K and K' points occur in k-space at $\vec{k}_K = \frac{2\pi}{3\sqrt{3}\tilde{a}}\left(\sqrt{3}\hat{i} + \hat{j}\right)$ and $\vec{k}_{K'} = \frac{2\pi}{3\sqrt{3}\tilde{a}}\left(\sqrt{3}\hat{i} - \hat{j}\right)$, respectively, and at these two points

$$\epsilon_{\vec{k}} = 0 \tag{4.54}$$

so that the upper and lower branches of the dispersion touch at K and K'. As a result of this touching, many of the interesting properties of graphene come from the dispersive properties of the system in the vicinity of the K and K' points. This is now discussed.

To study the dispersive properties of the electrons in the vicinity of the K and K' points, let us begin by considering the dispersion relation in Eqs. (4.53) and (4.45b) in terms of

$$\vec{k}_K + \vec{q} \text{ and } \vec{k}_{K'} + \vec{q} \tag{4.55}$$

where $|\vec{q}| \ll \left|\vec{k}_K\right|$. To leading order in $\vec{q}$, it is then found from Eq. (4.45b) that

$$\Delta_{\vec{k}_K + \vec{q}} = i\frac{3}{4}\left(1 + i\sqrt{3}\right)\left(q_x + iq_y\right)\tilde{a} \tag{4.56a}$$

and

$$\Delta_{\vec{k}_{K'} + \vec{q}} = i\frac{3}{4}\left(1 + i\sqrt{3}\right)\left(q_x - iq_y\right)\tilde{a}, \tag{4.56b}$$

where it is assumed that $|\vec{q}| \ll \left|\vec{k}_K\right|$. Consequently, from Eq. (4.53), it follows that [9–12,14]

$$\epsilon_{\vec{k}_K + \vec{q}} = \epsilon_{\vec{k}_{K'} + \vec{q}} = \pm\frac{3t\tilde{a}}{2}\sqrt{q_x^2 + q_y^2} \tag{4.57a}$$

expresses a linear relation between the energy and momentum of the system. This is a characteristic of relativistic, massless, particles where, for example, the energy dispersion of a massless particle of momentum p is expressed by $\varepsilon = cp$. Within this interpretation then

$$\epsilon_{\vec{k}_K + \vec{q}} = \epsilon_{\vec{k}_{K'} + \vec{q}} = \pm\frac{3t\tilde{a}}{2\hbar}\hbar\sqrt{q_x^2 + q_y^2} = \pm v_{el}\hbar\sqrt{q_x^2 + q_y^2} = \pm v_{el}\, p, \tag{4.57b}$$

where $p = \hbar\sqrt{q_x^2 + q_y^2}$ is an effective electron momentum measured from the K or K' points, and $v_{el} \approx 8 \times 10^5$ m/sec is the electron group velocity. This same expansion used to obtain Eq. (4.57) can also be used in Eq. (4.46c) to rewrite the Hamilton at that level of the formulation into the form of the Dirac equation for a massless fermion.

Returning to the first line of Eq. (4.46c), the Hamiltonian of the conduction electrons given by the form [9–12,14]

$$H = \sum_{\vec{k}} H_{\vec{k}} = -t \sum_{\vec{k}} \left(b_{\vec{k}}^{+}, \vec{a}_{\vec{k}}^{+}\right) \begin{vmatrix} 0 & \Delta_{\vec{k}} \\ \Delta_{\vec{k}}^{*} & 0 \end{vmatrix} \begin{vmatrix} b_{\vec{k}} \\ a_{\vec{k}} \end{vmatrix} \tag{4.58}$$

can be examined in the neighborhood of the k-space points K and K'. In these neighborhoods, it is shown to take a format related to that of the Weyl equation for a massless spin of one-half particle. (Note that the Weyl equation is obtained from the Dirac equation in the limit in which the particle masses become zero.) For example, in the vicinity of the K point located at $\vec{k}_K = \frac{2\pi}{3\sqrt{3}\tilde{a}}\left(\sqrt{3}\hat{i} + \hat{j}\right)$, consider taking $\vec{k} = \vec{k}_K + \vec{q}$ where $|\vec{q}| \ll |\vec{k}_K|$. Expanding the $H_{\vec{k}}$ terms in the Hamiltonian sum to leading order in $\vec{q}$, it follows that

$$H_{\vec{k}_K+\vec{q}} = -v_F \left(e^{i\delta} b_{\vec{k}_K+\vec{q}}^{+}, \vec{a}_{\vec{k}_K+\vec{q}}^{+}\right) \begin{vmatrix} 0 & (q_x + iq_y) \\ (q_x - iq_y) & 0 \end{vmatrix} \begin{vmatrix} e^{-i\delta} b_{\vec{k}_K+\vec{q}} \\ a_{\vec{k}_K+\vec{q}} \end{vmatrix}, \tag{4.59a}$$

where $v_F = \frac{3t\tilde{a}}{2}$ and $e^{i\delta} = i\frac{1+i\sqrt{3}}{2}$. Likewise, in the vicinity of the K' point located at $\vec{k}_{K'} = \frac{2\pi}{3\sqrt{3}\tilde{a}}\left(\sqrt{3}\hat{i} - \hat{j}\right)$, consider taking $\vec{k} = \vec{k}_{K'} + \vec{q}$ where $|\vec{q}| \ll |\vec{k}_{K'}|$. From a similar expansion to that in Eq. (4.59a) to leading order in $\vec{q}$, it is found that

$$H_{\vec{k}_{K'}+\vec{q}} = -v_F \left(e^{i\delta} b_{\vec{k}_{K'}+\vec{q}}^{+}, \vec{a}_{\vec{k}_{K'}+\vec{q}}^{+}\right) \begin{vmatrix} 0 & (q_x - iq_y) \\ (q_x + iq_y) & 0 \end{vmatrix} \begin{vmatrix} e^{-i\delta} b_{\vec{k}_{K'}+\vec{q}} \\ a_{\vec{k}_{K'}+\vec{q}} \end{vmatrix}, \tag{4.59b}$$

where $v_F = \frac{3t\tilde{a}}{2}$. (Note that in both of Eq. (4.59) the factors $e^{i\delta}$ and $e^{-i\delta}$ are part of a unitary transformation on the operator set $b_{\vec{k}}^{+}, \vec{a}_{\vec{k}}^{+}, b_{\vec{k}}, a_{\vec{k}}$ and do not affect the energy of the model.)

Due to the similarity of Eq. (4.59) to the relativistic equations of massless particles, many of the interesting relativistic properties of the relativistic Weyl systems have counterparts in the properties of the graphene conduction electrons near the K and K' points. Some of these properties of graphene are now discussed.

4.1.3 Graphene Conductivity

As in the case of the initial discussions of the one-dimensional tight-binding model, the above results in two dimensions were obtained for spinless fermions. The inclusion of spin into the problem can now be made, just as in the one-dimensional model, by noting that the spin of the conduction electrons enters only as an addition quantum number labeling the electron states. The spin-up (down) quantum numbers, however, do not enter into determining the energy of the particles. Even in the presence of spin, the Hamiltonian still only depends on the wavevector and the energy band of the solutions.

The spin quantum number, however, does enter into determining the occupancy of the energy levels, and the occupancy is regulated by the Pauli exclusion principle. In the spinless treatment, each energy level is occupied by one spinless particle, whereas each energy level is occupied by at most one spin-up and one spin-down particle in the presence of spin.

For graphene this means that, since there is one conduction electron for each carbon atom, at $T = 0$ only the lower $\epsilon_{\vec{k}} = -t|\Delta_{\vec{k}}|$ band of modes are completely occupied, and the upper $\epsilon_{\vec{k}} = t|\Delta_{\vec{k}}|$ band is completely vacant. As a result, the system is a semiconductor or semimetal, because the upper and lower energy bands only touch at the K and K' points [9–12,14].

4.1.4 Graphene Nanotubes

A geometry related to that of grapheme, which is of interest in nanoscience, is that of graphene nanotubes. These are tubes of circular cross section that are formed by cutting a finite-width infinite strip of graphene and joining the edges of its width together to form an infinite length of nanotube. The conduction properties of the nanotubes are of interest in certain nanotechnologies, as, depending on how the tube is wrapped, it can function as a metal or semiconductor. In this section, a brief presentation is given of an example of one such nanotube structure [2,9–12].

Consider the solution of the tight-binding model of graphene treated earlier in the context of an infinite plane of carbon atoms. From those considerations, the eigen energies of the planar modes were obtained in Eq. (4.53) by $\epsilon_{\vec{k}} = \pm t|\Delta_{\vec{k}}|$. In addition, the periodic boundary conditions over the infinite plane on the wavefunctions determined $\vec{k} = \frac{\pi l}{N}\vec{g}_1 + \frac{\pi p}{N}\vec{g}_2$, where $l, p = 0, \pm 1, \pm 2, \ldots, \pm N$ for $N \to \infty$ as the set of wavevectors at which Eq. (4.53) is evaluated. (Ultimately, the region of k-space summed over was deformed into the hexagon region shown in Figure 4.3c, but this transformation is not of interest here.) As a consequence, the properties of the energies and eigenmodes are determined to a large degree by the boundary conditions set over the infinite plane. In the case of the nanotube, the energies and eigenmodes are again obtained from Eq. (4.53) but now evaluated for periodic boundary conditions applied to the infinite strip of finite width [2,9–12].

In the case of the infinite tube of circular cross section, the values of the energies of the tube modes are still given from the Hamiltonian as $\epsilon_{\vec{k}} = \pm t|\Delta_{\vec{k}}|$ just as in the case of the infinite plane of graphene, but now the values of $\vec{k}$ for the modes are set by different boundary conditions. The components of $\vec{k}$ parallel to the infinite axis of the tube obey periodic boundary conditions over the infinite length of the tube, but the components of $\vec{k}$ perpendicular to the tube axis obey periodic boundary conditions along the circumference of the circular cross section of the tube. This circumference corresponds to the finite width of the strip from which the tube is generated.

As a result of the finite circumference of the tube the energy modes of the tube separate into spectra that are labeled in terms of the components of the wavevector perpendicular to the tube axis. For each discrete wavevector component perpendicular to the tube axis, there is a continuum density of states corresponding to the wavevector components parallel to the tube axis.

As with planar graphene, the behavior of the system in the neighborhood of the points K and K' is essential in determining the metallic or semiconducting properties of the materials. This comes as at $T = 0$ half of the available band modes are occupied in the nanotube, and the K and K' points are the points at which the upper (i.e., $\epsilon_{\vec{k}} = t|\Delta_{\vec{k}}|$) and lower (i.e., $\epsilon_{\vec{k}} = -t|\Delta_{\vec{k}}|$) energy bands in the system touch. Consequently, the focus in the following will be on the nature of the nanotubes in the vicinity of these two points [2,9–12].

As an example of a nanotube, consider a strip of the graphene plane that is rolled into a tube of circular cross section. When the nanotube is unrolled, its circumference sites are along the $\vec{a}_1$ direction, and the axis of the nanotube is perpendicular to the $\vec{a}_1$ direction. The sites on the finite circumference of the strip are subject to periodic boundary conditions so that they are translated into each other along the circumference by the so-called chiral translation vector defined by

$$\vec{C} = n\vec{a}_1, \tag{4.60}$$

where n is a fixed positive integer. The finite periodicity of the circular cross section of the tube is then along the $\vec{a}_1$ direction, and there is only a finite solution set of wavevector components k_1 in the $\vec{a}_1$ direction. Next, we determine the nature of the nanotube solutions in the vicinity of the K and K' points. These are related to the conductivity properties of the half-filled energy bands of the nanotubes through the electron dispersion relation.

To understand the nature of the solutions near K and K' note that

$$\vec{a}_1 = \frac{\tilde{a}\sqrt{3}}{2}\left(\sqrt{3}\hat{i} + \hat{j}\right), \tag{4.61a}$$

and the position of the K point in k-space is

$$\vec{k}_K = \frac{2\pi}{3\sqrt{3}\tilde{a}}\left(\sqrt{3}\hat{i} + \hat{j}\right) \tag{4.61b}$$

so that these two vectors are parallel. Consequently, the periodicity of the solutions along the $\vec{a}_1$ direction requires that if k_1 is the component of the wavevector of a mode along the $\vec{a}_1$ direction, then

$$n\left|\vec{a}_1\right|k_1 = 2\pi p, \tag{4.62}$$

where p is an integer. Using Eq. (4.61b) in Eq. (4.62), it follows that the components of wavevector along the circumference are of the form

$$k_1 = \frac{2\pi p}{n\left|\vec{a}_1\right|} = \frac{3}{2}\left|\vec{k}_K\right|\frac{p}{n}. \tag{4.63}$$

Consequently, the wavevectors of the modes in the system are then written in terms the component $k_1$that is in the direction of $\vec{a}_1$ forming the circumference of the circular cross section of the tube and a component $k_{\parallel}$ that is a continuum component along the tube axis.

It is seen from Eq. (4.63) that in the case that $n = 3q$ for some integer q, the component of wavevector along the circumference $k_1 = \frac{p}{2q}\left|\vec{k}_K\right|$, which for $p = \pm 2q$ is the length of the vector in Eq. (4.61b). As a result, when $k_{\parallel} = 0$ at these K, K' points $\epsilon_{k_1} = 0$, and the spectrum of energies exhibit a zero-energy mode. Specifically, at the values of $k_1 = \pm\left|\vec{k}_K\right|$, $k_{\parallel} = 0$, both the upper and lower branches of the dispersion $\epsilon_{\vec{k}} = \pm t\left|\Delta_{\vec{k}}\right|$ share a $\epsilon_{\vec{k}} = 0$ modal solution, and this is a requisite for a metallic behavior [9–12].

However, when $n = 3q+1$ or $n = 3q-1$ for an integer q in Eq. (4.63), the spectra of the electrons $\epsilon_{\vec{k}} = \pm t\left|\Delta_{\vec{k}}\right|$ under these conditions do not pass through the K, K' points, even when $k_{\parallel} = 0$. Consequently, in these cases the system does not exhibit a nonzero-energy gap between the upper $\epsilon_{\vec{k}} = t\left|\Delta_{\vec{k}}\right|$ and lower $\epsilon_{\vec{k}} = -t\left|\Delta_{\vec{k}}\right|$ energy bands. Only under the $n = 3q$ condition can the conduction electrons traveling down the nanotube exhibit metallic conductivity, as otherwise the system will exhibit a gap in the excitation spectrum and display semiconducting properties [9–12].

4.1.5 Some of the Interesting Properties of Graphene

Graphene has properties of unusually high thermal and electrical conductivities, in addition to exhibiting a great mechanical strength [9–12,15,16]. This makes it favorable for applications in a variety of diverse fields, including electronics, strengths of materials, photonics and optoelectronics, optics, and medicine. For example, electronics applications have included its use in the design of transistors, touchscreens and liquid crystal displays, anodes in photovoltaic devices, organic light-emitting diodes, and frequency multipliers. In optoelectronics, graphene has been utilized in the development of plasmonic devices, Hall sensors, quantum dots, solar cell devices, and in some spintronic applications. In addition, it has been applied in the building of optical modulators, ultraviolet hyperlenses for breaking the diffraction limits of optics, and in infrared light detection and photodetectors [9–12,15,16]. Graphene has also been useful in the construction of batteries and fuel cells, steam condenser coatings, and in the development of supercapacitors. In medicine, it has been applied to the design of sensors, nano–electro–mechanical (NEMS) configurations, and in drug-delivery devices. While many of these topics are in the initial developmental stages, their possibilities indicate that graphene will exhibit an interesting future in technological applications.

4.2 QUANTUM DOTS, QUANTUM WELLS, AND QUANTUM WIRES

In this section an introduction is given to the electronic properties of artificially engineered devices designed on mesoscopic scales to mimic some of the electrical and optical properties characteristically found in atoms, molecules, and semiconducting band structures [2,4–7,9–12,15,16,21]. These devices include quantum dots, quantum wells, and quantum wires. (See Figure 4.4 for a schematic representation of these three types of structures.) They are, respectively, designed to simulate the electronic features of atoms, two-dimensional electron gases, and one-dimensional electron gases with the goal of extending the range of electronic and optical phenomena available for technological applications. A quantum dot, as represented in Figure 4.4a, is a volume of metal or semiconductor that is much larger than an atom but small enough to exhibit discrete sets of energy levels that can be tuned to fit the requirements of engineering applications. Similarly, a quantum well, illustrated in Figure 4.4b, is a mesoscopic metal or semiconductor slab that confines an electron gas in a two-dimensional motion and is thin enough such that quantum mechanical effects associated with the slab thickness are observed. Finally, Figure 4.4c shows a quantum wire as basically a wire of small

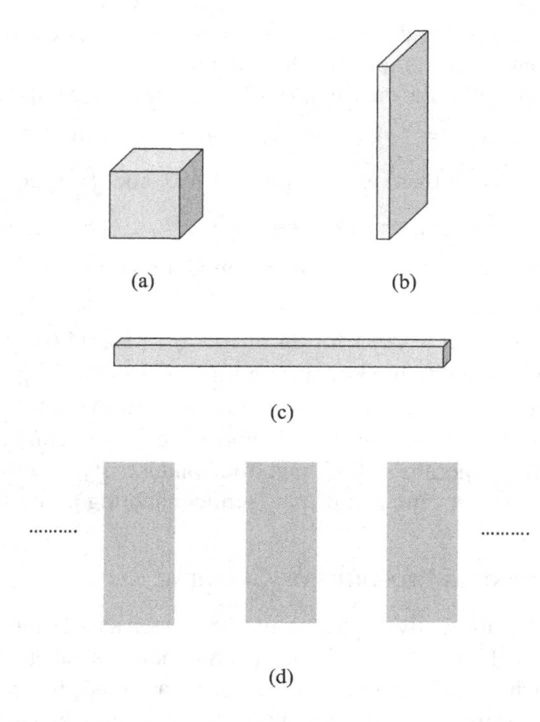

FIGURE 4.4 Schematic plots of: (a) a cube of material, which along with the media outside the cube models a quantum dot, (b) an infinite slab of material, which along with the media outside the slab forms a quantum well, (c) a wire of square cross section and a long length, and (d) an infinite array of layered infinite slabs forming a heterostructure.

enough cross section so that discrete quantum mechanical effects are observed in the electron transport along the wire.

For example, atoms exhibit discrete quantum energy levels of the electrons forming them, and this shows up in their spectroscopy. Quantum dots are mesoscopic analogues of atoms that display similar properties to atoms but on larger length scales [9–12]. They are formed as mesoscopic clumps of materials with conduction electrons occupying discrete energy levels dependent on the size and shape of the dot. In this regard, the dots are a kind of mesoscopic atom with the discrete optical spectra familiar from atoms, but the spectral properties of the dot can be engineering by changing the geometry of the dot. Similarly, quantum wells are slabs of conducting material that confine electrons within them to exhibit two-dimensional motion. When a one-dimensional array of such slabs is formed with insulating barrier materials layered between adjacent slabs, the analogy of a one-dimensional metal is created. Such an array of quantum wells is known as a heterostructure, and the electrons in the conducting slabs of the heterostructure can tunnel between the slabs of the array. (See Figure 4.4d for a schematic of a heterostructure.) Consequently, a hopping conductivity is established in the heterostructure in the direction perpendicular to the slab interfaces. The resulting electron dynamics leads to a series of conduction electron bands similar to those obtained in the one-dimensional tight-binding model, and the array of quantum wells is a mesoscopic analog of the electronic bands found in pure metals or semiconductors. Finally, wires of mesoscopic cross section can be designed to exhibit a current flow occurring in various states that are quantized by the small cross section of the wire. These quantum wires display some of the interesting properties associated with waveguides.

In the following, the ideas behind quantum wells, dots, and wires are introduced in the context of simple models [21]. Furthermore, for the discussions, illustrations are made in the context of $GaAs$ and $Ga_{1-x}Al_xAs$-based systems, as these are among the most commonly employed materials in the technology of device applications. To this end, first some remarks are made about the properties of $GaAs$ and $Ga_{1-x}Al_xAs$. The focus of the remarks is on the description of the electronic properties of $GaAs$ and $Ga_{1-x}Al_xAs$ within the effective mass and envelope function approximations. These approximations of the electrical properties of $GaAs$ and $Ga_{1-x}Al_xAs$ materials are appropriate and well suited to the pedagogical development to follow. More advanced discussions, however, generally require extensive computer simulation techniques. After the introduction of these aspects of material science, the simplest treatment of quantum wells, dots, and wires within the effective mass and envelope formulations are discussed.

4.2.1 Properties of $GaAs$ and $Ga_{1-x}Al_xAs$

A simple model for the electrical conduction properties of $GaAs$ and $Ga_{1-x}Al_xAs$ is based on treating the dynamics of the carriers in these materials within the effective mass and envelope function approximations [1,2,9–12,21]. Both the approximations represent great simplifications in developing the dynamics of the conduction electrons, and designs provide a means to a clear understanding of the basic mechanisms of the device functioning when applied to considerations of device.

In the absence of such approximations, the actual dynamics of the conduction electrons in $GaAs$ or $Ga_{1-x}Al_xAs$ arises from the detailed electron interactions with all of the many dynamical degrees of freedom in these materials. The details of most of these interactions, however, are found to be unnecessary in understanding and describing the conductivity properties of $GaAs$ or $Ga_{1-x}Al_xAs$ occurring in many applications. Most of these details can be averaged out, leading to the simplifications found in the effective mass and envelope function approximations [21].

In this regard, it is found that by removing the detailed short wavelength components of the conduction electron wavefunctions and their interactions within the crystalline medium, a model is obtained in which the motion of the conduction electrons is accounted for by an effective mass and an effective Schrodinger equation for an electron envelope wavefunction. The envelope function then represents the gross long-wavelength behavior of the conduction electrons. The averaging in the development of these approaches is done by the application of a series of reasonable approximations to the study of the conduction electron dynamics. For example, the short wavelength components in the system can be accounted for in terms of Wannier functions [1,2,8] or linear combinations of atomic orbits, which give a description of the wavefunction properties over individual atoms. When combined with longer wavelength envelope functions, these functions of the atomic scale features help remove the short-range behaviors in the wavefunctions, while leaving long-wavelength envelope functions described by an effective mass Schrodinger equation [1,2,8].

The effective mass approximation [2,3,9,10,21], then, represents a renormalization of the inertial properties of the conduction electron obtained through its average scattering with the medium in which it propagates. This averaging is similar to that of the development of the idea of a refractive index in optics as a means of replacing the details of the interaction of light with a crystal by an average interaction summarized by the index of refraction. An effective Schrodinger equation for the envelope wavefunctions of the conduction electrons is also developed by accounting for the short wavelength Fourier components of the electron scattering in an averaging of the wavefunction. As a consequence of this, the resulting envelope functions represent the dominant long-wavelength features of the electron propagations. For details of the development of these two approximations, the reader is referred to the standard texts [1–3]. Here the results of these two basic approximations will be brought into the theories developed in the following discussion as understood descriptions of the conduction elections in $GaAs$ and $Ga_{1-x}Al_xAs$.

4.2.2 Effective Mass Approximation

Within the effective mass approximations, the propagation of the electrons that participate in the conductivity of $GaAs$, $Ga_{1-x}Al_xAs$, and other semiconductors commonly used in technology are described by a set of simple dispersion relations of a standard parabolic form [1–3,8–11,21]. In the description of the conduction properties of these materials, three sets of electrons bands are commonly accounted for in most discussions. These include electrons propagating in a higher energy conduction band and two distinct lower energy valence bands. (A schematic of these bands is presented in Figure 4.5 showing plots of the energy versus wavevector of their electrons.) At zero temperatures, the two valence bands are completely occupied by electrons and the system behaves as an insulator. Upon introducing a nonzero temperature, electrons are promoted to the conduction band by thermal fluctuations, and the material behaves as a semiconductor.

The highest energy electrons in the materials illustrated in Figure 4.5 occur in the conduction band with energies described by parabolic dispersion relations of the form [21]

$$E_c = E_g + \frac{\hbar^2}{2m_c}\left(k_x^2 + k_y^2 + k_z^2\right) \tag{4.64a}$$

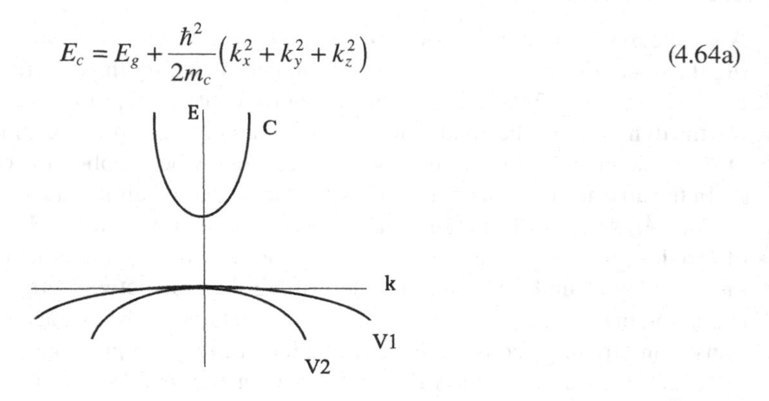

FIGURE 4.5 Schematic plot of an idealized semiconductor with band energies E presented versus wave vector k. The conduction band with energies E_c is labeled c, and the valence band with energies E_{v1}, E_{v2} are labeled $v1$ and $v2$. Here, we assume that the part of the dispersion shown has spherical symmetry. Consequently, the direction in which k is measured is not important.

where E_c is the energy of the electrons propagating in the conduction band, E_g is a constant gap energy, and m_c is the effective mass of the electrons in the conduction band. At lower energies, the parabolic forms of the dispersion relations of the two valence bands are given by [21]

$$E_{v1} = -\frac{\hbar^2}{2m_{v1}}\left(k_x^2 + k_y^2 + k_z^2\right) \tag{4.64b}$$

and

$$E_{v2} = -\frac{\hbar^2}{2m_{v2}}\left(k_x^2 + k_y^2 + k_z^2\right). \tag{4.64c}$$

Here, E_{v1} and E_{v2} are the energies of the first and second valence bands, respectively, and m_{v1} and m_{v2} are the effective masses of these two bands. Note that in this representation the zero of energy is at the top of the two valence bands.

A schematic representation of the three bands in Eq. (4.64) for a general system is presented in Figure 4.5. As a particular example of a parameterization of Eq. (4.64), the form for *GaAs* is often parameterized with values for the effective masses as $m_c = 0.067m_e$, $m_{v1} = 0.45m_e$, and $m_{v2} = 0.082m_e$, where m_e is the free electron mass, and the gap energy is $E_g = 1.424\ eV$. Another important system is that of $Ga_{1-x}Al_xAs$ which for $x < 0.41 - 0.45$ is parameterized by the effective masses $m_c = (0.063 + 0.083x)m_e$, $m_{v1} = (0.51 + 0.25x)m_e$, and $m_{v2} = (0.082 + 0.068x)m_e$ and an energy gap $E_g = 1.424 + 1.247x\ eV$. In addition, it should be noted here that not all effective mass representations of semiconductors are of the general form shown schematically in Figure 4.5. Important examples of system with different schematic representations for their effective mass dispersion relations are *Si*, *Ge*, and $Ga_{1-x}Al_xAs$ for $x > 0.41 - 0.45$.

4.2.3 Envelope Function Approximation

Next the envelope function approximation for the wavefunctions and its applications in treating boundary value problems is reviewed. In the envelope function approximation [2,8,21], it is found that the Schrodinger equation for the envelope wavefunction of the electrons contributing to the conduction properties of the material is easily obtained from the parabolic dispersion relations. This is effectively done by replacing the wavevectors in the dispersion relations by space derivatives so that $\hbar k_x \to -i\hbar\frac{\partial}{\partial x}$, $\hbar k_y \to -i\hbar\frac{\partial}{\partial y}$, etc. The differential equation resulting from this replacement is an effective Schrodinger equation for the envelope wavefunctions [1–3].

In this formulation, the Schrodinger equation for the envelope function of the conduction band in Eq. (4.64a) is given by [1–3,21]

$$E_c\psi_c = E_g\psi_c - \frac{\hbar^2}{2m_c}\nabla^2\psi_c \tag{4.65a}$$

where E_c is the energy of the electrons propagating in the conduction band, E_g is a constant gap energy, and m_c is the effective mass of the electrons in the conduction band. Similarly, at lower energies the two valence bands in Eqs. (4.64b) and (4.64c) are generalized into Schrodinger equations given by the forms [1–3,21]

$$E_{v1}\psi_{v1} = \frac{\hbar^2}{2m_{v1}}\nabla^2\psi_{v1} \tag{4.65b}$$

and

$$E_{v2}\psi_{v2} = \frac{\hbar^2}{2m_{v2}}\nabla^2\psi_{v2}. \tag{4.65c}$$

In bulk medium each of the Schrodinger equations in Eq. (4.65) are then solved by envelope wavefunctions with the planewave form

$$\psi_j(\vec{r}) \propto e^{i(k_x x + k_y y + k_z z)}, \tag{4.65d}$$

where $j = c,\ v1,\ v2$.

At surfaces or interfaces between two different semiconductors, sets of boundary conditions are required on the wavefunction between the different media. These are determined from the Schrodinger equations in Eq. (4.65), and the matchings are made between solutions in the two different media that are at the same energy. At an interface between two media described by Schrodinger equations with the forms in Eq. (4.65), the boundary conditions linking the wavefunctions at the surface between the two media are that [21]

$$\psi_1|_S = \psi_2|_s \tag{4.66a}$$

and

$$\frac{1}{m_{i1}}\hat{n}\cdot\nabla\psi_1\Big|_S = \frac{1}{m_{j2}}\hat{n}\cdot\nabla\psi_2\Big|_s \tag{4.66b}$$

where S is the interface of the two different semiconductors labeled as 1 and 2, and $\hat{n}$ is the unit normal to the interface. The effective masses m_{i1} and m_{i2} are for the $i, j = c, v1, v2$ solutions being connected across the surface between the two media 1 and 2, which are indicated by the end subscripts on the masses in Eq. (4.66b).

Before proceeding to examples of boundary value problems, we first offer some explanation of the boundary condition in Eq. (4.66b). In standard boundary value problems for the Schrodinger equation, the electron mass does not enter into the derivative conditions, because the electron mass does not change between media. This is not the case in the effective mass approximation, but some guidance as to the correct form of the derivative boundary conditions is found in treating the Hermitian properties of the kinetic operator at the interface between media.

Since the effective mass is a different constant in the two interfaced media, it must exhibit a position dependence in a small region centered about the interface. Consider a planar interface at $z = 0$ separating a medium with effective electron mass m_1 and a medium of effective electron mass m_2. In the small transition region containing the interface, the mass is represented by the position-dependent effective mass $m(z)$ interpolating between the effective masses m_1 and m_2. The derivative boundary conditions are now obtained by treating the effects of the position-dependent mass on the Hermeticity properties of the kinetic energy operator.

Consider the requirement that the kinetic energy operator in Eq. (4.65) is Hermitian. A statement of this condition in terms of the position-dependent mass going between the two media labeled 1 and 2 is that [21]

$$\int_1^2 dz\varphi^* \frac{d}{dz}\left(\frac{\hbar^2}{2m(z)}\frac{d\phi}{dz}\right) = \int_1^2 dz\frac{d}{dz}\left(\frac{\hbar^2}{2m(z)}\frac{d\varphi^*}{dz}\right)\phi, \tag{4.67}$$

and it is seen that for the integrand to exist it is necessary that the operator $\dfrac{1}{m(z)}\dfrac{d\phi}{dz}$ is continuous. This leads to the condition in Eq. (4.66b). In addition, note that in the limit $m_1 = m_2 = m(z)$ the above treatment reduces to the standard expression for kinetic energy operator and the standard boundary condition that $\hat{n}\cdot\nabla\psi_1\big|_S = \hat{n}\cdot\nabla\psi_2\big|_S$. The generalized boundary conditions in Eq. (4.66), known as ben Daniels–Duke boundary conditions, are now applied to the study of quantum wells and heterojunctions [21].

4.2.4 Quantum Wells and Heterostructures

A simple quantum well is formed by embedding an infinite slab of semiconductor into a surrounding medium composed of a different insulating or semiconducting material [9–12]. (See Figure 4.4b for a schematic.) Typically, the width of such a slab structure is of the order of nanometers or tens of nanometers. The importance of the structure is that it allows the conduction electron wavefunctions and dispersion relations of the slab to be modified from those of their bulk material. These modified dispersion properties and the associated modified electron wavefunctions have led to a variety of novel optical device applications [4–8,18–21].

In addition, a layered array of semiconducting slabs embedded in a background of insulator or of different semiconductor forms a heterostructure in which electrons may hop from layer to layer [9–12]. This is schematically illustrated in Figure 4.4d. Such arrays are also found to display new dispersive and wavefunction properties as well as exhibiting important properties for optical and transport applications. The design features offered by both wells and heterostructures have contributed to the design of lasers, sensors, thermoelectric devices, solar cells, and saturable absorbers.

In the following, a number of examples of quantum wells and heterostructures are studied. These are meant to afford an insight into the novel properties of engineered quantum wells and heterojunctions [21]. For the discussions, a focus is on systems composed of *GaAs*- and $Ga_{1-x}Al_xAs$-like materials.

4.2.4.1 Quantum Wells

The example of a quantum well treated in the following is that of a *GaAs* slab placed between two semi-infinite media [9–12]. In the first example, the semi-infinite media are represented by an infinite barrier that limits the electrons to be contained within the *GaAs* slab. In this highly idealized model, the *GaAs* electron motion is predominantly two dimensional in nature, exhibiting some aspects of a discrete nature that are highly dependent on the width of the slab. This is followed by a study of the *GaAs* slab embedded between two semi-infinite media composed of $Ga_{1-x}Al_xAs$.

In the case of a *GaAs* slab confined between two infinite barrier walls, the motion of the conduction electrons in the slab are again described by the effective Schrodinger equation in Eq. (4.61) but with boundary conditions that require the electron wavefunctions in the slab to vanish at the slab walls. (See the schematic in Figure 4.6a). If the walls are perpendicular to the *z*-axis and are located

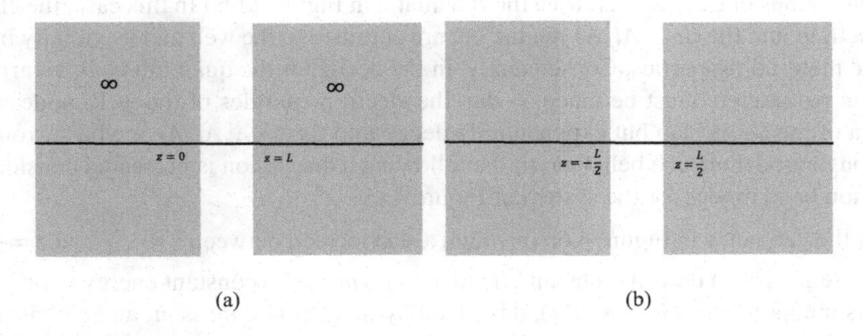

FIGURE 4.6 Schematic of a quantum well. In (a) the well is formed as an infinite slab between two regions of infinite potential energy. In (b) the well is formed as an infinite slab between two confining regions of finite potential energy.

at $z = 0$ and $z = L$, the hard wall boundary conditions require that the electron wavefunctions are of the form

$$\psi_j(\vec{r}) \propto e^{i(k_x x + k_y y)} \sin \frac{n\pi}{L} z, \tag{4.68}$$

where $j = c$, $v1$, $v2$ and $n = 1, 2, 3, \ldots$, and associated with these wavefunctions are the corresponding energy dispersion relations obtained from Eq. (4.65). From Eq. (4.65), it then follows that the higher energy conduction band electron energies are of the form

$$E_c = E_g + \frac{\hbar^2}{2m_c}\left(k_x^2 + k_y^2\right) + \frac{\hbar^2}{2m_c}\left(\frac{n\pi}{L}\right)^2, \tag{4.69a}$$

and for the two lower energy valence bands the electron energies are given by

$$E_{v1} = -\frac{\hbar^2}{2m_{v1}}\left(k_x^2 + k_y^2\right) - \frac{\hbar^2}{2m_{v1}}\left(\frac{n\pi}{L}\right)^2 \tag{4.69b}$$

and

$$E_{v2} = -\frac{\hbar^2}{2m_{v2}}\left(k_x^2 + k_y^2\right) - \frac{\hbar^2}{2m_{v2}}\left(\frac{n\pi}{L}\right)^2. \tag{4.69c}$$

These dispersion relations differ from the bulk electron dispersion relations in the treatment of the $z-$ component of wavevector. In the bulk system in Eqs. (4.64) and (4.65), the k_z is a continuum variable, but in the slab geometry (particularly for small L) the contribution of the $\frac{n\pi}{L}$ terms in Eq. (4.69) become large and highly discrete in nature.

The dispersion relations of the electrons in the slab system are seen to separate into a series of widely displaced parabolic dispersion relations in k_x, k_y. In this way, the three-dimensional isotropic dispersion relation in the bulk is sliced up into a discrete series of two-dimensional isotropic dispersion relations dependent on $k_x^2 + k_y^2$, which are shifted from one another in energy by large constants proportional to $\left(\frac{n\pi}{L}\right)^2$. This offers an engineering feature to be utilized in the design of systems with customized dispersion relations and electronic densities of states. The customization proceeds by changing L of the slab, providing a degree of freedom with which to tune the spectra of the electrons. In addition, the concentration of the wavefunction in space is helpful to engineer optical interactions in the system [9–12,21].

Next, consider replacing the infinite barrier media on either side of the *GaAs* slab of width L by semi-infinite regions of $Ga_{1-x}Al_xAs$. (See the schematic in Figure 4.6b.) In this case, the electrons in the slab penetrate into the $Ga_{1-x}Al_xAs$ media, but are confined to the well and its vicinity by the mismatch in the material properties. Consequently, in the design of the quantum well, an arrangement of the system parameters must be made so that the electron energies of the well modes propagate in the media of the *GaAs* slab but exponentially decay into the $Ga_{1-x}Al_xAs$ media surrounding the slab. As a rough model of such behavior, in the following a discussion is presented considering only the conduction band modes for the system of Figure 4.6b.

Consider the geometry in Figure 4.6b in which a slab located between $z = -\frac{L}{2}$ and $z = \frac{L}{2}$ is made of material a (e.g., *GaAs*) described by an effective mass m_a and a constant energy gap E_{ga}. Outside of the slab is media b (e.g., $Ga_{1-x}Al_xAs$), described by an effective mass m_b and a constant energy gap E_{gb}. In the following, the focus on this system is to present a rough theory for the conduction electron eigenmodes bound to the well. This is done within the context of the effective mass–envelope function theory, in which the dynamics is approximated by Eq. (4.65a), and the solutions are matched together with appropriate boundary conditions. The results for the eigenvalues and eigenmodes lead to a qualitative understanding of the system, but the solution is not accurate for technological applications.

For an accurate quantitative solution of the dynamics, an account must also be made of the mixing of the modes of the conduction and valence bands states, as they are coupled at the slab interfaces. It is generally found that the energies of the conduction and valence bands are close enough that the boundary conditions at the interface between different media couples the conduction and valence modes, and this coupling must be accounted for to arrive at a fully accurate solution. This is especially true for the case of modes with nonzero components of wavevector parallel to the slab interfaces. These band mixing effects are ignored here.

From Eq. (4.65a) for the dynamics of the conduction band electrons, it follows that the Schrodinger equation of the slab media a is of the form [9–12,21]

$$E_a\psi_a = E_{ga}\psi_a - \frac{\hbar^2}{2m_a}\nabla^2\psi_a \tag{4.70a}$$

where E_a is the energy of the electrons propagating in the conduction band of material a, and $-\frac{L}{2} \le z \le \frac{L}{2}$. Similarly, it follows that the Schrodinger equations for the conduction band electrons of the surrounding media b is of the form

$$E_b\psi_b = E_{gb}\psi_b - \frac{\hbar^2}{2m_b}\nabla^2\psi_b \tag{4.70b}$$

where E_b is the energy of an electron propagating in the conduction band of material b where $z < -\frac{L}{2}$ or $\frac{L}{2} < z$. In the treatment of the quantum well, the solutions of the well modes are composed from the constant energy E solutions in both the well and the confining media. As a result, in Eq. (4.70) it is required that $E_a = E_b = E$.

The solutions of Eq. (4.70) in the three spatial regions of interest are of the form

$$\psi_a(x,y,z) = Ae^{i(k_x x + k_y y)}\begin{cases}\cos kz \\ \sin kz\end{cases} \text{for } -\frac{L}{2} \le z \le \frac{L}{2}, \tag{4.71a}$$

where $k^2 = \frac{2m_a}{\hbar^2}(E - E_{ga}) - (k_x^2 + k_y^2)$,

$$\psi_b(x,y,z) = \pm Be^{i(k_x x + k_y y)}e^{\kappa z} \text{ for } z < -\frac{L}{2}, \tag{4.71b}$$

and

$$\psi_b(x,y,z) = Be^{i(k_x x + k_y y)}e^{-\kappa z} \text{ for } \frac{L}{2} < z \tag{4.71c}$$

where $\kappa^2 = \frac{2m_b}{\hbar^2}(E_{gb} - E) + (k_x^2 + k_y^2)$ for $\kappa > 0$. Note that for a bound state solution in the well it is required that $k^2, \kappa^2 > 0$. In addition, due to the symmetry of the potential of the quantum well, the solutions separate into even- and odd-parity eigenmodes.

Applying the requirements that $k^2, \kappa^2 > 0$ and the boundary conditions on ψ and $\frac{1}{m}\frac{\partial\psi}{\partial z}$ obtained from Eq. (4.66) to the solution forms in Eq. (4.71) generates the conditions on E needed for a solution to exist. For the $\cos kz$ modes in Eq. (4.71), solutions exist at energies determined from the requirement that [21]

$$\frac{k}{m_a}\tan k\frac{L}{2} = \frac{\kappa}{m_b}. \tag{4.72a}$$

Similarly, the conditions for odd-parity solutions of the $\sin kz$ form to exist is that their E are determined from

$$\frac{k}{m_a}\cot k\frac{L}{2} = -\frac{\kappa}{m_b}. \tag{4.72b}$$

These are common types of transcendental equations arising in well boundary value problems and lead to a finite set of discrete E solutions for each set of the continuum variables k_x, k_y.

4.2.4.2 Quantum Heterostructures

Quantum heterostructures are formed as periodic arrays of quantum wells made from different semiconducting materials. The periodicity is one-dimensional in a direction perpendicular to the surfaces of the parallel interfaced slabs comprising the array. In the array the layering is designed so that the electron states are generally confined within the wells but can tunnel between the different wells through the regions of barrier material separating the individual wells. General heterostructures of this type are of importance for their electronic and optical properties.

Due to the tunneling processes the conduction electrons propagate throughout the heterostructure with a greatly modified dispersion relation and with wavefunctions concentrated within the regions of the wells. The dispersion relation of the electrons is described by a series of bands similar to those found in semiconductors and metals, but additional bands and band gaps occur in the system due to the periodicity of the array of slabs in the layering in the case of heterostructures. This new band structure is highly dependent on the material mismatches, and geometry of the wells and array of wells in the systems. It allows for the dispersion relation to be tuned by engineering the heterostructure design [9–12,21].

In addition, the wavefunctions of the electronic modes tend to be localized in the well regions, and this changes the nature of the coupling of these modes upon the introduction of perturbations to the system. Such modifications of the spatial array of the wavefunctions is another feature that is subject to engineering considerations for the enhancement of applications, including those involving laser, solar cell, and transistor designs.

There are generally two different types of heterostructures, depending on the offset of the valence and conduction bands of the materials comprising the heterostructures. These are illustrated in Figure 4.7, where it is shown that the type of the heterostructure depends on the alignment of the active modes in the conduction band and in the valence band. In a type I heterostructure, the conduction band and valence band states of the wells both occur in the slab layers, and the barrier media

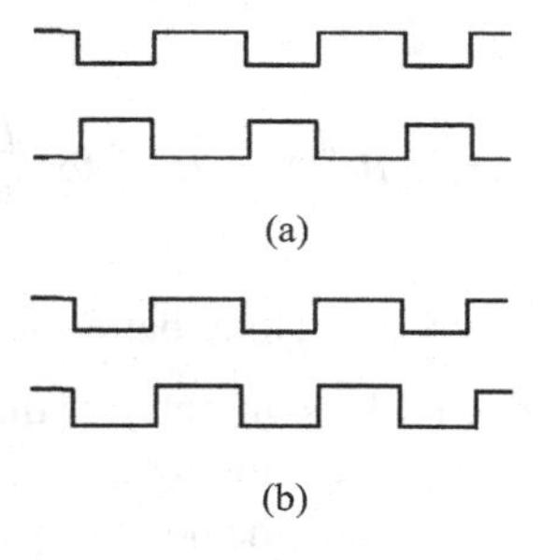

FIGURE 4.7 Schematic of the lower energy of the upper band and the upper energy of the lower band in a heterostructure. In (a) is a type I model, and in (b) is a type II model.

surrounding the slabs primarily serve as a tunneling medium for the electrons confined in the slab wells. In a type II heterostructure, the conduction band and valence band states of the well occur in different layers of the heterostructure. In these structures, the one layer serves as a well for the conduction band states, while the other layer acts as a barrier medium for the conduction band states. Conversely, for the valence band states the layer that acts as a barrier for the conduction electron states acts as a well for the valence band states, and the layer that is a well for conduction electron states acts as a barrier for the valence band states. In the following example, discussions are limited to treat Type I heterostructures formed of *GaAs* and $Ga_{1-x}Al_xAs$ like materials, and the focus is on the modes in the conduction band for simplicity.

Consider a sample of $Ga_{1-x}Al_xAs$ into which an array of N identical, equally spaced quantum wells of *GaAs* are introduced. This structure that is illustrated schematically in Figure 4.8a provides a basic example of a Type I heterostructure in which the *GaAs* contains the propagating conduction and valence electron states, and the $Ga_{1-x}Al_xAs$ acts as the tunneling barriers. In the treatment to follow, the slabs of the array are created by locating *GaAs* in the regions $na \le z \le \left(n + \frac{1}{2}\right)a$ where n is an integer ranging between 1 and N. Outside of these slabs in the regions, $\left(n - \frac{1}{2}\right)a \le z \le na$ the embedding medium of the system is composed of $Ga_{1-x}Al_xAs$. On either side of the layering are semi-infinite regions of $Ga_{1-x}Al_xAs$.

It is seen from Eq. (4.65a) that the conduction band in the region of *GaAs* is described by [21]

$$E_a\psi_a = E_{ga}\psi_a - \frac{\hbar^2}{2m_a}\nabla^2\psi_a \tag{4.73a}$$

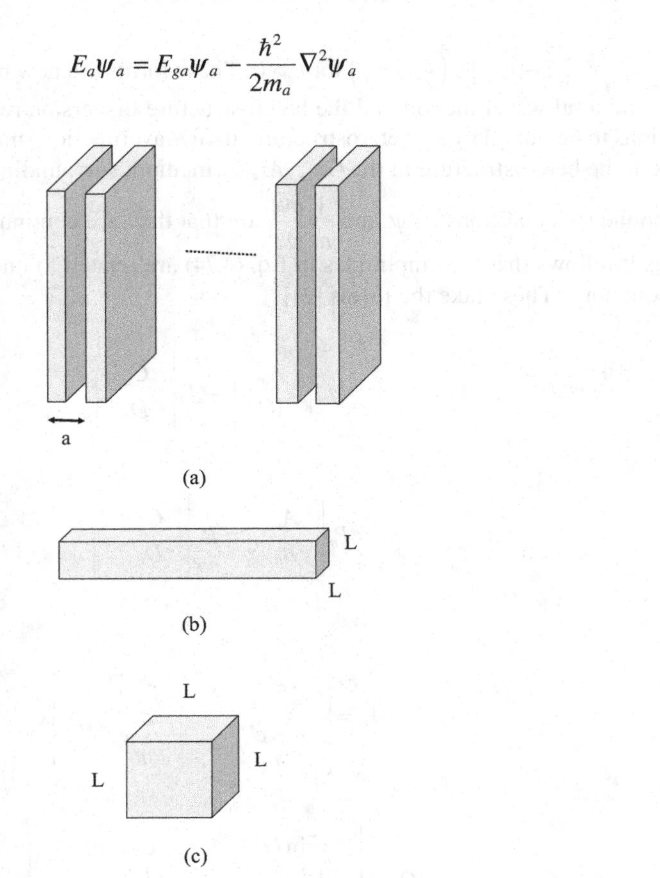

FIGURE 4.8 Schematic geometries of: (a) a heterostructure composed as an infinite layering of infinite slabs that are periodically stacked along the z-axis with the smallest periodicity a, (b) an infinitely long wire of square cross section with square sides of length L, and (c) a quantum dot formed as a cube of material of side L.

and, in the region of $Ga_{1-x}Al_xAs$, it is obtained from

$$E_b\psi_b = E_{gb}\psi_b - \frac{\hbar^2}{2m_b}\nabla^2\psi_b. \tag{4.73b}$$

A simplified study is obtained by treating the propagating modes in of $GaAs$ and matching them through the boundary conditions to the evanescent solution in $Ga_{1-x}Al_xAs$.

In this regard, the evanescent modes in the regions $\left(n-\frac{1}{2}\right)a \le z \le na$ have the general form

$$\psi(x,y,z) = \left(A_n e^{\kappa z} + B_n e^{-\kappa z}\right)e^{i\left(k_x x + k_y y\right)}, \tag{4.74a}$$

where the decay in the barrier region is given by $\kappa^2 = \frac{2m_b}{\hbar^2}\left(E_{gb} - E\right) + \left(k_x^2 + k_y^2\right)$ for $\kappa > 0$. Within the well regions $na \le z \le \left(n+\frac{1}{2}\right)a$, the propagating solutions have the general form

$$\psi(x,y,z) = (C_n \sin kz + D_n \cos kz)e^{i\left(k_x x + k_y y\right)}, \tag{4.74b}$$

where $k^2 = \frac{2m_a}{\hbar^2}\left(E - E_{ga}\right) - \left(k_x^2 + k_y^2\right)$ for $k > 0$. These forms are now matched at the well interfaces to obtain the total wavefunction and the heterostructure dispersion relations. In this matching, for the solutions to be bound to the heterostructure, their wavefunctions must go to zero at infinite separation from the heterostructure in the $Ga_{1-x}Al_xAs$ medium surrounding it.

The boundary conditions on ψ and $\frac{1}{m}\frac{\partial\psi}{\partial z}$ are that they are continuous over the heterostructure. From this it follows that the amplitudes in Eq. (4.74) are related to one another through a series of matrix equations. These take the forms [21]

$$M_n\begin{vmatrix} A_n \\ B_n \end{vmatrix} = Q_n\begin{vmatrix} C_n \\ D_n \end{vmatrix} \tag{4.75a}$$

and

$$P_n\begin{vmatrix} A_n \\ B_n \end{vmatrix} = R_n\begin{vmatrix} C_{n-1} \\ D_{n-1} \end{vmatrix} \tag{4.75b}$$

where

$$M_n = \begin{vmatrix} e^{\kappa na} & e^{-\kappa na} \\ \frac{\kappa}{m_b}e^{\kappa na} & -\frac{\kappa}{m_b}e^{-\kappa na} \end{vmatrix} \tag{4.76a}$$

$$Q_n = \begin{vmatrix} \sin kna & \cos kna \\ \frac{k}{m_a}\cos kna & -\frac{k}{m_a}\sin kna \end{vmatrix} \tag{4.76b}$$

$$P_n = \begin{vmatrix} e^{\kappa\left(n-\frac{1}{2}\right)a} & e^{-\kappa\left(n-\frac{1}{2}\right)a} \\ \frac{\kappa}{m_b} e^{\kappa\left(n-\frac{1}{2}\right)a} & -\frac{\kappa}{m_b} e^{-\kappa\left(n-\frac{1}{2}\right)a} \end{vmatrix} \tag{4.76c}$$

$$R_n = \begin{vmatrix} \sin k\left(n-\frac{1}{2}\right)a & \cos k\left(n-\frac{1}{2}\right)a \\ \frac{k}{m_a}\cos k\left(n-\frac{1}{2}\right)a & -\frac{k}{m_a}\sin k\left(n-\frac{1}{2}\right)a \end{vmatrix}. \tag{4.76d}$$

Within this notation it follows that

$$\begin{vmatrix} A_n \\ B_n \end{vmatrix} = M_n^{-1} Q_n R_{n+1}^{-1} P_{n+1} \begin{vmatrix} A_{n+1} \\ B_{n+1} \end{vmatrix}, \tag{4.77}$$

and, consequently, for a heterostructure consisting of N quantum wells

$$\begin{vmatrix} A_1 \\ B_1 \end{vmatrix} = \prod_{n=1}^{N} \left(M_n^{-1} Q_n R_{n+1}^{-1} P_{n+1}\right) \begin{vmatrix} A_{n+1} \\ B_{n+1} \end{vmatrix}. \tag{4.78}$$

In order that the solutions obtained from Eq. (4.78) represent states bound to the array of heterostructure wells, solutions on either side of the well must become zero at an infinite separation along the z-axis from the wells. For this to be the case the wavefunction coefficients $B_1 = 0$ and $A_{N+1} = 0$.

A numerical solution of Eq. (4.78) for the modes bound to the heterostructure is obtained under these conditions by choosing

$$\begin{vmatrix} A_{n+1} \\ B_{n+1} \end{vmatrix} = \begin{vmatrix} 0 \\ B \end{vmatrix}, \tag{4.79a}$$

where B is a nonzero fixed constant. A guess is then made for the energy E, and the product on the right in Eq. (4.78) is computed. If the result on the left of the equation is of the form

$$\begin{vmatrix} A_1 \\ B_1 \end{vmatrix} = \begin{vmatrix} A \\ 0 \end{vmatrix}, \tag{4.79b}$$

the energy guess was corrected, and the result for E and the amplitudes in Eq. (4.79b) is an answer for Eq. (7.79a). Note, in this regard that the equation in Eq. (4.78) is linear so that both Eq. (4.79), and a multiple of it by a nonzero constant are answers, i.e., both represent the same answer to the problem. In this way a computer search can generate the dispersion relation of the heterostructure.

For a finite array (i.e., for $N \ll \infty$) the solutions of Eq. (4.78) for the $k_x = k_y = 0$ tunneling results in a finite discrete set of propagating modes $\{E_m\}$ in the conduction and valence bands. Relaxing this restriction so that nonzero k_x or k_y are allowed, a continuum of energy modes is developed based on each of the $\{E_m\}$.

4.2.5 Quantum Wires and Dots

Next is treated an example of a simple quantum wire [21]. The wire considered, illustrated in Figure 4.8b, involves a geometry of square cross section and infinite length. For simplicity, it is assumed that the region outside the wire is a barrier medium that restricts the modes to remain inside the wire. The barrier medium then acts as a hard wall. The axis of the wire is along the z-direction, and in the plane perpendicular to the axis of the wire the walls of the wire are located at $x = 0$ and $x = L$, and at $y = 0$ and $y = L$.

Applying boundary conditions requiring the wavefunctions to vanish at the walls of the wire determines the conduction and valence band dispersion relations and wavefunctions. In this way from Eq. (4.65), the dispersion relation of the conduction electrons is of the form [21]

$$E_c = E_g + \frac{\hbar^2}{2m_c}k_z^2 + \frac{\hbar^2}{2m_c}\left[\left(\frac{n\pi}{L}\right)^2 + \left(\frac{m\pi}{L}\right)^2\right], \tag{4.80a}$$

and the two lower energy valence bands electron energies are given by

$$E_{v1} = -\frac{\hbar^2}{2m_{v1}}k_z^2 - \frac{\hbar^2}{2m_{v1}}\left[\left(\frac{n\pi}{L}\right)^2 + \left(\frac{m\pi}{L}\right)^2\right] \tag{4.80b}$$

and

$$E_{v2} = -\frac{\hbar^2}{2m_{v2}}k_z^2 - \frac{\hbar^2}{2m_{v2}}\left[\left(\frac{n\pi}{L}\right)^2 + \left(\frac{m\pi}{L}\right)^2\right]. \tag{4.80c}$$

Here k_z is the continuum wavevector component parallel to the axis of the wire, and $n, m = 1, 2, 3\ldots$ are the indices of the discrete transverse components of the modes. The wavefunctions corresponding to the dispersion relations are given by

$$\psi_j(x,y,z) \propto e^{ik_z z}\sin\left(\frac{n\pi}{L}x\right)\sin\left(\frac{m\pi}{L}y\right) \tag{4.81}$$

for the band indices $j = c, v1, v2$.

The motion of the electrons in the wire is reminiscent of the propagation of waves in a waveguide. In the absence of impurities, it is a simple ballistic transport involving sets of discrete modes characterized by the quantum numbers n, m. The various modes of n, m represent channels for the transport of electrons, and the current flow in the wire increases as these various channels become occupied [21].

Unlike the well and wire features discussed earlier, quantum dots do not support currents but give a total confinement of their electron modes. Similar to their atomic counterparts, the energy levels in quantum dots are discrete and support an optical spectroscopy involving electronic transitions between the discrete electron levels. An advantageous feature of the dots is that the discrete electron levels can be engineered by varying the geometry, size, and composition of the dots. This has found an application in the design of displays and televisions.

A simple model of a quantum dot is illustrated in Figure 4.8c. It involves a dot designed as a cube of semiconductor material. Outside of the dot is a hard wall barrier medium that confines the electrons to the volume of the dot. The walls of the dot are at $x = 0$ and $x = L$, at $y = 0$ and $y = L$, and at $z = 0$ and $z = L$.

From discussions paralleling those made earlier for the quantum wire, it follows that the wavefunctions of the dot are of the form [21]

$$\psi_j(x,y,z) \propto \sin\left(\frac{l\pi}{L}y\right)\sin\left(\frac{n\pi}{L}x\right)\sin\left(\frac{m\pi}{L}z\right), \tag{4.82}$$

where $l, n, m = 1, 2, 3\ldots$ for the band indices $j = c, v1, v2$. The corresponding energy levels of the conduction band electrons are

$$E_c = E_g + \frac{\hbar^2}{2m_c}\left[\left(\frac{l\pi}{L}\right)^2 + \left(\frac{n\pi}{L}\right)^2 + \left(\frac{m\pi}{L}\right)^2\right], \tag{4.83a}$$

and for the two lower energy valence bands the electron energies are given by

$$E_{v1} = -\frac{\hbar^2}{2m_{v1}}\left[\left(\frac{l\pi}{L}\right)^2 + \left(\frac{n\pi}{L}\right)^2 + \left(\frac{m\pi}{L}\right)^2\right] \tag{4.83b}$$

and

$$E_{v2} = -\frac{\hbar^2}{2m_{v2}}\left[\left(\frac{l\pi}{L}\right)^2 + \left(\frac{n\pi}{L}\right)^2 + \left(\frac{m\pi}{L}\right)^2\right]. \tag{4.83c}$$

The energy spectrum is found to be a series of discrete energy levels. These depend on the materials and geometry of the dot.

4.3 EXCITONS

In addition to the single particle excitations considered in the earlier discussions, another set of excitations with technological applications exist in engineered materials [2,9–12]. These are the so-called excitons that are bound states consisting of an electron excited into the conduction band and a positive charged valance band mode known as a hole. In bulk materials typical exciton-bound state energies are of order of 5 meV, and this can be increased in quantum well structures to be of order 10 meV. In both bulk and well structures, the excitons appear as energy levels just below the conduction band single-electron excitations.

Excitons can have interesting physical effects on the optical properties of materials, including some resulting in technological applications. These include design applications in solar cell and photovoltaic devices, photo emitters and detectors, antennas, LED's, optical modulators, and in other electrooptical functions. In the following, the nature of excitons is discussed and illustrated in terms of a crude theory for the exciton-bound state. Some of the modifications of excitations in quantum wells and heterostructures are indicated, and the interest of these modifications to device applications are described.

Excitons are essential bound states formed from an electron that has been promoted into the conduction band and a positive charge that has been left in a valence band when its neutralizing electron is raised into the conduction band [1–3]. Specifically, in a semiconductor at zero temperature, before the promotion of an electron to the conduction band the positive ions and valence band electrons form a charge-neutral system. When an electron in one of the valence bands is promoted into the conduction band, the part of the semiconductor composed from the positive ion lattice and remaining valence electrons acquires a net positive charge. The net positive charge of the ion lattice and remaining valence electrons is associated with the absence of the neutralizing electron that is now in the conduction band. As a consequence, the dynamics of the net positive charge in the valence band is determined by that of the remaining electrons in the valence band. This dynamics mimics that of a single positive charged electron, and the excitation essentially behaves as a positive charged particle known as a hole. In this scheme the conduction band electron and the valence band hole interact through the Coulomb force to form an electron–hole-bound state known as an exciton.

From these discussions, it is seen that the theoretical treatment of a hole is similar to that of a positron in particle physics. In particle physics, the positron is the antiparticle of the electron, while a hole is a type of antiparticle to the band electron in semiconductor physics. Continuing with the analogy, an exciton is the condensed matter physics analog of the high-energy configuration known as positronium, which is an electron–positron bound state.

To develop a simple model of an exciton in a bulk material, consider the system described in Eq. (4.60) for the dispersion relation of the electrons in the conduction and valence bands of a bulk semiconductor. For the following discussions, only one valence band is needed, and $v1$ is arbitrarily chosen for the presentation.

When an electron in a state with wavevector (k_x, k_y, k_z) is removed from the valence band and placed into a conduction band mode having a wavevector (k_x', k_y', k_z'), the change in energy is given by

$$\Delta E = E_c - E_{v1} = E_g + \frac{\hbar^2}{2m_c}\left(k_x'^2 + k_y'^2 + k_z'^2\right) + \frac{\hbar^2}{2m_{v1}}\left(k_x^2 + k_y^2 + k_z^2\right). \tag{4.84}$$

As in the discussions of Eq. (4.65) in the effective mass–envelope function approximations, the form in Eq. (4.84) can be used to obtain an approximating Schrodinger equation for the exciton energies. Rewriting Eq. (4.84) by replacing the wavevectors by differential operators, it follows that in terms of the operator form in Eq. (4.65) this gives [21]

$$\Delta E \psi(\vec{r}', \vec{r}'') = [E_c - E_{v1}]\psi(\vec{r}', \vec{r}'') = \left[E_g - \frac{\hbar^2}{2m_c}\nabla'^2 - \frac{\hbar^2}{2m_{v1}}\nabla''^2\right]\psi(\vec{r}', \vec{r}'') \tag{4.85}$$

for the now two-particle wavefunction where the particles are located at $\vec{r}'$ and $\vec{r}''$ and the wavefunction is $\psi(\vec{r}', \vec{r}'')$.

Going further, the two-particle Coulomb interaction can be included in Eq. (4.85), resulting in the Schrodinger equation

$$\Delta E \psi(\vec{r}', \vec{r}'') = [E_c - E_{v1}]\psi(\vec{r}', \vec{r}'') = \left[E_g - \frac{\hbar^2}{2m_c}\nabla'^2 - \frac{\hbar^2}{2m_{v1}}\nabla''^2 - \frac{1}{\varepsilon}\frac{e^2}{r}\right]\psi(\vec{r}', \vec{r}'') \tag{4.86}$$

where $r = |\vec{r}' - \vec{r}''|$, and ε is the dielectric constant of the bulk medium. This is of the form of a Schrodinger equation for a hydrogen atom and includes the center of mass motion of the atom. By a standard treatment, the expression for the two-particle Schrodinger equation is separated into relative and center of mass coordinates giving the following form

$$\Delta E \psi(\vec{r}_{CM}, \vec{r}) = \left[E_g - \frac{\hbar^2}{2M}\nabla_{CM}^2 - \frac{\hbar^2}{2\mu}\nabla^2 - \frac{1}{\varepsilon}\frac{e^2}{r}\right]\psi(\vec{r}_{CM}, \vec{r}). \tag{4.87}$$

Here, ∇_{CM}^2 is the Laplacian in center of mass coordinates $\vec{r}_{CM}$, which are defined by $M\vec{r}_{CM} = m_c\vec{r}' + m_{v1}\vec{r}''$, ∇^2 is the Laplacian in the relative coordinates $\vec{r} = \vec{r}' - \vec{r}''$, $M = m_c + m_{v1}$ is the total mass, and $\mu = \dfrac{m_c m_{v1}}{m_c + m_{v1}}$ is the reduced mass.

The resulting problem in Eq. (4.87) is essentially that of an effective hydrogen atom translating in space. Its energy eigenvalues for the s excitonic states are given by the form

$$\Delta E = E_g + \frac{\hbar^2}{2M}\mathrm{k}_{CM}^2 - \frac{\mu e^4}{2\varepsilon^2\hbar^2}\frac{1}{n^2}, \tag{4.88}$$

where $n = 1, 2, 3, \ldots.$, and $\vec{k}_{CM}$ is the wavevector of the center of mass of the exciton. In Eq. (4.88) on the right side of the equation, the first term is the energy gap between the top of the valence bands and the bottom of the conduction band, the second term is the center of mass kinetic energy, and the third term are familiar as the binding energies of the hydrogen-like electron–hole-bound states. The exciton modes are found as a series of bands located just below the electronic conduction energy band.

The results in Eqs. (4.84)–(4.88) can also be modified to study the bound state properties in two-dimensional systems. These are useful in understanding aspects of the properties of surfaces, quantum wells and heterostructures where the exciton becomes confined along one direction in space. To make the modification, the Laplacians in Eq. (4.87) are changed from spherical coordinates to polar coordinates, and the radial variable is now not that of a sphere but that of the polar radian vector in the plane of motion. The result from such an analysis for the energies of the s states of the two-dimensional exciton give

$$\Delta E = E_g + \frac{\hbar^2}{2M}\mathrm{k}_{CM}^2 - \frac{2\mu e^4}{\varepsilon^2\hbar^2}\frac{1}{n^2}. \tag{4.89}$$

It is seen that the third terms on the right of Eqs. (4.88) and (4.89), representing the binding energy contributions from the Coulomb interaction between the two particles, is in magnitude four times greater in the two-dimensional system over that in the three-dimensional system.

The models presented in Eqs. (4.88) and (4.89) are for Wannier excitons [1,2,21]. These excitons typically have binding energies of order 0.01 eV so that they are mostly observed in low temperature systems. In Wannier excitons, the separation between the electron and hole is large compared to inter-atomic separations. This is why the dielectric constant of the medium enters into the calculation. The Wannier excitons are most often found in semiconductor crystals [1,2]. Another type of exciton is the Frenkel excitons [1,2], which have binding energies of $0.1-1.0$ eV. These types of excitons are formed of electron and holes that are spatially close together so that the background medium sometimes cannot be treated by a dielectric constant. They are found in alkali halides and molecular crystals.

The earlier results for three- and two-dimensional excitons were both discussed for bulk materials in which the excitons are not spatially restricted by the geometry of the medium. The introduction of the geometry of quantum wells, heterostructures, dots, and wires, however, allows for geometric constraints to be placed on the wavefunction of the excitons along with consequent modifications to their binding energies. For example, in a quantum well of width much greater than that of the bulk exciton wavefunction, the well has little effect on the three-dimensional exciton-binding energies. However, in a quantum well of width much less than that of the bulk exciton wavefunction, the well has a greater effect on the exciton mode and its binding energies. In particular, for a well embedded in a hard wall medium, which does not permit the electrons and holes to leave the well, wide wells have binding energies given by the three-dimensional bulk result, $-\frac{\mu e^4}{2\varepsilon^2\hbar^2}\frac{1}{n^2}$, whereas narrow wells have binding energies that approach the two-dimensional bulk result, $-\frac{2\mu e^4}{\varepsilon^2\hbar^2}\frac{1}{n^2}$, as the width of the well collapses.

In this regard, some further variation of exciton properties is found in the heterostructure geometry. Type I and Type II heterostructures make for different spatial configurations of the excitons created within them. In Type I systems the electron and hole bound in the exciton are formed in conduction and valence bands within the same slab. However, in Type II systems the electron and hole bound in the exciton are formed in conduction and valence bands within neighboring slabs. These differences allow for further possibilities in the design of systems with tailored electrooptical properties and responses.

4.4 PHOTONIC AND PHONONIC CRYSTALS

In this section, the physical properties of periodic systems described by classical mechanical wave equations are treated [4–7,18–20]. These properties commonly are used in the design and study of photonic and phononic crystals that have device applications in developing filtering, confinement, and circuit features for the processing of classical radiation fields. For simplicity in the presentation of the basic ideas, the focus of the discussions emphasizes systems described by two-dimensional Helmholtz wave equations. Nevertheless, three-dimensional designs and applications of periodic materials also have found important functions in filtering and antenna problems involving the emission and reception of radiation. All applications of photonic and phononic crystals arise from the band structure of the frequency versus wavevector dispersion relation developed from the periodicity of the systems sustaining the radiation fields. This band structure of the periodic materials provides them with the ability to block or allow the propagation of select frequency bands of radiation into the bulk of the crystal.

In the following, first some discussions are made of the general nature of the solutions of periodic systems for classical continuum radiation fields. This is followed by treatments of examples from the classical electrodynamics of photonic crystals and the classical acoustics of phononic crystals. Finally, general remarks are given regarding applications involving photonic and phononic crystals, summarizing their properties found of interest to technology.

4.4.1 Properties of Waves in Periodic Media

In an isotropic, homogeneous medium, the solutions for the propagation of waves through space are generally proportional to planewave forms represented by [4–7]

$$e^{i(\vec{k}\cdot\vec{r}-\omega t)}, \tag{4.90a}$$

where $\vec{k}$ is the wavevector, and ω the frequency of the excitation. The planewave in Eq. (4.90a) is essentially a statement of the conservation of energy and momentum for motion within a linear, nonscattering, nondissipative medium. During this motion, it is found that the wave propagates everywhere with the same energy and momentum density. Consequently, the intensity of the wave in the medium, which is proportional to the modulus square of its planewave form, is related through a constant by the condition

$$\left|e^{i(\vec{k}\cdot\vec{r}-\omega t)}\right|^2 = 1. \tag{4.90b}$$

Both forms in Eq. (4.90) then represent statements of the complete translational invariance of the medium in space, i.e., throughout space and time the medium is the same everywhere.

When spatial periodicity is introduced into the medium, the medium supporting the wave is no longer everywhere the same, and the form of the wavefunction must change. In particular, it is found that the waves in the medium become proportional to the form [4–7]

$$e^{i(\vec{k}\cdot\vec{r}-\omega t)}u_{\vec{k},j}(\vec{r}) \tag{4.91}$$

where $u_{\vec{k},j}(\vec{r})$ is a periodic function with the periodicity of the medium through which the wave propagates. The function $u_{\vec{k},j}(\vec{r})$ depends on the wavevector $\vec{k}$ so that the nature of the periodic function is different for different wavevectors, and j labels the different components of the wave in the case of a vector field, e.g., $j = x, y, z$. The reason for the introduction of the periodic form $u_{\vec{k},j}(\vec{r})$ in Eq. (4.91) is that the intensity of the radiation in space should exhibit the translational symmetry in space of the periodic medium. This is a basic result of the symmetry properties of the material.

As in the case of Eq. (4.90b), in the expression for the spatial intensity of the wavefunction in Eq. (4.91) the planewave form $e^{i(\vec{k}\cdot\vec{r}-\omega t)}$ drops out. It is removed during the process of computing the modulus of the wavefunction. The remaining terms in the intensity display only a periodic spatial variation of the wave intensity with the periodicity of the material.

On the other hand, the planewave form $e^{i(\vec{k}\cdot\vec{r}-\omega t)}$ in Eq. (4.91) remains a significant part of the wavefunction in two ways. First, the time-dependent part of the planewave indicates that the energy is conserved in a nondissipative system, i.e., the intensity of the wave is independent of time. Secondly, the spatial wavevector part of the planewave is a statement that the crystal momentum (proportional to $\vec{k}$) is a constant of the motion. The crystal momentum is not the same as the momentum of the system, but it is a distinguishing label of the modal solutions and has important conservation properties. Like the momentum, the crystal momentum characterizes the translational symmetry of the material, which is now restricted to a countably infinite set of discrete spatial translations. In this regard, the crystal momentum is important in computing the scattering arising when impurities are introduced into the periodic media. In the course of scattering interactions, it serves a similar function to that of the momentum (proportional to $\vec{k}$) in Eq. (4.90a).

The solutions in a periodic medium exhibit a variety of important properties related to the symmetry of the medium. A very important set of properties come from the translational symmetries that translate the medium into itself. These translational invariants and their effects on the modes of the medium are now discussed.

Consider a vector $\vec{T}_{nm}$ that translates the periodic structure of the material into itself and the actions of the translation on the wavefunction in Eq. (4.91). In a medium with two-dimensional periodicity, $\vec{T}_{nm}$ has the general form $\vec{T}_{nm} = n\vec{a}_1 + m\vec{a}_2$, where $\vec{a}_1$ and $\vec{a}_2$ are two linearly independent vectors in the two-dimensional symmetry plane, and n and m are integers. Shifting each point of the medium by $\vec{T}_{nm}$ maps the medium onto itself without changing its appearance.

For the translation vector it then follows that the periodic part of the wavefunction in Eq. (4.91) satisfies the periodicity condition [4–7]

$$u_{\vec{k},j}\left(\vec{r}+\vec{T}_{nm}\right) = u_{\vec{k},j}\left(\vec{r}\right), \tag{4.92a}$$

because $u_{\vec{k},j}(\vec{r})$ has the same translational symmetry as that of the periodic medium. Looking at the planewave $e^{i\vec{k}\cdot\vec{r}}$ in Eq. (4.91), however, it is seen that, in general, the planewave form changes under a space translation of the periodic material so that

$$e^{i\vec{k}\cdot\left(\vec{r}+\vec{T}_{nm}\right)} = e^{i\vec{k}\cdot\vec{T}_{nm}}e^{i\vec{k}\cdot\vec{r}}. \tag{4.92b}$$

The planewave form only is invariant under the translation if

$$e^{i\vec{k}\cdot\vec{T}_{nm}} = 1, \tag{4.92c}$$

and this condition is met only for a particular set of wavevectors. Specifically, Eq. (4.92c) is satisfied for wavevectors $\vec{k}$ with the property that $\vec{k}\cdot\vec{T}_{nm} = 2\pi M$, where M is an integer.

The set of wavevectors that leaves the planewave form invariant under translation has the general form $\vec{k} = \vec{G}_{pq}$ where

$$\vec{G}_{pq} = p\vec{b}_1 + q\vec{b}_2, \tag{4.93a}$$

p and q are integers, and $\vec{b}_1$ and $\vec{b}_2$ are solutions of

$$\vec{a}_i\cdot\vec{b}_j = 2\pi\delta_{i,j}. \tag{4.93b}$$

Under these conditions

$$\vec{G}_{pq} \cdot \vec{T}_{nm} = 2\pi(np + mq) = 2\pi M, \tag{4.94}$$

where M is an integer. It then follows that

$$e^{i\vec{G}_{pq} \cdot \vec{T}_{nm}} = e^{i2\pi M} = 1, \tag{4.95a}$$

so that the condition in Eq. (4.92c) is realized. Consequently, for these wavevectors the planewave form has a translation symmetry in k-space given by

$$e^{i\vec{G}_{pq} \cdot (\vec{r} + \vec{T}_{nm})} = e^{i\vec{G}_{pq} \cdot \vec{r}}. \tag{4.95b}$$

Both of the two symmetries in Eqs. (4.92a) and (4.95b) are now useful in determining the periodicity properties of the wavefunctions and their dispersion relations.

The symmetries in Eqs. (4.92a) and (4.95b) have important consequences for the form of the wavefunction in Eq. (4.91). In particular, for a modal solution in the medium with a wavevector of the form $\vec{k} + \vec{G}_{pq}$, it follows from Eq. (4.91) that

$$e^{i\left[(\vec{k} + \vec{G}_{pq}) \cdot \vec{r} - \omega t\right]} u_{\vec{k} + \vec{G}_{pq}, j}(\vec{r}) = e^{i(\vec{k} \cdot \vec{r} - \omega t)} e^{i\vec{G}_{pq} \cdot \vec{r}} u_{\vec{k} + \vec{G}_{pq}, j}(\vec{r}) = e^{i(\vec{k} \cdot \vec{r} - \omega t)} v_{\vec{k}, j}(\vec{r}), \tag{4.96}$$

where both $u_{\vec{k},j}(\vec{r})$ and $v_{\vec{k},j}(\vec{r})$ represent periodic space solutions of the system. An important point of the transformation is that both $e^{i\left[(\vec{k} + \vec{G}_{pq}) \cdot \vec{r} - \omega t\right]} u_{\vec{k} + \vec{G}_{pq}, j}(\vec{r})$ and $e^{i(\vec{k} \cdot \vec{r} - \omega t)} v_{\vec{k},j}(\vec{r})$ correspond to the same modal solutions of a wave in the medium. Specifically, the equivalence of these two wavefunctions implies that the sets of modal solutions generated for $\vec{k}$ and $\vec{k} + \vec{G}_{pq}$ are identical sets corresponding to the same eigenfrequencies.

The equivalence of the $\vec{k}$ and $\vec{k} + \vec{G}_{pq}$ solutions implies that the dispersion relation at these two different wavevectors are the same so that the dispersion relation of the eigenmodes is invariant under translations in k-space by the set of $\vec{G}_{pq}$. Consequently, the dispersion relation is a periodic function in k-space.

From the previous discussions, it is found that the set of all eigenmodes and eigenfrequencies are uniquely specified in a small neighborhood in k-space centered about the k-space origin, i.e., the $\vec{k} = 0$ point. The points in this set have the property that no two k-space points in the set are translated into one another through an application of a nonzero $\vec{G}_{pq}$. However, all the duplicate solutions at the points in k-space outside the neighborhood centered on $\vec{k} = 0$ can be generated from those in the neighborhood by an application of a nonzero $\vec{G}_{pq}$. These properties are now illustrated by examples.

4.4.2 Examples of Periodic Media in One and Two Dimensions

As a first example of a periodic medium, consider a one-dimensional periodic material in which the translation vectors are of the form $\vec{T}_n = n\vec{a}_1$, where $\vec{a}_1 = a\hat{i}$. A scalar wavefunction in such a medium is then from Eq. (4.91) of the form [4–7]

$$\psi_k(x,t) = e^{i(kx - \omega t)} u_k(x), \tag{4.97a}$$

with the spatial translational properties

$$\psi_k(x + na, t) = e^{ikna} e^{i(kx - \omega t)} u_k(x) = e^{ikna} \psi_k(x,t). \tag{4.97b}$$

In k-space, it then follows from Eq. (4.93) that $\vec{G}_n = n\frac{2\pi}{a}\hat{i}$

$$\psi_{k+n\frac{2\pi}{a}}(x,t) = e^{i\left[\left(k+n\frac{2\pi}{a}\right)x-\omega t\right]}u_{k+n\frac{2\pi}{a}}(x) = e^{i[kx-\omega t]}v_k(x) = \psi_k(x,t), \tag{4.98}$$

where $v_k(x) = e^{in\frac{2\pi}{a}x}u_{k+n\frac{2\pi}{a}}(x)$ is a periodic function in x with period a. A unique set of wavefunctions and their dispersion relations are located within the k-space region $-\frac{\pi}{a} \leq x \leq \frac{\pi}{a}$, and the k-space solutions outside this region are duplicates of those within this region.

In the case of a periodic medium in two dimensions with square lattice symmetries, the translation vectors are of the form $\vec{T}_{nm} = n\vec{a}_1 + n\vec{a}_2$, where $\vec{a}_1 = a\hat{i}$, $\vec{a}_2 = a\hat{j}$. From Eq. (4.93), $\vec{G}_{nm} = n\frac{2\pi}{a}\hat{i} + m\frac{2\pi}{a}\hat{j}$ defines the periodic properties of the system in k-space, and the region of unique solutions and dispersion relations in k-space is the square region defined by $-\frac{\pi}{a} \leq x \leq \frac{\pi}{a}$, $-\frac{\pi}{a} \leq y \leq \frac{\pi}{a}$. Outside this k-space square are duplicate solutions.

Next these considerations are applied to the solution of a two-dimensional problem.

4.4.2.1 Examples of Two-Dimensional Photonic Crystals

A common type of equation for waves propagating in a periodic two-dimensional medium is that of the two-dimensional Helmholtz equation expressed as [4–8]

$$\left[\frac{\partial^2}{\partial x^2} + \frac{\partial^2}{\partial y^2}\right]\psi(\vec{r},t) - \frac{1}{v^2(\vec{r})}\frac{\partial^2}{\partial t^2}\psi(\vec{r},t) = 0. \tag{4.99}$$

Here $\vec{r} = x\hat{i} + y\hat{j}$, $v(\vec{r})$ is a periodic function of space representing the speed of the wave, and $\psi(\vec{r},t)$ is the modal amplitude for a scalar wave or one of the components of a vector wave. For example, for electron problems $\psi(\vec{r},t)$ is the wavefunction, for photonic crystal problems $\psi(\vec{r},t)$ is the wavefunction of an electric or magnetic field component, and for phononic crystals $\psi(\vec{r},t)$ is the wavefunction of a pressure or elastic wave component. The general form of the solutions of Eq. (4.99) are from Eq. (4.91) of the form [4–8]

$$\psi(\vec{r},t) = A_{\vec{k}}\, e^{i(\vec{k}\cdot\vec{r}-\omega t)}u_{\vec{k}}(\vec{r}), \tag{4.100}$$

where $A_{\vec{k}}$ is the normalizing factor of the wavefunction.

The solution of Eq. (4.99) for an interesting problem of two-dimensional photonic crystals composed of semiconductor materials is now discussed.

4.4.2.2 Two-Dimensional Semiconducting Photonic Crystal

The system treated is a square lattice array of infinitely long semiconductor cylinders [22,23] of radii R with the nearest-neighbor separation $a > 2R$. The cylinders are perpendicular to the $x-y$ plane with the cylinder axes in the z-direction, and the electromagnetic modes of interest propagate in the $x-y$ plane. Outside the cylinders is vacuum. Within the cylinders, the semiconductor dielectric constant is described by a standard form given by [22,23]

$$\varepsilon(\omega) = \varepsilon_\infty\frac{\omega_L^2 - \omega^2}{\omega_T^2 - \omega^2}, \tag{4.101}$$

where ω_L and ω_T are, respectively, the frequencies of the zero wavevector longitudinal and transverse optical phonon modes, and ε_∞ is the $\omega \to \infty$ dielectric constant.

The translation vectors of the square lattice for the array of cylinders are of the form

$$\vec{T}_{nm} = na\hat{i} + ma\hat{j}, \tag{4.102}$$

where n, m are integers. These are the translations that leave the periodic medium invariant. The dielectric function of the photonic crystal is then given by the periodic form [22,23]

$$\varepsilon(\vec{r}) = 1 + [\varepsilon(\omega) - 1]\sum_{n,m} S(\vec{r} - \vec{T}_{nm}), \tag{4.103}$$

where $\vec{r} = x\hat{i} + y\hat{j}$ is the position vector in the $x - y$ plane, and

$$\begin{aligned} S(\vec{r} - \vec{T}_{nm}) &= 1 \text{ if } \left|\vec{r} - \vec{T}_{nm}\right| < R \\ &= 0, \text{ otherwise} \end{aligned} \tag{4.104}$$

Consequently, $\varepsilon(\vec{r} + \vec{T}_{lk}) = \varepsilon(\vec{r})$ signifies the periodicity of the photonic crystal.

Related to the $\vec{T}_{nm}$ are the sets of k-space translation vectors, which are defined by

$$\vec{G}_{pq} = p\frac{2\pi}{a}\hat{i} + q\frac{2\pi}{a}\hat{j} \tag{4.105}$$

where p, q are integers. These are translation vectors that through their application define a periodic k-space lattice known as the reciprocal lattice. The reciprocal lattice enters into representing periodic space functions that are invariant under the spatial translations in Eq. (4.102) in terms of Fourier series. This is now discussed.

In terms of translation vectors $\vec{G}_{pq}$ of the reciprocal lattice, the periodic dielectric function becomes

$$\varepsilon(\vec{r}) = \sum_{\vec{G}} \hat{\varepsilon}(\vec{G}) e^{i\vec{G}\cdot\vec{r}}, \tag{4.106}$$

where the sum is over all $\vec{G}$ of the form given in Eq. (4.105). It follows from Eqs. (4.95) to (4.106) that $\varepsilon(\vec{r} + \vec{T}_{lk}) = \varepsilon(\vec{r})$. Here, in Eq. (4.106)

$$\hat{\varepsilon}(\vec{G}) = \frac{1}{a_c}\int_{a_c} d^2r\, \varepsilon(\vec{r}) e^{-i\vec{G}\cdot\vec{r}}, \tag{4.107}$$

where $a_c = a^2$ is one of the unit squares composing the lattice. By direct substitution, it is readily seen that Eqs. (4.107) and (4.106) are inverse relations. In addition, for the following discussions, a useful function is the reciprocal of the dielectric function. This is given by the form

$$\frac{1}{\varepsilon(\vec{r})} = \sum_{\vec{G}} \hat{\kappa}(\vec{G}) e^{i\vec{G}\cdot\vec{r}} \tag{4.108}$$

with

$$\hat{\kappa}(\vec{G}) = \frac{1}{a_c}\int_{a_c} d^2r \frac{1}{\varepsilon(\vec{r})} e^{-i\vec{G}\cdot\vec{r}}. \tag{4.109}$$

In Eq. (4.108) again for the sum over all $\vec{G}$ in Eq. (4.105), it follows from Eq. (4.95) that $\frac{1}{\varepsilon\left(\vec{r}+\vec{T}_{lk}\right)}=\frac{1}{\varepsilon(\vec{r})}$.

That Eqs. (4.108) and (4.109) are inverse relations is shown by direct substitution [22,23].

Now, as a first consideration, the case in which the dielectric function does not depend on frequency is studied. In particular, in this limit Eq. (4.101) then becomes

$$\varepsilon(\omega)=\varepsilon \tag{4.110}$$

and from Eq. (4.103) the periodic dielectric function is given by

$$\varepsilon(\vec{r})=1+[\varepsilon-1]\sum_{n,m}S\left(\vec{r}-\vec{T}_{nm}\right). \tag{4.111}$$

For the case of electromagnetic modes moving in the $x-y$ plane with a single electric field component polarized along the z-direction, the single electric field component written as $\vec{E}(\vec{r},t)=\hat{k}E(\vec{r},t)$ is described by a Helmholtz equation of the form [22,23]

$$\left[\frac{\partial^2}{\partial x^2}+\frac{\partial^2}{\partial y^2}\right]E(\vec{r},t)-\frac{\varepsilon(\vec{r})}{c^2}\frac{\partial^2}{\partial t^2}E(\vec{r},t)=\left[\frac{\partial^2}{\partial x^2}+\frac{\partial^2}{\partial y^2}\right]E(\vec{r},t)+\frac{\varepsilon(\vec{r})\omega^2}{c^2}E(\vec{r},t)=0, \tag{4.112a}$$

where a harmonic form

$$E(\vec{r},t)=E(\vec{r})e^{-i\omega t} \tag{4.112b}$$

has been taken for the time dependence.

The solution for the electric field as a function of position can be written in the general form of a planewave times a periodic function of space. Specifically, it is assumed that

$$E(\vec{r})=e^{i\vec{k}\cdot\vec{r}}\sum_{\vec{G}}a\left(\vec{k}+\vec{G}\right)e^{i\vec{G}\cdot\vec{r}}, \tag{4.113}$$

where $\sum_{\vec{G}}a\left(\vec{k}+\vec{G}\right)e^{i\vec{G}\cdot\vec{r}}$ from Eq. (4.95) is seen to be periodic in space. Rewriting Eq. (4.112a) into the form [20–22]

$$\frac{1}{\varepsilon(\vec{r})}\left[\frac{\partial^2}{\partial x^2}+\frac{\partial^2}{\partial y^2}\right]E(\vec{r},t)+\frac{\omega^2}{c^2}E(\vec{r},t)=0 \tag{4.114}$$

and applying Eqs. (4.108), (4.109) and (4.113) in Eq. (4.114), the eigenvalue problem can be rewritten in a matrix form as

$$\sum_{\vec{G}'}\hat{\kappa}\left(\vec{G}-\vec{G}'\right)\left(\vec{k}+\vec{G}'\right)^2 a\left(\vec{k}+\vec{G}'\right)=\frac{\omega^2}{c^2}a\left(\vec{k}+\vec{G}\right). \tag{4.115}$$

Defining

$$a\left(\vec{k}+\vec{G}\right)=\frac{b\left(\vec{k}+\vec{G}\right)}{\left|\vec{k}+\vec{G}\right|} \tag{4.116a}$$

Eq. (4.115) takes the symmetric form [20,22,23]

$$\sum_{\vec{G}'} \left|\vec{k}+\vec{G}\right| \hat{\kappa}\left(\vec{G}-\vec{G}'\right) \left|\vec{k}+\vec{G}'\right| b\left(\vec{k}+\vec{G}'\right) = \frac{\omega^2}{c^2} b\left(\vec{k}+\vec{G}\right) \tag{4.116b}$$

and aids in the diagonalization of the matrix.

Turing to the evaluation of Eq. (4.109) for the constant uniform dielectric constant of the cylinders, it follows from the planewave expansion in cylinder coordinates and with the further application of a Bessel function identity that [20,22,23]

$$\hat{\kappa}\left(\vec{G}\right) = \begin{array}{ll} \frac{1}{\varepsilon} f + (1-f), & \text{if } \vec{G} = 0 \\ \left(\frac{1}{\varepsilon} - 1\right) f \frac{2J_1(GR)}{GR}, & \text{otherwise} \end{array}, \tag{4.117}$$

where $f = \frac{\pi R^2}{a^2}$ is the cylinder filling fraction. In Figure 4.9, theoretical results are now presented for an example of a photonic crystal that has been studied both theoretically and experimentally [20,22,23].

The theory in Figure 4.9 is a plot of the frequencies of the electromagnetic modes versus wavevector presented for a photonic crystal formed of dielectric cylinders arranged on a square lattice [20]. In the photonic crystal the cylinders are of dielectric constants $\varepsilon = 9$, are embedded in a vacuum, and have a filling fraction $f = 0.4488$. The system has been realized experimentally using an aluminum composite, and the agreement of theory and experiment is excellent. The results display a sequence of pass and stop band regions in which the electromagnetic modes, respectively, propagate or fail to propagate into the bulk of the photonic crystal.

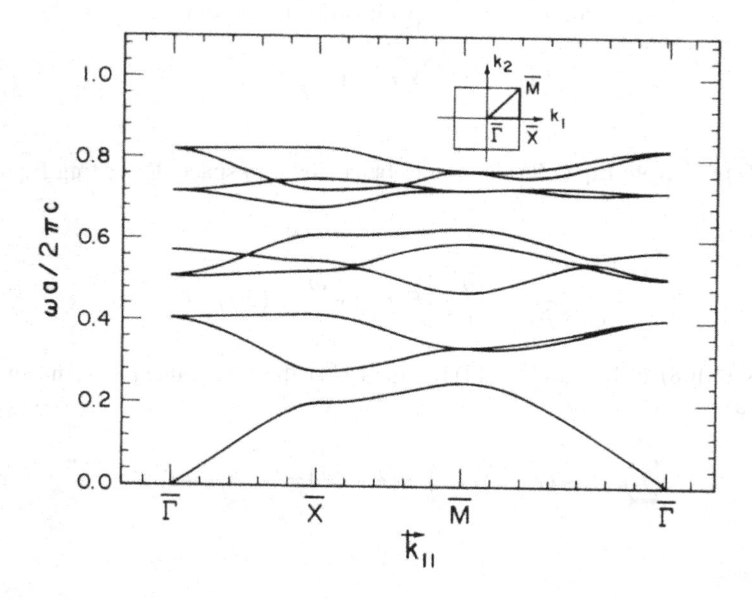

FIGURE 4.9 Plot of the frequency versus wavevector for a square lattice photonic crystal of infinite parallel axis dielectric cylinders arrayed in vacuum [20]. The optical modes propagate in the plane of the square lattice array with electric fields polarized parallel to the cylinder axes. An inset to the figure shows the wavevectors lines considered in making the plot. (Reprint with permission from Ref. [20], Optical Society of America.)

In the case of a frequency-dependent dielectric constant, a different approach to determining the excitation spectrum is needed [22–24]. The reason for this is that the dielectric function of the system depends on the eigenvalue frequencies that are being computed. One method, however, that is effective for solving for the modes in the presence of frequency-dependent materials is a determinantal approach. This approach is now discussed by treating a photonic crystal composed of cylinders with a dielectric constant of the form in Eq. (4.101). The geometry of the present system is the same as that for the earlier study of a frequency-independent dielectric function; the only change in the problem is that the dielectric function is now frequency dependent.

For frequency- dependent materials now considered, the formulation in Eq. (4.112a) is used. Substituting the general form for the electric field [22–24],

$$E(\vec{r}) = e^{i\vec{k}\cdot\vec{r}} \sum_{\vec{G}} B(\vec{k} \mid \vec{G}) e^{i\vec{G}\cdot\vec{r}} \tag{4.118}$$

into Eq. (112a), it then follows that

$$(\vec{k}+\vec{G})^2 B(\vec{k} \mid \vec{G}) = \frac{\omega^2}{c^2}\hat{\varepsilon}(0) B(\vec{k} \mid \vec{G}) + \frac{\omega^2}{c^2} \sum_{\vec{G}'}{}' \hat{\varepsilon}(\vec{G}-\vec{G}') B(\vec{k} \mid \vec{G}'), \tag{4.119}$$

where the second sum on the right is only over $\vec{G}' \neq \vec{G}$. From Eqs. (4.101), (4.103)–(4.106), it then follows that

$$\hat{\varepsilon}(0) = 1 + f \frac{(\varepsilon_\infty \omega_L^2 - \omega_T^2) - (\varepsilon_\infty - 1)\omega^2}{\omega_T^2 - \omega^2} \tag{4.120a}$$

$$\hat{\varepsilon}(\vec{G} \neq 0) = f \frac{(\varepsilon_\infty \omega_L^2 - \omega_T^2) - (\varepsilon_\infty - 1)\omega^2}{\omega_T^2 - \omega^2} \frac{2J_1(GR)}{GR}, \tag{4.120b}$$

where $f = \frac{\pi R^2}{a^2}$ is the cylinder filling fraction so that Eq. (4.119) can be rewritten as [22–24]

$$(\omega_T^2 - \omega^2)\left[(\vec{k}+\vec{G})^2 - \frac{\omega^2}{c^2}\right] B(\vec{k} \mid \vec{G})$$

$$- f\left[(\varepsilon_\infty \omega_L^2 - \omega_T^2) - (\varepsilon_\infty - 1)\omega^2\right] \frac{\omega^2}{c^2} \sum_{\vec{G}'} \frac{2J_1(|\vec{G}-\vec{G}'|R)}{|\vec{G}-\vec{G}'|R} B(\vec{k} \mid \vec{G}') = 0. \tag{4.121}$$

The resulting system of equations in Eq. (4.121) is homogeneous. Consequently, the determinant of the system must be zero for a solution to exist. A determination of the frequencies $\frac{\omega^2}{c^2}$ at which the determinant is zero can be obtained numerically by studying the determinant as a function of $\frac{\omega^2}{c^2}$. In this way, for example, the positions of the zeros of the determinant are found in a plot of the determinant as a function of $\frac{\omega^2}{c^2}$.

Results are shown in Figure 4.10 for a study of a photonic crystal composed of GaAs cylinders with dielectric properties modeled by the frequency-dependent dielectric [22,24] form in Eq. (4.101). The photonic crystal is a square lattice array of semiconductor cylinders with very small filling fractions. For the frequency-dependent dielectric, a series of flat bands are introduced into the spectrum of electromagnetic modes [23]. These arise in the vicinity of the poles and zeros of

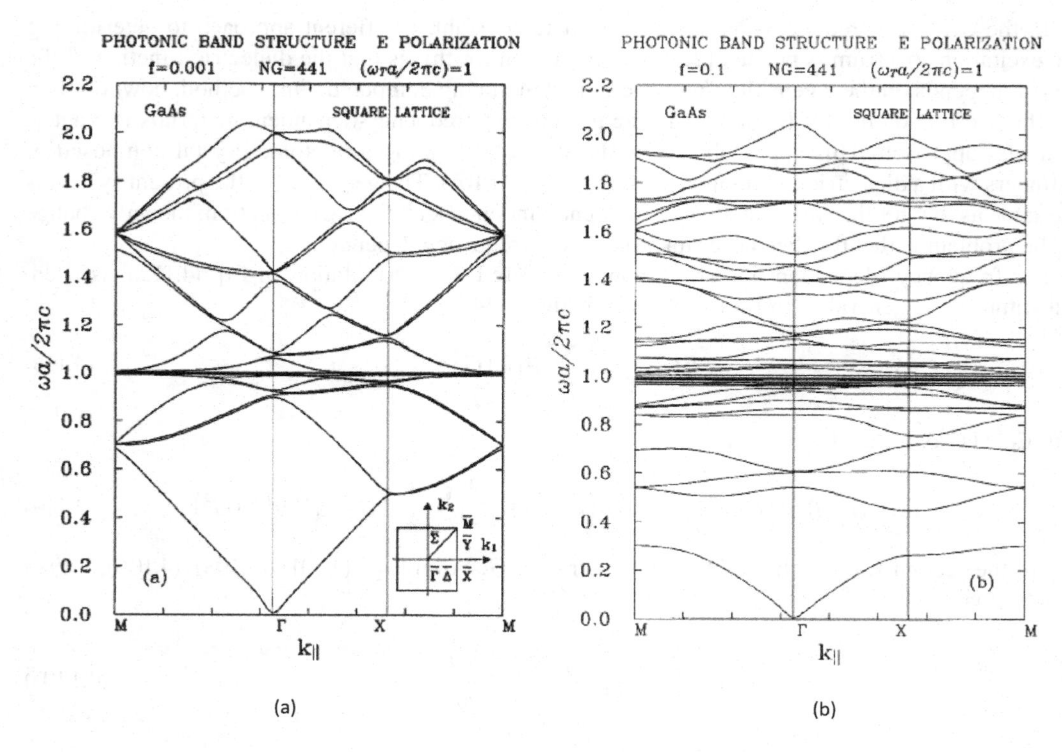

FIGURE 4.10 Plots of the frequency versus wavevector for a square lattice photonic crystal of infinite parallel axis cylinders arrayed in vacuum [23]. The optical modes propagate in the plane of the square lattice array with electric fields polarized parallel to the cylinder axes. An inset to the figure shows the wavevectors lines considered in making the plot, and the form of the dielectric function of the cylinders is given in Eq. (4.101). An interesting sequence of flat bands are observed at $\omega \approx \omega_T$. Results are shown for filling fractions (a) $f=0.001$ and (b) $f=0.1$. (Reprinted figure with permission form [23]. Copyright 1997, American Physical Society.)

the dielectric function. Similar considerations have also been made in three-dimensional arrays of metals and semiconductors [24].

4.4.3 Applications of Photonic and Phononic Crystals

Many of the applications of photonic and phononic crystals come from their ability to block stop band frequencies of radiation from entering their bulk [4–7,18–23]. These stop bands can be engineered by varying the sizes and dielectric properties of the features forming the crystal. An advantage of the photonic and phononic crystal approach is that often the materials used in these technologies lead to lower loss systems than could otherwise be achieved by more conventional methods employed in the design of, for example, electromagnetic and acoustic resonators, and filters.

The ability to exclude radiations allows for photonic and phononic crystal to be used as filters, resonators, waveguides, and in the design of optical and acoustic circuits. There is then a potential for the replacement of electronic technologies by optical components leading to the field of optoelectronics and acoustic components applied in conjunction with electronic or optical systems [4–7,25,26]. In addition, with the introduction of nonlinear optical and acoustic media, in which the response of the material to stimuli is not linearly related in the intensity of the input, more sophisticated diode and transistor effects can be obtained for applications [25,26]. In this regard, further applications have included antennas, repression of emissions, sensors, and the generation of higher harmonics [4–7].

REFERENCES

1. Kittle, C. 1996. *Introduction to solid state physics, 7th Edition*, 393–410. New York: John Wiley & Sons, Inc.
2. Marder, M. P. 2000. *Condensed matter physics*. New York: John Wiley & Sons, Inc.
3. Martin, R. M. 2004. *Electronic structure: Basic theory and practical methods*. Cambridge: Cambridge University Press.
4. McGurn, A. R. 2020. *Introduction to photonic and phononic crystals and metamaterials*. San Rafael: Morgan & Claypool Publishers.
5. McGurn, A. R. 2021. *Introduction to nonlinear optics of photonic crystals and metamaterials, 2nd Edition*. Bristol: IOP Publishing Ltd.
6. McGurn, A. R. 2015. *Nonlinear optics of photonic crystals and meta-materials*. San Rafael: Morgan & Claypool Publishers.
7. McGurn, A. R. 2018. *Nanophotonics*. Cham: Springer.
8. McGurn, A. R. 2005. Impurity mode techniques applied to the study of light sources. *Journal of Physics D: Applied Physics* 38: 2338–2349.
9. Streetman, B. G. and Banerjee, S. 2000. *Solid state electronic devices, 5th Edition*, New Jersey: Prentice Hall.
10. Ouisse, T. 2008. *Electron transport in nanostructures and mesoscopic devices: An introduction*. New York: Wiley.
11. Barnham, K. and Vvedensky D. 2001. *Low-dimensional semiconductor structures: Fundamentals and device applications*. Cambridge: Cambridge University Press.
12. Datta, S. 2005. *Quantum transport: Atom to transistor*. Cambridge: Cambridge University Press.
13. Kroemer, H. 1994. *Quantum mechanics: For engineering, materials science, and applied physics*. New Jersey: Prentice Hall.
14. Wallace, P. R. 1947. The band theory of graphite. *Physical Review* 71: 622–634.
15. Katsnelson, M. I. 2020. *The physics of graphene, 2nd Edition*. Cambridge: Cambridge University Press.
16. Sharon, M. and M. Sharon 2015. *Graphene: An introduction to the fundamentals and industrial applications*. New York: John Wiley & Sons, Inc.
17. Gerry, C. and P. Knight 2005. *Introductory quantum optics*. Cambridge: Cambridge University Press.
18. Joanopoulos J. D., Vilenueve, P. R. and Fan, S. 1995. *Photonic crystals*. Princeton: Princeton University Press.
19. Favennec, P. N. 2005. *Photonic crystals: Towards nanoscale photonic devices*. Berlin: Springer.
20. Maradudin, A. A. and A. R. McGurn 1993 The photonic band structure of a truncated, two-dimensional, periodic medium. *Journal of the Optical Society of America* B 10: 307–313.
21. Harrison, P. and P. Harrison 2000. *Quantum wells, wires, and dots*. New York: John Wiley & Sons, Inc.
22. Maradudin, A. A. Kuzmiak, V. and A. R. McGurn 1995. Photonic band structures of systems with components characterized by frequency-dependent dielectric functions, in *Photonic band gap materials* Ed. C.M. Soukoulis, NATO advanced studies series. Vol. 315, 271–280, Dordrecht: Kluver Academic Press.
23. Kuzmiak, V., Maradudin, A. A. and A. R. McGurn 1997. Photonic band structures of two-dimensional systems fabricated from rods of a cubic polar crystal. *Physical Review* B 55: 4298–4311.
24. McGurn, A. R., and A. Maradudin 1992. Photonic band structure of two- and three-dimensional periodic metal and semiconductor arrays. *Physical Review B* 48: 17576–17579.
25. Wang, C-Y., Chen, C-W. and T.-L. Lin 2016. Diode-action and logic gates based on anisotropic nonlinear responsive liquid crystals, *Scientific Reports* 6: 30873. https://doi.org/10.1038/srep30873
26. Almeida, V. R., Barrios, C. A., Panepuccu, R.R. and M. Lipson 2004. All-optical control of light on a silicon chip. *Nature* 431: 1081–1084.

5 Basic Properties of Light and Its Interactions with Matter

In this chapter, an introduction to the quantization of the electromagnetic fields and some of the properties of their interactions with bound electronic systems are presented [1–7]. This involves a treatment along the lines of the Planck and Einstein hypotheses in which the field energies are taken to occur in integer multiples of $\hbar\omega$ and in which the degrees of freedom in classical electrodynamics are replaced by quantum mechanical operators characterizing photon excitations of the fields. Specifically, the classical formulation of electrodynamics in terms of a Hamiltonian and its dynamics generated by Poisson brackets is replaced by a quantum formulation in terms of a Hamiltonian operator and its commutator-generated dynamics [8,9]. This formulation of electrodynamics is then developed to treat the quantum mechanical interaction of the fields with a two-level quantum mechanical model of a single bound electron [3–7]. In this regard, the basic dynamics of the quantum electron-field system is discussed in the context of the so-called Jaynes–Cumming model [3,6,7]. This model provides a basic treatment of interest for electrons bound in atoms and molecules confined within electromagnetic cavities [3,7].

In addition, discussions are also presented of the basic ideas related to the coherence of the electromagnetic fields [3,7]. These are made in terms of an infinite set of coherence functions that quantitatively characterize the extent to which the fields are correlated in space and time. Specific considerations are given to the coherence properties of Fock states, coherent states, and thermal distributions of radiation fields [3–7]. These characterizations will later be of interest in the context of the treatment presented in Chapter 6 on lasers and in the discussion of problems related to the interactions of light with matter.

5.1 QUANTIZED ELECTROMAGNETIC WAVES

The quantization of electromagnetic waves is intimately related to the quantization of the mechanical harmonic oscillator [3,7–9]. This was discussed by both Planck and Einstein early in the development of quantum mechanics. Planck originally noted the nature of quantum phenomena in his quantization scheme, which postulated that the energy changes within the electromagnetic modes of a system occur in integer multiples of $\hbar\omega$. Einstein latter extended this idea to treat the energy contained in the vibrational modes of materials. Both of these results are ultimately statements of the linearity in the nature of the excitations of electromagnetic and acoustic systems, and similarly are applied in a variety of magnetic, ferroelectric, and general media supporting Bose–Einstein excitations. Consequently, as a starting point a review is given of the properties of a one-dimensional mechanical harmonic oscillator. This is then extended to the problem of the quantization of the electromagnetic fields.

In the simplest formulation of classical mechanics, the one-dimensional harmonic oscillator is described by a linear differential equation given by [3,8,9]

$$\ddot{x} + \omega^2 x = 0 \tag{5.1}$$

where ω is the angular frequency of the oscillator. The general harmonic solution is then written as a linear combination of a set of exponential harmonic solutions, i.e., $x(t) = x_0 e^{-i\omega t} + x_0^* e^{i\omega t}$. For the extension of the theory to a consideration of the quantum mechanical oscillator, however, a description of the system in terms of the Hamiltonian approach is taken. In this formulation, the Hamiltonian in both classical and quantum mechanics is a quadratic form written as

DOI: 10.1201/9781003031987-5

$$H = \frac{1}{2m} p^2 + \frac{1}{2} m\omega^2 x^2. \tag{5.2}$$

Within the context of the quantum mechanical treatment of Eq. (5.2), an addition quantization condition must also be introduced that has the form of the commutation relation [3,7–9]

$$[x, p] = i\hbar, \tag{5.3a}$$

where the momentum is now expressed by a differential operator

$$p = -i\hbar \frac{\partial}{\partial x}. \tag{5.3b}$$

In this regard, the commutation condition in Eq. (5.3a) replaces the Poisson brackets, $\{x, p\} = 1$, which is used to solve Eq. (5.2) in the context of classical mechanics [8,9]. Both the commutator solution of the quantum mechanical oscillator and the Poisson brackets solution of the classical mechanic oscillator closely parallel one another.

Proceeding with the quantum mechanical solution, it is useful to define the operators [3,9]

$$a = \left(\frac{\omega m}{2\hbar}\right)^{1/2} \left[x + \frac{i}{\omega m} p\right] \tag{5.4a}$$

and

$$a^+ = \left(\frac{\omega m}{2\hbar}\right)^{1/2} \left[x - \frac{i}{\omega m} p\right]. \tag{5.4b}$$

From Eq. (5.3a), it follows that these operators have an important commutation relation given by

$$[a, a^+] = 1, \tag{5.5}$$

which shall be useful in generating the solutions of the Hamiltonian. In this regard, the Hamilton is written in terms of the operators a and a^+, becoming of the form [3,9]

$$H = \left(a^+ a + \frac{1}{2}\right)\hbar\omega, \tag{5.6a}$$

where later it is shown that the operator

$$N = a^+ a \tag{5.6b}$$

is the number operator with eigenvalues 0, 1, 2,…. Within the Einstein model of lattice vibrations, the number operator corresponds to the number of phonons present in the harmonic system.

As an additional important consideration, the factor of 1/2 in Eq. (5.6a) is the zero-point energy of the oscillator. This has its origin in the Heisenberg uncertainty principle [3,9]. Specifically, it represents a statement that the energy of the oscillator cannot be zero, as this would require both the position and momentum of the oscillator to be simultaneously specified.

In the context of the Heisenberg formulation, the equations of motion of a and a^+ are given by [3,9]

$$-i\hbar\dot{a} = [H, a] = -\hbar\omega a \tag{5.7a}$$

and

$$-i\hbar\dot{a}^{+} = \left[H, a^{+}\right] = \hbar\omega a^{+}. \tag{5.7b}$$

Consequently, the solutions of the equations of motion in Eq. (5.7) are obtained by an integration so that

$$a(t) = a(0)e^{-i\omega t} \tag{5.8a}$$

and

$$a^{+}(t) = a^{+}(0)e^{i\omega t}. \tag{5.8b}$$

It is seen from these solutions that when they are applied to eigenfunctions of the harmonic oscillator a and a^{+}, respectively, lower and raise the energy of the eigenfunction of the system by $\hbar\omega$. As a result, an application of $\left(a^{+}\right)^{n}$ to an eigenfunction raises its energy so that it is an eigenstate of energy increased by $n\hbar\omega$, and an application of $(a)^{n}$ to an eigenfunction lowers its energy by $n\hbar\omega$. In this regard, a and a^{+} are often, respectively, referred to as destruction and creation operators that remove or introduce a quantum of energy of the oscillator.

Consequently, if $|0\rangle$ is the wavefunction corresponding to the ground state energy of the system, then [3,9]

$$a|0\rangle = 0, \tag{5.9}$$

as the system cannot go to a lower energy state. From Eq. (5.6a), it then also follows that

$$H|0\rangle = \left(a^{+}a + \frac{1}{2}\right)\hbar\omega|0\rangle = \frac{1}{2}\hbar\omega|0\rangle \tag{5.10}$$

and consequently $N|0\rangle = 0$. Considering the generation of higher energy vibrational states from the ground state $|0\rangle$, it is seen that

$$\left(a^{+}\right)^{n}|0\rangle = \sqrt{n!}\,|n\rangle \tag{5.11a}$$

where $|n\rangle$ is an eigenfunction with eigenenergy $\left(n + \frac{1}{2}\right)\hbar\omega$ and for $m > n$

$$(a)^{n}|m\rangle = \sqrt{\frac{m!}{(m-n)!}}\,|m-n\rangle, \tag{5.11b}$$

so that from Eq. (5.6a)

$$H\frac{1}{\sqrt{n!}}\left(a^{+}\right)^{n}|0\rangle = \left(a^{+}a + \frac{1}{2}\right)\hbar\omega|n\rangle = \left(n + \frac{1}{2}\right)\hbar\omega|n\rangle \tag{5.12}$$

and $N|n\rangle = n|n\rangle$. It should be noted that the relationship $N|n\rangle = n|n\rangle$ accounts for the factorials occurring in Eqs. (5.11a) and (5.11b), i.e., the factorials adjust for the required normalization condition $\langle n|n\rangle = 1$ on the wavefunction. In this sense, often $|n\rangle$ is referred to as the wavefunction for a system containing n phonons.

These results are now translated into the context of the study of the harmonic modes of an electromagnetic wave propagating in vacuum. Specifically, it is shown that the modes of the harmonic oscillator and the electromagnetic fields both exhibit a harmonic variation in time and are subject to the same Planck quantization postulate. This linkage between the two physical systems arises as the theories governing the physics of the two systems are isomorphic to each other so that the development of the properties of one system is intimately related to that of the other.

The classical electrodynamics of electromagnetic radiation in vacuum is described by the Maxwell equations. In the absence of charges and currents, these take the particular forms given in SI units by [3,7]

$$\nabla \times \vec{E} = -\frac{\partial \vec{B}}{\partial t} \tag{5.13a}$$

$$\nabla \times \vec{H} = \frac{\partial \vec{D}}{\partial t} \tag{5.13b}$$

$$\nabla \cdot \vec{D} = 0 \tag{5.13c}$$

$$\nabla \cdot \vec{B} = 0 \tag{5.13d}$$

where $\vec{D} = \varepsilon_0 \vec{E}$ and $\vec{B} = \mu_0 \vec{H}$. The focus in the following discussions is on first studying the classical electrodynamic planewave solutions of these Maxwell equations, followed by a treatment of the quantization of the radiation fields. The presentation for the harmonic electromagnetic waves proceeds directly along the lines of the earlier discussions of the mechanical oscillator.

A simplification of the classical field equations in Eq. (5.13) occurs upon the introduction of a scalar potential ϕ and a vector potential $\vec{A}$, which are defined such that [3,7,10]

$$\vec{B} = \nabla \times \vec{A} \tag{5.14a}$$

and

$$\vec{E} = -\nabla \phi - \frac{\partial \vec{A}}{\partial t}. \tag{5.14b}$$

As a consequence of the introduction of these new fields, Eq. (5.14a) causes Eq. (5.13d) to be satisfied, and, similarly, Eq. (5.14b) assures that Eq. (5.13a) is satisfied. Further simplification is now made to the two remaining Maxwell equations in Eqs. (5.13b) and (5.13c). This simplification arises from certain gauge symmetries that are associated with the definitions in Eq. (5.14).

The forms in Eq. (5.14) are invariant under a gauge transformation, which states that for a scalar function χ

$$\vec{B} = \nabla \times \vec{A} = \nabla \times \left(\vec{A} + \nabla \chi \right) \tag{5.15a}$$

where the identity $\nabla \times \nabla \chi = 0$ is used, and [10]

$$\vec{E} = -\nabla \phi - \frac{\partial \vec{A}}{\partial t} = -\nabla \left(\phi - \frac{\partial \chi}{\partial t} \right) - \frac{\partial}{\partial t} \left(\vec{A} + \nabla \chi \right). \tag{5.15b}$$

Consequently, adding $\nabla\chi$ to $\vec{A}$ and subtracting $\frac{\partial\chi}{\partial t}$ from ϕ leaves the electromagnetic fields unchanged. The nature of the gauge symmetry is such that it can be used to eliminate ϕ from Eq. (5.14) so that within the context of the gauge transformation the fields take the form [3,10]

$$\vec{B} = \nabla\times\vec{A} \tag{5.16a}$$

and

$$\vec{E} = -\frac{\partial\vec{A}}{\partial t}. \tag{5.16b}$$

For this to be the case, it is required that

$$\phi - \frac{\partial\chi}{\partial t} = 0 \tag{5.17a}$$

or

$$\chi = \int^t \frac{\partial\chi}{\partial t}dt. \tag{5.17b}$$

Under these conditions, the selection of the condition in Eq. (5.17a) along with the additional requirement that $\nabla\cdot\vec{A} = 0$ is known as the Coulomb gauge [10].

In this context, note that for our treatment of gauge symmetry $\nabla\cdot\vec{A} = \nabla\cdot\left(\vec{A}+\nabla\chi\right) = 0$ and the Poison equation $\nabla^2\chi = -\nabla\cdot\vec{A}$ is always solvable for χ. Consequently, a χ can always be determined so that $\nabla\cdot\vec{A} = 0$ is satisfied. It is also found in the present problem that the $\nabla\cdot\vec{A} = 0$ condition assures that Eq. (5.13c) is satisfied, and, as a result, only the Maxwell equation in Eq. (5.13b) remains to be studied.

From Eq. (5.13b), within the Coulomb gauge, it then follows upon the application of Eq. (5.16) and the vector identity $\nabla\times\nabla\times\vec{A} = \nabla\left(\nabla\cdot\vec{A}\right) - \nabla^2\vec{A}$ that the vector potential satisfies a Helmholtz wave equation of the form [3,10]

$$\nabla^2\vec{A} - \frac{1}{c^2}\frac{\partial^2\vec{A}}{\partial t^2} = 0. \tag{5.18}$$

Consequently, $\vec{A}$ is the solution of Eq. (5.18) which in vacuum can be taken to be of the form of a planewave. In addition, the condition of the Coulomb gauge setting

$$\nabla\cdot\vec{A} = 0 \tag{5.19}$$

requires that the planewave solution for $\vec{A}$ must be a transverse wave.

As an example of a planewave solution of Eq. (5.18), consider $\vec{A}$ in the form of a planewave propagating along the x-axis. In this case, the general solution is given by

$$\vec{A}(x,t) = \left(A_y\hat{j} + A_z\hat{k}\right)e^{i(kx-\omega t)} + \left(A_y^*\hat{j} + A_z^*\hat{k}\right)e^{-i(kx-\omega t)}, \tag{5.20}$$

where $\omega = ck$ is required as a condition for the solution to exist. (Note that the solution in Eq. (5.20) has a form that is similar to that of the solution $x(t) = x_0e^{-i\omega t} + x_0^*e^{i\omega t}$ of Eq. (5.1) for the classical harmonic oscillator.) From Eqs. (5.16b) to (5.20), it then follows that the electric field of the planewave is given by

$$\vec{E}(x,t) = i\omega\left(A_y\hat{j} + A_z\hat{k}\right)e^{i(kx-\omega t)} - i\omega\left(A_y^*\hat{j} + A_z^*\hat{k}\right)e^{-i(kx-\omega t)}, \tag{5.21a}$$

and from Eqs. (5.16a) to (5.20) the magnetic induction is

$$\vec{B}(x,t) = ik\left(-A_z\hat{j} + A_y\hat{k}\right)e^{i(kx-\omega t)} - ik\left(-A_z^*\hat{j} + A_y^*\hat{k}\right)e^{-i(kx-\omega t)}. \tag{5.21b}$$

The resulting electromagnetic planewave propagates along the x-axis with electric and magnetic induction polarized in the $y-z$ plane and perpendicular to each other.

Substituting Eq. (5.20) into Eq. (5.18), it follows that [3,10]

$$-k^2\vec{A} - \frac{1}{c^2}\frac{\partial^2\vec{A}}{\partial t^2} = 0 = \ddot{\vec{A}} + \omega^2\vec{A}. \tag{5.22}$$

Comparing the far right of Eq. (5.22) with Eq. (5.1), it is found that the resulting equation of motion for the vector potential in classical electrodynamics is just the harmonic oscillator equation of classical mechanics. The quantization of the vector potential in electrodynamics should then proceed by a process that is the analogy of that for the quantization of the classical mechanical oscillator. As with the mechanical problem, the quantization of the electromagnetic waves takes place by considering the total energy in the system.

Consider the energy in the fields of Eq. (5.21). In classical electrodynamics, the time averaged energy of the fields is given by the form [3,7,10]

$$U = \frac{1}{2}\int dxdydz\left(\varepsilon_0\left|\vec{E}\right|^2 + \frac{1}{\mu_0}\left|\vec{B}\right|^2\right), \tag{5.23}$$

where it is assumed that the modes exist in a cubic box of sides of length $L \to \infty$ and are subject to periodic boundary conditions over the box. Upon substituting Eq. (5.21) into Eq. (5.23), it follows that

$$\begin{aligned} U &= \varepsilon_0 L^3\omega^2\left[A_yA_y^* + A_zA_z^* + A_y^*A_y + A_z^*A_z\right] \\ &= 2\varepsilon_0 L^3\omega^2\left[\left|A_y\right|^2 + \left|A_z\right|^2\right] \end{aligned}, \tag{5.24}$$

where the classical nature of the fields are used in the second line of Eq. (5.24).

In order to perform the quantization step, it is good to rewrite Eq. (5.24) in the form

$$U = \frac{1}{2}\hbar\omega\left\{\frac{2\varepsilon_0 L^3\omega}{\hbar}\left[A_yA_y^* + A_zA_z^* + A_y^*A_y + A_z^*A_z\right]\right\}. \tag{5.25}$$

Here a factor of $\hbar\omega$ has been separated out from the terms contained within the brackets. The separation is made in anticipation of the Planck quantization postulate that the discrete photon quanta of the radiation fields carry a quantum of energy $\hbar\omega$. For the case of a planewave with the electric field polarized along the y-direction $A_z = 0$, Eq. (5.25) becomes

$$\begin{aligned} U &= \frac{1}{2}\hbar\omega\left\{\frac{2\varepsilon_0 L^3\omega}{\hbar}\left[A_yA_y^* + A_y^*A_y\right]\right\} \\ &= \frac{1}{2}\hbar\omega\left\{a_ya_y^* + a_y^*a_y\right\} \end{aligned}, \tag{5.26}$$

where

$$a_y = \sqrt{\frac{2\varepsilon_0 L^3 \omega}{\hbar}} A_y \tag{5.27a}$$

and

$$a_y^* = \sqrt{\frac{2\varepsilon_0 L^3 \omega}{\hbar}} A_y^*. \tag{5.27b}$$

At this point the fields can be quantized by introducing the quantization condition in Eq. (5.5). This is done in analogy with the quantization of the classical mechanical oscillator. Specifically, the condition in Eq. (5.5) is taken over to apply to new quantum operators based on the forms in Eq. (5.27), i.e., the following commutator is adopted

$$\left[a_y, a_y^+\right] = 1, \tag{5.28}$$

where the associations $a_y \leftrightarrow a_y$ and $a_y^* \leftrightarrow a_y^+$ match the classical coefficients in Eq. (5.27) with their counterpart quantum mechanical operators. From Eq. (5.26), it then follows that the quantum mechanical Hamiltonian of the fields are rewritten in terms of the quantum operators as

$$\begin{aligned} H &= \frac{1}{2}\hbar\omega\left\{a_y a_y^+ + a_y^+ a_y\right\} \\ &= \hbar\omega\left\{a_y^+ a_y + \frac{1}{2}\right\}. \end{aligned} \tag{5.29}$$

This expresses the wave as a quantized mode of frequency ω with an electric field polarized along the y-direction [3,7,10].

From the properties of Eqs. (5.28) and (5.29) and following the same arguments as in Eq. (5.7)–(5.11) for the mechanical harmonic oscillator, it is seen that a_y is the destruction operator, and a_y^+ the creation operator for the frequency ω quantum mechanical electromagnetic modes of the system. In addition, by similar arguments as those for the quantum harmonic oscillator, it follows that

$$N_y = a_y^+ a_y \tag{5.30}$$

is the number operator with eigenvalues 0, 1, 2, 3,... of the frequency ω modes. It gives the number of photons of energy $\hbar\omega$ in the electromagnetic mode. In terms of Eq. (5.30), Eq. (5.29) then becomes

$$H = \hbar\omega\left\{N_y + \frac{1}{2}\right\}. \tag{5.31}$$

Both the electric and magnetic induction can also be quantized by writing them in terms of the quantum operators a_y and a_y^+. In the context of the quantum formulation, the field operators are from Eqs. (5.27) to (5.21) then given by [2,4,7,10]

$$\vec{E}(x,t) = i\omega\sqrt{\frac{\hbar}{2\varepsilon_0 L^3 \omega}}\hat{j}\left[a_y e^{i(kx-\omega t)} - a_y^+ e^{-i(kx-\omega t)}\right], \tag{5.32a}$$

and

$$\vec{B}(x,t) = ik\sqrt{\frac{\hbar}{2\varepsilon_0 L^3 \omega}}\hat{k}\left[a_y e^{i(kx-\omega t)} - a_y^+ e^{-i(kx-\omega t)}\right], \tag{5.32b}$$

and the vector potential from Eq. (5.20) becomes

$$\vec{A}(x,t) = \sqrt{\frac{\hbar}{2\varepsilon_0 L^3 \omega}}\hat{j}\left[a_y e^{i(kx-\omega t)} + a_y^+ e^{-i(kx-\omega t)}\right]. \tag{5.32c}$$

These form the fields for a mode propagating along the x-direction in the case of a transverse mode with the electric field polarized along the y-axis. A second transverse mode of frequency ω exists with the electric field polarized along the z-axis. The quantization of the z-component of electric field is done by taking $A_y = 0$ and $A_z \neq 0$, and following through the calculations similar to those in Eqs. (5.25)–(5.32).

5.1.1 General Form of 3D Quantized Electromagnetic Waves

The earlier discussions only considered a single planewave mode of frequency ω, which propagates in the x-direction. These results are directly generalized to treat modes of arbitrary frequency propagating in general directions in space. The argument relies on the isotropy and homogeneity of vacuum. In addition, due to the linearity of the theory of electrodynamics in vacuum, the modes of the system exist independently of one another, and the electromagnetic fields of a general solution are formed as a sum over the different modal solutions.

In addition, the electromagnetic fields within an infinite box of vacuum for each wavevector state are composed of the sum of transverse planewave modes with two separate polarizations. In the planewave sum, the length and direction of the wavevectors of the modes are set by the periodic boundary conditions at the surface of the infinite box and by the isotropic dispersion relation $\omega_{\vec{k}} = c\left|\vec{k}\right|$. In the case of periodic boundary conditions, the modal wavevectors have the general form

$$\vec{k} = k_x\hat{i} + k_y\hat{j} + k_z\hat{k}, \tag{5.33}$$

where $k_l = \frac{2\pi}{L}n_l$ and $l = x, y, z$; n_l is an integer; and the plane of the electromagnetic polarization of each mode is perpendicular to its wavevector. These results yield planewave solutions useful in the study of the transport properties within the box.

Since the operators for the $\vec{A}, \vec{E}, \vec{B}$ fields are linear operators, the total fields in the box are composed as sums over their wavevectors and polarization states. In this regard, the vector potential in three dimensions for a planewave mode of wavevector $\vec{k}$ takes the general form [3,4,7,10]

$$\vec{A}_{\vec{k}s}(\vec{r},t) = \sqrt{\frac{\hbar}{2\varepsilon_0 L^3 \omega_{\vec{k}}}}\hat{e}_{\vec{k}s}\left[a_{\vec{k}s}e^{i(\vec{k}\cdot\vec{r}-\omega_{\vec{k}}t)} + a_{\vec{k}s}^+ e^{-i(\vec{k}\cdot\vec{r}-\omega_{\vec{k}}t)}\right], \tag{5.34}$$

where $\hat{e}_{\vec{k}s}$ is a unit vector in the direction of the electric field polarization, s labels the two possible polarizations in the plane perpendicular to the wavevector, and we have now explicitly indicated on the left of the equation that $\vec{A}_{\vec{k}s}(\vec{r},t)$ is a mode of wavevector $\vec{k}$ and polarization s. Consequently, the vector potential operator for a general electromagnetic field composed of a sum of many modes in the infinite cube is then a sum over the wavevector components and their polarizations. The resulting sum for the total field is expressed as

$$\vec{A}(\vec{r},t) = \sum_{\vec{k}s}\vec{A}_{\vec{k}s}(\vec{r},t), \tag{5.35}$$

and corresponding $\vec{E}$, $\vec{B}$ fields are obtained from the relations in Eq. (5.16).

Similarly, corresponding to the field operator in Eq. (5.35), the general expression for the energy of the total fields in the box is the sum over the energies of the individual wavevectors and polarization modes. This is given by

$$H = \sum_{\vec{k}s} \hbar\omega \left\{ N_{\vec{k}s} + \frac{1}{2} \right\}, \tag{5.36}$$

where $N_{\vec{k}s} = a^{+}_{\vec{k}s} a_{\vec{k}s}$ is the number operator for the $\vec{k}s$ mode.

5.1.2 Cavity Modes

As an additional example of a quantized electromagnetic field, consider the case in which a mode propagating along the x-axis is placed between two perfectly conducting parallel plates of area $L^2 \to \infty$, one located at $x = 0$ and the other at $x = d$. The solution for the field operators now involves a mixture of right and left propagating planewaves. These waves add to form standing waves matching the electromagnetic boundary conditions at the perfect conducting plates. (See Figure 5.1 for a schematic of the cavity geometry.)

Specifically, the electric fields of the modes must vanish at $x = 0$ and $x = d$. Under these conditions, from Eq. (5.21) and considering both the $\pm k$ solutions, it follows that the classical electrodynamic fields of the modal solutions form a discrete set of modes between the plates. These discrete modes, for the case of an electric field polarized along the y-direction, are of the general form [4]

$$\vec{A}_n(x,t) = 2i\hat{j}\left[A_n e^{-i\omega_n t} - A_n^* e^{i\omega_n t}\right]\sin(k_n x), \tag{5.37a}$$

$$\vec{E}_n(x,t) = -2\omega_n \hat{j}\left[A_n e^{-i\omega_n t} + A_n^* e^{i\omega_n t}\right]\sin(k_n x), \tag{5.37b}$$

and

$$\vec{B}_n(x,t) = i2k_n \hat{k}\left[A_n e^{-i\omega_n t} - A_n^* e^{i\omega_n t}\right]\cos(k_n x), \tag{5.37c}$$

where $k_n = \dfrac{n\pi}{L}$, $\omega_n = ck_n$, and n is a positive integer. (Note: An addition set of modes with the electric field polarized along the z-directions also exists and are similarly obtained from the solutions

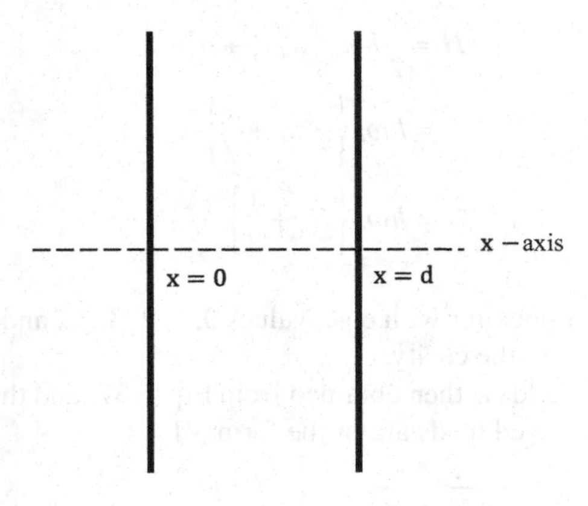

FIGURE 5.1 Two perfect conducting parallel plates separated along the x-axis by a distance d.

of the planewave modes with electric fields polarized along the z-directions. These standing wave modes are not addressed here.)

The energy within the classical fields represented in Eq. (5.37) is then obtained from an application of Eq. (5.23). In this way for the fields in Eq. (5.37), it is found that [3,4,7]

$$\begin{aligned} U &= 2\varepsilon_0 L^2 d\omega_n^2 \left[A_n A_n^* + A_n^* A_n \right] \\ &= \frac{1}{2}\hbar\omega_n \left\{ \frac{4\varepsilon_0 L^2 d\omega_n}{\hbar} \left[A_n A_n^* + A_n^* A_n \right] \right\}. \end{aligned} \tag{5.38}$$

Following the same quantization procedure as with the planewave modes discussed in Eqs. (5.23)–(5.29), we define

$$a_n = 2\sqrt{\frac{\varepsilon_0 L^2 d\omega_n}{\hbar}} A_n \tag{5.39a}$$

and

$$a_n^* = 2\sqrt{\frac{\varepsilon_0 L^2 d\omega_n}{\hbar}} A_n^*. \tag{5.39b}$$

In terms of Eq. (5.39), the energy becomes

$$U = \frac{1}{2}\hbar\omega_n \left\{ a_n a_n^* + a_n^* a_n \right\}, \tag{5.40}$$

and is in a form that can now be quantized.

The quantization follows by introducing quantum mechanical operators in place of a_n and a_n^*, along with a quantization hypothesis. Specifically, the association $a_n \leftrightarrow a_n$ and $a_n^* \leftrightarrow a_n^+$ is made between the classical coefficients (on the left) and quantum operators (on the right). For the quantization, the quantum operators are postulated to obey the commutation relations

$$\left[a_n, a_n^+ \right] = 1. \tag{5.41}$$

From Eqs. (5.40) and (5.41), it then follows that the quantum mechanical Hamiltonian of the fields is rewritten in terms of the quantum operators as [3,7]

$$\begin{aligned} H &= \frac{1}{2}\hbar\omega_n \left\{ a_n a_n^+ + a_n^+ a_n \right\} \\ &= \hbar\omega_n \left\{ a_n^+ a_n + \frac{1}{2} \right\} \\ &= \hbar\omega_n \left\{ N_n + \frac{1}{2} \right\}. \end{aligned} \tag{5.42}$$

Here $N_n = a_n^+ a_n$ is the number operator with eigenvalues 0, 1, 2, 3, …, and the number operator represents the number of photons in the cavity.

The quantized form of the fields is then obtained from Eq. (5.37) and the quantization conditions. Proceeding this way, the quantized fields are of the form [4]

$$\vec{A}_n(x,t) = i\sqrt{\frac{\hbar}{\varepsilon_0 L^2 d\omega_n}}\,\hat{j}\left[a_n e^{-i\omega_n t} - a_n^+ e^{i\omega_n t} \right] \sin(k_n x), \tag{5.43a}$$

$$\vec{E}_n(x,t) = -\omega_n \sqrt{\frac{\hbar}{\varepsilon_0 L^2 d\omega_n}} \hat{j}\left[a_n e^{-i\omega_n t} + a_n^+ e^{i\omega_n t}\right]\sin(k_n x), \tag{5.43b}$$

and

$$\vec{B}_n(x,t) = ik_n \sqrt{\frac{\hbar}{\varepsilon_0 L^2 d\omega_n}} \hat{k}\left[a_n e^{-i\omega_n t} - a_n^+ e^{i\omega_n t}\right]\cos(k_n x). \tag{5.43c}$$

In Eq. (5.42), the eigenstates of $N_n = a_n^+ a_n$ and, consequently of the Hamiltonian, are known as Fock or number states. In these states, while the Hamiltonian is diagonal, the fields in Eq. (5.43) are not diagonal.

The Fock states are not the only important states occurring in quantum optics. Next, another type of photon state of the system, known as a coherent state, is discussed. This has important statistical properties that are useful in later discussions.

5.1.3 Coherent States

While the Fock states are the eigenmodes of the electrodynamic Hamiltonians in vacuum and of their number operators, the electrodynamic field operators in Eqs. (5.32) and (5.43) are not diagonal in their respective Fock state bases [3–7]. In fact, the expectation values of the field operators of Eqs. (5.32) and (5.43) in their $|n\rangle$ respective Fock states are zero. This follows directly from the expectation values of the operators a_y and a_y^+ of Eq. (5.32)

$$\langle n|a_y|n\rangle = \langle n|a_y^+|n\rangle = 0 \tag{5.44a}$$

in their Fock basis, and for the expectation values of the operators a_n and a_n^+ of Eq. (5.43) in their Fock states, i.e.,

$$\langle n|a_n|n\rangle = \langle n|a_n^+|n\rangle = 0. \tag{5.44b}$$

Note that no matter how large the number of photons in $\langle n|$, the expectation value of the fields observed from Eqs. (5.32) or (5.43) is zero. Specifically, raising the number of photons n in the state does not make the expectation value approach a classical observable electromagnetic field. A way of arriving at the classical (correspondence principle) limit of the fields is through a different set of states known as the coherent states [3–7].

To understand the coherent state approach, first consider its application to the one-dimensional mechanical harmonic oscillator problem developed in Eqs. (5.1)–(5.12). For this problem, the position operator of the oscillator from Eq. (5.4) is

$$x = \frac{1}{2}\left(\frac{2\hbar}{m\omega}\right)^{\frac{1}{2}}\left[a + a^+\right], \tag{5.45}$$

and, similar to the considerations of the electric field, a Fock state $|n\rangle$ of the harmonic oscillator is defined as an eigenstate of the mechanical energy $U = \left(N + \frac{1}{2}\right)\hbar\omega$ and the number operator $N = a^+ a$. The expectation value of the position operator in Eq. (5.45) in a Fock state, however, gives

$$\langle n|x|n\rangle = 0, \tag{5.46}$$

no matter how large is n. In this regard, the expectation value of the position operator of the oscillator in any of the Fock states does not exhibit the expected periodic motion of the classical limit of the system [3–7,10].

One way of obtaining the periodic motion of an oscillator in the classical limit of large fields is to form an oscillator wavefunction as a linear combination of Fock states. This is done by looking for linear combinations of Fock states that are eigenstates of the a operator in Eq. (5.45). These eigenstates have nonzero values when acted on by the a operator and are known as coherent states. (Note that the a^+ operator has no eigenstates and that is why only the eigenstates of the a operator are studied in the present context.) It is subsequently shown that, within the context of the coherent states defined in this way, the expectation value of x in the coherent states exhibits periodic oscillations in space with a frequency ω of the harmonic oscillator. In the limits of a large occupancy of the Fock modes in the coherent states, the limiting behavior of a classical oscillator is achieved for the expectation value of x.

The eigenstates of the a operator are obtained as the solutions of an eigenvalue problem written as

$$a|\alpha\rangle = \alpha|\alpha\rangle. \tag{5.47}$$

Solving the eigenvalue problem, it is found that the normalized eigenfunctions are [3–7]

$$|\alpha\rangle = e^{-\frac{|\alpha|^2}{2}} \sum_{n=0}^{\infty} \frac{\alpha^n}{\sqrt{n!}} |n\rangle, \tag{5.48a}$$

where α is a complex number, and the related eigenvalue problem $\langle\alpha|a^+ = \langle\alpha|\alpha^*$ similarly has the wavefunction solution

$$\langle\alpha| = e^{-\frac{|\alpha|^2}{2}} \sum_{n=0}^{\infty} \frac{(\alpha^*)^n}{\sqrt{n!}} \langle n|. \tag{5.48b}$$

Note that for the wavefunctions in Eq. (5.48), it follows that $\langle\alpha|\alpha\rangle = 1$ so that the coherent states are normalized to one. Also note that both of these wavefunctions are in the form of a Poison distribution over the Fock states of the system.

Since a and a^+ are not Hermitian, the coherent state eigenmodes are not orthogonal to one another. Specifically, for general $\alpha \neq \beta$ it follows from Eq. (5.48) that $\langle\beta|\alpha\rangle \neq 0$. In fact, the states in Eq. (5.48) are over complete, i.e., they form a basis of states, which in total are not linearly independent. This is not a problem for the present discussions as the coherent states are only being used here to show how Fock states can be combined into a wavefunction with a limiting behavior approaching that of the motion of a classical harmonic oscillator.

To see that the coherent states in Eq. (5.48) can reproduce classical mechanical oscillator motion, the expectation values of the position and momentum operators in the coherent states are computed. First consider the expectation value of the position operator in the coherent state $|\alpha\rangle$ for which

$$\begin{aligned}\langle\alpha|x(t)|\alpha\rangle &= \sqrt{\frac{\hbar}{2m\omega}}\left[\langle\alpha|a(t)|\alpha\rangle + \langle\alpha|a^+(t)|\alpha\rangle\right] = \sqrt{\frac{\hbar}{2m\omega}}\left[\alpha e^{-i\omega t} + \alpha^* e^{i\omega t}\right] \\ &= \sqrt{\frac{\hbar}{2m\omega}}\, 2|\alpha|\cos(\omega t - \vartheta)\end{aligned}, \tag{5.49a}$$

where $\alpha = |\alpha| e^{i\vartheta}$. The position expectation value is seen to display the sinusoidal dependence on time of a classical oscillator of frequency ω. Similarly, for the expectation value of the momentum operator

$$\langle\alpha|p(t)|\alpha\rangle = -i\sqrt{\frac{\hbar m\omega}{2}}\left[\langle\alpha|a(t)|\alpha\rangle - \langle\alpha|a^{+}(t)|\alpha\rangle\right] = -i\sqrt{\frac{\hbar m\omega}{2}}\left[\alpha e^{-i\omega t} - \alpha^{*}e^{i\omega t}\right]$$
$$= -\sqrt{\frac{\hbar m\omega}{2}}2|\alpha|\sin(\omega t - \vartheta) \quad (5.49b)$$

This again displays the correct behavior of the momentum in the classical oscillator limit.

Preceding along these lines, it is also found that [3–7]

$$\langle\alpha|[x(t)]^2|\alpha\rangle = \frac{\hbar}{2m\omega}\left[2|\alpha|^2\cos(2\omega t - 2\vartheta) + 2|\alpha|^2 + 1\right] \quad (5.50a)$$

and

$$\langle\alpha|[p(t)]^2|\alpha\rangle = \frac{\hbar m\omega}{2}\left[-2|\alpha|^2\cos(2\omega t - 2\vartheta) + 2|\alpha|^2 + 1\right]. \quad (5.50b)$$

Computing the statistical variance of the position and momentum from Eqs. (5.49) and (5.50), it then follows that

$$\sqrt{\langle\alpha|[p(t)]^2|\alpha\rangle - \langle\alpha|p(t)|\alpha\rangle^2}\sqrt{\langle\alpha|[x(t)]^2|\alpha\rangle - \langle\alpha|x(t)|\alpha\rangle^2} = \frac{\hbar}{2}. \quad (5.51)$$

This is a statement of the minimum Heisenberg uncertainty in the context of the coherent state averages. The coherent states exhibit a minimum uncertainty motion. This means that they generate states that give the correct correspondence principle limit between the quantum and classical mechanic problems.

It is seen from Eq. (5.49) that the average position and momentum in the coherent states behave like classical harmonic oscillations, and, in addition, from Eq. (5.51) these states have the minimum uncertainty in position and momentum. Consequently, the development of the coherent state wavefunctions in the large α limit corresponds to the limit of the classical oscillator solutions. Since the modes of the quantum electrodynamic fields are essentially oscillator modes, in an appropriate limit the coherent states of these quantum fields should reproduce the behaviors of the fields in classical electrodynamics. This is now discussed.

Consider the quantized fields in Eq. (5.32) for the electric and magnetic induction in vacuum. These describe planewave propagation along the x-direction with wavevector k and frequency $\omega = ck$ for the case of electric field polarization along the y-axis, and are represented by

$$\vec{E}(x,t) = E_y(x,t)\hat{j} = i\omega\sqrt{\frac{\hbar}{2\varepsilon_0 L^3\omega}}\hat{j}\left[a_y e^{i(kx-\omega t)} - a_y^{+}e^{-i(kx-\omega t)}\right], \quad (5.52a)$$

and

$$\vec{B}(x,t) = B_z(x,t)\hat{k} = ik\sqrt{\frac{\hbar}{2\varepsilon_0 L^3\omega}}\hat{k}\left[a_y e^{i(kx-\omega t)} - a_y^{+}e^{-i(kx-\omega t)}\right]. \quad (5.52b)$$

It is seen from Eq. (5.52) that at each point x the time development of the fields is harmonic, just like the x amplitude of the one-dimensional mechanical oscillator. As with the mechanical oscillator, in terms of the field operators a_y and a_y^{+} of the electrodynamic problem, the Fock states (denoted by $|n\rangle$) are again eigenstates of the number operator $a_y^{+}a_y$, with the properties $a_y|n\rangle = \sqrt{n}|n-1\rangle$ for $n > 0$ and $a_y^{+}|n\rangle = \sqrt{n+1}|n+1\rangle$ for $n \geq 0$. Consequently, in terms of these Fock states, the coherent states of the a_y and a_y^+ electromagnetic field operators are again given by the standard forms $|\alpha\rangle = e^{-\frac{|\alpha|^2}{2}}\sum_{n=0}^{\infty}\frac{\alpha^n}{\sqrt{n!}}|n\rangle$ and $\langle\alpha| = e^{-\frac{|\alpha|^2}{2}}\sum_{n=0}^{\infty}\frac{(\alpha^*)^n}{\sqrt{n!}}\langle n|$.

As in the earlier discussion of the mechanical oscillator, the expectation value of the fields in the Fock modes are [3–7]

$$\langle n|E_y(x,t)|n\rangle = 0 \tag{5.53a}$$

and

$$\langle n|B_z(x,t)|n\rangle = 0, \tag{5.53b}$$

so that the Fock states are not appropriate forms for investigating the approach to the limiting fields of classical electrodynamics. In the coherent states, however, it follows that

$$\langle\alpha|E_y(x,t)|\alpha\rangle = -\omega\sqrt{\frac{\hbar}{2\varepsilon_0 L^3\omega}}2|\alpha|\sin(kx-\omega t+\vartheta) \tag{5.54a}$$

and

$$\langle\alpha|B_z(x,t)|\alpha\rangle = -k\sqrt{\frac{\hbar}{2\varepsilon_0 L^3\omega}}2|\alpha|\sin(kx-\omega t+\vartheta). \tag{5.54b}$$

From Eq. (5.54), it is seen that both the fields exhibit the harmonic variations in space and time, which are familiar from classical electrodynamics. Consequently, the correspondence principle limits between the quantum and classical systems are reached within the representation of the quantum system in terms of the coherent states.

Similar to the treatment of the mechanical oscillator in Eq. (5.50), the moments of the field operators in the coherent states are calculated and found to be of the forms

$$\langle\alpha|E_y^2(x,t)|\alpha\rangle = \omega^2\frac{\hbar}{2\varepsilon_0 L^3\omega}\left[1+2|\alpha|^2-2|\alpha|^2\cos 2(kx-\omega t+\vartheta)\right] \tag{5.55a}$$

and

$$\langle\alpha|B_z^2(x,t)|\alpha\rangle = k^2\frac{\hbar}{2\varepsilon_0 L^3\omega}\left[1+2|\alpha|^2-2|\alpha|^2\cos 2(kx-\omega t+\vartheta)\right]. \tag{5.55b}$$

From Eqs. (5.54) and (5.55), it then follows that the statistical variances of the two fields are computed, and their product seems to obey the relation

$$\sqrt{\langle\alpha|E_y^2(x,t)|\alpha\rangle-\langle\alpha|E_y(x,t)|\alpha\rangle^2}\sqrt{\langle\alpha|B_z^2(x,t)|\alpha\rangle-\langle\alpha|B_z(x,t)|\alpha\rangle^2} = \frac{\hbar}{2\varepsilon_0 L^3}\frac{\omega}{c}. \tag{5.56}$$

From Eqs. (5.54) to (5.56), it is then found that the fields are well defined in the $|\alpha|\to\infty$ classical limit of the quantum fields and approach the limits of classical electrodynamics. Specifically, the amplitudes of both the fields are proportional to $|\alpha|\to\infty$, while the product of their variances is independent of $|\alpha|\to\infty$.

Note that in the treatment in Eqs. (5.52)–(5.56) only a single mode of wavevector k and frequency $\omega = ck$ has been considered, but the electrodynamics of the total vacuum fields can be composed of many modes of different wavevectors and frequencies. As discussed in Section 5.1.1, these different modes are linear modes of the system that exist independently of one another. As a result, the total fields in the system add, and the total wavefunctions of the system are just a direct product of the independent modes possible. Specifically, the vector potential operator of a single mode is represented in Eq. (5.34), and the total vector potential field operator, given as a sum over the individual field operators, is in Eq. (5.35). For the correspondence principle limit, the wavefunction of the total electromagnetic field operator is a direct product of the coherent states of all of the different modes of the system.

Up to now only the quantization of the fields have been treated. In the following sections, the interaction of the fields with atoms and electronic systems is considered.

5.2 FIELD INTERACTIONS WITH ATOMS AND ELECTRONS

In this section, simple models for the interaction of the electromagnetic fields with atoms and electronic systems are considered. These include discussions of the Jaynes–Cumming model for the case of an electromagnetic field interaction with a two-level atom and some basic treatment of atoms confined within an electromagnetic cavity [3–7]. While the Jaynes–Cumming model is of general interest in condensed matter physics, the cavity problem has been of particular interest in the development of various ideas of quantum computing [11] and in the study of fundamental problems in the nature of quantum systems.

First a general discussion of the Jaynes–Cumming problem is given, and this is followed by its application in a treatment of the problem of an atom in an electromagnetic cavity.

5.2.1 JAYNES–CUMMING MODEL

The Jaynes–Cumming model is a simple model for the description of the interaction of a bound electron with an externally applied electromagnetic field [3–7]. It treats a single electron that occupies one of two energy levels (i.e., a ground state or an excited state) of a two-level electronic system while interacting with a single mode of an applied electromagnetic field. The two quantum mechanical systems—the bound electron and the electromagnetic fields are coupled to one another through an electric dipole interaction [12] so as to transfer energy back and forth between the two. In the following, the Jaynes–Cumming model Hamiltonian is developed and illustrated with a number of example problems.

To begin, consider an electron that occupies one of the two different energy levels. The lower energy level is the ground state with an electron wavefunction denoted by $|g\rangle$, and the upper energy level is the excited state with an electron wavefunction $|e\rangle$. Both of these electron states are separated from one another by an energy $\hbar\omega_{eg}$ so that the electron Hamiltonian in the absence of external interactions is given by [3–7]

$$H^E = \frac{1}{2}\hbar\omega_{eg}|e\rangle\langle e| - \frac{1}{2}\hbar\omega_{eg}|g\rangle\langle g|. \tag{5.57}$$

The applied electromagnetic field interacting with the electron is described as a single mode of frequency ω. For specificity, the mode considered here is taken to be of the form of a cavity mode propagating along the x-direction with an electric field polarized in the y-direction. (See Figure 5.2

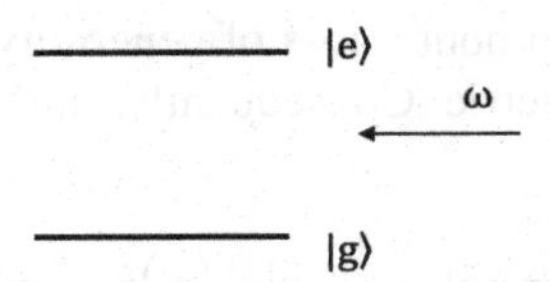

FIGURE 5.2 A two-level electron system with a ground state $|g\rangle$ and an excited $|e\rangle$ interacting with a resonant cavity field of frequency $\omega \approx \omega_{eg}$. The atom is contained within the cavity and interacts with the cavity fields by the Jaynes–Cumming Hamiltonian.

for a schematic of the cavity geometry.) The general form of the modal fields of such a mode is given in Eq. (5.43), and an abbreviated notation is adopted in which the distinguishing modal index n is omitted in the present discussions. Under these conditions, the electric field of the single cavity mode has the basic representation

$$E(x,t) = -\epsilon_\omega \left(a + a^+\right)\sin(kx) \tag{5.58a}$$

where $a = a_0 e^{-i\omega t}$ and $a^+ = a_0^+ e^{i\omega t}$, the subscript 0 is the index of the cavity mode chosen, and

$$\epsilon_\omega = \omega\sqrt{\frac{\hbar}{\varepsilon_0 L^2 d\omega}}, \tag{5.58b}$$

with $\omega = ck$ for k set by the cavity boundary conditions.

In this notation, the Hamiltonian of the field in Eq. (5.58) is from Eq. (5.42) given by

$$H^F = \hbar\omega a^+ a, \tag{5.59}$$

where the constant zero-point energy $\frac{1}{2}\hbar\omega$ term is dropped. The zero-point energy does not affect the dynamics of interest in the following considerations so is not included in the Jaynes–Cumming Hamiltonian.

The two systems represented by the Hamiltonians in Eqs. (5.57) and (5.59) exist independent of one another, each displaying its own separate dynamics. The final consideration in the formulation of the Jaynes–Cumming Hamiltonian is to include the interaction of the electron with the electromagnetic field. This comes from the coupling of the electric field with the dipole matrix for the electron transition between its ground and excited states. In the formulation of the coupling, it is assumed that the wavelength of the electromagnetic mode is much larger than the size of the atom, molecule, or other feature described by the two-level electron Hamiltonian. Under these considerations the potential characterizing the interaction of the electron with the field takes the form

$$V^I = -PE(x_0) \tag{5.60}$$

where P is the projection of the electron dipole moment on the electric field vector, and x_0 is the location of the electron in the cavity.

For a transition between the ground and excited state of the two-level system, the dipole moment of the transition is written as [3–7]

$$\begin{aligned} P &= |g\rangle\langle g|\tilde{P}|e\rangle\langle e| + |e\rangle\langle e|\tilde{P}|g\rangle\langle g| \\ &= \tilde{P}_{ge}|g\rangle\langle e| + \tilde{P}_{eg}|e\rangle\langle g| \end{aligned} \tag{5.61}$$

where $\tilde{P}_{ge} = \langle g|\tilde{P}|e\rangle$ and $\tilde{P}_{eg} = \langle e|\tilde{P}|g\rangle$ are the dipole transition matrices of the electronic feature, and $E(x_0)$ is the intensity of the electric field at the position x_0 of the electron. A useful simplification occurs in Eq. (5.61) by noting that, without a loss of generality, the transition coefficients $\tilde{P}_{ge} = \tilde{P}_{eg}$ can be assumed to be real and symmetric. Consequently, in terms of Eqs. (5.58) and (5.52) it then follows that Eq. (5.60) becomes

$$V^I = \hbar\tilde{g}\left(|g\rangle\langle e| + |e\rangle\langle g|\right)\left(a + a^+\right), \tag{5.62}$$

where $\tilde{g} = \dfrac{\tilde{P}_{ge}\epsilon_\omega \sin(kx_0)}{\hbar}$ is real.

Assembling these results, the total Hamiltonian of the electron, the fields, and their interaction from Eqs. (5.57), (5.59), and (5.61) is given by [4]

$$\begin{aligned} H &= H^E + H^F + V^I \\ &= \frac{1}{2}\hbar\omega_{eg}|e\rangle\langle e| - \frac{1}{2}\hbar\omega_{eg}|g\rangle\langle g| + \hbar a^+ a + \hbar\tilde{g}\big(|g\rangle\langle e| + |e\rangle\langle g|\big)\big(a + a^+\big) \end{aligned} \quad (5.63)$$

This expresses the complete dynamics of the Jaynes–Cumming problem in terms of the Schrodinger equation based on Eq. (5.63), which takes the form [4]

$$\frac{\partial}{\partial t}|\psi\rangle = -\frac{i}{\hbar}\left[H^E + H^F + V^I\right]|\psi\rangle \quad (5.64)$$

where $|\psi\rangle$ is the wavefunction of the electron and fields.

An additional simplification of Eq. (5.64) can be made by assuming a solution for $|\psi\rangle$ of the form

$$|\psi\rangle = e^{\frac{i}{\hbar}\left(H^E + H^F\right)t}|\psi_I\rangle. \quad (5.65)$$

Upon substituting Eq. (5.65) into Eq. (5.64), it then follows that [3–7]

$$\frac{\partial}{\partial t}|\psi_I\rangle = -\frac{i}{\hbar}\mathcal{V}|\psi_I\rangle \quad (5.66)$$

where

$$\begin{aligned} \mathcal{V} &= e^{\frac{i}{\hbar}\left(H^E + H^F\right)t} V^I e^{-\frac{i}{\hbar}\left(H^E + H^F\right)t} \\ &= e^{\frac{i}{\hbar}\left(H^E + H^F\right)t}\left\{\hbar\tilde{g}\big(|g\rangle\langle e| + |e\rangle\langle g|\big)\big(a + a^+\big)\right\}e^{-\frac{i}{\hbar}\left(H^E + H^F\right)t}. \end{aligned} \quad (5.67)$$

Within this reformulation of the problem, known as the interaction formulation, it is seen from Eqs. (5.57) to (5.59) that the coupling between the fields and the electron is completely contained in the single interaction potential

$$\mathcal{V} = \hbar\tilde{g}\left(|g\rangle\langle e|e^{-i\omega_{eg}t} + |e\rangle\langle g|e^{i\omega_{eg}t}\right)\left(ae^{-i\omega t} + a^+e^{i\omega t}\right), \quad (5.68)$$

and terms related solely to either the fields or to the electron are removed from the considerations in Eq. (5.66).

Treating Eq. (5.66) in the case that $\omega \approx \omega_{eg}$ (i.e., the transitions of the electron and electromagnetic modes are assumed to be near resonance.), the dominant terms in Eq. (5.68) take the form [4]

$$\begin{aligned} \mathcal{V} &\approx \hbar\tilde{g}\left(|g\rangle\langle e|a^+e^{i\omega t}e^{-i\omega_{eg}t} + |e\rangle\langle g|ae^{-i\omega t}e^{i\omega_{eg}t}\right) \\ &= \hbar\tilde{g}\left(|g\rangle\langle e|a^+e^{-i(\omega_{eg}-\omega)t} + |e\rangle\langle g|ae^{i(\omega_{eg}-\omega)t}\right). \end{aligned} \quad (5.69)$$

These terms change slowly in time, while those that have been deleted in going from Eqs. (5.68) to (5.69) have rapid time variations proportional to the time factors

$$e^{\pm i(\omega_{eg}+\omega)t} \approx e^{\pm i2\omega_{eg}t}. \quad (5.70)$$

On average, only the slowing varying terms in time, given in Eq. (5.69), contribute a significant contribution to the dynamics of the problem. On the scale of the dynamics governed by the slow terms in Eq. (5.69), the dynamics of the rapidly varying terms tend to quickly average to zero.

In addition, from Eq. (5.68), the rapidly varying terms are seen not to conserve the energy of the system so that they become less important at increasing time scales. The approximation in Eq. (5.69) is known as the Rotating Wave Approximation [3,4,7] and is often applied in quantum optics, magnetic resonance, and a variety of other problems in quantum mechanics.

5.2.2 Jaynes–Cumming Model: Example of Fock States

As an example of the Jaynes–Cumming problem, the case of a resonant solution for which $\omega \approx \omega_{eg}$ can be directly obtained from the formulation in Eq. (5.65)–(5.69). Specifically, writing $|\psi_I\rangle$ in Eq. (5.65) in the form [3,4,7]

$$|\psi_I\rangle = c_1(t)|g\rangle|n\rangle + c_2(t)|e\rangle|n-1\rangle, \tag{5.71}$$

a set of equations for the unknown coefficients $c_1(t)$ and $c_2(t)$ is generated for the dynamics between the ground state in the presence of n phonons and the excited state in the presence of $n-1$ phonons. This is done by substituting Eq. (5.71) into Eq. (5.66) for the interaction potential in Eq. (5.69).

Upon making the substitution and using the orthogonality of the basis $|g\rangle|n\rangle$ and $|e\rangle|n-1\rangle$, it is found that the unknown coefficients are the solutions of the set of two differential equations [4]

$$\dot{c}_1(t) = -i\sqrt{n}\tilde{g}c_2(t) \tag{5.72a}$$

and

$$\dot{c}_2(t) = -i\sqrt{n}\tilde{g}c_1(t) \tag{5.72b}$$

for appropriate boundary conditions on $c_1(t=0)$ and $c_2(t=0)$. From Eqs. (5.72a) and (5.72b), it then follows for the case in which $\omega = \omega_{eg}$ that

$$\ddot{c}_1(t) + n\tilde{g}^2 c_1(t) = 0, \tag{5.73}$$

and $c_2(t)$ is obtained from Eq. (5.72b).

For a specific example consider the case of the initial conditions $c_1(0)=1$ and $c_2(0)=0$. From Eqs. (5.72) and (5.73), the coefficients of Eq. (5.71) have the solution [3,7]

$$c_1(t) = \cos(\Omega t) \tag{5.74a}$$

$$c_2(t) = -\mathrm{i}\sin(\Omega t), \tag{5.74b}$$

where $\Omega = \sqrt{n}\tilde{g}$ is the Rabi frequency. It then follows from Eqs. (5.65) to (5.71) that

$$\begin{aligned} |\psi\rangle &= e^{-i[\omega_g + n\omega]t}\cos(\Omega t)|g\rangle|n\rangle - \mathrm{i}\, e^{-i[\omega_e+(n-1)\omega]t}\sin(\Omega t)|e\rangle|n-1\rangle \\ &= e^{-i[\omega_e+(n-1)\omega]t}\left\{e^{i[\omega_{eg}-\omega]t}\cos(\Omega t)|g\rangle|n\rangle - \mathrm{i}\sin(\Omega t)|e\rangle|n-1\rangle\right\} \\ &= e^{-i[\omega_e+(n-1)\omega]t}\left\{\cos(\Omega t)|g\rangle|n\rangle - \mathrm{i}\sin(\Omega t)|e\rangle|n-1\rangle\right\}, \end{aligned} \tag{5.75}$$

where $\omega_g = -\frac{1}{2}\omega_{eg}$ and $\omega_e = \frac{1}{2}\omega_{eg}$ are, respectively, the frequencies of the ground and excited electron states. Consequently, the transition frequency $\omega_{eg} = \omega_e - \omega_g = \omega$ is reproduced.

In the more general off resonance case for which $\omega \approx \omega_{eg}$, the formulation in Eqs. (5.63) and (5.64) is now used in order to provide an example of the direct application of the Schrodinger equation formulation to the Jaynes–Cumming model. This avoids the assumption of a wavefunction of the form in Eq. (6.65) made in the previous example. In this regard, assuming $|\psi\rangle = |\psi_E\rangle e^{-\frac{i}{\hbar}Et}$ is an eigenstate of energy E for the Jaynes–Cumming problem, the Schrodinger equation in Eq. (5.64) becomes an eigenvalue problem of the form [3,4,7]

$$E|\psi_E\rangle = H|\psi_E\rangle \tag{5.76a}$$

where

$$H = \frac{1}{2}\hbar\omega_{eg}|e\rangle\langle e| - \frac{1}{2}\hbar\omega_{eg}|g\rangle\langle g| + \hbar a^{+}a + \hbar\tilde{g}\big(|g\rangle\langle e| + |e\rangle\langle g|\big)\big(a + a^{+}\big). \tag{5.76b}$$

Considering Eq. (5.76) in the basis of states defined by $|1\rangle = |g\rangle|n\rangle$ and $|2\rangle = |e\rangle|n-1\rangle$, it follows that the Hamiltonian matrix takes the form

$$\langle 1|H|1\rangle = -\frac{1}{2}\hbar\omega_{eg} + n\hbar\omega = \epsilon + \hbar\Delta \tag{5.77a}$$

$$\langle 2|H|2\rangle = \frac{1}{2}\hbar\omega_{eg} + (n-1)\hbar\omega = \epsilon \tag{5.77b}$$

$$\langle 1|H|2\rangle = \hbar\sqrt{n}\tilde{g} \tag{5.77c}$$

$$\langle 2|H|1\rangle = \hbar\sqrt{n}\tilde{g} \tag{5.77d}$$

where $\Delta = (\omega - \omega_{eg})$ is the detuning and $\epsilon = \frac{1}{2}\hbar\omega_{eg} + (n-1)\hbar\omega$. In this notation, the eigenvalue problem in Eq. (5.76) becomes a matrix eigenvalue problem of the form

$$\begin{vmatrix} \langle 1|H|1\rangle - E & \langle 1|H|2\rangle \\ \langle 2|H|1\rangle & \langle 2|H|2\rangle - E \end{vmatrix} \begin{vmatrix} |1\rangle \\ |2\rangle \end{vmatrix} = 0, \tag{5.78}$$

where E is the eigenvalue.

The problem in Eq. (5.78) has two eigenvalue energies $E_\pm$ corresponding to two orthogonal eigenstates $|\psi\rangle_\pm$. From Eqs. (5.77) and (5.78), it is then found that the eigenvalues are [3–7]

$$E_\pm = \epsilon + \frac{1}{2}\hbar\Delta \pm \sqrt{\frac{1}{4}\hbar^2\Delta^2 + n(\hbar\tilde{g})^2}, \tag{5.79}$$

and these correspond to the eigenfunctions

$$|\psi\rangle_\pm = \frac{1}{\sqrt{N_\pm}}\left\{\sqrt{n}\hbar\tilde{g}|g\rangle|n\rangle - \left(\frac{1}{2}\hbar\Delta \mp \sqrt{\frac{1}{4}\hbar^2\Delta^2 + n\hbar^2\tilde{g}^2}\right)|e\rangle|n-1\rangle\right\}e^{-\frac{i}{\hbar}E_\pm t}, \tag{5.80}$$

where $N_\pm = 2\hbar^2\left(\frac{1}{4}\Delta^2 + n\tilde{g}^2\right) \mp \hbar^2\Delta\sqrt{\frac{1}{4}\Delta^2 + n\tilde{g}^2}$.

Note that the solution generated from Eq. (5.80), which matches the $t = 0$ boundary conditions of the wavefunction of Eq. (5.75), has the form

$$|\psi\rangle = \frac{1}{\sqrt{2}}\left(|\psi\rangle_+ + |\psi\rangle_-\right). \tag{5.81}$$

In the limit that $\Delta = (\omega - \omega_{eg}) \to 0$, the wavefunction in Eq. (5.81) reduces to the resonance solution in Eq. (5.75).

5.2.3 Jaynes–Cumming Model: Example of Coherent States

The earlier results were for Fock states of the electromagnetic fields, but another type of important field configuration to consider are those in the form of coherent states. These are now discussed for the particular dynamics of an initial $t = 0$ coherent state mode of the form [3,4,7]

$$|g,\alpha\rangle = e^{-\frac{|\alpha|^2}{2}} \sum_{n=0}^{\infty} \frac{\alpha^n}{\sqrt{n!}} |g\rangle|n\rangle. \tag{5.82}$$

The discussion is heavily based on our treatment of the Fock modes in which the $|g\rangle|n\rangle$ and $|e\rangle|n-1\rangle$ states of the system were shown to be the only states that are coupled by the Jaynes–Cumming Hamiltonian. In this regard, it is also noted that the $|g\rangle|0\rangle$ configuration is not coupled by the Hamiltonian to any other degrees of freedom. We now consider the $\omega = \omega_{eg}$ case.

Consequently, the dynamics of the mode in Eq. (5.82) follows directly from the dynamics of each of the individual $|g\rangle|n\rangle$ composed in the sum of Eq. (5.82). From the results in Eqs. (5.74) and (5.75), it then follows that at general time [3,4,7]

$$|g,\alpha\rangle = e^{-\frac{|\alpha|^2}{2}} e^{-i\omega_g t}|g\rangle|0\rangle$$

$$+ e^{-\frac{|\alpha|^2}{2}} \sum_{n=1}^{\infty} \frac{\alpha^n}{\sqrt{n!}} \left[e^{-i[\omega_g + n\omega]t} \cos\left(\tilde{g}\sqrt{n}t\right)|g\rangle|n\rangle - ie^{-i[\omega_e + (n-1)\omega]t} \sin\left(\tilde{g}\sqrt{n}t\right)|e\rangle|n-1\rangle \right]. \tag{5.83}$$

Here no photons are present in $|g\rangle|0\rangle$, so there is an absence of transitions, while the other terms of the sum develop in time independent of one another.

5.2.4 Jaynes–Cumming Model: Temperature Effects

As a final consideration of the Jaynes–Cumming model, the effects on the dynamics of a thermal distribution of photons are now discussed. The calculations for the dynamics of the thermal system proceed similar as in the case of the coherent states. In both systems, the Jaynes–Cumming dynamics between the $|g\rangle|n\rangle$ and $|e\rangle|n-1\rangle$ modes are combined as a weighted sum involving the initial probability that n photons are in the system. The total dynamics is then given as a sum of many possible independent dynamics in the system subject to the initial photon thermal distribution [3–7]. Again, we consider the $\omega = \omega_{eg}$ case.

To obtain the electron–photon wavefunction, it is important to remember the nature of the thermal photon distribution. Specifically, the distribution of photons of frequency ω at a temperature T is given by the Einstein occupation formula, and from this distribution the probability $P(n)$ of finding n photons in the system is described in the standard form

$$P(n) = e^{-n\hbar\omega/k_BT}\left(1 - e^{-\hbar\omega/k_BT}\right). \tag{5.84}$$

From Eq. (5.84), the mean number of photons in the system as a function of temperature is then given in terms of $P(n)$ by

$$\bar{n} = \sum_{n=0}^{\infty} nP(n) = \frac{1}{e^{\hbar\omega/k_BT} - 1}. \tag{5.85}$$

Inverting Eq. (5.85), $e^{\hbar\omega/k_BT}$ is obtained as an expression in terms of $\bar{n}$, which when combined with Eq. (5.84) relates $P(n)$ to the average number of photons in the system by [4,7]

$$P(n) = \frac{\bar{n}^n}{(\bar{n}+1)^{n+1}}. \tag{5.86}$$

Later it is found that the form in Eq. (5.86) is most useful in writing the general form of the Jaynes–Cumming wavefunction averages in temperature-dependent systems.

Once the distribution of photon modes is determined, the results in Eqs. (5.74) and (5.75) can be assembled along with the weighting in Eq. (5.86) to determine the total electron–photon operator averages in the system. For this assembly, it is assumed that the electron is in its ground state $|g\rangle$ at $t = 0$. Consequently, under these conditions for a thermally distributed photon field of frequency ω, the average of the operator O in the Schrodinger representation is given by [4]

$$\langle O \rangle = \sum_{n=0}^{\infty} \frac{\bar{n}^n}{(\bar{n}+1)^{n+1}} \langle \tilde{n} \,|\, O \,|\, \tilde{n} \rangle, \tag{5.87a}$$

where

$$|\tilde{0}\rangle = e^{-i\omega_g t}|g\rangle|0\rangle \tag{5.87b}$$

and for $\tilde{n} > 0$

$$\begin{aligned} |\tilde{n}\rangle &= \left[e^{-i[\omega_g + n\omega]t} \cos(\tilde{g}\sqrt{n}t)|g\rangle|n\rangle - ie^{-i[\omega_e + (n-1)\omega]t} \sin(\tilde{g}\sqrt{n}t)|e\rangle|n-1\rangle \right] \\ &= e^{-i[\omega_e + (n-1)\omega]t} \left\{ \cos(\Omega t)|g\rangle|n\rangle - \mathrm{i}\, \sin(\Omega t)|e\rangle|n-1\rangle \right\}. \end{aligned} \tag{5.87c}$$

As with the coherent states, again the $|g\rangle|0\rangle$ term remains uncoupled to the other degrees of freedom of the system, and the $|g\rangle|n\rangle$ and $|e\rangle|n-1\rangle$ degrees of freedom transition between one another. The average operator $\langle O \rangle$ in Eq. (5.87a) then involves the average of the operator $\langle \tilde{n} \,|\, O \,|\, \tilde{n} \rangle$ in each of the quantum states of the system weighted with the initial photon distribution $P(n)$. In this regard, the thermal average is taken after the quantum averages.

Next, the nature of the coherence of light is investigated. This is involved with the quantification of how correlated in space and time are the optical wavefunctions and the properties they represent.

5.3 OPTICAL CORRELATIONS AND COHERENCE

An important property of light is its coherence. Coherence has to do with the correlation of the physical properties of light in space and time [7,10]. Specifically, based on a knowledge of its behavior in a small space–time region, how predictable is the behavior of the light outside that region. In the following, an assessment of the predictability of the nature of various types of light is made through the evaluation of a systematically defined series of field coherence functions. These functions provide a quantification of the behavior of light in space and time.

For a discussion of the coherence functions of light in vacuum, we revert to the three-dimensional formulation of the field operators in terms of electromagnetic planewave modes. This formulation was discussed in Section 5.1.1, where the general expression for the vector field operator $\vec{A}(\vec{r},t)$ was obtained in Eq. (5.34). From the relationship $\vec{E} = -\frac{\partial \vec{A}}{\partial t}$, it is then found that the modal electric field operator in the planewave basis has the form [3,7]

$$\begin{aligned}\vec{E}_{\vec{k}s}(\vec{r},t) &= i\omega_{\vec{k}}\sqrt{\frac{\hbar}{2\varepsilon_0 L^3\omega_{\vec{k}}}}\hat{e}_{\vec{k}s}\left[a_{\vec{k}s}e^{i(\vec{k}\cdot\vec{r}-\omega_{\vec{k}}t)} - a^{+}_{\vec{k}s}e^{-i(\vec{k}\cdot\vec{r}-\omega_{\vec{k}}t)}\right] \\ &= \vec{E}^{(+)}_{\vec{k}s}(\vec{r},t) + \vec{E}^{(-)}_{\vec{k}s}(\vec{r},t)\end{aligned} \tag{5.88}$$

where

$$\vec{E}^{(+)}_{\vec{k}s}(\vec{r},t) = i\omega_{\vec{k}}\sqrt{\frac{\hbar}{2\varepsilon_0 L^3\omega_{\vec{k}}}}\hat{e}_{\vec{k}s}a_{\vec{k}s}e^{i(\vec{k}\cdot\vec{r}-\omega_{\vec{k}}t)} \tag{5.89a}$$

and

$$\vec{E}^{(-)}_{\vec{k}s}(\vec{r},t) = -i\omega_{\vec{k}}\sqrt{\frac{\hbar}{2\varepsilon_0 L^3\omega_{\vec{k}}}}\hat{e}_{\vec{k}s}a^{+}_{\vec{k}s}e^{-i(\vec{k}\cdot\vec{r}-\omega_{\vec{k}}t)}. \tag{5.89b}$$

In this notation, the modal field operator is separated into a collection of the terms $\vec{E}^{(+)}_{\vec{k}s}(\vec{r},t)$ that remove $\vec{k}s$ photons from the optical fields, and terms $\vec{E}^{(-)}_{\vec{k}s}(\vec{r},t)$ that add $\vec{k}s$ photons to the optical fields. In this regard, $\vec{E}^{(+)}_{\vec{k}s}(\vec{r},t)$ enters into the description of photon absorption processes, while $\vec{E}^{(-)}_{\vec{k}s}(\vec{r},t)$ enters into the treatment of photon emission processes. The total electric field operator is then composed as a sum over all the various planewave eigenmodes and is given by

$$\vec{E}(\vec{r},t) = \sum_{\vec{k}s}\left[\vec{E}^{(+)}_{\vec{k}s}(\vec{r},t) + \vec{E}^{(-)}_{\vec{k}s}(\vec{r},t)\right] = \vec{E}^{(+)}(\vec{r},t) + \vec{E}^{(-)}(\vec{r},t) \tag{5.90}$$

so that the total field operator $\vec{E}^{(+)}(\vec{r},t)$ removes photons from the total fields, and the terms $\vec{E}^{(-)}(\vec{r},t)$ add photons to the total fields.

The formulation in Eqs. (5.88)–(5.90) is then the most general formulation of the optical fields in three dimensions. For simplicity, in the following discussions of the field correlations, however, it is assumed that the modes of interest propagate in the $x-y$ plane and are polarized with their electric fields in the z-direction. This allows us to treat the electric field and its correlator as scalars, eliminating the full vector fields and tensor correlation functions considered in more general problems.

In terms of the fields in Eq. (5.90) a first coherence function, providing a measure of the phase coherent properties of the electromagnetic field, is defined. The coherence function is of the form of a statistical correlation function given by the specific form [3–7]

$$g^{(1)}(\vec{r}_1,t_1 \mid \vec{r}_2,t_2) = \frac{G^{(1)}(\vec{r}_1,t_1 \mid \vec{r}_2,t_2)}{\sqrt{G^{(1)}(\vec{r}_1,t_1 \mid \vec{r}_1,t_1)G^{(1)}(\vec{r}_2,t_2 \mid \vec{r}_2,t_2)}}, \tag{5.91a}$$

where for the electric field along the z-direction

$$G^{(1)}(\vec{r}_1,t_1 \mid \vec{r}_2,t_2) = \sum_{i,f} P_i \langle i|E^{(-)}(\vec{r}_1,t_1)|f\rangle\langle f|E^{(+)}(\vec{r}_2,t_2)|i\rangle, \tag{5.91b}$$

and $|f\rangle$ and $|i\rangle$ are planewave eigenmodes of the fields. Both the i and f sums in Eq. (5.91b) are over the possible modes of the system; however, note that the $|i\rangle$ modes are weighted by the probability P_i that they are present in the equilibrium radiation fields of the system. It follows from Eq. (5.91b) that $G^{(1)}(\vec{r}_1,t_1 \mid \vec{r}_1,t_1)$ and $G^{(1)}(\vec{r}_2,t_2 \mid \vec{r}_2,t_2)$ are both positive real functions, so that the argument of the square root in Eq. (5.91a) is real.

The coherence function has a number of important general properties arising from its mathematical structure. Specifically, from the Schwartz inequality it is seen that [3–7]

$$\left|G^{(1)}(\vec{r}_1,t_1 \mid \vec{r}_2,t_2)\right| \le \sqrt{G^{(1)}(\vec{r}_1,t_1 \mid \vec{r}_1,t_1)G^{(1)}(\vec{r}_2,t_2 \mid \vec{r}_2,t_2)}, \tag{5.92}$$

And, applying this to the form in Eq. (5.91a), it is found that the coherence function is bounded by the inequality

$$\left|g^{(1)}(\vec{r}_1,t_1 \mid \vec{r}_2,t_2)\right| \le 1. \tag{5.93}$$

In addition, as $|t_2 - t_1| \to \infty$ from Eq. (5.91b), it follows that for random fields with a zero time average, the time separation of the fields results in the limit form $\left|g^{(1)}(\vec{r}_1,t_1 \mid \vec{r}_2,t_2)\right| \to 0$.

The general form of $g^{(1)}(\vec{r}_1,t_1 \mid \vec{r}_2,t_2)$ arises from the discussions of the interference effects measured in the Young demonstration of the wave nature of light. (The two-slit setup of the Young experiment is illustrated in Figure 5.3.) A planewave of light of frequency ω propagating in the plane of the page is incident from the left onto the screen containing two parallel infinite line slits. For the simplification, in the following it is also assumed that the light is polarized with the electric field parallel to the slits that are perpendicular to the plane of the page. Upon incidence on the screen, some of the light passes through the slits at $\vec{r}_1$ and $\vec{r}_2$, so that the slits act as sources for light propagating into the region to the right of the screen. The widths of the slits are small enough that single-slit diffraction effects can be neglected.

Light from the two silts is ultimately absorbed at $\vec{r}$ by the screen on the far right of the figure after traversing paths of length $s_1 = |\vec{r} - \vec{r}_1|$ and $s_2 = |\vec{r} - \vec{r}_2|$, respectively. The fields are absorbed at $\vec{r}$ so that (since an absorption process is govern by the $a_{\vec{k}s}$ operators) at this position the field absorption processes are managed by the $E_\omega^{(+)}(\vec{r},t)$ operator, where the subscript ω indicates that only modes of frequency ω in Eq. (5.90) are involved. For the Young geometry, it follows that

$$E_\omega^{(+)}(\vec{r},t) = E_{1,\omega}^{(+)}(\vec{r},t) + E_{2,\omega}^{(+)}(\vec{r},t). \tag{5.94}$$

and $E_{1,\omega}^{(+)}(\vec{r},t)$ and $E_{2,\omega}^{(+)}(\vec{r},t)$ are for the fields arriving from the slits 1 and 2, respectively.

In this notation, the transition matrix for the absorption at the far-right screen in Figure 5.3 is obtained from the standard form of the Fermi Gold Rule [9]. The transition matrix is given by

$$T_{i \to f} = \langle f | E_\omega^{(+)}(\vec{r},t) | i \rangle, \tag{5.95}$$

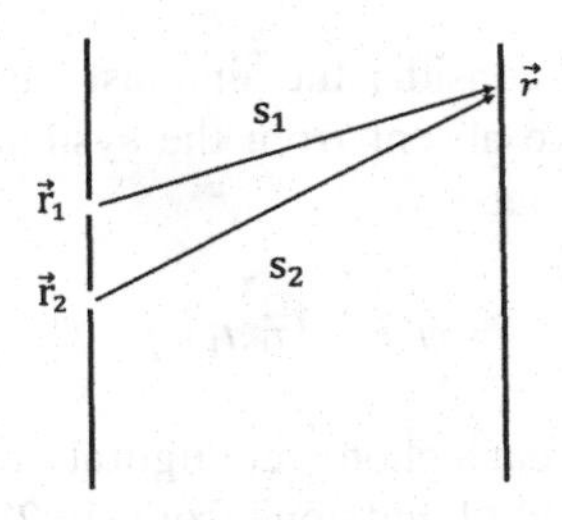

FIGURE 5.3 Double-slit geometry in which a planewave of light passes through two slits (i.e., $\vec{r}_1$ and $\vec{r}_2$) located in a planar screen, and an interference pattern is created at $\vec{r}$ on a second screen on the far right. The distances s_1 and s_2 are, respectively, the shortest separations $|\vec{r} - \vec{r}_1|$ and $|\vec{r} - \vec{r}_2|$.

where now $|i\rangle$ is the initial state of the incoming field at the right-most screen, and $|f\rangle$ is the final state of the fields at the right-most screen. In terms of Eq. (5.95), the rate of transition from $|i\rangle$ to $|f\rangle$ is proportional to

$$\sum_{i,f} P_i |T_{i\to f}|^2 = \sum_{i,f} P_i \langle i|E_\omega^{(-)}(\vec{r},t)|f\rangle\langle f|E_\omega^{(+)}(\vec{r},t)|i\rangle = G^{(1)}(\vec{r},t\,|\,\vec{r},t). \tag{5.96}$$

From this it is seen that the coherence function in Eq. (5.91) is just a generalization of the transition rate determining the interference pattern in the Young experiment.

The operator $E_\omega^{(+)}(\vec{r},t)$ in Eq. (5.94) is written in terms of the fields arriving at the screen on the far right from slits 1 and 2 at time t. These fields can now be expressed in terms of their past values when they left their respective slits. Specifically, expressed in the fields at the slits, it is found that

$$E_\omega^{(+)}(\vec{r},t) \propto \frac{e^{iks_1}}{\sqrt{s_1}} E_{1,\omega}^{(+)}\left(\vec{r}_1, t-\frac{s_1}{c}\right) + \frac{e^{iks_2}}{\sqrt{s_2}} E_{2,\omega}^{(+)}\left(\vec{r}_2, t-\frac{s_2}{c}\right), \tag{5.97}$$

where $t-\frac{s_1}{c}$ is the time that the wave left slit 1 to arrive at $\vec{r}$ at time t, and $t-\frac{s_2}{c}$ is the time that the wave left slit 2 to arrive at $\vec{r}$ at time t. The factors $\frac{e^{iks_1}}{\sqrt{s_1}}$ and $\frac{e^{iks_2}}{\sqrt{s_2}}$ account for the decrease in field amplitude and change in phase with distance during the transit between the two screens.

Considering Eq. (5.97) for the case that $|\vec{r}_2-\vec{r}_1| \ll s_1 \approx s_2 = R$, it follows that [3–7]

$$\frac{1}{\sqrt{s_1}} E_1^{(+)}\left(\vec{r}_1, t-\frac{s_1}{c}\right) \approx \frac{F}{\sqrt{R}} a_1, \tag{5.98a}$$

where $F \propto e^{-i\omega t}$ and, similarly,

$$\frac{1}{\sqrt{s_2}} E_2^{(+)}\left(\vec{r}_2, t-\frac{s_2}{c}\right) \approx \frac{F}{\sqrt{R}} a_2. \tag{5.98b}$$

From Eqs. (5.94), (5.97), and (5.98), it is then found that

$$E^{(+)}(\vec{r},t) \approx \frac{F}{\sqrt{R}}\left(a_1 e^{iks_1} + a_2 e^{iks_2}\right), \tag{5.99}$$

and this can be used to obtain simplified forms for the evaluation of the Eqs. (5.91) and (5.95). Examples of these applications in the evaluation of the coherence functions of light are now given for some simple cases.

As an example of the formalism, consider the simplest case of Eq. (5.91) in which thermal or other types of distribution effects are absent from the system, and only one term in the sum is needed. In this limit, it then follows that

$$G^{(1)}(\vec{r}_1,t_1\,|\,\vec{r}_2,t_2) = \langle i|E^{(-)}(\vec{r}_1,t_1)|f\rangle\langle f|E^{(+)}(\vec{r}_2,t_2)|i\rangle. \tag{5.100}$$

As a specific case of Eq. (5.100), we treat a photon configuration generated at the slits and composed of an initial state of one photon from slit 1 and none from slit 2. This has the form

$$|i\rangle = |1\rangle_1 |0\rangle_2. \tag{5.101}$$

Upon substituting the state Eq. (5.101) into Eq. (5.100), it follows that

$$G^{(1)}(\vec{r}_1,t_1 \mid \vec{r}_2,t_2) \propto \left(e^{-iks_1}e^{i\omega t_1}\right)\left(e^{iks_1}e^{-i\omega t_2}\right) = e^{i\omega(t_1-t_2)}, \tag{5.102a}$$

and from Eq. (5.91a), the first-order coherence function is

$$g^{(1)}(\vec{r}_1,t_1 \mid \vec{r}_2,t_2) = e^{i\omega(t_1-t_2)}. \tag{5.102b}$$

The coherence function in Eq. (5.102b) is consistent with the lack of spatial correlation for a single photon, which only passes through one slit. In this case, interference fringes are absent on the far-right screen.

On the other hand, for an initial state of the form

$$|i\rangle = \frac{1}{\sqrt{2}}\left[|1\rangle_1|0\rangle_2 + |0\rangle_1|1\rangle_2\right], \tag{5.103}$$

the slit through which the photon passes is not determined. In this case, the function in Eq. (5.100) is given by

$$\begin{aligned} G^{(1)}(\vec{r}_1,t_1 \mid \vec{r}_2,t_2) &\propto \left(e^{-iks_1}e^{i\omega t_1}\right)\left(e^{iks_1}e^{-i\omega t_2}\right) + \left(e^{-iks_2}e^{i\omega t_1}\right)\left(e^{iks_2}e^{-i\omega t_2}\right) \\ &\quad + \left(e^{-iks_1}e^{i\omega t_1}\right)\left(e^{iks_2}e^{-i\omega t_2}\right) + \left(e^{-iks_2}e^{i\omega t_1}\right)\left(e^{iks_1}e^{-i\omega t_2}\right) \\ &= 2e^{i\omega(t_1-t_2)}\left[1+\cos\left[k(s_2-s_1)\right]\right], \end{aligned} \tag{5.104a}$$

and the coherence function becomes

$$g^{(1)}(\vec{r}_1,t_1 \mid \vec{r}_2,t_2) = e^{i\omega(t_1-t_2)}\left[1+\cos\left[k(s_2-s_1)\right]\right]. \tag{5.104b}$$

As it is uncertain through which slit the photon goes, a spatial interference is observed following from the presence of the cosine form.

A higher order coherence function is obtained by considering the problem in which two photons are absorbed due to their interaction with a medium. For these processes, the transition matrix in Eq. (5.95) characterizing a single-photon absorption process changes to the form [3–7]

$$T_{i\to f} = \langle f | E_\omega^{(+)}(\vec{r}_1,t_1) E_\omega^{(+)}(\vec{r}_2,t_2) | i \rangle. \tag{5.105}$$

The presence of two destruction operators in Eq. (5.105) now removes or absorbs two different photons from the fields at two different space–time points. This multiple photon process represents a directed generalization of the single-photon process treated in the first-order coherence function. From the transition matrix in Eq. (5.105), a higher order coherence function is now formulated.

Related to Eq. (5.105) is the general second-order coherence function generalization of Eq. (5.91), accounting for the two-photon transitions in Eq. (5.105). Specifically, we define the function of four field operators

$$G^{(2)}(\vec{r}_1,t_1,\vec{r}_2,t_2,\mid \vec{r}_3,t_3,\vec{r}_4,t_4) = \sum_f \langle i \mid E^{(-)}(\vec{r}_1,t_1)E^{(-)}(\vec{r}_2,t_2) \mid f\rangle\langle f \mid E^{(+)}(\vec{r}_3,t_3)E^{(+)}(\vec{r}_4,t_4) \mid i\rangle, \tag{5.106}$$

which follows naturally from its two-photon counterpart. Similar to the treatment of Eq. (5.91), the second-order coherence function is now defined in terms of Eq. (5.106) by [3–7]

$$g^{(2)}(\vec{r}_1,t_1,\vec{r}_2,t_2,|\vec{r}_3,t_3,\vec{r}_4,t_4) = \frac{G^{(1)}(\vec{r}_1,t_1,\vec{r}_2,t_2,|\vec{r}_3,t_3,\vec{r}_4,t_4)}{\sqrt{G^{(1)}(\vec{r}_1,t_1|\vec{r}_1,t_1)G^{(1)}(\vec{r}_2,t_2|\vec{r}_2,t_2)}\sqrt{G^{(1)}(\vec{r}_3,t_3|\vec{r}_3,t_3)G^{(1)}(\vec{r}_4,t_4|\vec{r}_4,t_4)}}. \tag{5.107}$$

In this way, adding more $E^{(-)}$ fields to the product in the left transition matrix and an equal number of $E^{(+)}$ to the product in the right transition matrix, an infinite sequence of higher order coherence functions can be generated through initial considerations of the matrix elements of higher order photon absorption processes. Here we stop at the second-order correlator, and finish with some illustrations of the properties of the first- and second-order coherence functions. The reader is referred to the literature for a further consideration of the problem of higher coherence functions [1–7].

As examples of the second-order coherence function, consider the particular form of

$$g^{(2)}(\vec{r},t,\vec{r},t+\tau|\vec{r},t,\vec{r},t+\tau) \tag{5.108}$$

for a single planewave of frequency ω and wavevector $\vec{k}$ propagating in the $x-y$ plane with an electric field polarized in the z-direction. An expression for this type of field configuration is obtained from the general fields found in Eq. (5.100). For the evaluations of Eq. (5.108), from Eq. (5.106) the field operator for a single planewave mode written in terms of creation and destruction operators is used, generating a four-field product that is written as

$$E_{\vec{k}}^{(-)}(\vec{r},t)E_{\vec{k}}^{(-)}(\vec{r},t+\tau)E_{\vec{k}}^{(+)}(\vec{r},t)E_{\vec{k}}^{(+)}(\vec{r},t+\tau) \propto \\ \propto a_{\vec{k}}^{+}a_{\vec{k}}^{+}a_{\vec{k}}a_{\vec{k}}. \tag{5.109}$$

Applying Eq. (5.109) to light of a single frequency ω generated in the form of the coherent state $|\alpha\rangle_{\vec{k}}$, it follows from Eqs. (5.107)–(5.109) that

$$g^{(2)}(\vec{r},t,\vec{r},t+\tau|\vec{r},t,\vec{r},t+\tau) = 1. \tag{5.110a}$$

This coherent state result contrasts with the result for a single-mode Fock state $|n_{\vec{k}}\rangle$ for $n \geq 2$, which has a coherence function given by

$$g^{(2)}(\vec{r},t,\vec{r},t+\tau|\vec{r},t,\vec{r},t+\tau) = 1 - \frac{1}{n}. \tag{5.110b}$$

The second-order correlation function for the single quantum modes in Eq. (5.110) are found to exhibit a range of values $0 \leq g^{(2)}(\vec{r},t,\vec{r},t+\tau|\vec{r},t,\vec{r},t+\tau) \leq 1$. In general, only quantum mechanical single modes exhibit values of $g^{(2)}(\vec{r},t,\vec{r},t+\tau|\vec{r},t,\vec{r},t+\tau)$ in this range. However, for the case in which there are a thermal distribution of quantum states, as in Eqs. (5.84)–(5.86), the system is in a chaotic thermal mixture of states. For this limit, the single-mode chaotic light is found to give

$$g^{(2)}(\vec{r},t,\vec{r},t+\tau|\vec{r},t,\vec{r},t+\tau)|_{\tau=0} = 2. \tag{5.110c}$$

It is interesting to compare the quantum results for $g^{(1)}(\vec{r}_1,t_1|\vec{r}_2,t_2)$ and $g^{(2)}(\vec{r},t,\vec{r},t+\tau|\vec{r},t,\vec{r},t+\tau)$ in single planewave modes with those for classical electromagnetic radiation based on planewave modes. In the classical theory, the operators

$$E^{(-)}(\vec{r}_1,t)E^{(+)}(\vec{r}_2,t+\tau) \tag{5.111a}$$

and

$$E_{\vec{k}}^{(-)}(\vec{r},t)E_{\vec{k}}^{(-)}(\vec{r},t+\tau)E_{\vec{k}}^{(+)}(\vec{r},t)E_{\vec{k}}^{(+)}(\vec{r},t+\tau) \tag{5.111b}$$

are replaced by classical number so that $E_{\vec{k}}^{(-)}(\vec{r},t)=\epsilon_{\vec{k}}^{*}e^{-i(\vec{k}\cdot\vec{r}-\omega t)}$ and $E_{\vec{k}}^{(+)}(\vec{r},t)=\epsilon_{\vec{k}}e^{i(\vec{k}\cdot\vec{r}-\omega t)}$ for the complex amplitude $\epsilon_{\vec{k}}$. In computing the coherence function for these fields, the field amplitudes are then distributed by a classical probability distribution $P(\epsilon_{\vec{k}})$, which is used to perform the weighted average of Eq. (5.111) over amplitudes or by a classical probability distribution over the frequency components given by the form $P(\omega)$. In addition, in computing the coherence functions, the quantum mechanical matrix averages $\langle i\,|\,AB\,|\,i\rangle$ for quantum operators A and B are replaced by time averages of the form

$$\lim_{T\to\infty}\frac{1}{T}\int dtA(t)B(t+\tau), \tag{5.111c}$$

where T is the length of time over which the averages are computed. In the following, to make a comparison, some discussions are made for both the $g^{(1)}$ and $g^{(2)}$ coherence functions in classical electrodynamics systems.

In the case of single planewave modes distributed over frequency by $P(\omega)=\dfrac{1}{1+(\omega-\omega_c)^2\tau_c^2}$ in the form of a Lorentzian chaotic light distribution, it is found for the treatment of first-order coherence that [3–7]

$$g^{(1)}(\vec{r},t\,|\,\vec{r},t+\tau)=e^{-i\omega_c\gamma\tau}e^{-|\tau|/\tau_c}. \tag{5.112a}$$

Similarly, for a Gaussian chaotic distribution $P(\omega)$ it is found that [3–7]

$$g^{(1)}(\vec{r},t\,|\,\vec{r},t+\tau)=e^{-i\omega_c\gamma\tau}e^{-\frac{\pi}{2}\left(\frac{\tau}{\tau_c}\right)^2}. \tag{5.112b}$$

where $\tau_c=(8\pi\ln 2)^{1/2}/\Delta\omega$ and $\Delta\omega$ is the full width at the half-maximum of the Gaussian frequency distributions. In both of these expressions, ω_c is the frequency maximum of the distribution, and τ_c is a characteristic time scale over which the coherence function approaches zero as time increases.

In the classical system, the second-order coherence functions are simply related to the first-order coherence functions through the identity [3–7]

$$g^{(2)}(\vec{r},t,\vec{r},t+\tau\,|\,\vec{r},t,\vec{r},t+\tau)=1+\left|g^{(1)}(\vec{r},t\,|\,\vec{r},t+\tau)\right|^2. \tag{5.113a}$$

In this regard, from Eq. (5.112a) for the Lorentzian distribution, it follows that [3–7]

$$g^{(2)}(\vec{r},t,\vec{r},t+\tau\,|\,\vec{r},t,\vec{r},t+\tau)=1+e^{-2|\tau|/\tau_c}. \tag{5.113b}$$

It is seen from Eq. (5.113) that $g^{(2)}(\vec{r},t,\vec{r},t+\tau\,|\,\vec{r},t,\vec{r},t+\tau)$ of the classical systems are always greater than 1. On the other hand, from Eq. (5.110), the quantum single-mode states are generally equal to or less than 1. This is a fundamental difference between the classical and quantum theories of optics, and provides a distinguishing criterion to differential between the two theories [3].

REFERENCES

1. Kittle, C. 1996. *Introduction to solid state physics, 7th Edition*, 393–410. New York: John Wiley & Sons, Inc.
2. Marder, M. P. 2000. *Condensed matter physics*. New York: John Wiley & Sons, Inc.
3. Gerry, C. and P. Knight 2005. *Introductory quantum optics*. Cambridge: Cambridge University Press.
4. Lambropoulos, P. and D. Petrosyan 2007. *Fundamentals of quantum optics and quantum information*. Berlin: Springer-Verlag.
5. M. Fox 2006. *Quantum optics: An introduction*. Oxford: Oxford University Press.
6. Band, Y. B. 2006. *Light and matter: Electromagnetism, optics, spectroscopy and lasers*. Chichester: John Wiley & Sons, Ltd.
7. Walls, D.F. and G. J. Milburn 1995. *Quantum optics*. Berlin: Springer-Verlag.
8. Goldstein, H., Poole, C. and J. Safko 2002. *Classical mechanics, 3rd Edition*. San Francisco: Addison Wesley.
9. Kroemer, H. 1994. *Quantum mechanics: For engineering, materials science, and applied physics*. New Jersey: Prentice Hall.
10. Jackson, J. D. 1975. *Classical electrodynamics, 2nd Edition*. New York: John Wiley & Sons, Inc.
11. McGurn, A. R. 2018. *Nanophotonics*. Cham: Springer.
12. Streetman, B. G. and S. Banerjee 2000. *Solid state electronic devices, 5th Edition*. New Jersey: Prentice Hall.

6 Basic Properties of Lasers, Masers, and Spasers

In this chapter, the basic operations of a laser are discussed [1–3]. The components entering into laser designs are each treated with a focus on generality and the development of the physics ideas rather than on the design details of engineering concern. Specifics as to the requirements needed to achieve a threshold for laser operation, the mechanisms of gain creation, and the sources of energy dissipation in the laser are addressed. In addition, a focus is on the nature of the radiation fields that are emitted by a laser, i.e., what is the distribution of photons in the laser fields and how are these correlated with the stimulating radiation. Current generalizations of the ideas of laser physics to the enhancement of plasmons arising in the design of spaser devices are also discussed as well as certain laser ideas applied to atomic beams, which form the basis of maser operations [4–6].

The focus of laser operation is on the amplification of stimulated emissions of light, as they are generated in a lasing medium [1–3,7–10]. In this process, photons of stimulating light at the frequency of an electronic energy-level transition of the lasing medium, through the process of stimulated emission, create additional photons at the transition frequency. This provides a basis for the amplification of the light, as it passes through the medium. By placing the lasing medium within a cavity resonating at the transition frequency, the stimulating and stimulated light are made to build up within the cavity. In this way, the stimulating and stimulated light develop a phase relation to one another, which is set by the Hamiltonian describing the process of stimulated emission. Consequently, the fields of the generating and generated radiations tend to add to one another in a total phase-coherent enhancement. Within the cavity is then generated an increasing intensity of coherent radiation fields [11,12].

In addition to the gain mechanism of the laser, there is dissipation in the system. The dissipation processes come from energy losses in the lasing medium and cavity mirrors, spontaneous emissions to the cavity fields, and the emission of laser fields from the cavity in the form of laser radiation [1–3,7–12]. For the system to generate a net amplification of the light, the amplification of the fields from the stimulated emissions must dominate the losses within the system and the emissions from the system. Specifically, the intensity of generated light needs to reach a threshold of generated energy density known as the laser threshold.

The threshold for lasing is attained by creating an imbalance of states in the initial electron energy occupancy of the states of the medium involved in stimulated transition events [7–10]. In particular, this imbalance is accomplished by pumping the laser medium to have an excess of states in the higher energy electron level of the stimulated transition. Consequently, after the medium is pumped, the occupancy of its electron energy levels is no longer characterized by a thermal distribution.

A variation of the process outlined is that of the maser process. In the maser, a beam of excited atoms travels through a resonant cavity, and each atom deposits a photon to the resonant cavity fields through the mechanism of stimulated emission. In this way, a gain is introduced into the development of the cavity fields, which is eventually emitted from the cavity as maser radiation. The generation of the maser fields is easier to treat theoretical than those of the standard laser operations and are treated here in discussions of the nature of the configuration of the photon field present in the radiation emitted from the cavity. Consequently, the maser model is used to determine the distribution of photons in the maser fields, which also models the fields generated in a laser [7–12].

In the following, first a brief review of stimulated emission is given, followed by a discussion of the lasing process in terms of rate equations for the various processes involved in lasing. After this,

DOI: 10.1201/9781003031987-6

a full quantum optics treatment of the fields generated in the maser is presented. Finally, at the end of the chapter, some brief discussions of the spaser and atomic lasers are introduced. These have been of recent interest in the application of ideas of lasers operating in surface physics.

6.1 STIMULATED EMISSION

In this section, the basic theory of stimulated emission is briefly reviewed in the context of presenting a study of the thermal equilibrium of a two-level electron system interacting with an applied electromagnetic field [3,7–12]. The elementary discussions focus on a treatment of the processes involved in transitioning between the two electronic levels, with a determination of the transition rates acting between the electron levels, and the distribution of the electrons and photons in the system. The results of the outlined system are developed from the simplest model allowing for the determination of the most general basic features of the theory of spontaneous emission and of the electron occupancies.

Specifically, in the context of a simple model, a single electron occupies an upper energy state called level 2 with energy ε_2, or a lower energy state called level 1 with energy ε_1 and transitions between the two levels by emitting or absorbing a photon of energy $h\nu = \varepsilon_2 - \varepsilon_1$, where ν is the photon frequency [3,7,10,11]. As a simplification in the discussions, the applied radiation fields are assumed to also be at the frequency ν. The object of the discussions is then to obtain the various transition rates that operate between the electronic levels, as they interact with the photons and a thermal reservoir. From these transition rates, the occupation probabilities of the electron and photon modes are then computed. As a reference in the proceeding treatments, Figure 6.1 gives a schematic of the system and its various transitions.

As the electron transitions between the two energy levels, there are three types of processes involved. These include two types of stimulated processes involving the applications of photons to the two-level electron system. One process is an emission of a photon, as the electron lowers its energy, and another is an absorption of a photon, as the electron raises its energy. (Note that such stimulated processes only occur in the presence of a preexisting photon in the system.) In addition to these two, there is a spontaneous emission process in which the electron lowers its energy by a photon emission even in the absence of a preexisting photon in the system. Note that there is no spontaneous process in which an electron is raised to a higher energy level [3,7,10,11].

To treat the equilibrium state of the system, consider a statistical array of $N \gg 1$-independent two-level systems, which each carry a single electron and separately interact with the photon field of frequency ν. In the array of N two-level systems at thermal equilibrium, let N_1 and N_2 be the number of two-level electrons in the energy level 1 or energy level 2, respectively, where $N = N_1 + N_2$. For the electronic array and fields to exist in thermal equilibrium, it is then required that the rate of transitions in the system from level 1 to level 2 must be the same as the rate of transitions from level 2 to level 1. In addition, each of these transitions must include all of the three types of basic transitions mentioned earlier [3,7,10,11].

Per each two-level system, the probability per second of a transition from level 2 to level 1 can be represented as [3,7,10,11]

$$T_{2\to1} = B_{2\to1}\rho(\nu) + A_{2\to1}, \tag{6.1}$$

ϵ_2

ϵ_1

FIGURE 6.1 Two-level state system for the occupancy of a single electron. The ground state of the electron is ϵ_1, and the excited state of the electron is ϵ_2. The thermodynamic array is composed of N two-level states system—N_1 of which with electrons in energy states ϵ_1 and N_2 of which with electrons in energy states ϵ_2.

where $\rho(\nu)$ is the energy density per unit frequency in the photon fields, and $A_{2\to1}$ and $B_{2\to1}$ are rate coefficients that are to be determined. The first term on the right depends on the energy density of photons in the fields and represents stimulated emission processes, while the second term represents spontaneous emission processes that are independent of the photon energy density in the field. Note that increasing the energy density of photons increases the transition rate of stimulated processes, while the spontaneous processes are independent of the energy density of photons.

Per two-level system, the probability per second of a transition from level 1 to level 2 can be represented as [3,7,9,10,11]

$$T_{1\to2} = B_{1\to2}\rho(\nu) \tag{6.2}$$

where $\rho(\nu)$ is the energy density per unit frequency in the photon fields, and $B_{1\to2}$ is a rate coefficient, which is to be determined. Here, only a stimulated absorption process occurs as a spontaneous absorption process would involve promoting an electron to a higher energy in the absence of photons in the system. This would not conserve energy.

When the array of N two-level systems are in equilibrium, the number of transitions per second from level 2 must equal the number of transitions per second from level 1, otherwise there would be a net change in the occupancy in time. From Eqs. (6.1) and (6.2), this requires that in the array of N two-level systems [3,7,9,10,11]

$$N_2\left[B_{2\to1}\rho(\nu)+A_{2\to1}\right]=N_1B_{1\to2}\rho(\nu). \tag{6.3}$$

In order to determine the coefficients $B_{2\to1}$, $B_{1\to2}$, $A_{2\to1}$, considering the special case in which the radiation fields are in a state of blackbody equilibrium, the photon occupancy in this arrangement is now governed by the Planck distribution [7,9,11]

$$\rho(\nu)=\frac{8\pi h\nu^3}{c^3}\frac{1}{e^{h\nu/k_BT}-1}. \tag{6.4}$$

In addition, at thermal equilibrium the electron occupancy of the two energy levels, each of which containing a single electron, is given by the Boltzman condition [3,7,9,10,11]

$$\frac{N_2}{N_1}=e^{-h\nu/k_BT}. \tag{6.5}$$

The conditions in Eqs. (6.3)–(6.5) on the photons and electrons characterize the occupancy in the system and completely set the coefficients $B_{2\to1}$, $B_{1\to2}$, $A_{2\to1}$.

Introducing Eqs. (6.4) and (6.5) into Eq. (6.3), it is found that [9,11]

$$B_{2\to1}=B_{1\to2}=B=\frac{c^3}{8\pi h\nu^3}A_{2\to1} \tag{6.6}$$

and [9]

$$A_{2\to1}=A=\frac{1}{t_{spon}}. \tag{6.7}$$

where t_{spon} is the average time of a spontaneous emission transition. Consequently, the equilibrium condition in Eq. (6.3) becomes [3,7,9,10,11]

$$N_2\left[B\rho(\nu)+A\right]=N_1B\rho(\nu) \tag{6.8a}$$

or upon substituting for A and B

$$N_2\left[\frac{c^3}{8\pi h\nu^3}\rho(\nu)+1\right]=N_1\frac{c^3}{8\pi h\nu^3}\rho(\nu) \tag{6.8b}$$

6.1.1 Deviations of the System from Thermal Equilibrium

If one were to impose on the equilibrium system, a sudden increase in occupancy of the electrons in level 2 to a new number $N_2' > N_2$, then [3,7,10,11]

$$N_2'\left[B\rho(\nu)+A\right]>N_2\left[B\rho(\nu)+A\right]=N_1B\rho(\nu). \tag{6.9}$$

The resultant number of transitions per second from level 2 in the imbalanced state, $N_2'\left[B\rho(\nu)+A\right]$, is then greater than the number of transitions from level 1, $N_1B\rho(\nu)$. Consequently, the transitions of the electron levels 2 will dump a net increased number of photon states into the system, as they seek to transition and reestablish equilibrium. This causes an amplification of the number of ν photons in the system and is an essential element in the laser amplification.

In this view of lasing or light amplification, an essential condition for laser operation is that the array of two-level systems be configured so that $N_2 > N_1$. When the array is then subjected to stimulating radiation of frequency ν, the rate of stimulated emission from the upper level is given from Eqs. (6.1), (6.6), and (6.7) by [3,4,7,9,10,11]

$$B_{2\to1}\rho(\nu)=\frac{c^3}{8\pi h\nu^3t_{spon}}\rho(\nu)=\frac{c^2}{8\pi h\nu^3t_{spon}}I(\nu) \tag{6.10}$$

where $I(\nu)=c\rho(\nu)$ is the energy current density [9] of the simulating field of frequency ν.

Note, however, that in the last equation in Eq. (6.10) the expression for the transition rate is now written in terms of the intensity of the energy current in the lasing medium. In turn, the stimulated emission transition rate, set by the energy current, generates additional photons in the simulating field.

It is also found that the phase of the radiation simulated by the fields bears a definite relationship to that of the stimulating light so that the intensity generated at a point in space tends to be additive to the stimulating fields [9]. This is due to the fact that the stimulated emission process arises from the dynamics of the quantum mechanical Hamiltonian of the system. Consequently, if the two-level system is interacting with a planewave traveling in the z-direction, the increase in the energy current as a function of propagation distance is [9]

$$\frac{d}{dz}I(z,\nu)=\frac{c^2}{8\pi\nu^2t_{spon}}g(\nu)\frac{(N_2-N_1)}{V}I(z,\nu)=\gamma I(\nu,z), \tag{6.11a}$$

where V is the volume of the medium, and $g(\nu)$ accounts for the frequency line shape of the planewave radiation of frequency ν so that [9]

$$\int_{-\infty}^{\infty}g(\nu)d\nu=1, \tag{6.11b}$$

which for our purposes is considered to be narrow. In addition, here [9]

$$\gamma=\frac{c^2}{8\pi\nu^2t_{spon}}g(\nu)\left(\frac{N_2-N_1}{V}\right) \tag{6.11c}$$

is the net number of transitions from level 2 to level 1 per second [11], and $(N_2 - N_1)h\nu$ is the net energy transferred to the cavity fields by these transitions. In this regard, it is assumed in Eq. (6.11) that the energy current generated by the stimulated emission adds directly to the energy current of the generating fields [11].

If Eq. (6.11a) is integrated for the energy current as a function of distance, it is found for fixed $N_2 - N_1$ that [9]

$$I(z,\nu) = I(0,\nu)e^{\gamma z} \tag{6.12}$$

where [9] $\gamma = \frac{c^2}{8\pi\nu^2 t_{spon}} g(\nu) \frac{(N_2 - N_1)}{V}$. In the case that $(N_2 - N_1) > 0$, it is seen that the intensity of the planewave increases as it travels through the lasing medium. On the other hand, in the case that $(N_2 - N_1) < 0$, it is found that the intensity of the planewave decreases as it travels through the lasing medium. This is the fundamental mechanism in the laser, but there are a number of other considerations that go into developing its full implementation.

In order to intensify and collect the amplified radiation of frequency ν, the array of two-level electron systems is placed within a resonant cavity, which partially transmits the amplified radiation to the outside world. The cavity is tuned to the frequency ν so as to enhance the spontaneous emission process and to help the development of the fields needed for the generation of spontaneous emission. It also acts to output the laser radiation. The cavity is an important component in maintaining a steady state operation of the laser and is treated in detail later.

In addition to the mechanism of stimulated emission, the pumping or overpopulation of the upper electronic level from its equilibrium occupancy is one of the essential elements in the laser amplification of light. These processes, however, are not the only important electronic functions in the operation of lasers. There are a variety of other mechanisms of energy loss and energy fluctuation that enter into the working of the laser. The treatment of these additional considerations requires the introduction of more electron levels into the model as well as considerations of the emitted radiation from the laser cavity. In the next section, it is shown how the pumping of excited states and their stimulated emissions fit into the rate equation of a more complete model for the general operation of a laser.

6.2 RATE EQUATION MODEL OF LASER OPERATIONS

In the earlier discussions, the idea of the amplification of light through its stimulation of an excited array of two-level electron systems was treated [3,7,10–12]. Two other considerations, however, are required in the functioning of a laser. These include a means of generating the excited array of two-level electron systems, and a means of creating a steady state output of coherent amplified light accounting for loss mechanisms and emission of laser radiation from the cavity containing the lasing medium. Creating the excited array is done by a mechanism known as pumping, and the output of the laser is modulated by a resonant cavity tuned to the lasing frequency ν. In the following discussions, these two additional features are treated and inserted into the laser mechanism.

The addition of these features necessitates the introduction of a more general system composed as an array of single electrons systems with four distinct electron energy levels. This array is located within the resonant cavity in place of the two-level systems of our earlier discussions. The four levels are composed of the two energy levels of our earlier discussion, which directly enter into the generation of laser light, and two additional levels involved with a pumping mechanism and the modeling of a variety of loss mechanisms.

In the following, a rate equation description of laser operation is developed. In this description, the laser is composed of an array of N electronic units (e.g., atoms, molecules, etc.), and each unit is comprised of four different electronic energy levels. In a specific unit, a single electron transitions between the four different energy levels generating or absorbing laser or loss photon modes, depending on which energy-level transitions are involved. The electron transitions generate light at either the resonant frequency of the resonant cavity containing the array or at loss frequencies that differ from the resonant cavity frequency.

Specifically, the photonic laser modes arise from stimulated emission processes involved with the resonant cavity modes, but there are additional electron transitions yielding photons at frequencies different than that of the resonant cavity mode and which contribute to energy losses in the system. Eventually, the laser light in the resonant cavity modes is allowed to escape the cavity, making the cavity a source of laser light. This is also a loss to the system.

A four-level electronic unit is represented schematically in Figure 6.2a for a single electron transiting between four different electron energy levels and a resonant cavity for photons of frequency ν. In the schematic, there are two lasing electronic energy levels denoted as 2 for the highest electron energy with energy ϵ_2 and 1 for the lower energy state at an energy ϵ_1. The energy difference between the two levels is $\epsilon_2 - \epsilon_1 = h\nu$ so that the electron transition is in resonance with the photon cavity modes. Below these two energy levels are two electron levels 4 and 3 of even lower energies ϵ_4 and ϵ_3, so that the hierarchy of energy relations is $\epsilon_2 > \epsilon_1 > \epsilon_4 > \epsilon_3$. While the levels 2 and 1 are involved in the generation of coherent laser light, the levels 4 and 3 are involved with internal losses representing dissipative effects in the generation of laser light.

An external pumping mechanism exists in the system, which allows for the electronic occupancies of the levels 1 and 2 to be artificial adjusted [12]. The rate that electrons are pumped into the levels 1 and 2 are denoted R_1 and R_2, and the occupancies of these levels are denoted N_1 and N_2. The electrons in the levels 1 and 2 can, respectively, decay into levels 3 and 4 with the respective transitions per second given by γ_1 and γ_2. In terms of the average time of spontaneous transitions between

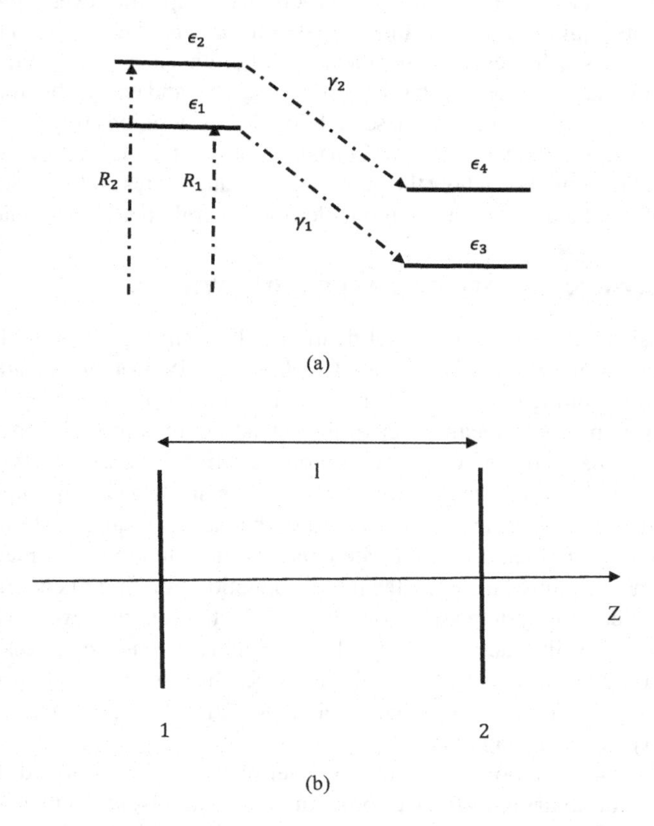

FIGURE 6.2 Energy schematic for: (a) the four electron energy levels for the discussion of the laser rate equations, and (b) the resonant cavity that contains the lasing medium.

levels 2 and 4, denoted t_{spon}, and the average time of other lossy processes resulting in transitions between levels 2 and 4, denoted t_{20},

$$\gamma_2 = \frac{1}{t_2} = \frac{1}{t_{spon}} + \frac{1}{t_{20}}. \tag{6.13}$$

Similarly, in computing the average time of nonstimulated transition processes between levels 1 to 3, denoted t_1,

$$\gamma_1 = \frac{1}{t_1}. \tag{6.14}$$

In addition to the pumping and loss processes to other electron transitions, there are stimulated emission and absorption processes between the electronic levels 1 and 2. The rate of stimulated emission is equal to the rate of stimulated absorption, and is denoted by T. The rate of change in the N_1 and N_2 occupancies in terms of all of the earlier mention rates is then given by [3,7,9–12]

$$\frac{dN_2}{dt} = R_2 - \frac{N_2}{t_2} - (N_2 - N_1)T \tag{6.15a}$$

and

$$\frac{dN_1}{dt} = R_1 - \frac{N_1}{t_1} + \frac{N_2}{t_{spon}} + (N_2 - N_1)T. \tag{6.15b}$$

In Eq. (6.15a) the first term (i.e., R_2) on the right accounts for the pumping into the level, the second term accounts for transitions outside the resonant cavity modes, and the third term is the generation of ν resonant cavity modes. Similarly, in Eq. (6.15b), the first term on the right (i.e., R_1) accounts for the pumping into the level, the second term accounts for transitions outside the resonant cavity modes, the third term involves spontaneous emissions from level 2 to level 1, and the fourth term involves the generation of ν resonant cavity modes.

In the stead state configuration $\frac{dN_2}{dt} = 0$ and $\frac{dN_1}{dt} = 0$. Consequently, solving the resulting Eq. (6.15) under these conditions, it is found that [9]

$$N_1 - N_2 = \frac{R_2 t_2 - \left(R_1 + \frac{t_2}{t_{spon}} R_2\right) t_1}{1 + \left[t_2 + \left(1 - \frac{t_2}{t_{spon}}\right) t_1\right] T}. \tag{6.16}$$

This gives the population imbalance required to overcome losses to the medium and leads to a net amplification in the array. In addition to these losses, however, an account needs to be made of the losses due to emission of laser radiation from the resonant cavity. This is the next consideration.

6.3 RESONATOR CAVITY

An important component of the laser is the resonant cavity. This contains the lasing medium that generates the contributions to the field from stimulated emission. The resonant cavity is tuned to the frequency of light emitted by the medium so as to reinforce the accumulation of energy in the resonant mode, which is eventually radiated from the cavity as the laser emission. In this regard, the cavity allows the radiation to build up within it, as more and more stimulated emissions are collected from the lasing medium.

To understand the resonant function of the cavity, consider a simple model of the cavity formed from two partially reflecting infinite parallel plates that are separated along the z-axis by a distance l. (See the schematic in Figure 6.2b.) Light moving along the z-axis causes the stimulated emission of radiation from the lasing medium contained between the plates, and the stimulated light, in turn, travels along the z-axis. The total component of radiation then resonantly bounces back and forth between the two plates [3,4,7,1–12].

In these travels between the plates, a series of processes occur. Specifically, a component of radiation generates more radiation through the stimulation process described by Eq. (6.12), a component of radiation is lost in the cavity due to dissipative effects, and a component is radiated from the cavity in emitted laser light. These processes represent a sequence of gains and losses involving the cavity radiation fields.

Considering the geometry in Figure 6.2b, light at the resonant frequency enters the left edge of the cavity and begins to stimulate emissions from the lasing medium. In this configuration, the total component of light within the cavity is described by a planewave form

$$\vec{E}(z,t)=\vec{E}_0e^{i(kz-\omega t)}, \tag{6.17a}$$

where the complex wavevector is given by [9,12]

$$k=k_R-i\frac{\gamma-\alpha}{2} \tag{6.17b}$$

Here from Eq. (6.12), γ describes the increase in the field amplitude with distance due to the generation of radiation from stimulated emission, and α represents loss processes within the cavity due to dissipation in the cavity–medium system that decrease the field amplitude with distance traveled. As the wave in Eq. (8.17) travels the length of the cavity, the amplitude of the field then changes by a factor of

$$e^{ikl}. \tag{6.18}$$

The total transmission of a wave from the left to the right side of the system of parallel plates can be written in terms of the transmission and refection coefficients at the two plates and the phase change of the waves between the plates. For this formulation, the transmission and reflection coefficients at each plate are denoted t_1,t_2 and r_1,r_2, respectively, where the left (right) plate is denoted $1(2)$. Here t_1 is the transmission into the cavity at the left interface, and t_2 is the transmission out of the cavity at its right interface. Similarly, r_1 is the reflection back into the cavity at the left interface, and r_2 is the reflection back into the cavity at its right interface.

Specifically, the total transmission from the left to the right of the parallel plates is [3,7,10–12]

$$T=t_1t_2e^{ikl}+t_1r_2r_1t_2e^{3ikl}+t_1r_2r_1r_2r_1t_2e^{5ikl}+\ldots. \tag{6.19a}$$

Here, the first term on the right represents transmissions at plate 1 and plate 2, and a phase change through the length l. The second term on the right includes an intermediate reflection process at surface 2 and surface 1 between the two transmissions, where now the length traveled in the region between the plates is $3l$. Similarly, the third term represents four intermediate reflections between the two transmissions and a transit through a length $5l$.

Summing all the infinite set of processes in Eq. (6.19a) gives [12]

$$T=\frac{t_1t_2e^{ikl}}{1-r_2r_1e^{2ikl}}=\frac{t_1t_2e^{ik_Rl}e^{\frac{\gamma-\alpha}{2}l}}{1-r_2r_1e^{2ik_Rl}e^{(\gamma-\alpha)l}}, \tag{6.19b}$$

and for the mode to be a resonant mode of the cavity, in addition, requires $e^{2ik_R l} = 1$.

The condition for a sustained oscillation within the cavity, balancing the losses of the system with the gains in the cavity, occurs at the singularity of T. This is the threshold of the laser operation and is found at the condition [9,12]

$$1 - r_2 r_1 e^{(\gamma-\alpha)l} = 0. \tag{6.20}$$

From this and Eq. (6.11c), it then follows that [9,12]

$$\gamma = \frac{c^2}{8\pi v^2 t_{spon}} g(v)\left(\frac{N_2 - N_1}{V}\right) = \alpha - \frac{1}{l}\ln r_1 r_2. \tag{6.21}$$

Consequently, a threshold population inversion is defined by the requirement that [3,7,10,11]

$$\left(\frac{N_2 - N_1}{V}\right) = 8\pi v^2 t_{spon} \frac{\alpha - \frac{1}{l}\ln r_1 r_2}{c^2 g(v)}. \tag{6.22}$$

6.4 MASER

In this section, discussions are made for a type of lasing system known as a maser, and the object is to present a quantum mechanical treatment that determines the distribution of photon modes in the maser when it generates a coherently amplified output of radiation [1,4,10–12]. Specifically, it will be shown that the maser has a threshold of operation below which it generates an incoherent thermal distribution of photons somewhat like that of a lightbulb, and above which it outputs amplified coherent radiation. The distribution of photons in the coherent radiative state are shown to be in a Poisson distribution [12], similar to the coherent modes of light discussed in the previous chapter.

In the operation of the maser [1,4,10–12], a beam of atoms in an excited energy state is run through a resonant cavity, which contains resonant photons of frequency ω. The resonant frequency ω of the cavity is tuned to an atomic transition from the atomic excited state in the beam. Most of the atoms passing through the cavity deposit a photon of frequency ω to the photon distribution in the cavity, as the atom transitions to a lower energy state. This amplifies the cavity resonant photon mode that is eventually allowed to escape from the cavity as maser radiation. The interest here is in determining the distribution of the Fock modes [7,12] in the cavity, and the maser light emitted from the cavity both above and below the lasing threshold.

In the model, each atom is assumed to be represented by four electronic energy levels. The excited state of the atom is the highest energy level, and there is a lower energy state to which it decays to contribute a photon of frequency ω to the cavity. In addition, there are two lower energy atomic levels that are used to model atomic losses to other internal atomic relaxation processes. The light emitted from the cavity is modeled by a term representing losses of the photon modes to the regions outside the cavity.

For the discussions, an idealized system is addressed, and, for convenience, it is assumed that the excited state atoms in the beam pass through the cavity one at a time.

6.4.1 THE MODEL

The model is composed of a single atom represented by four different electron energy levels [7,10,12] denoted by the atomic eigenstates $\{|1\rangle_a, |2\rangle_a, |3\rangle_a, |4\rangle_a\}$. A single electron of the atom is shuttled between the atomic energy levels as the atom passes through the photon cavity, causing the electron to interact with the electromagnetic cavity fields. In the cavity, the fields are of a single

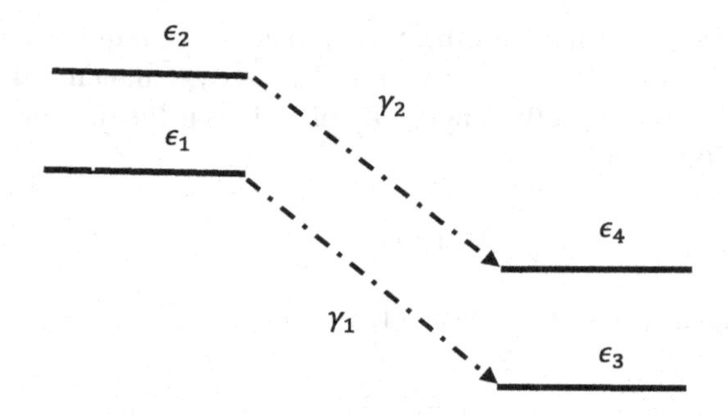

FIGURE 6.3 Schematic of the energy levels of a maser. The transitions between the energy levels $|1\rangle_a, |2\rangle_a$ generate the coherent radiation, while energy losses and fluctuations are modeled by transitions to the lower energy states labels $|3\rangle_a, |4\rangle_a$. The rate of transitions between the states $|1\rangle_a$ and $|3\rangle_a$ is γ_1, and the rate of transitions between the states $|2\rangle_a$ and $|4\rangle_a$ is γ_2. Ultimately, it is assumed that $\gamma_1 = \gamma_2 = \gamma$.

frequency ω and are represented by Fock states $\{|n\rangle\}$, where n is the number of photons present in the cavity [7,10,12].

A schematic of the electron levels is indicated in Figure 6.3 showing the possible transitions [12] between the composite states of the electron and photon modes, which are denoted by $|2,n\rangle = |2\rangle_a |n\rangle$, $|1,n+1\rangle = |1\rangle_a |n+1\rangle$, $|4,n\rangle = |4\rangle_a |n\rangle$, and $|3,n+1\rangle = |3\rangle_a |n+1\rangle$. The energy separation of the levels $|1\rangle_a, |2\rangle_a$ is the same as the energy separation of the levels $|3\rangle_a, |4\rangle_a$, and both of the $|3\rangle_a, |4\rangle_a$ energy levels occur at lower energies that the $|1\rangle_a, |2\rangle_a$. The electron is transferred between its various energy levels by exchanging energy with the cavity fields and the internal atomic degrees of freedom.

The total wavefunctions of the atom–cavity system are made up from the basis set of states $\{|2,n\rangle, |1,n+1\rangle, |4,n\rangle, |3,n+1\rangle\}$ designating the configuration of the single atomic electron and the corresponding photon occupancy of the cavity photons. The processes linking the transitions between these states of the atom–cavity system are energy exchanges between the electron states $|1\rangle_a, |2\rangle_a$ and the cavity photons, decay processes into the electron states $|3\rangle_a, |4\rangle_a$ within the atom, a Markovian energy extraction form the cavity representing the emission of maser light from the cavity, and Markovian fluctuations between the radiation emission and the atomic states [12].

One loss of energy from the atom–field modes is through processes involving the transitions of the atomic electron into either of the states $|3\rangle_a, |4\rangle_a$, where the decay rate of the electron states $|1\rangle_a, |2\rangle_a$ into the states $|3\rangle_a, |4\rangle_a$ is characterized by a rate parameter γ. In this regard, the $|3\rangle_a, |4\rangle_a$ atomic levels act as part of internal dissipative processes that removes energy from the system and also accounts for energy fluctuations into the system. A second energy loss from the atom–cavity field system is in the actual transmission of field energy out of the maser cavity in the form of emitted radiation. This is characterized by a cavity loss rate parameter κ, which shall be discussed later [12].

In the maser process, atoms are introduced into the cavity fields in the excited state $|2\rangle_a$. These excited states decay to $|1\rangle_a$ atomic states and in the process contribute a photon of energy $\hbar\omega$ to the photon cavity modes. In addition, some of the energy of the excited state is also lost internally to the atomic and cavity modes. Eventually, light of frequency ω is emitted by the cavity in the form of the maser emission of light.

6.4.2 The Hamiltonian: Absence of Maser Fields

To begin the treatment of the dynamics of the atom–cavity–field system, first consider the case in which no energy is extracted from the cavity [12], i.e., no maser emissions are allowed, $\gamma = 0$, and the atom–field system merely shuttles energy elastically between the atom and the cavity photons. In this limit, only the states $\{|2,n\rangle,\ |1,n+1\rangle\}$ enter into the considerations, and the system is essentially reduced to a consideration of the Jaynes–Cumming model [7,10,12].

The Jaynes–Cumming Hamiltonian representing the atom–cavity fields and their interaction in the absence of emitted radiation is then given by [7,10–16]

$$H_{JC}=\frac{\hbar\epsilon}{2}\left[b_2^+b_2-b_1^+b_1\right]+\hbar\omega a^+a+\hbar g\left[a^+b_1^+b_2+ab_2^+b_1\right], \tag{6.23}$$

where $\hbar\epsilon$ is the energy of the electron transitions between the states $|1\rangle_a$ and $|2\rangle_a$, $\hbar\omega=\hbar\epsilon$ is the photon energy of the cavity mode, and g is the coupling parameter of the electron states with the photons of the cavity. In the Hamiltonian H_{JC}, the b_i^+ and b_i operators create and destroy electrons in the various atomic electron levels, i.e., if $|0\rangle_a$ represents the ionized atom in the absence of the single electron, then

$$b_1^+|0\rangle_a=|1\rangle_a,\quad b_2^+|0\rangle_a=|2\rangle_a\quad b_1|1\rangle_a=|0\rangle_a,\quad b_2|2\rangle_a=|0\rangle_a. \tag{6.24a}$$

In addition, because of the nature of $|0_a\rangle$ and the requirements of the Pauli principle

$$b_1|0\rangle_a=0,\quad b_2|0\rangle_a=0,\quad b_1^+|1\rangle_a=0,\quad b_2^+|2\rangle_a=0 \tag{6.24b}$$

Similar properties exist for the operators a^+ and a of the photon fields, which are defined by

$$a^+|n\rangle=\sqrt{n+1}|n\rangle+1\rangle,\quad a|n+1\rangle=\sqrt{n+1}|n\rangle,\quad a|0\rangle=0. \tag{6.25}$$

Later, in the discussion of the electron states $|3_a\rangle,|4_a\rangle$, sets of operators $\{b_3^+,\ b_3,\ b_4^+,\ b_4\}$ are similarly introduced as creation and destruction operators of these electron states.

It should also be noted that in Eq. (6.23), the possibility of nonenergy conserving processes has been excluded from the Hamiltonian, and, in addition, it is required that $\hbar\epsilon=\hbar\omega$ so that the interaction between the electron and the fields is resonant. These conditions on the form of Eq. (6.23) have also been included in our earlier considerations of the Jaynes–Cumming Hamiltonian of which the present system is an example [12–16].

The eigenvalue problem for the Hamiltonian of the two-level conservative atomic system in Eq. (6.23) is written as [12]

$$H_{JC}|\psi\rangle=i\hbar\frac{\partial}{\partial t}|\psi\rangle=E_0|\psi\rangle, \tag{6.26a}$$

with the formal time solution given by

$$|\psi(t)\rangle=e^{-i\frac{H_{JC}}{\hbar}t}|\psi(0)\rangle. \tag{6.26b}$$

Here, the wavefunction in Eq. (6.26b) is found to be a complex mixture of electron and phonon eigenstates that are mixed by the coupling of the last term on the righthand side of Eq. (6.23).

Of interest, however, in our discussions is the distribution of photon cavity modes in the system as electron–cavity states develops in time. This involves considerably less information than that contained within the entire wavefunction of electron and photon modes. To facilitate the development of the photon probability distribution, it is helpful to transform the problem in Eq. (6.26) into a simplifying notation. In this regard, a useful reformulation of the eigenvalue problem is made by rewriting Eq. (6.26) in the context of the density matrix formulation of quantum mechanics. This is now introduced and explained [12].

6.4.3 Density Matrix

Consider forming from the solution of Eq. (6.26) the density matrix, which is defined as [7,12–16]

$$\rho_{JC}(t) = |\psi(t)\rangle\langle\psi(t)| = e^{-i\frac{H_{JC}}{\hbar}t}|\psi(0)\rangle\langle\psi(0)|e^{i\frac{H_{JC}}{\hbar}t}. \tag{6.27}$$

This matrix is useful in determining the probabilities with which certain configurations of field and electron occupancies are found in the eigenstate $|\psi(t)\rangle$ of H_{JC}. For example, what if we were to ask the question of the probability of finding the solution in a state such as $|2,n\rangle$ or $|1,n+1\rangle$ within the electron–photon system.

To answer this question, it would be necessary to compute the matrix elements

$$\begin{aligned} |\langle 2,n||\psi(t)\rangle|^2 &= \langle 2,n|\rho_{JC}(t)|2,n\rangle \\ \text{or } |\langle 1,n+1||\psi(t)\rangle|^2 &= \langle 1,n+1|\rho_{JC}(t)|1,n+1\rangle, \end{aligned} \tag{6.28}$$

respectively. From Eq. (6.28), it is seen that questions regarding the nature of the distribution over the basis set of modes of the electrons and photons in the noninteraction (i.e., $g = 0.$) model are directly related to the matrix elements of the density matrix. Specifically, as shall be seen, these matrix elements are very helpful in realizing the goal of determining the nature of the distribution of photon occupancy states in the maser fields generated within the cavity resonator. It is even found to be the best to study the dynamics of the density matrix rather than to directly treat the wavefunction of the problem in Eq. (6.26). We now turn to a treatment of the dynamics of the density matrix for Eq. (6.23).

Taking the time derivative of the density matrix in Eq. (6.27), it is found to have an equation of motion of the form [7,12–16]

$$\frac{\partial}{\partial t}\rho_{JC}(t) = \frac{i}{\hbar}\left[\rho_{JC}(t), H_{JC}\right]. \tag{6.29}$$

The equation of motion obtained in Eq. (6.29) not only includes terms from the electron–photon interaction on the far right of Eq. (6.23) but also terms from the independent electron and photon dynamics in the first two terms on the right of Eq. (6.23). In this form, the problem looks complicated, but a simplification is made by applying a transformation related to the first two terms on the right of Eq. (6.23). This is now considered.

A helpful transformation on the density matrix in Eq. (6.27) is to apply a unitary transformation of the form [7,12–16]

$$\rho_{JC}^{(I)}(t) = e^{i\frac{H_{AF}}{\hbar}t}\rho_{JC}(t)e^{-i\frac{H_{AF}}{\hbar}t} = e^{i\frac{H_{AF}}{\hbar}t}|\psi(t)\rangle\langle\psi(t)|e^{-i\frac{H_{AF}}{\hbar}t}, \tag{6.30a}$$

where

$$H_{AF} = \frac{\hbar\epsilon}{2}\left[b_2^+b_2 - b_1^+b_1\right] + \hbar\omega a^+a \tag{6.30b}$$

is the Hamiltonian of the noninteracting electron and fields. With this transformation, the resulting equation of motion for $\rho_{JC}^{(I)}(t)$ then becomes [12]

$$\frac{\partial}{\partial t}\rho_{JC}^{(I)}(t) = \frac{i}{\hbar}\left[\rho_{JC}^{(I)}(t), H_{JC}^{(I)}(t)\right]. \tag{6.31a}$$

where

$$H_{JC}^{(I)}(t) = e^{i\frac{H_{AF}}{\hbar}t}\hbar g\left[a^{+}b_{1}^{+}b_{2} + ab_{2}^{+}b_{1}\right]e^{-i\frac{H_{AF}}{\hbar}t}, \tag{6.31b}$$

and the system has been chosen to be at resonance so that $\hbar\epsilon = \hbar\omega$. Now, only the third term in Eq. (6.23) enters the commutator in Eq. (6.31a), and this leads to the generation of fewer terms in the equations of motion for $\rho_{JC}^{(I)}(t)$.

Since the number of terms generated by the commutator in Eq. (6.31) is much less than that in Eq. (6.29), a significant reduction occurs to the dynamics of the problem. Viewed in the context of the reformulation in Eqs. (6.30) and (6.31), the system is said to be stated in the interaction picture. The interaction picture has the advantage that the phase information from the noninteracting electron and field degrees of freedom have been removed from the equations of motion. In the following, we proceed in the context of the interaction picture and consider the addition of phenomenological terms that account for internal and external losses of the cavity.

6.4.4 Hamiltonian: Phenomenological Dissipative Terms

The object in the following discussions is to study the photon field distributions of the system in Eqs. (6.30) and (6.31) for the case in which they are modified to include losses within the atom and also to represent a radiating maser light source [12–15]. In order to do this, a phenomenological account is developed of the losses and the radiation of the photon modes exiting from the cavity. In both the cases of energy loss from the system, this is done with the inclusion of terms generated based on general loss considerations for the simulation of the withdrawal of energy from the cavity.

These phenomenological terms represent Markovian stochastic processes [12,13] involving losses from the cavity into electron modes, the emission of maser fields, and the interactions of the electron–cavity states with fluctuations in these processes. In the maser process, the light emitted from the cavity is initially generated by the introduction into the cavity of excited state atoms (i.e., atoms in a $|2\rangle_a$ state), which then transitions to a lower atomic energy level (i.e., the $|1\rangle_a$ atomic state). These transitions supply energy to the fields of the cavity, and the supplied field energy, in turn, escapes the cavity as maser light. The exit of energy from the cavity is again modeled by general statistical considerations, imposing losses on the modes of the quantum mechanical dynamics of the cavity. On the other hand, the loss processes to electronic degrees of freedom present in the system involve the $|3\rangle_a, |4\rangle_c$ atomic states. These electronic states have not as yet been introduced into our formulation.

Under a steady state operation of the maser system, the resulting formulation of the Jaynes–Cumming dynamics with included loss terms is used to determine the distribution of Fock state photons $|n\rangle$ in the cavity fields. This distribution represents the nature of the fields of maser light, which ultimately escapes from the cavity. In the following discussions, the addition to the equations of motion in Eq. (6.31) of dissipative terms simulating the energy loss to photons and electronic degrees of freedom are made followed by a treatment of the density matrix obtained in the interaction picture [7,12–16].

The dissipative losses are introduced phenomenologically by the addition to the equations of motion in Eq. (6.31) of a so-called Lindbaden operator [7,12,13] accounting for the energy exchanges with the world outside and inside the cavity. The operator is designed to represent the most general Markovian stochastic effects of an interaction with a source of random energy exchanges. In the present problem, however, a derivation of the Lindbaden operator will not be given, and the reader is referred to the literature for these detains [7,12,13]. Some qualitative discussions of the processes involved, nevertheless, are presented. To understand the Lindbaden operator and its effects on the electron–cavity system, it is first necessary to start with the dynamics of the closed system of an atom interacting with the cavity fields.

For the present problem, begin by evaluating Eq. (6.31) to obtain the energy conserving equations of motion for the density matrix in the interaction formulation. This gives the dynamical form [7,12]

$$\frac{\partial}{\partial t}\rho_{JC}^{(I)} = -ig\left[a^{+}b_1^{+}b_2 + ab_2^{+}b_1, \rho_{JC}^{(I)}\right], \tag{6.32}$$

describing an isolated electron–cavity–field system at resonance and in the absence of losses. The conservative processes represented in Eq. (6.32) are purely dynamical quantum mechanical processes, which do not account for the communication of the cavity with the outside world. In the following, first the addition to Eq. (6.32) of electronic losses will be discussed within the Lindbaden formulation. This will be followed by considerations of the losses from the fields emitted from the cavity.

When energy is lost to the internal electronic degrees of freedom in the cavity, the energy removed from the system is described in terms of a rate at which energy is shifted within the atom. These processes are represented in Figure 6.3 by the γ_1 and γ_2 decay processes between the $|1\rangle_a, |3\rangle_a$ atomic states and the $|2\rangle_a, |4\rangle_a$ atomic states, respectively. The general expression for the stochastic shuttling of energy in the Lindbaden formulation is found in terms of the general density matrix ρ being considered for the schematic process in Figure 6.3 to be of the form [7,12–16]

$$\mathcal{L}_L\rho = \frac{\gamma_1}{2}\left[2\sigma_1^{-}\rho\sigma_1^{+} - \sigma_1^{+}\sigma_1^{-}\rho - \rho\sigma_1^{+}\sigma_1^{-}\right] + \frac{\gamma_2}{2}\left[2\sigma_2^{-}\rho\sigma_2^{+} - \sigma_2^{+}\sigma_2^{-}\rho - \rho\sigma_2^{+}\sigma_2^{-}\right] \tag{6.33}$$

Here, $\sigma_1^{+} = b_1^{+}b_3$ transfers the electron from energy level 3 and places it in level 1, $\sigma_1^{-} = b_3^{+}b_1$ transfers the electron from energy level 1 and places it in level 3, $\sigma_2^{+} = b_2^{+}b_4$ transfers the electron from energy level 4 and places it in level 2, and $\sigma_2^{-} = b_4^{+}b_2$ transfers the electron from energy level 4 and places it in level 2. The expression for the electronic losses in Eq. (6.33) actually involves two Lindbaden operators. The first operator $\frac{\gamma_1}{2}\left[2\sigma_1^{-}\rho\sigma_1^{+} - \sigma_1^{+}\sigma_1^{-}\rho - \rho\sigma_1^{+}\sigma_1^{-}\right]$ is for the losses associated with the electron in level 1, and the second operator $\frac{\gamma_2}{2}\left[2\sigma_2^{-}\rho\sigma_2^{+} - \sigma_2^{+}\sigma_2^{-}\rho - \rho\sigma_2^{+}\sigma_2^{-}\right]$ is for the losses associated with the electron in level 2.

Applying Eq. (6.33) to the rate equation in Eq. (6.32) gives dissipative terms of the form [12,13]

$$\mathcal{L}_L\rho_{JCRJ}^{(I)} = -\frac{1}{2}\left[R\rho_{JCRJ}^{(I)} + \rho_{JCRJ}^{(I)}R\right] + J\rho_{JCRJ}^{(I)}, \tag{6.34a}$$

where γ_1 and γ_2 are rate parameters for the shuttling of energy in and out of the $|1\rangle_a, |2\rangle_a$ electronic states,

$$R = \gamma_1 b_1^{+}b_3 b_3^{+}b_1 + \gamma_2 b_2^{+}b_4 b_4^{+}b_2 = \gamma_1 b_1^{+}b_1 + \gamma_2 b_2^{+}b_2 \tag{6.34b}$$

and

$$J\rho_{JCRJ}^{(I)} = \gamma_1 b_3^{+}\langle 1, n+1|\rho_{JCRJ}^{(I)}|1, n+1\rangle b_3 + \gamma_2 b_4^{+}\langle 2, n|\rho_{JCRJ}^{(I)}|2, n\rangle b_4. \tag{6.34c}$$

(Note, here the notation that $\rho_{JCRJ}^{(I)}$ is the density matrix of the system, including the addition of the terms in Eq. (6.34a) to the right side of Eq. (6.32), and, in this, composition of terms $\rho_{JC}^{(I)}$ in Eq. (6.32) is replaced by $\rho_{JCRJ}^{(I)}$.) The first term in Eq. (6.34a) is thought of as a rate at which energy is extracted from the states $|2, n\rangle$ and $|1, n+1\rangle$ when the system has occupancy in configurations composed from these elements. The second term represents a rate at which energy can be drawn from the $|4, n\rangle$ and $|3, n+1\rangle$ states and reintroduced in to the $|2, n\rangle$ and $|1, n+1\rangle$ states. In this regard, the $|4, n\rangle$ and $|3, n+1\rangle$ modes act as a reservoir into which energy is either dissipated or from which it is stochastically received as a fluctuation.

For a simplification in the following treatment, it is assumed that $\gamma_1 = \gamma_2 = \gamma$ and some qualitative discussions as to the nature of the processes in Eq. (6.34) are given. This is done by separately introducing first the stochastic term in Eq. (6.34b) followed by the stochastic term in Eq. (6.34c).

6.4.5 Development of the Solutions with Dissipative Effects

Upon the introduction into Eq. (6.32) of the first stochastic term in Eq. (6.34b), the new dynamical equation, including these particular stochastic effects, takes the form [7,12–16]

$$\frac{\partial}{\partial t}\rho_{JCR} = -ig\left[a^+b_1^+b_2 + ab_2^+b_1, \rho_{JCR}\right] - \frac{1}{2}\left[R\rho_{JCR} + \rho_{JCR}R\right]. \tag{6.35}$$

(Note, with the inclusion of the terms in R, the density matrix in denoted as ρ_{JCR}.) The solution of Eq. (6.35) is now given formally in terms of the $\rho_{JCR}(0)$ initial condition as [13]

$$\rho_{JCR}(t) = B(t)\rho_{JCR}(0)B^+(t), \tag{6.36}$$

where

$$B(t) = \exp\left[-i\frac{H_{JCR}}{\hbar}t\right] \tag{6.37a}$$

is a time translation operator. The solution in Eq. (6.36) introduces a dissipative Hamiltonian of the form [13]

$$H_{JCR} = H_{JC}^{(I)} - i\frac{\hbar}{2}R = \hbar g\left[a^+b_1^+b_2 + ab_2^+b_1\right] - i\frac{\hbar}{2}R, \tag{6.37b}$$

in which complex variable terms now appear [12–16].

It is important to note that the Hamiltonian in Eq. (6.37b) includes a real part based on the original dynamics of the conservative system and an imaginary part representing the energy losses in the system from the stochastic events associated with atomic degrees of freedom. The loses in the system are observed as an exponential decay of the time translation operator in Eq. (6.37a) with increasing time. This ultimately shows up as a decrease in the cavity energy.

The remaining second term in Eq. (6.34a) is more difficult to handle, as it cannot be rewritten into the effective Hamiltonian formulation in Eqs. (6.36) and (6.37). Nevertheless, it can be treated as a source of energy transfer in the system and handled in a formal perturbation approach. However, before this is done, it is useful to look at the properties of the effective Hamiltonian and the time translation operator in Eq. (6.36).

Consider the eigenvalue problem for the effective Hamiltonian in Eq. (6.37b). This takes the standard form [7,12–16]

$$H_{JCR}|\varphi\rangle = \lambda|\varphi\rangle. \tag{6.38}$$

Expressing Eq. (6.38) in terms of the basis states $|1,n+1\rangle$ and $|2,n\rangle$, Eq. (6.38) reduces to an algebraic 2×2 matrix eigenvalue problem given by [12,13]

$$\left|\begin{matrix} -i\frac{\gamma}{2} & g\sqrt{n+1} \\ g\sqrt{n+1} & -i\frac{\gamma}{2} \end{matrix}\right| \left|\begin{matrix} u_1 \\ u_2 \end{matrix}\right| = \left|\frac{\lambda}{\hbar}\right| \left|\begin{matrix} u_1 \\ u_2 \end{matrix}\right|, \tag{6.39}$$

where $|\varphi\rangle = u_1|1,n+1\rangle + u_2|2,n\rangle$. From considerations of linear algebra, this problem has two eigenvalue and eigenvector solutions for $\frac{\lambda}{\hbar}$ and $\begin{vmatrix} u_1 \\ u_2 \end{vmatrix}$, which may be listed as:

$$\frac{\lambda_1}{\hbar} = -i\frac{\gamma}{2} + g\sqrt{n+1} \quad \text{with eigenvector } |n,+\rangle = \frac{1}{\sqrt{2}}\left[|1,n+1\rangle + |2,n\rangle\right] \tag{6.40a}$$

and

$$\frac{\lambda_2}{\hbar} = -i\frac{\gamma}{2} - g\sqrt{n+1} \quad \text{with eigenvector } |n\rangle, - = \frac{1}{\sqrt{2}}\left[|1,n+1\rangle - |2,n\rangle\right]. \tag{6.40b}$$

These eigenvectors are then a natural basis in which to study the dynamical properties of the system.

Specifically, the results of the eigenvectors and eigenvalues of H_{JCR} provide a basis for studying the dynamics represented in Eqs. (6.35)–(6.37) and simplify the treatment of the perturbation represented by the second term in Eq. (6.34a). These eigenvector solutions are now used to obtain the effects on the dynamics of the perturbation represented by the second term in Eqs. (6.34a) and (6.34c).

6.4.6 Density Matrix of the Maser

The complete equation of motion for the density matrix describing the passage of a single atom through the cavity ρ_{JCRJ} is given from Eqs. (6.32) to (6.34) by [7,12–16]

$$\frac{\partial}{\partial t}\rho_{JCRJ} = -ig\left[a^+ b_1^+ b_2 + ab_2^+ b_1, \rho_{JCRJ}\right] - \frac{1}{2}\left[R\rho_{JCRJ} + \rho_{JCRJ} R\right] + J\rho_{JCRJ}. \tag{6.41}$$

For the passage of the first atom through the cavity, Eq. (6.41) is integrated in time for an initial density matrix of the form

$$\rho_{JCRJ}(0) = |2,n\rangle\langle 2,n|. \tag{6.42}$$

(Note: A more general initial photon state can be made [12,13], but, for simplicity, we assume this configuration to illustrate the method.) In the treatment of the pass through the cavity, the limits of integration beginning at $t = 0$ and carry over to the passage time τ of the atom through the cavity, and the initial density matrix in Eq. (6.42) represents an atom in its highest excited state $|2\rangle_a$, entering a cavity containing a Fock state $|n\rangle$.

As time progresses, the initial state in Eq. (6.42) develops in time as a mixture of the states $|2,n\rangle = |2\rangle_a|n\rangle$, $|1,n+1\rangle = |1\rangle_a|n+1\rangle$, $|4,n\rangle = |4\rangle_a|n\rangle$, and $|3,n+1\rangle = |3\rangle_a|n+1\rangle$. For each atomic passage through the cavity, Eq. (6.41) is again integrated using initial boundary conditions received as the solution from the passage of the previous atom. The totality of all such atomic passages sets the final mixture of photon states found within the maser cavity.

The problem for $\rho_{JCRJ}(t)$ in Eqs. (6.41) and (6.42) appears complicated but is simplified when rewritten in terms of the eigenvalues and eigenvectors discussed in Eqs. (6.39) and (6.40). In addition, as is now discussed, an application of a canonical transformation reduces the number of terms needed to be treated in the problem. Specifically, consider the new density function defined by the transformation [7,12–16]

$$\rho_{JCRJ}^{(I)}(t) = e^{i\frac{H_{JCR}}{\hbar}t}\rho_{JCRJ}(t)e^{-i\frac{H_{JCR}}{\hbar}t}. \tag{6.43}$$

The dynamics of $\rho^{(I)}_{JCRJ}(t)$ is obtained through an application of Eq. (6.41), which yields the dynamical equation [12]

$$\frac{\partial}{\partial t}\rho^{(I)}_{JCRJ}(t) = J\rho^{(I)}_{JCRJ}(t). \tag{6.44}$$

The dynamics including the effects from the second term in Eq. (6.34a) is obtained by formally integrating Eq. (6.44). In this way, it follows that

$$\rho^{(I)}_{JCRJ}(t) = \rho^{(I)}_{JCRJ}(0) + \int_0^t dt_1 J\rho^{(I)}_{JCRJ}(t_1). \tag{6.45}$$

Equation (6.45) is an integral equation for $\rho^{(I)}_{JCRJ}(t)$ from which an iterative solution in the form of a Neumann series involving successive powers of the small dissipation parameter γ can be generated.

Upon the first iteration, Eq. (6.45) yields the lowest order correction in J. This correction to $\rho^{(I)}_{JCRJ}(0)$ is given by [12,13]

$$\rho^{(I)}_{JCRJ}(t) = \rho^{(I)}_{JCRJ}(0) + \int_0^t dt_1 J\rho^{(I)}_{JCRJ}(0). \tag{6.46}$$

Applying Eqs. (6.36), (6.37a) and (6.43) in Eq. (6.46), it then follows that to the lowest order of approximation (Note: To simplify the notation, the superscript (I) in Eq. (6.46) is dropped in the following discussions.)

$$\rho_{JCRJ}(t) = \mathcal{L}(t)\rho_{JCRJ}(0) + \int_0^t dt_1 \mathcal{L}(t - t_1) J\mathcal{L}(t_1)\rho_{JCRJ}(0), \tag{6.47}$$

where here in operator notation

$$\mathcal{L}(t)\rho = B(t)\rho B^+(t) \tag{6.48}$$

is defined as the time translation operator in the case of noninteracting electron fields.

For an atom passing through the cavity in a time τ, the distribution of the electron-field states at the end of the passage is

$$\rho_{JCRJ}(\tau) = \exp\left[-i\frac{H_{JCR}}{\hbar}\tau\right]\rho_{JCRJ}(0)\exp\left[i\frac{H_{JCR}^{\ *}}{\hbar}\tau\right] + \int_0^\tau dt_1 \mathcal{L}(\tau - t_1) J\mathcal{L}(t_1)\rho_{JCRJ}(0), \tag{6.49a}$$

where Eq. (6.48) has been explicitly used in the first term on the right. In the limit that the atomic passage time is much greater that the decay time of the excited atomic state, so that $\gamma\tau \gg 1$, it then follows that [7,12–16]

$$\rho_{JCRJ}(\tau) \approx \int_0^\infty dt_1 \mathcal{L}(\tau - t_1) J\mathcal{L}(t_1)\rho_{JCRJ}(0), \tag{6.49b}$$

where the first term in Eq. (6.49a) exponential decays to zero in time τ. In addition, because the scale of time variation is set by $\frac{1}{\gamma} \ll \tau$, the upper limit of the integration has been taken to be infinity.

Consider the evaluation of the integral in Eq. (6.49b) for the case of the passage through the cavity of the first atom described by the initial density matrix in Eq. (6.42). Begin by evaluating the operator

$$J\mathcal{L}(t_1)\rho_{JCRJ}(0) = J\mathcal{L}(t_1)|2,n\rangle\langle 2,n| \tag{6.50}$$

in the integrand of Eq. (6.49b), where from Eq. (6.42) $\rho_{JCRJ}(0)=|2,n\rangle\langle 2,n|$. In terms of the eigenvectors in Eq. (6.40), it follows that [12–16]

$$|2,n\rangle\langle 2,n| = \frac{1}{\sqrt{2}}\left[|n,+\rangle - |n,-\rangle\right]\frac{1}{\sqrt{2}}\left[\langle n,+| - \langle n,-|\right] \tag{6.51}$$

so that

$$\mathcal{L}(t_1)|2,n\rangle\langle 2,n| = e^{-\gamma t_1}\frac{1}{\sqrt{2}}\left[e^{-i\Omega t_1}|n,+\rangle - e^{i\Omega t_1}|n,-\rangle\right]\frac{1}{\sqrt{2}}\left[\langle n,+|e^{i\Omega t_1} - \langle n,-|e^{-i\Omega t_1}\right], \tag{6.52}$$

where $\Omega = g\sqrt{n+1}$. Finally, operating on Eq. (6.52) with J, it is found that

$$J\mathcal{L}(t_1)\rho_{JCRJ}(0) = \frac{1}{2}\gamma e^{-\gamma t_1}\left[(1-\cos 2\Omega t_1)|3\rangle\langle 3||n+1\rangle\langle n+1| + (1+\cos 2\Omega t_1)|4\rangle\langle 4||n\rangle\langle n|\right]. \tag{6.53}$$

Substituting the result in Eq. (6.53) into Eq. (6.49b) and tracing over the electron states leaves a density matrix describing the distribution of photon fields generated in the cavity by the atomic passage.

In this way, it follows that [12,13]

$$\begin{aligned}\rho_{ph}(\tau) &= Tr\rho_{JCRJ}(\tau) \approx Tr\int_0^\infty dt_1\mathcal{L}(\tau - t_1)J\mathcal{L}(t_1)\rho_{JCRJ}(0) \\ &= \frac{1}{2}\gamma\int_0^\infty dt_1 e^{-\gamma t_1}\left[(1-\cos 2\Omega t_1)|n+1\rangle\langle n+1| + (1+\cos 2\Omega t_1)|n\rangle\langle n|\right] \\ &= \frac{2g^2(n+1)}{\gamma^2+4g^2(n+1)}|n+1\rangle\langle n+1| + \frac{\gamma^2+2g^2(n+1)}{\gamma^2+4g^2(n+1)}|n\rangle\langle n|\end{aligned} \tag{6.54a}$$

From the result in Eq. (6.54a), it is seen that an operator $\wp$ can be introduced, which advances the system through the time interval τ for the atomic passage through the cavity. It then follows that [12,13]

$$p_{ph}(\tau) = \wp p_{ph}(0) = \frac{2g^2(n+1)}{\gamma^2+4g^2(n+1)}|n+1\rangle\langle n+1| + \frac{\gamma^2+2g^2(n+1)}{\gamma^2+4g^2(n+1)}|n\rangle\langle n|. \tag{6.54b}$$

Here, $p_{ph}(\tau)$ represents the distribution of photons in the system after a single atom has gone through the cavity in an initial electron–cavity state described by the density matrix $|2,n\rangle\langle 2,n|$ with an initial photon density of states $p_{ph}(0)$. After this passage, the next atom going through the cavity will then involve the development of the states $|2,n\rangle\langle 2,n|$ and $|2,n+1\rangle\langle 2,n+1|$, and these are of the same form as the $|2,n\rangle\langle 2,n|$ state treated in the first atom to pass the cavity.

As a consequence of these, note that the result in Eq. (6.54b) can be generalized to treat an atomic passage involving an initial density matrix expressed as a linear combination of terms of the form $|2,m\rangle\langle 2,m|$. In this view, the advancement of a general term of the form $|2,m\rangle\langle 2,m|$ during the passage of a single atom through the cavity is accomplished by the mapping [12,13]

$$\begin{aligned}\wp|m\rangle\langle m| &= \frac{2g^2(m+1)}{\gamma^2+4g^2(m+1)}|m+1\rangle\langle m+1| + \frac{\gamma^2+2g^2(m+1)}{\gamma^2+4g^2(m+1)}|m\rangle\langle m| \\ &= A_{m+1,m+1}|m+1\rangle\langle m+1| + B_{m,m}|m\rangle\langle m|\end{aligned} \tag{6.55a}$$

of the photon fields, and, after the successive passage of k separate atoms, the mapping of $|m\rangle\langle m|$ in Eq. (6.55a) is given by

$$\wp^k |m\rangle\langle m|. \tag{6.55b}$$

Consequently, the operator $\wp$ (when applied to the density matrix of the photons for an atomic excited state $|2_a\rangle$ entering a cavity containing a diagonal distribution of photon cavity modes) advances the cavity photon distribution to that found in the cavity following the atomic passage. Starting from a diagonal distribution of cavity photons $\rho_{ph}(0)$ after k atomic passages, the photonic distribution is given as [12,13]

$$\rho_{ph}(k) = \wp^k \rho_{ph}(0), \tag{6.56}$$

where $\wp$ generates the mapping of the photon number modes during the passage of a single atom. In Eq. (6.56), the time dependence of $\rho_{ph}(\tau)$ has now been replaced by the number of distinct passages k through the cavity.

To understand the distribution of fields within the maser in its steady state operation, it is necessary to consider the effects on the cavity fields of a passage of a steady state sequence of atoms through the maser cavity. This is now addressed.

6.4.7 The General Atomic Passage Maser Processes

The earlier discussions were for the passage of a single atom through the cavity in the time interval τ. For the maser operation [12], however, a stream of atoms goes through the cavity so that in an infinitesimal time $\Delta t > \tau$ several atoms may individually contribute energy to the cavity fields. An account of this contribution must be made to determine the amount of energy given to the cavity photons in Δt. From this, the rate of energy transfer into the cavity fields is obtained.

Consider that the atoms enter the cavity at the rate r measured in the probability of an atomic passage per second, and that this rate is small enough that there is at most one atom in the cavity at any given time. During the time interval Δt, the probability that an atom passes through the cavity is $r\Delta t$, and the probability that an atom does not pass through the cavity during Δt is $1 - r\Delta t$. If in the time t a number k of excited atoms have successively past through the cavity, then from Eq. (6.56), the density matrix of the photon modes is $p_{ph}(t) = \wp^k \rho_{ph}(0)$.

In the next infinitesimal time interval Δt, an additional single atom passing through the cavity would change the density matrix to $p_{ph}(t+\Delta t) = \wp p_{ph}(t)$. Taking into account the probability that an atom may or may not pass through the cavity in time Δt, the average change experienced by the photon density matrix during the time Δt is then [12,13]

$$p_{ph}(t+\Delta t) = r\Delta t \,\wp p_{ph}(t) + (1 - r\Delta t)\, p_{ph}(t). \tag{6.57}$$

Here, the first term on the right accounts for the change in the density matrix when an atoms passes through the cavity and releases its excited state energy to the fields, and the second term accounts for the case in which no atom passes through the cavity. From Eq. (6.57), the rate of change of the photon density function in time is then expressed as

$$\frac{p_{ph}(t+\Delta t) - p_{ph}(t)}{\Delta t} = \frac{d}{dt} p_{ph}(t) = r[\wp - 1]\, p_{ph}(t) = rUp_{ph}(t). \tag{6.58}$$

Of particular interest in the discussion of the distribution of photon modes contained in the cavity is [12]

$$\frac{d}{dt}\langle n|\rho_{ph}(t)|n\rangle = \langle n|rU\rho_{ph}(t)|n\rangle. \tag{6.59}$$

This is the rate of change of the probability of finding a mode $|n\rangle$ within the cavity. This rate can be evaluated by applying Eq. (6.55a) in Eq. (6.59) for a density matrix of the general form

$$\rho_{ph} = \sum_{n,m=0}^{\infty} \rho_{n,m}|n\rangle\langle m|, \tag{6.60a}$$

In this way, it is found that

$$\langle n|\wp\rho_{ph}|n\rangle = \rho_{n-1,n-1}A_{n,n} + \rho_{n,n}B_{n,n}. \tag{6.60b}$$

Consequently, combining Eqs. (6.58)–(6.60) gives the rate equations [12]

$$\begin{aligned}\frac{d}{dt}\rho_{n,n} &= r\left[\rho_{n-1,n-1}A_{n,n} + \rho_{n,n}\left(B_{n,n}-1\right)\right]\\ &= r\left[\frac{2g^2n}{\gamma^2+4g^2n}\rho_{n-1,n-1} - \frac{2g^2(n+1)}{\gamma^2+4g^2(n+1)}\rho_{n,n}\right].\end{aligned} \tag{6.61}$$

This expresses the rate of change of the probability of finding photon modes in the Fock state $|n\rangle$ inside the cavity in terms of the diagonal terms of the photon density matrix.

More generally proceeding as above in the derivation of Eq. (6.61), it can be shown that for the various element of the photon density matrix [12,13]

$$\frac{d}{dt}\rho_{n,m} = r\left[\frac{2g^2\sqrt{mn}}{\gamma^2+2g^2(n+m)}\rho_{n-1,m-1} - \frac{g^2(m+n+2)+g^4(m-n)^2\gamma^{-2}}{\gamma^2+2g^2(m+n+2)}\rho_{n,m}\right]. \tag{6.62}$$

In the steady state operation of the maser, the terms in Eq. (6.61) must be balanced by a term accounting for the energy radiated from the cavity as maser light. It is now needed to obtain an expression for the rate of loss of the photonic cavity due to the emission of maser light.

6.4.8 The Loss Terms in the Maser Processes

The reduction of the density matrix due to maser emissions from the cavity is represented, as with the internal cavity losses, by a dissipation term of the Lindbaden form [7,12]. This term is directly added to the photon density matrix rate equation in Eq. (6.61). Specifically, upon its addition to Eq. (6.61), the rate equation of the electron-field density matrix including all energy losses has the form [12,13]

$$\begin{aligned}\frac{d}{dt}\rho_{n,n} &= r\left[\frac{2g^2n}{\gamma^2+4g^2n}\rho_{n-1,n-1} - \frac{2g^2(n+1)}{\gamma^2+4g^2(n+1)}\rho_{n,n}\right] + \langle n|\frac{\kappa}{2}\left(2a\rho a^+ - a^+a\rho - \rho a^+a\right)|n\rangle\\ &= r\left[\frac{2g^2n}{\gamma^2+4g^2n}\rho_{n-1,n-1} - \frac{2g^2(n+1)}{\gamma^2+4g^2(n+1)}\rho_{n,n}\right] + \kappa\left((n+1)\rho_{n+1,n+1} - n\rho_{n,n}\right),\end{aligned} \tag{6.63a}$$

where κ is a parameter characterizing the rate at which energy is extracted from the cavity through the radiation emissions of the maser cavity. Note that in Eq. (6.63a) the term $-\kappa n\rho_{n,n}$ contributes an exponential decay in the time dependence of the density matrix, whereas the $\kappa(n+1)\rho_{n+1,n+1}$ term introduces a much slower loss with time into the system.

The Lindbaden loss term in Eq. (6.63a) contains the operator $\frac{\kappa}{2}\left(2a\rho a^{+} - a^{+}a\rho - \rho a^{+}a\right)$, which is the analogy of the operator $\frac{\gamma_1}{2}\left[2\sigma_1^{-}\rho\sigma_1^{+} - \sigma_1^{+}\sigma_1^{-}\rho - \rho\sigma_1^{+}\sigma_1^{-}\right]$ in Eq. (6.33) for the electronic losses. Note that, for the emission losses, the a^{+} and a operators of the photons introduce and remove photon modes from the system, and $a^{+}a$ counts the number of photons in the cavity. Similarly, in the electron operator, σ_1^{+} and σ_1^{-} shift an electron from the 3 state to the 1 state and from the 1 state to the 3 state, respectively. Also, in the electron loss term, $\sigma_1^{+}\sigma_1^{-}$ essentially counts the number of electrons in the 1 state. For the photon modes, there is only one Lindbaden operator for the single photon cavity mode, but, for the electron modes, there is one Lindbaden operator for the 1-state electrons and another for the 2-state electrons.

Following some algebra, Eq. (6.63a) is rewritten into the compact form [12,13]

$$\frac{d}{dt}\rho_{n,n} = G\left[\frac{n}{1+\frac{n}{n_s}}\rho_{n-1,n-1} - \frac{(n+1)}{1+\frac{n+1}{n_s}}\rho_{n,n}\right] + \kappa\left((n+1)\rho_{n+1,n+1} - n\rho_{n,n}\right), \tag{6.63b}$$

where $G = \frac{r}{2n_s}$ and $n_s = \frac{\gamma^2}{4g^2}$. By defining

$$t_{+}(n) = G\frac{(n+1)}{1+\frac{n+1}{n_s}} \tag{6.64a}$$

and

$$t_{-}(n) = \kappa n \tag{6.64b}$$

Eq. (6.63b) then becomes

$$\frac{d}{dt}\rho_{n,n} = -\left[t_{+}(n) + t_{-}(n)\right]\rho_{n,n} + t_{+}(n-1)\rho_{n-1,n-1} + t_{-}(n+1)\rho_{n+1,n+1}. \tag{6.64c}$$

Note that in Eq. (6.64c) $t_{-}(n+1)\rho_{n+1,n+1}$ represents the rate at which $\rho_{n+1,n+1}$ transitions into $\rho_{n,n}$, and $t_{+}(n-1)\rho_{n-1,n-1}$ represents the rate at which $\rho_{n-1,n-1}$ transitions into $\rho_{n,n}$.

In this same sense, $t_{-}(n)\rho_{n,n}$ represents the rate at which $\rho_{n,n}$ transitions into $\rho_{n-1,n-1}$, and $t_{+}(n)\rho_{n,n}$ represents the rate at which $\rho_{n,n}$ transitions into $\rho_{n+1,n+1}$. For the condition of steady state operation of the maser, the rate at which the system leaves a given photon occupancy state, going to another state, must equal the rate for the reversed process. Consequently, in the steady state it follows that [12,13]

$$t_{-}(n+1)\rho_{n+1,n+1} = t_{+}(n)\rho_{n,n} \tag{6.65a}$$

and similarly

$$t_{+}(n-1)\rho_{n-1,n-1} = t_{-}(n)\rho_{n,n}. \tag{6.65b}$$

These are a set of recursion relations that can now be solved for $\rho_{n,n}$.

Making this solution, from Eq. (6.65b), it is found that

$$\rho_{n,n} = \frac{t_{+}(n-1)}{t_{-}(n)}\rho_{n-1,n-1} \tag{6.66a}$$

and, in particular,

$$\rho_{1,1} = \frac{t_+(0)}{t_-(1)}\rho_{0,0}. \tag{6.66b}$$

From the two relations in Eq. (6.66), it then follows iteratively that [12,13]

$$\rho_{n,n} = \rho_{0,0}\prod_{k=1}^{n}\frac{t_+(k-1)}{t_-(k)} = \rho_{0,0}\left(\frac{G}{\kappa}\right)^n\prod_{k=1}^{n}\left(\frac{n_s}{n_s+k}\right). \tag{6.67}$$

The result in Eq. (6.67) relates the general photonic occupancy probability $\rho_{n,n}$ to that of the $\rho_{0,0}$ mode in terms of the ratio $\frac{G}{\kappa}$ and the parameter n_s. In this regard, $\rho_{0,0}$ acts as a normalization.

Note that in Eq. (6.67), the function $\prod_{k=1}^{n}\left(\frac{n_s}{n_s+k}\right)$ decreases with increasing values of n. On the other hand, $\left(\frac{G}{\kappa}\right)^n$ for $\frac{G}{\kappa}>1$ is an increasing function of increasing n, while, for $\frac{G}{\kappa}<1$, it is a decreasing function of increasing n. In this regard, it is expected that $\rho_{n,n}$ has a maximum about $n=0$ in the case that $\frac{G}{\kappa}<1$, whereas in the case that $\frac{G}{\kappa}>1$, the $\rho_{n,n}$ distribution should be peaked at some $n \neq 0$. Specifically, for $\frac{G}{\kappa}<1$, Eq. (6.67) generates a distribution of photons characteristic of an incoherent thermally generated distribution of photon Fock modes. For $\frac{G}{\kappa}>1$, Eq. (6.67) generates a distribution of photons approximating a Poisson distribution of Fock modes. This last case is approximated by the coherent state of photons discussed in the previous chapter. In Figure 6.4, a representative plot of $\rho_{n,n}$ versus n is shown for these cases of the photon distribution.

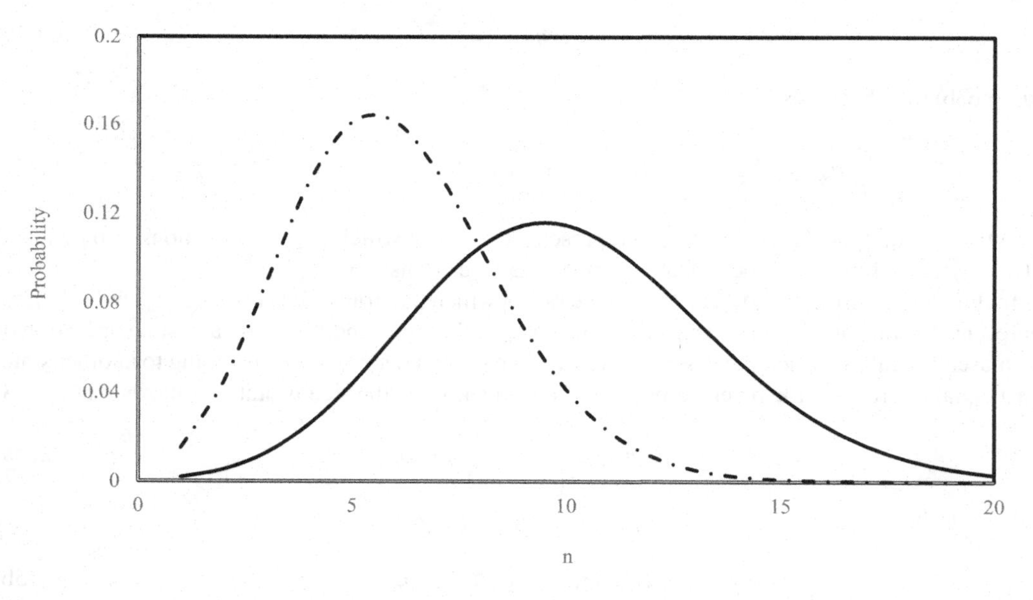

FIGURE 6.4 Plot of the probability $\rho_{n,n}$ for the cavity occupancy of the photons versus the number of photons n in the cavity maser field of frequency $\nu=\omega/2\pi$. The solid curve is for the cavity fields given in Eq. (6.67), and the dash–dot curve is for the Poison distribution in Eq. (6.71). The plots are made for $G/\kappa=6$, $n_s=2$, and $\alpha=6$. Note that the plot for the Poison distribution is made to give a crude comparison showing that, in this case for the parameters of the cavity field (solid curve), the poison distribution (dash–dot curve) has a similar form to that of the cavity fields.

Consequently, the light generated for $\frac{G}{\kappa} > 1$ is a coherent maser light, and the laser threshold of the system is obtained from the condition that $\frac{G}{\kappa} = 1$. Below this threshold the light is not sustained as coherent maser light. Its properties are closer to the light generated by an incoherent thermal source [12,13].

6.4.9 Statistical Properties of Maser Radiation

The steady state of the maser in which energy is pumped into the system and emitted from the system at a constant rate from Eq. (6.63b) requires that [12,13]

$$\frac{d}{dt}\rho_{n,n} = 0 = G\left[\frac{n}{1+\frac{n}{n_s}}\rho_{n-1,n-1} - \frac{(n+1)}{1+\frac{n+1}{n_s}}\rho_{n,n}\right] + \kappa\left((n+1)\rho_{n+1,n+1} - n\rho_{n,n}\right). \tag{6.68}$$

This condition of the maser is now studied to obtain details of the nature of the distribution of photon Fock modes in the cavity, as reflected in the fields emitted by the maser.

In the case that $\frac{n}{n_s} = n\frac{4g^2}{\gamma^2} \ll 1$, Eq. (6.68) has the form

$$\frac{d}{dt}\rho_{n,n} = 0 = -G\left[(n+1)\rho_{n,n} - n\rho_{n-1,n-1} + \frac{1}{n_s}\left(n^2\rho_{n-1,n-1} - (n+1)^2\rho_{n,n}\right)\right] + \kappa\left((n+1)\rho_{n+1,n+1} - n\rho_{n,n}\right). \tag{6.69}$$

Multiplying by n and then summing over n, it is found after some algebra that [12,13]

$$\frac{d\bar{n}}{dt} = 0 = (G-\kappa)\bar{n} - \frac{G}{n_s}\overline{n^2} - \frac{G}{n_s}(2\bar{n}) - \frac{G}{n_s} + G. \tag{6.70a}$$

Considering the limit $n_s \gg G$, it is seen that in the region in which maser radiation is absent (i.e., for $G < \kappa$), the first term on the right of Eq. (6.70) is negative so that a fluctuation in the system that increases $\bar{n}$ in this region of operation would be damped out. For this case the mean number of photons is

$$\bar{n} \approx \frac{G}{\kappa - G}. \tag{6.70b}$$

On the other hand, in the region in which maser radiation is present (i.e., for $G > \kappa$), it can be shown [13] from Eq. (6.69) that $\bar{n}$ would be amplified. In this region of operation, it is found that [12,13]

$$\bar{n} \approx n_s\frac{G-\kappa}{\kappa}. \tag{6.70c}$$

From the distribution in Eq. (6.67), it can be shown [12,13] that above the laser threshold and for $\bar{n} \gg n_s$, the variance of the distribution approaches $\bar{n}$. This is a characteristic of the Poisson distribution. Consequently, the coherent state modes discussed in the previous chapter are closely associated with a Poisson distribution of Fock modes, such as those generated by the maser. In this regard, for a comparison with Eq. (6.67), the general form of the Poisson distribution $P(\alpha, n)$ over integers is given by [12,13]

$$P(\alpha,n)=e^{-\alpha}\alpha^{n}\prod_{k=1}^{n}\left(\frac{1}{k}\right), \tag{6.71}$$

where α is a parameter characterizing the width of the distribution. Indeed, it is seen that as $n_s \to 0$, the distribution in Eq. (6.67) becomes exactly a Poisson distribution.

For further discussions of these points, the reader is referred to Refs. [7,12–16].

6.5 SPASERS AND ATOM LASERS

The ideas of optical lasers have been extended to the development of laser-like applications in other systems of bosonic excitations [4–6]. These include the development of spasers and atom lasers. In the spaser, a laser-like production of surface plasmons is made, based on a mechanism involving many of the features found in the functioning of an optical laser [4–6]. Similarly, atom lasers are studied that operate on Bose–Einstein condensates of atoms. The atom lasers produce a beam of atoms that display some of the properties of the radiation generated by a laser, including an emission of a phase-coherent stream of atoms [17–21].

A typical spaser design involves a dye that is pumped so that the electrons in the excited state of the dye molecules can be stimulated to down transition to lower molecular energy levels [4–6]. During this process, energy is transferred to a resonant cavity in the form of plasmon oscillations of the cavity. Here the plasmons are collective excitations formed from electromagnetic fields and the electronic polarization of the resonator material.

Typically, the resonant cavity of a spaser would be a metal nanoparticle. For these particles, the dye molecules may be positioned as an outer coating of the particle or may be included within a hollow of the particle. Polaritons emitted from the cavity would act as a supply source of surface plasmons for various plasmonic applications on surfaces. In this regard, the spaser could function as a more efficient source of surface plasmons [4–6].

As an example [3] of a resonant cavity for plasmon generation, consider a spherical dielectric of radius a and dielectric constant $\frac{\varepsilon}{\varepsilon_0}$ placed in a surrounding vacuum. It is a standard result that if a uniform electric field E_0 is applied to the sphere it becomes polarized. The induced polarization of the sphere, in turn, generates a field outside of the sphere, which is a dipolar field. Specifically, in the region outside the sphere, the field is that of a point dipole [3]

$$p=4\pi\varepsilon_0\left(\frac{\varepsilon-\varepsilon_0}{\varepsilon+2\varepsilon_0}\right)a^3E_0. \tag{6.72a}$$

In the quasi-static limit of a low-frequency harmonic applied field

$$E_0=E_{0,0}e^{-i\omega t}, \tag{6.72b}$$

the induced polarization is also given by Eq. (6.72a), with E_0 now replaced by the harmonic form in Eq. (6.72) and ε a frequency-dependent $\varepsilon(\omega)$. In this limit, note that for the case that $\varepsilon(\omega)+2\varepsilon_0=0$ the dipole moment becomes infinite, independent of the nonzero E_0. This instability of the system is due to the onset of a plasmon excitation in the sphere. In this regard, independent of the applied field, the sphere acts as a resonant cavity supporting a polarization oscillation, i.e., plasmon excitations.

Another type of bosonic mode that has been of interest is that of Bose–Einstein condensates of atoms [17–21]. These low-temperature condensates have been shown to exhibit atomic matter waves with some of the interesting phase-coherent properties of laser light. However, they do not exhibit an amplification of the matter waves. In this regard, unlike photons, atoms are not created in the condensate of the atom laser, but they only develop a coherent wave propagation. Such atomic waves have been used to demonstrate imagining capabilities of atomic beams that are like those exhibited by laser light [17–19].

In the extreme limit of superfluid He^4, due to the strong interactions between the atoms, a highly correlated fluid is developed that can be related to the ideas of the generation of a Bose–Einstein condensate [2,17,18]. This superfluidity of He^4 arises, in part, from the strong interactions of the helium atoms, so that the state of the system differs from that of a true Bose–Einstein condensate. More recently, however, atomic systems with weaker interactions have been developed into true Bose–Einstein condensates using various atomic trapping schemes, and these have provided the basis of atomic lasers [17–19]. Early studied systems in this respect were on gases of sodium and rubidium atoms [20,21].

REFERENCES

1. Kittle, C. 1996. *Introduction to solid state physics, 7th Edition*, 393–410. New York: John Wiley & Sons, Inc.
2. Marder, M. P. 2000. *Condensed matter physics*. New York: John Wiley & Sons, Inc.
3. Streetman, B. G. and S. Banerjee 2000. *Solid state electronic devices, 5th Edition*. New Jersey: Prentice Hall.
4. McGurn, A. R. 2018. *Nanophotonics*. Cham: Springer.
5. Stochman, M. I. 2013. *Spaser, plasmonic amplification, and loss compensation*. In Active plasmonics and tuneable plasmonic metamaterials Ed A. V. Zayats and S. A. Maier. New York: John Wiley & Sons, Ltd.
6. Noginov, M. A., Zhu, G., Belgrave, A. M., et al. 2009. Demonstration of a spaser-based nanolaser. *Nature* 460: 1110–1113.
7. Gerry, C. and P. Knight 2005. *Introductory quantum optics*. Cambridge: Cambridge University Press.
8. Lambropoulos, P. and D. Petrosyan 2007. *Fundamentals of quantum optics and quantum information*. Berlin: Springer-Verlag.
9. Yariv, A. 1971. *Introduction to optical electronic*. New York: Holt, Rinehart and Winston, Inc.,
10. Band, Y. B. 2006. *Light and matter: Electromagnetism, optics, spectroscopy and lasers*. Chichester: John Wiley & Sons, Ltd.
11. Solymar, L. and D. Walsh 2010. *Electrical properties of materials, 8th Edition*. Oxford: Oxford University Press.
12. Walls, D.F. and G. J. Milburn 1995. *Quantum optics*. Berlin: Springer-Verlag.
13. Scully, M. O. and W. E. Lamb 1967. Quantum theory of an optical maser. I. General theory. *Physical Review* 159, 208–226.
14. Scully, M. O. and W. E. Lamb 1968. Quantum theory of an optical maser II. Spectral profile. *Physical Review* 166: 246–249.
15. Scully, M. O. and W. E. Lamb 1969. Quantum theory of an optical maser. III. Theory of photoelectron counting statistics. *Physical Review* 179: 368–374.
16. Lax, M. and W. H. Louisell 1969. Quantum noise XII density operator treatment of field and population fluctuations. *Physical Review* 185: 568–591.
17. Martellucci, S., Chester, A. N., Aspect, A. and M. Inguscio 2002. *Bose-Einstein condensates and atomic lasers*. New York: Kluwer Academic Publishers.
18. V. S. Letokhov 2007. *Laser control of atoms and molecules*. Oxford: Oxford University Press.
19. Martellucci, S., Chester A. N., Aspect, A. and M. Inguscio 2002. *Bose-Einstein condensates and atom lasers*. New York: Kluwer Academic Publisher.
20. Davis, K. B., Mewes, M. O., Andrews, M. R., van Druten, J., Durfee, D.S., Kurn, D. M. and W. Ketterle 1995. Bose-Einstein condensation in a gas of sodium atoms, *Physical Review Letters* 75, 3969–3973.
21. Anderson, M. H., Ensher, J. R. Matthews, M.R., Wieman, C. E. and E. A. Cornell 1995. Observation of Bose-Einstein condensation in a dilute atomic vapor. *Science* 269: 198–201.

7 Semiconductor Junctions

In this chapter a review of some of the basic properties of semiconducting materials is presented, followed by an application of these properties to the discussion of semiconductor junctions [1–4]. For our studies, a semiconductor junction in its basic formulation is designed as a simple planar interface between two different types of semi-infinite semiconductor materials, and its importance resides in the interesting nonlinear electronic properties it displays. The focus of our treatment is on the fundamental principles needed to understand the elementary functioning of standard electronic devices that are composed of semiconductor junctions. In this regard, the engineering details required to finesse the characteristics or maximize the performance of such electronic devices are not a consideration. Only the most general principles in device operations are a consideration.

The chapter begins with a consideration of the conduction properties of semiconductor materials, in general [1,2]. The semiconductor is described by an energy band model in which two pass bands of electron energy states are separated by a stop band. The higher energy pass band of electron solutions is known as the conduction band, and the lower energy pass band of electron solutions is known as the valence band. The stop band between these bands represents energies at which no propagating electron solutions exist and which do not support a current in the material. At zero temperature, all of the electron states in the valence band of the semiconductor are occupied by electrons, and the electron states in the conduction band are completely vacant. Consequently, the system behaves as an insulator. This is due to the fact that electrons in the completely filled valence band require significant energy to be moved to the conduction band where they can form a current.

At nonzero temperatures, electrons are thermally promoted from the valence to the conduction band, and the electronic modes in the two bands both contribute to give the material a nonzero conductivity. The conductivity of the electrons in the conduction bands is handled in a formulation that is similar to that used to treat electrons in a metallic conductor. On the other hand, the conductivity of the electrons in the valence band is handled by introducing the idea of holes and their conduction properties. In this formulation, the holes are effectively positive-charge carriers, and their motion contributes to the current supported by the semiconductor. Consequently, both types of charge carriers enter into the determination of the conductivity of the material.

Holes are treated in a formulation that is similar to that of the treatment of electrons in metals but which is based on the ideas of the transport of positive charges [1–3]. In effect holes are absences of electrons in the valence band, and enter into the study of semiconductor band structure in a similar manner to that of the discussions of electrons and positrons in the theory of quantum electrodynamics. In this regard, a hole represents an electron vacancy in the valence band, and the motion of the vacancy arises from and is highly correlated with that of the valence band electrons. Once the dynamics of all of the electrons and holes in the material is determined, the conductivity is set from the sum of both the electron and hole currents.

The above description is that for a so-called intrinsic semiconductor [1–3]. This is a material that is a pure sample without chemical impurities. If a small number of chemical impurities are added to the semiconductor, a type of semiconductor known as an extrinsic semiconductor is formed. In an extrinsic semiconductor, there are again a conduction and a valence pass band separated by a stop band. Now, however, spatially localized impurity electron energy levels can occur within the stop band. These arise from the chemical nature of the impurities introduced into the material [1–3] and their abilities to bind electrons to them. In an extrinsic material, the valence band is fully occupied at zero temperature, and the conduction band is vacant so that the material is an insulator. The impurity states in this limit do not affect the insulating properties of the material.

DOI: 10.1201/9781003031987-7

At nonzero temperatures, however, because of thermal interaction with their environment, the impurity states within the stop band can become ionized. In this way, their ionization also contributes either electrons to the conduction band or holes to the valence band. Consequently, depending on the types of impurities present, the impure materials exhibit conductivities involving an unequal number of conduction band electrons and valence band holes. (In this regard, note that in an intrinsic semiconductor the number of conduction band electrons is equal to the number of valence band holes, i.e., the holes are formed by the promotion of valence band electrons to the conduction band.) As a result, two types of extrinsic semiconductors can be made. An extrinsic material in which the conduction is dominated by electrons in the conduction band is known as an n-type semiconductor, and an extrinsic material in which the conduction is dominated by valence band holes is known as a p-type semiconductor [1].

We shall see that the n- and p- types materials are very important for their applications in semiconductor junctions and devices [1–3]. In this regard, the basis of most of these applications is centered in an arrangement known as a p–n semiconductor junction. This is idealized as a junction formed as a planar interface between two semi-infinite regions of extrinsic semiconductor: one region is of p-type semiconductor, and the other is of n-type semiconductor. Its importance arises from the nonlinear electrical properties the junction exhibits for currents crossing the interface between the two materials. Such nonlinearity is required for a variety of switching and rectification effects, and for the interactions between different currents flowing in complex electrical circuits.

In summary, the basis of junction applications is the interesting nonlinear conductivity arising at the junction when it is subject to a potential difference applied across the interface. These types of applications will be elaborated upon in discussions given in this and later chapters, and forms the basis of most of the important developments of electronic technologies. Specifically, it is a topic entering into the design of various rectifier and transistor applications, which are the focus of our interests.

In this chapter, first a general review of basic semiconductor properties is given. This is followed by a discussion of the characteristics of semiconductor junctions. The elaboration of these properties into the discussions of semiconductor devices will occupy the later chapters of the text.

7.1 SEMICONDUCTOR MODEL

In its basic model, a semiconductor is described by an electronic energy band structure composed of two electronic pass bands that are separated by an electronic stop band [1,2]. The two pass bands are a low-energy valence band that has a high electron occupancy and a high energy conduction band which is only sparsely occupied with electrons. These two pass bands are separated by an intermediate energy stop band that contains no propagating electron solutions. The energy stop band [1,2] typically has an energy width less than 2 eV. (A schematic of these bands in the electronic energy spectrum is shown in Figure 7.1.) For a comparison of the energy scales involved, the Boltzmann energy at room temperature is $k_B T \approx \frac{1}{40}$ eV. Consequently, the room temperature energy distribution of electrons appears to be little changed from that of the zero-temperature system. Nevertheless, the small changes that do occur in the electron distribution in the semiconductor between these two temperatures have significant effects on the conductivity of the semiconductor.

At zero temperature all of the electrons fill the lower energy valence band, while the high-energy pass band is vacant. In this configuration, the material is an insulator. As the temperature increases, electrons in the valence band are promoted by thermal fluctuations through the stop band separating the two pass bands, and begin to occupy the energy levels of the upper energy conduction band. The material now exhibits a conductivity that arises from the motion of the promoted electrons in the conduction band, and the motion of the holes or absence of electrons that are created in the valence band. Both of these currents combine to contribute to a net total current in the material.

While the conductivity of the electrons in the conduction band can be handled within the context of a treatment similar to that for electrons within a metal, the conductivity of holes created in the

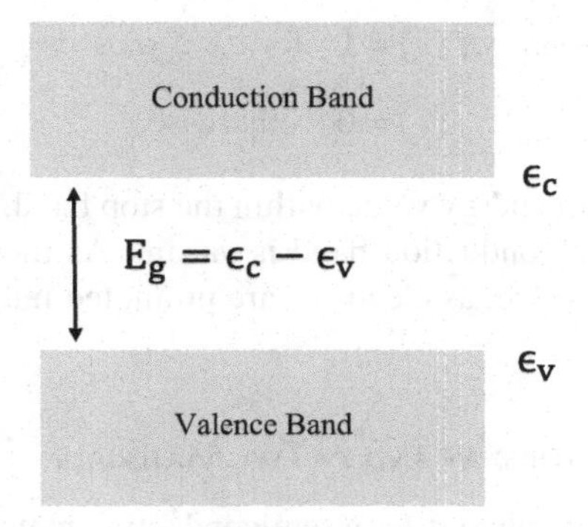

FIGURE 7.1 The conduction pass band and valence pass band of a semiconductor where the energy is measured on the vertical axis. The lowest energy of the conduction band is denoted ϵ_c, and the highest energy of the valence band is denoted ϵ_v. The stop band separating the two pass bands has an energy width $E_g = \epsilon_c - \epsilon_v$.

valence band is a little different. When an electron is removed from the valence band, it is found that the conductivity arising from the remaining electrons is the same as that of the vacancy left within the valence band by the removed electron. Specifically, the current carried by all of the remaining electrons is equal to the current carried by the electron vacancy when it is treated as a positively charged particle. In this context, the positive-charged vacancy is termed a hole, and the current from the valence band is characterized as arising from the motion of the single-particle hole vacancy.

To see the relationship of holes to vacancies in bands of valence electrons, consider a material with a full valence band of electrons. The current density of the completely filled band is obtained from the sum over electron states and is given by [1,2]

$$J = -\sum_i e\vec{v}_i = 0, \tag{7.1a}$$

where $e > 0$ is the fundamental unit of electric charge, the sum is over all the valence band states, and $\vec{v}_i$ is the velocity of the ith valence band electron. If the jth electron is removed from the sum, the new current flow of the remaining electrons in the band is given by [1–3]

$$J = -\sum_i{}' e\vec{v}_i = -\sum_i e\vec{v}_i + e\vec{v}_j = e\vec{v}_j, \tag{7.1b}$$

where the prime on the first sum indicates the absence of the contribution of the jth electron. Consequently, the current contribution of the jth electron is $-e\vec{v}_j$, and removing the jth electron introduces a net current to the system of $e\vec{v}_j$. This is the current of a positive charge with velocity $\vec{v}_j$.

7.1.1 Thermal Occupancy

Specifically, the energy occupancy at temperature T of the conduction and valence bands in a semiconductor is governed by the Fermi–Dirac distribution. This is given by [1]

$$f(\epsilon) = \frac{1}{e^{\beta(\epsilon-\mu)} + 1}, \tag{7.2a}$$

where $0 \le f(\epsilon) \le 1$ is the probability that an electron state of energy ϵ is occupied, $\beta = \dfrac{1}{k_B T}$ for the temperature T, and μ is the chemical potential. At zero temperature, Eq. (7.2a) becomes [1]

$$f(\epsilon) = 1 \quad \text{for} \quad \epsilon \le \mu$$
$$f(\epsilon) = 0 \quad \text{otherwise,} \tag{7.2b}$$

and the chemical potential has an energy value within the stop band. In this limit, the valence band is completely occupied, and the conduction band is vacant. As the temperature is increased, the chemical potential increases in value, as electrons are promoted into the conduction band, leaving holes in the valence band.

7.1.2 Extrinsic Semiconductors: *N*- and *P*- Type Materials

The two pass band model just described is for a semiconducting material of a type termed an intrinsic semiconductor. These materials are chemically pure and are characterized by the property that the number of conduction electrons in their conduction band is equal to the number of holes in their valence band. A second type of semiconductor, also of great technological importance, is termed an extrinsic semiconductor. An extrinsic semiconductor is formed by the addition of a small concentration of chemical impurities to an otherwise intrinsic semiconductor. The addition of chemical impurities allows for the preferential introduction into the system of electrons in the conduction band or holes into the valence band. These adjustments provide for the creation of extrinsic materials with electrical currents that are dominated by either electron or hole transport [1].

Specifically, there are two type of extrinsic materials known as n- and p-type semiconductors. In an n-type material, an impurity atom (termed a donor) is introduced into an intrinsic semiconductor so that it binds an electron to the impurity, existing at an energy level located within the stop band of the intrinsic semiconductor. The bound electron is localized at the impurity atom within the semiconductor, and the impurity is chosen such that the energy of the bound impurity electron is close to the uppermost energy of the stop band. As a consequence of its close proximity to the uppermost energy of the stop band, the impurity is easily ionized by thermal fluctuations to contribute its bound electron to the electron carriers in the conduction band. This creates a net increase in the number of electrons in the conduction band over that of the material in the absence of the impurities. In the following discussions, the concentration of donor impurity atoms in the n-type material will be denoted as n_+.

In a p-type material an impurity atom (termed an acceptor) is introduced into an intrinsic semiconductor so that it can bind an electron already present in the valence band [1,2]. This creates a hole or absence of an electron in the valence band, which contributes to the number of holes available in the valence band to carry electrical current. The electron removed from the valence band is localized about the impurity atom within the semiconductor, and the impurity is chosen such that the energy of the electron bound to the impurity atom is greater than but close to the lowermost energy of the stop band. Due to its close proximity to the lowermost energy of the stop band, the valence band electron is easily promoted in energy by thermal fluctuations to bind to the atomic site impurity. This creates a net increase in the number of holes in the valence band over that of the material without impurities. In the following discussions, the concentration of acceptor impurity atoms in the p-type material will be denoted as n_-.

7.1.3 Positioning of the Chemical Potential

In the presence of a nonzero temperature [1,2], the detailed properties of the semiconductor conductivity depend on the stop band energy parameters, and the effective masses of the electrons m_e and holes m_h. In this regard, if the bottom of the conduction band is located at the energy ϵ_c, and the top of the valence band is denoted ϵ_v, then the width of the stop band energy region $E_g = \epsilon_c - \epsilon_v$ forms a general characterization of the important basic electron properties of the material. With reference to these fixed energy and mass parameters, it is found that the position of the chemical potential $\mu(m_e, m_h\ \epsilon_c, \epsilon_v, T)$ within the stop band changes with temperature to reflect the occupancy of conduction band electrons and valence band holes. These occupancies, in turn, determine the current density in the material, which is a fundamental property for semiconductor applications, and the stop band energy width also sets the overall nature of the conductive features displayed by the material as a function of temperature.

In addition, in the case of extrinsic semiconductors, the position of the chemical potential within the stop band is also adjusted by the concentration of chemical impurities within the material and whether these contribute to *n*- or *p*-type behaviors. By changing the concentration of impurities in impurity-doped materials, the concentration of conduction band electrons and valence band holes is adjusted. Consequently, these changes show up as modifications to the chemical potentials of the materials that characterize the adjustment in conduction band electron and valence band hole concentrations.

Added to the considerations of the current density of a bulk semiconductor, in a second major consideration the chemical potential fixes the nature of the chemical equilibrium of the semiconductor with its environment and its interaction with other electronic components, e.g., by the introduction of externally applied electric potentials and the creation of internal electric potential gradients. This is an important factor in setting engineering designs for technology. Such reactive properties are found to form the basis of operation of device applications in sensor, rectifier, and transistor mechanisms [1–5].

A number of factors can be manipulated to affect the position of the chemical potential of the semiconductor. Such changes involve both permanent intrinsic modifications to the semiconducting material with or without impurities and/or temporary reversible changes applied to adjust the state of the material. These include shifts in fields or other environmental factors applied to the semiconductor and also the introduction of impurities that modify the semiconductor chemistry. The details of both temporary and permanent effects are discussed more fully later. For now, however, it is assumed that the position of the chemical potential within the stop band can be adjusted at will by applications of external potentials and by the introduction of chemical impurities, and the focus will turn to the study of the *p*–*n* junction interface between two different extrinsic semiconductors. As we shall see, the operation of the *p*–*n* junction is based on the chemical potential effects due to impurity concentration gradients and to applied electric potentials [1–3].

7.2 SEMICONDUCTOR JUNCTION MODEL

A basic device that enters into a variety of applications is the so-called *p*–*n* semiconductor junction [1]. This is composed as a planar interface between two different types of semiconductors known as *p*- and *n*-type semiconductors. In these types of semiconducting materials, small traces of chemical impurities are added in order to adjust the chemical potentials of the materials. Specifically, the chemical potentials are modified so that either electrons or holes dominate the current densities in one or the other of the materials. In *p*-type semiconductors, the arrangement is made so that positive-charged holes dominate the current density, whereas, in *n*-type semiconductors, the arrangement is such that negative charged electrons dominate the current density.

When the *p*- and *n*-type semiconductors are considered separately in the absence of the formation of a *p*–*n* junction, the *p*-type material has a conduction band with the lowest energy $\epsilon_{c,p}$, a valance band with the greatest energy $\epsilon_{v,p}$, and a chemical potential μ_p located within the semiconductor stop band. Similarly, when the *n*-type material is considered by itself, the *n*-type material has a conduction band with the lowest energy $\epsilon_{c,n}$, a valance band with the greatest energy $\epsilon_{v,n}$, and a chemical potential μ_n located within the semiconductor stop band. (This is shown schematically in Figure 7.2a and b for the two isolated bulk materials.) It should be noted that the chemical potential of the *n*-type materials tends to be located just below the conduction band. This favors the introduction of electrons into the conduction band [1,2]. Similarly, the chemical potential tends to be located just above the top of the valence band in *p*-type materials, and in this configuration the electron occupancy of the conduction band is less favored, while holes are preferentially occupied in the valence band.

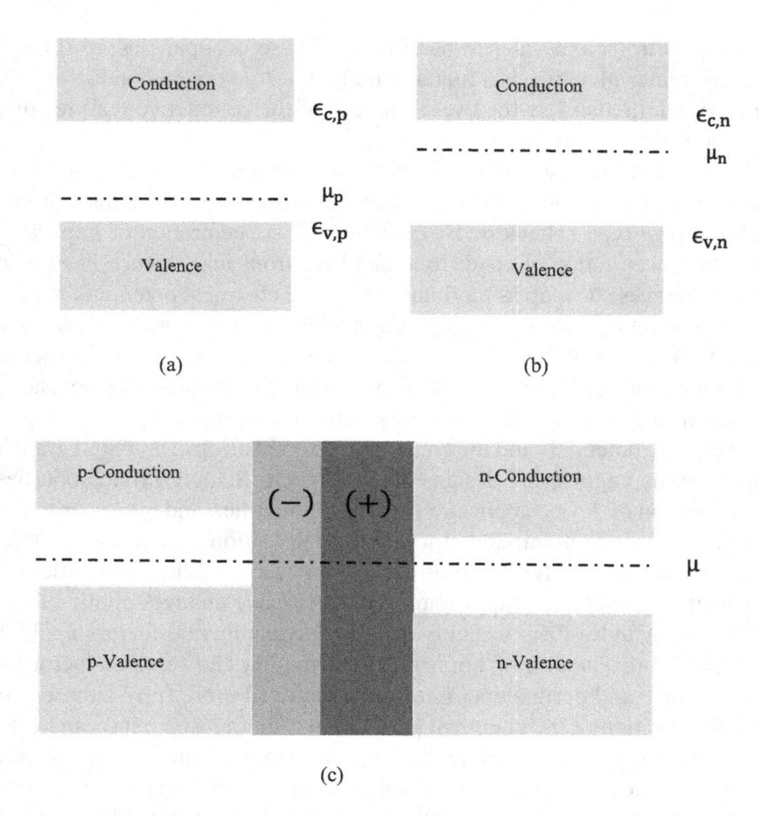

FIGURE 7.2 The energy band configuration of: (a) a p semiconductor and (b) an n semiconductor. The parameters characterizing the p-type material are $\epsilon_{c,\,p}$, $\epsilon_{v,\,p}$, μ_p, and the parameters characterizing the n-type material are $\epsilon_{c,\,n}$, $\epsilon_{v,\,n}$, μ_n. In (c) is the p–n junction, with the p-type material on the left, and the n-type material on the right. The p- and n-type materials are separated by an infinite slab of negative charge denoted (–) and an infinite slab of positive charge denoted (+). The energy is measured on the vertical axis.

If the two materials are now formed into a p–n junction, their band parameters are changed as the junction seeks an equilibrium configuration of the charge carriers at the interface. The reason for this change is the existence of different types of mobile charge carriers within the two materials, i.e., electrons in the n-type material and holes in the p-type material. Due to this difference, the two types of carriers are now subject to a process of osmosis at the junction interface. This is essentially the same type of osmosis process that occurs in the biological transfers at cellular membranes in which chemicals are exchanged between the cell and its environment. As a result of the osmosis force at the interface of the p–n junction, the charge carriers diffuse across the junction, i.e., holes move to the n side of the junction n and electrons move into the p side of the junction. An additional feature of the diffusion of charges across the p–n junction, however, is that the osmosis at the junction interface now is limited by the electrostatic interactions of the electron and hole carriers, as they separate at the interface. This imposes a limit on the energy configuration formed at the junction interface, and an equilibrium configuration is eventually attained, representing a balance of the osmotic and electrostatic forces around the junction region [1,2].

In the course of osmosis at the p–n junction, interface holes are passed from the p-type material into the n-type material, while electrons are passed between the n-type material into the p-type material. A consequence of this is that at the junction interface a small region of positive charge density is developed in the n-type material, and a small region of negative charge density is developed in the p-type material. These two charged regions essentially form a charged parallel-plate capacitor around the interface, and the electric field from the capacitor acts so as to limit the extent of the osmotic effects to a small region located about the interface. In Figure 7.2c, the developed regions of net charge about the interface are represented schematically.

In Figure 7.2c, the regions of net charge density have been approximately represented as infinite slabs of uniform positive- or negative charge density. This is done to facilitate a simple calculation of the changes in the band structures of the p- and n-type materials arising from the charge layer developed about the planar junction interface. A treatment of the electrostatics associated with the interface charges will soon follow [1,2].

First, however, note that across the interface of the p–n junction, the electrons and holes of the combined systems are allowed to circulate throughout the two different materials forming the junction. This creates a single state of thermodynamic equilibrium characterized by a Fermi–Dirac distribution in the total electronic system [1] of the form given in Eq. (7.2). Specifically, in this characterization of the equilibrium state of the junction, a single chemical potential μ fixes the occupation of the electrons throughout the material in place of the bulk μ_p, μ_n associated separately with the noninteracting bulk p- and n-type semiconductors. In this regard, the chemical potential μ must be the same throughout the equilibrium system, in order for the system to be in a steady state everywhere in space. The determination of μ and its relationship to the conduction and valence bands of the p- and n-type semiconductors is central to the following considerations of the electrostatics at the junction interface.

Consequently, the modifications at the p–n junction leading to the development of the charge densities at the junction interface arise in order to develop the uniform μ within the junction and the materials forming it. To understand the nature of these considerations, a treatment of the charge configuration in Figure 7.2c is made for a determination of the electrostatic potential it creates everywhere. For the study of the electric potential of the charge separation at the interface, only the charge distribution as represented in Figure 7.3a needs to be considered.

7.2.1 Electrostatics at the Junction

In Figure 7.3a two slabs of opposite charge are located about the junction interface at $x = 0$. In the region, $0 < x < l_n$ is a positive charge density $en_+ > 0$ arising from the removal of the mobile electron charge carriers from the n-type material in that region, and in the region $-l_p < x < 0$ is a negative charge density $-en_- < 0$ arising from the removal of the mobile hole charge carriers from the p-type material in that region. (Note: For simplicity and as a reasonable first approximation both en_+, where n_+ is the density of donor impurities, and $-en_-$, where n_- is the density of acceptor impurities, are taken to be constants over their regions of definition. In addition, it is also assumed that the junction materials are at a temperature at which the donor and acceptor impurities are highly ionized.) For the condition of total charge neutrality in the system, it then follows that $n_+ l_n = n_- l_p$, so that the two uniform charge densities are related to one another through the lengths of their regions of definition along the x-axis. The focus in the following is now on the determination of the electric potential from this charge distribution. This will cause a shift of the energy band structures of the p- and n-type materials in the regions outside the $-l_p < x < l_n$ transition region between the two materials [1–3].

The Poisson equation for the electrostatic potential throughout space is then given by the form [2]

$$\frac{d^2V}{dx^2} = -\frac{\rho(x)}{\varepsilon}, \tag{7.3}$$

where

$$\begin{aligned} \rho(x) &= en_+ > 0, \quad 0 < x < l_n \\ \rho(x) &= -en_- < 0, \quad -l_p < x < 0 \\ \rho(x) &= 0, \quad \text{otherwise.} \end{aligned} \tag{7.4}$$

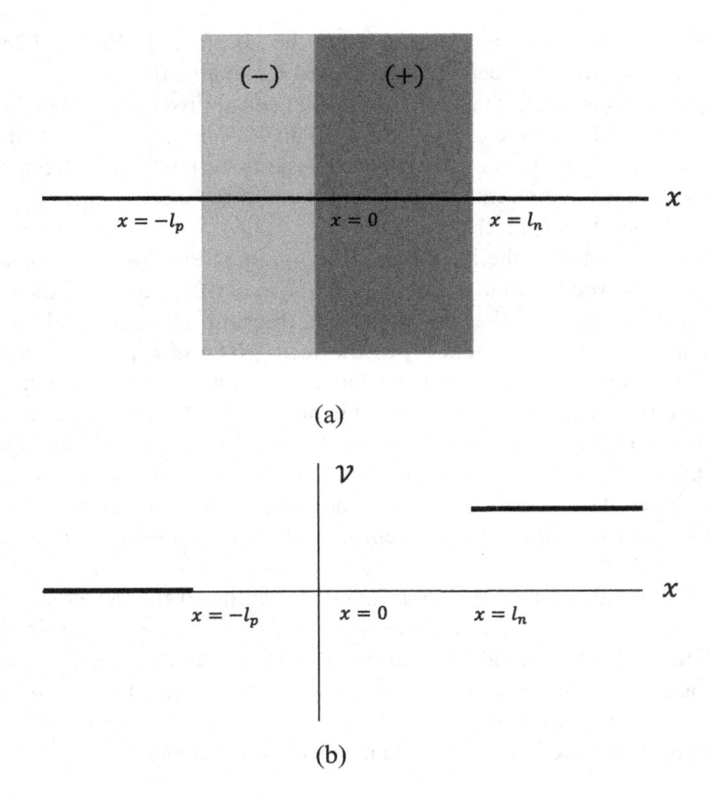

FIGURE 7.3 In (a) the two infinite slabs of charge at the p–n junction interface are shown with the infinite slab of net negative charge denoted by (–) and the infinite slab of net positive charge denoted by (+). In (b), the electric potential is shown as a function of x in the p- and n-type materials outside of the charged slabs at the junction interface.

Here, as a simplification, it is assumed that the dielectric constant of the n- and p-type materials are the same. Solving Eq. (7.4) in conjunction with the condition $n_+ l_n = n_- l_p$ of total charge neutrality determines the change in potential between $x = -l_p$ and $x = l_n$. This potential difference is given by

$$\mathcal{V} = V(l_n) - V(-l_p) = \frac{e}{2\varepsilon}\left[\frac{n_-}{n_+}(n_- + n_+) l_p^2\right]. \tag{7.5}$$

The change in electric potential in Eq. (7.5) across the junction ultimately shifts the band structure of the n- and p-type materials relative to each other outside the transition region of the junction.

From Eq. (7.5), it is found that the n-type material is shifted upward in electric potential by $\mathcal{V}$ from that of the p-type material. Consequently, the potential energy of the electrons in the n-type material is shifted downward by the factor $U = -e\mathcal{V}$, where the charge of the electron is $-e < 0$.

This shift of the bulk band structure of the n-type material is illustrated schematically in Figure 7.3b. Note that, as was discussed earlier, in the equilibrium configuration of the junction, the chemical potential μ must now be the same value throughout the system. This is the condition of thermal equilibrium and will be used in the following discussions of the equilibrium state.

Under the equilibrium conditions of the junction, there are a number of mechanisms of current flow acting at the junction interface. Specifically, there is an osmotic flow of electrons from the n-type material to the p-type material. This is known as a recombinant current and is denoted as J_{nr}. The electrons in the recombinant current eventually enter the p-type material, where they combine with holes in the p-type material, causing both charges to disappear from the system with an associated emission of a photon. In addition, there is a generative current arising from electrons in the p-type material, which have been thermally promoted into the conduction band of the p-type materials. These electrons enter the junction transition region of positive and negative charged interface slabs. Once in this region, they flow from the p-type to the n-type material in the form of a generative current denoted J_{ng}. In the equilibrium system, these two currents sum to zero so that [3]

$$J_{nr} + J_{ng} = 0, \tag{7.6a}$$

and there is no net current in the system due to electron transport.

Similarly, for the holes in the system, there is a recombinant current J_{pr} of holes from the p-type to the n-type material that is osmotic in nature, a generative current J_{pg} of holes from the n to the p-type material that is from thermal promotion in the n-type material, and the equilibrium sum of these two currents

$$J_{pr} + J_{pg} = 0. \tag{7.6b}$$

Again, it is found that no net current flows in the junction due to these hole transport processes at equilibrium.

While the junction is in the equilibrium state, the recombinant and generative currents are seen to cancel one another. This changes, however, with the introduction of an externally applied potential. In the case when a net external potential is applied across the junction interface, a net steady state current is supported across the junction. This new aspect of the junction is now discussed.

7.2.2 Application of a Potential across the Junction

In addition to the introduction of impurities to the system, the p- and n-type materials can have their electric potentials manipulated by the application of a battery acting between the materials forming the p–n junction [1–4]. In Figure 7.4a, a schematic is shown of the application of a battery with its positive terminal connected to the n-type material and its negative terminal connected to the p-type material.

Before the application of the battery across the junction interface, a positive electric potential $\mathcal{V}$ given by Eq. (7.5) already exists at the junction interface. The electric potential $\mathcal{V}$ shifts the potential of the n-type material relative to the p-type material, with a consequent negative potential energy shift of the electron energies in the n-type material relative to the electron energy levels in the p-type material. With the additional application of the positive electric potential of the battery $|V_B|$,

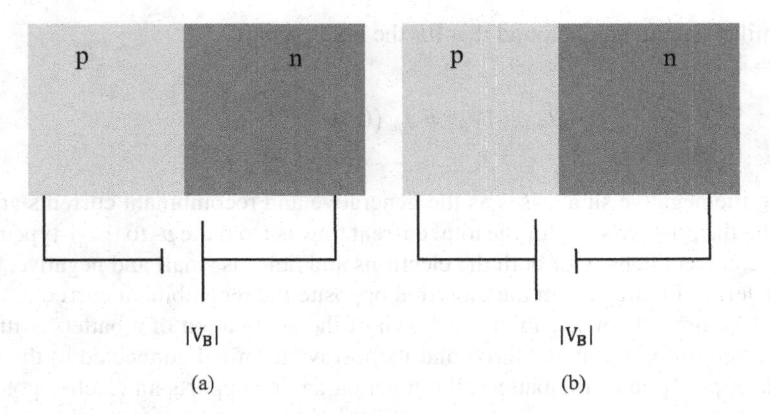

FIGURE 7.4 In (a) is the reversed biased p–n junction and in (b) is the forward-biased p–n junction. Note that the positive flow of current is from the left to the right, and the magnitude of the battery voltage is $|V_B|$.

the n-type materials will experience a total positive electric potential shift of $\mathcal{V}+|V_B|$ relative to the p-type material. This results in a total downward shift of the electronic energy levels of the n-type material by $-e(\mathcal{V}+|V_B|)$ relative to the electron levels of the p-type material. In this configuration, the junction is said to be reverse biased [1–4].

To understand the effects of $|V_B|$ on the recombinant and generative currents, begin with a consideration of the generative current of electrons [1–4]. These electrons are generated in the conduction band of the p-type material and travel to the conduction band of the n-type material. The battery potential has shifted the conduction band of the n-type material downward in energy. In this regard, it has not blocked the electrons in the p-type material from entering electron states in the n-type material. They still freely transfer to the n-type material. As a rough approximation, we expect little change to occur in the generative current, so that

$$J_{ng}(V_B) \approx J_{ng}(0). \tag{7.7a}$$

The case of the recombinant current is more difficult, because the electrons in the n-type material faced a potential energy barrier $e\mathcal{V}$ in the absence of the battery, given from Eq. (7.5), opposing their entry into the p-type material. With the addition of the battery potential, this barrier is now increased by the additional barrier potential energy $e|V_B|$, so that the total potential energy barrier opposing the entry of the n-type electrons to the p-type material becomes $e(\mathcal{V}+|V_B|)$. Consequently, as a rough approximation [1–4]

$$J_{nr}(V_B) \approx J_{nr}(0)e^{-\frac{e|V_B|}{k_BT}}, \tag{7.7b}$$

where the exponential factor is a Boltzmann factor representing the effects of the increase in the potential energy barrier that the n-type material electrons experience in making their thermal transition into the p-type material. (Note that this Boltzmann factor is a standard factor commonly encountered in the treatment of thermal transitions through a barrier.) Upon the introduction of the electric potential $|V_B|$, the recombinant current consequently suffers a decrease by the Boltzmann factor $e^{-\frac{e|V_B|}{k_BT}}$ arising from the increased difficulty in traveling into the p-type material.

Applying Eqs. (7.7a) and (7.7b) for the total electron current

$$J_{n,\text{Total}}(V_B) = J_{nr}(0)\left[e^{-\frac{e|V_B|}{k_BT}} - 1\right], \tag{7.8a}$$

and from a similar treatment it is found that for the hole current

$$J_{p,\text{Total}}(V_B) = J_{pr}(0)\left[e^{-\frac{e|V_B|}{k_BT}} - 1\right]. \tag{7.8b}$$

(Note here that the negative sign arises as the generative and recombinant currents are opposite to one another, and the positive sense of the total current flow is from the p- to the n-type material.) For positive $|V_B|$, the current density of both the electrons and holes is small and negative, flowing from the right to the left in Figure 7.4a in the direction opposite the recombinant current.

Similarly, in Figure 7.4b, a schematic is shown of the application of a battery with its negative terminal connected to the n-type material and its positive terminal connected to the p-type material. Before the application of the battery, the junction again supports an electric potential $\mathcal{V}$ shift in the n-type material relative to the p-type material, with a consequent negative potential energy shift of the electron energies in the n-type material relative to those in the p-type material. Upon the additional application to the n-type material of the negative electric potential of the battery $-|V_B|$, the n-type materials will experience a total electric potential shift of $\mathcal{V}-|V_B|$ relative to the p-type material. This results in a total downward shift by $-e(\mathcal{V}-|V_B|)$ of the electronic energy levels of the n-type material relative to the electron energy levels in the p-type material. This configuration, which reduces the potential energy barrier between the n-type and p-type materials, is known as forward biasing.

To understand the effects of the application to the n-type material of the $-|V_B|$ from the battery on the recombinant and generative currents, begin with the generative current of electrons. These electrons are generated in the conduction band of the p-type material and travel to the conduction band of the n-type material. The small battery potential (i.e., $\mathcal{V} \gg |V_B|$) has shifted the conduction band of the n-type material upward in energy relative to the p-type material energies, but it has not blocked the electrons in the p-type material from entering electron states in the n-type material. The electrons still pass easily from the p-type to the n-type material. Consequently, as a rough approximation, we expect little change to occur in the generative current so that [1–4]

$$J_{ng}(V_B) \approx J_{ng}(0). \tag{7.9a}$$

The case of the recombinant current is more difficult, because in the absence of the battery the n-type material electrons face a potential energy barrier $e\mathcal{V}$, given from Eq. (7.5), to enter the p-type material. With the application of the small battery potential, this barrier is now decreased by $e|V_B|$ to the total barrier potential energy $e(\mathcal{V}-|V_B|)$. However, the potential energy barrier to enter the p-type material is only reduced not eliminated. Consequently, as a rough approximation

$$J_{nr}(V_B) \approx J_{nr}(0)e^{\frac{e|V_B|}{k_B T}}, \tag{7.9b}$$

where the exponential factor is a Boltzmann factor representing the effects of the contribution of the battery potential. In this configuration, the recombinant current is found to increase with the increased battery potential $|V_B|$.

Applying Eqs. (7.9a) and (7.9b) for the total electron current [1–4]

$$J_{n,\text{Total}}(V_B) = J_{nr}(0)\left[e^{\frac{e|V_B|}{k_B T}} - 1\right] \tag{7.10a}$$

and for the total hole current

$$J_{p,\text{Total}}(V_B) = J_{pr}(0)\left[e^{\frac{e|V_B|}{k_B T}} - 1\right], \tag{7.10b}$$

where the positive sense of the total current is from the p-type to the n-type material. For negative $-|V_B|$ applied to the n-type material, the current density can be made large and positive, flowing from the left to the right in Figure 7.4b in the direction of the recombinant current.

7.2.3 Resulting Current versus Voltage Relationship of the Junction

As a summary of the results from Eqs. (7.8) to (7.10) relating the current flow through the junction to its applied bias, in Figure 7.5 a schematic plot of a typical behavior of current versus voltage of a p–n junction is presented. In the plot, the forward-biased junction has a positive voltage with a positive current flow, which initially for small biases is approximately increasing Ohmic in behavior. For the case of the reverse-biased junction, a negative voltage is applied to the junction. This results in a negative current flow that quickly approaches a small constant value. The highly

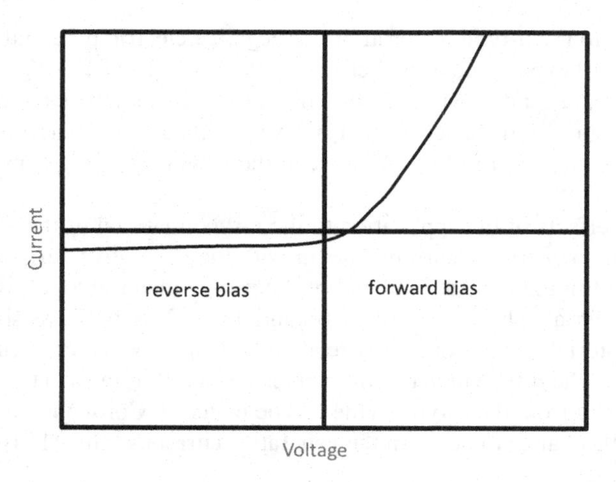

FIGURE 7.5 Schematic representing the typical current versus voltage characteristics of a *p–n* junction. In the forward-biased region of positive voltage, the positive battery terminal is applied to the *p*-type material and the negative terminal to the *n*-type material, and the current flows, in general, from the *p*- to the *n*-type material. In the reversed biased region of negative voltage, the positive battery terminal is applied to the *n*-type material and the negative terminal to the *p*-type material, and the current flows from the *n*- to the *p*-type material. In the figure, the condition $V \gg |V_B|$ is assumed.

nonlinear current–voltage behavior in Figure 7.5 forms the basis of many of the important technological behaviors of the junction [1–5].

The discussions in this chapter have now described the basic functioning of *p–n* junctions in the regulation of the flow of electron and holes. The considerations have treated the two possible cases of forward and reverse biasing, and the nonlinear current response of the junction. In the next chapter, these properties of junctions will be used in various junction combinations to discuss switching, rectifying, and other related transistor device applications [1–5].

REFERENCES

1. Kittle, C. 1996. *Introduction to solid state physics, 7th Edition*, 393–410. New York: John Wiley & Sons, Inc.
2. Marder, M. P. 2000. *Condensed matter physics*. New York: John Wiley & Sons, Inc.
3. Streetman, B. G. and S. Banerjee 2000. *Solid state electronic devices, 5th Edition*. New Jersey: Prentice Hall.
4. Schilling, D. L. and C. Belove 1968. *Electronic circuits: Discrete and integrated*. New York: McGraw-Hill Book Company.
5. Datta, S. 2005. *Quantum transport: Atom to transistor*. Cambridge: Cambridge University Press.

8 Rectifiers and Transistors

In this chapter, discussions are given of some devices based on *p–n* junctions that are designed to manipulate the flow of electrical currents [1–4]. The devices of interest include diode rectifiers and transistors, and a consideration of some of their basic circuit applications. In the formulations of these electrical systems, the *p–n* junction is an important mechanism, as it allows for the introduction of nonlinearity into electrical circuits. In this way, junctions and their derivatives differ from the simpler linear circuit elements, i.e., resistors, capacitors, inductors, etc. Due to their nonlinear properties, junctions allow for the design of various circuits, which can act as signal rectifiers, amplifiers, switches, light sources, and sensing mechanisms. All these aspects of junction behavior come from the non-Ohmic properties of the *p–n* junction. A specific focus of the chapter is on the basic theoretical ideas of the physics behind rectifier and transistor technology, and some of their technological nanoscience applications. The specifics of the engineering implementation of the theory in the detailed design of devices and the experimental processes involved in making them are not a consideration.

Rectifiers and transistors allow for the introduction of nonlinearity and directionality into electrical circuits [1–5]. A rectifier is the simplest form of a device based on *p–n* junctions. In its elementary form, it involves a single *p–n* junction and uses the directional characteristics of the current versus voltage function of the junction to control the current flow through the rectifier circuit [1,2,4]. Specifically, the junction only permits a large current to flow in one direction through the interface between the two semiconducting media from which it is composed. At the next level of complexity is a transistor [1–4]. The transistor is based on the interaction of two *p–n* junctions and is involved with operations composed from a combination of three potentials or currents. It mediates a modulation between the various currents and voltages in a branched circuit through the directionality properties of the junctions and their interactions with each other. In this regard, the rectifier only regulates the flow of current within a single branch of an electrical circuit, while the transistor causes currents in different branches of a complex circuit to interact with one another.

In an ideal world for engineering applications, a rectifier would be a device that only allows current to pass through a wire or circuit in one direction [3]. In this regard, it would act as a short circuit for current flow in one direction but offers an infinite resistance to current flow in the opposite direction. The *p–n* junction considered in our discussion is an approximation to this behavior, involving a small current in one direction and an Ohmic current in the opposite direction. Nevertheless, the asymmetric current–voltage behavior of the junction is useful in the regulation of currents in a single branch of a circuit and is used, for example, in the conversion process of transforming AC current into DC current or in the generation of current pulses [4]. In these functions, however, the rectifier does not allow two different currents in two different branches of a circuit to interact with each other. These interactive applications require the application of a device of greater complexity known as a transistor.

The second device to be considered is a transistor [3–5]. The transistor is a device based on the interaction of two *p–n* junctions arising from their spatial proximity. It allows for electronic circuits to connect with the two junctions of the transistor by means of three circuit terminals that act to feed currents and potentials into the interacting junctions for processing. Consequently, unlike the rectifier, the transistor permits the interaction of potentials and currents in different branches of a complex system of circuits so that the potentials and currents in the different branches modulate one another in a nonlinear interaction. In this regard, there are basically two general types of transistors that shall be considered [3,4]. These are the field effect transistor and the bipolar transistor. Both of these transistor designs are focused on different sets of arrangements between the transistor's

DOI: 10.1201/9781003031987-8

two p–n junctions, and the input and output terminals of the transistor, which feed currents and potentials into the device. Consequently, the field effect and bipolar systems involve two distinctly different physical mechanisms in their methods of processing electrical inputs into output electrical signals. The details of these interactions are a focus in the course of the chapter.

Before we continue on to the development of the basic models of rectifier and transistor devices, it should be noted here that in the treatment of the p–n junctions involved in this chapter's discussions, the model of the junction developed in the previous chapter is used. This is a simplified model that accounts for the gross functioning of the p–n junction but leaves out details that the reader can find in the more advanced literature [1–5]. For example, in the model we have focused mainly on the flow of the so-called majority carriers in the systems. In the p-materials, these are the holes, and, in the n-materials, they are electrons [1,2]. In a more detailed treatment, an account must also be made of small effects arising from the so-called minority carriers. These involve some, for our purposes, small quantities of thermally created valence band holes in the n-material and valence band electrons in the p-materials. In addition, in the consideration of interacting junctions, the detailed treatment of designed systems generally involves computer simulation techniques that fully account for the system geometry and the detailed band structures of the materials involved. These considerations are not discussed here, but we have opted for the treatment of a simplified model accessible to analytic methods so as to provide for students an illustration of the basic operations. While this approach gives the general ideas of the device functions, it is an idealization of what exists in the real world, and must accordingly be modified in the engineering and design of devices for applications. For further details of all the finesses required in device design and the experimental implementation of these ideas, the reader is referred to the literature [1–5].

In the following, first a discussion of rectifiers is presented. This is followed by a consideration of the field effect transistor and, in turn, to discussions of the bipolar transistor. During the course of the development of the theory, a treatment of the operation of the simplest transistor circuits is given.

8.1 RECTIFIERS AND TRANSISTORS

The simplest electronic device with a design based on a single p–n junction is a current–voltage regulator known as a rectifier [2–4]. In its idealized form, the rectifier is a device that only allows an electric current to pass in one direction through an electronic circuit. Consequently, when introduced into a circuit, it functions as a short in the circuit for current flow in one direction but allows no current to flow in the opposite direction. For example, in the ideal rectifier the current is passed unchanged for a positive applied voltage, but no current is allowed to flow through the rectifier for a negative voltage. In the real world, this type of on–off behavior with the applied voltage is simulated by a junction that allows a large current to flow when it is biased with a positive voltage but only a small current to flow when biased with a negative voltage.

In this regard, the p–n junction approximates an ideal rectifier, essentially behaving as a resistor for current flow in its forward biased direction while passing only a small current in its reverse biased direction [3,4]. (See Figure 8.1 for a schematic representing the current versus voltage behavior of a p–n junction.) The goal of engineers is then to design an operational system with the approximate properties of an ideal rectifier using the imperfect behaviors of the p–n junction. This can be accomplished through a combination of material science and electrical engineering design methods.

A rectifier is a very simple device that affects the flow of a single current passing through a particular branch of a circuit. It does not, however, allow through its direct actions for the current flowing in a given circuit branch to affect the properties of the currents flowing in other branches of a complex circuit. To create such a device for modulations within multiple branched circuits requires device designs with a more complex structure than that provided by single-junction rectifiers. This leads to the ideas of transistors and arrays of transistors.

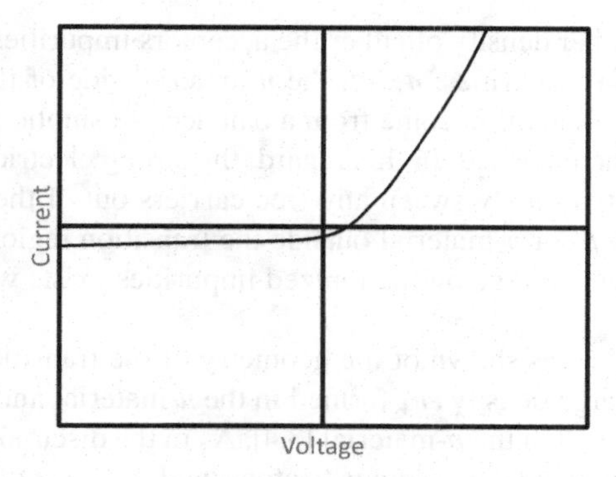

FIGURE 8.1 Schematic representation of the typical current versus voltage characteristics of a *p–n* junction. In the forward biased region of positive voltage, the positive battery terminal is applied to the *p*-type material and the negative terminal to the *n*-type material, and the current flows in general from the *p*-type to the *n*-type material. In the reversed biased region of negative voltage, the positive battery terminal is applied to the *n*-type material, and the negative terminal is applied to the *p*-type material, and the current flows from the *n*-type to the *p*-type material.

As shall be seen later, the rectifying properties of multiple *p–n* junctions can be combined to create a means of more complex current manipulations than just rectification [3,4]. Such combinations of junctions allow for currents and potentials in various parts of complex circuits to interact with one another, providing the mechanisms for these currents to modulate each other's properties. In this respect, multiple current interactive devices in their simplest forms are based entirely on the interactions between arrays of different *p–n* junctions placed so as to manipulate the flow of current in multiple circuits.

A simplest basic example of a device formed from at least two interacting junctions and designed to modulate currents and voltages between different circuit branches is known as a transistor. Later it will be shown how transistors and transistor arrays for specific technological applications can be made to create general current interactions for processing electron signals. The transistor is then a basic nonlinear device of current and voltage interactions in electronic circuits, and all other applications are essentially formed as combinations of transistors and rectifiers and linear electron devices.

In order to develop the ideas at the foundation of transistor technology, it is first necessary to revisit the discussion of the properties of *p–n* junctions introduced in the last chapter and to consider in more detail the nature of the rectifying behaviors occurring around the interface of the *p–n* junction [3,4]. In the previous chapter, the primary focus was on obtaining the nonlinear current versus voltage behavior, which is essential to understand the rectification properties of a single *p–n* junction. In the design of more complex combinations of junctions, however, care is needed to understand how the spatial proximity of closely placed junctions might affect the operations of the set of junctions. This interaction between junctions is closely related to the nature of the charge accumulations found at the interface of the *p–n* materials of the two proximate junctions.

Specifically, the proximity of the junctions is found to lead to interactions between junctions, which, in turn, affect their ability to pass current. This shows up in a dependence of one junction's potentials, and passed currents on the potentials and passed currents of its neighboring junctions. Let us then return to a reconsideration of the charge separation created at a single-junction interface.

8.1.1 The Transition Region in a *p–n* Junction

In the previous chapter, an isolated junction was shown to display subregions of net charge on either side of the chemical interface of the junction [3,4]. In these regions of net charge, the charge densities arise from the number density of either the acceptors impurities, n_-, on the p-material side of the interface or the donor impurities, n_+, on the n-material side of the interface. Each of these subregions of net charge accumulation come from a balance of osmotic and electromagnetic forces in the transition region of the interface. In this regard, the strong electric fields within the two subregions of net charge density quickly sweep any free carriers out of the junction transition region and into the regions of bulk p- or n-material outside the transition region. Consequently, to a good approximation only the bound charge on the ionized impurities exists within the subregions of net charge.

In Figure 7.3, a schematic was shown of the geometry of the transition region composed of the subregion of net positive charge density en_+ formed in the n-material and the subregion of net negative charge density $-en_-$ formed in the p-material [3,4]. As in the discussion presented there, the two charged subregions (one of ionized doner impurity atoms and the other of charged accepter impurity atoms) are formed so that the sum of their net positive and net negative charges is zero. Outside the transition region, the materials on either side of the junction interface revert to their bulk charge-neutral properties. The two subregions of net charge are now shown to be the mechanism for the development of the interactions between closely placed junctions and through this interaction to contribute to an important application in transistor technology.

When two junctions are placed in proximity to one another, the currents and voltages passing through their transition regions become a source of interaction between the two junctions, leading to means of current and voltage manipulation [3,4]. To see how this arises, let us first return to the details of the mathematical consideration of the subregions of charge accumulation at the interface of a single junction and the potential generated across the transition region between the two different bulk materials. This provides the foundation for the development of the theory of the interaction between two closely placed junctions.

8.1.2 The Transition Region Characteristics

Returning to the considerations of Chapter 7, the reader is reminded of the charge configurations found in a simple p–n junction. In Figure 7.3a, the transition region between the n- and p- bulk materials is composed of a subregion $0 < x < l_n$ of a net uniform positive charge density $en_+ > 0$ from the ionized doner impurities and a subregion $-l_p < x < 0$ of a net uniform negative charge density $-en_- < 0$ from the charged acceptor impurities [3,4]. The two charge densities are approximated as being constant in space and related by

$$n_+ l_n = n_- l_p, \tag{8.1a}$$

which is the condition that the transition region is charge-neutral on average. (Specifically, in the transition region the net amount of positive charge in $0 < x < l_n$ is equal to the net amount of negative charge in $-l_p < x < 0$.) This configuration defines a basic problem in electrostatics relating the electric potential developed across the transition region to the charge configuration within the transition region.

For this charge geometry the potential developed across the transition region is then described by a Poisson equation of the form [3,4]

$$\frac{d^2V}{dx^2} = -\frac{\rho(x)}{\varepsilon}, \tag{8.1b}$$

with a dependence only on the coordinate perpendicular to the junction interface. In addition, note that in the formulation of Eq. (8.1), to facilitate the calculation, it is assumed that both the n- and p-type bulk materials are formed by doping the same undoped semiconductor material. The n- and p-nature of the materials are then determined by doping with small amounts of, respectively, different appropriate impurities. This accounts for taking the same dielectric constant ε of the semiconductor in both the n- and p- regions.

Solving Eq. (8.1) for the change in potential between $x = -l_p$ and $x = l_n$, it is then found that [3,4]

$$\mathcal{V} = V(l_n) - V(-l_p) = \frac{e}{2\varepsilon}\left[\frac{n_-}{n_+}(n_- + n_+)l_p^2\right] = \frac{e}{2\varepsilon}\left[\frac{n_+}{n_-}(n_- + n_+)l_n^2\right] = \frac{e}{2\varepsilon}\left[\frac{n_+ n_-}{n_- + n_+}W^2\right], \quad (8.2a)$$

where the width of the transition region is expressed by

$$W = l_n + l_p = \frac{n_- + n_+}{n_-} l_n = \frac{n_- + n_+}{n_+} l_p. \quad (8.2b)$$

These formulae relate the width of the transition region and the potential across it. An important geometric result, then, for future considerations is obtained from Eq. (8.2) by inverting these relationships giving [3,4]

$$l_n^2 = \frac{2\varepsilon}{e}\mathcal{V}\frac{n_-}{n_+}\frac{1}{n_+ + n_-}, \quad (8.3a)$$

$$l_p^2 = \frac{2\varepsilon}{e}\mathcal{V}\frac{n_+}{n_-}\frac{1}{n_+ + n_-}, \quad (8.3b)$$

and

$$W^2 = \frac{2\varepsilon}{e}\mathcal{V}\frac{n_+ + n_-}{n_+ n_-}. \quad (8.3c)$$

From these expressions the lengths of the transition region are found to depend explicitly both on the charge densities and the electric potential across the interface. Consequently, by manipulating the potential across the transition region its geometry is found to be easily changed.

While the charge densities in Eqs. (8.2) and (8.3) depend only on the fixed chemistry of the p- and n- materials, the electrical potential $\mathcal{V}$ can be manipulated by the application of an external potential across the transition region. This dependence of $\mathcal{V}$ on an external factor is an important characteristic for the development of junction interactions. It is now examined and related to the geometry of the transition region and the manipulations that can be performed upon it.

From Eq. (8.3), the geometry of the transition region is found to be strongly dependent on the electric potential applied across the transition region. In this regard, we next consider three particularly interesting cases of Eq. (8.3) for technological applications and how these cases determine the junction geometry. The first is for the case in which the interface involves only a contact with no externally applied potential. In this case, $\mathcal{V}$ is just the contact potential. This consideration is followed by a treatment of an externally applied forward biased potential and a consideration of an externally applied reversed biased potential across the interface. These three configurations later form the basis for transistor design.

In the case that no external potential is applied across the p–n interface (i.e., the p- and n- materials are only placed in contact at the planar interface), the potential difference [3,4]

$$\mathcal{V} = V_0 \quad (8.4)$$

where V_0 is known as the contact potential and is a property that depends on the chemical nature of the p- and n- materials. In this regard, only by adjusting the values of n_+ and n_- can the contact potential be changed. This is accomplished by changing the doping of the materials.

If a battery is now connected to the bulk p- and n- materials across the transition region, the potential across the interface is changed from V_0 as are the lengths W, l_n, and l_p. To understand how these changes occur, it is necessary to begin by considering the effects of applying an external potential across the isolated junction. Specifically, from Eq. (8.2a), it follows for $\varepsilon > 0$ that in the absence of an externally applied potential the electric potential

$$V(l_n) > V(-l_p), \tag{8.5}$$

and the p-type material is lower in electric potential than is the n-type material. This is the case in which only the contact potential is present. In the case that a battery of potential $|V_B| > 0$ is now connected with the positive terminal applied to the p-material and the negative terminal attached to the n-materials, the new potential difference across the interface is changed to [3,4]

$$\mathcal{V} = V(l_n) - V(-l_p) = V_0 - |V_B|. \tag{8.6}$$

From Eq. (8.6) it is seen that the battery has made $V(-l_p)$ more positive by raising its potential relative to the n-material, and this shift is represented by the added $-|V_B|$ term.

Alternatively, in the case that a battery of potential $|V_B| > 0$ is connected with the positive terminal applied to the n-material and the negative terminal attached to the p-materials, the new potential difference across the interface is changed to

$$\mathcal{V} = V(l_n) - V(-l_p) = V_0 + |V_B|. \tag{8.7}$$

Now, the battery has made $V(-l_p)$ less positive by lowering its potential relative to the n-material. This shift is accounted for by the added $+|V_B|$ term in Eq. (8.7).

Earlier it was shown in Eq. (8.3) that the geometry of the junction transition region is directly related to $\mathcal{V}$. Consequently, the geometry of the region is easily changed by applying a battery in forward bias (i.e., $\mathcal{V} = V_0 - |V_B|$) or reverse bias (i.e., $\mathcal{V} = V_0 + |V_B|$). These biases not only affect the current flows through the individual junctions but also allow for the development of an interaction between two junctions that are placed in close proximity to one another. This interaction is readily seen in the operation of a single-field-effect transistor and is the next topic of discussion.

8.2 FIELD EFFECT TRANSISTOR

The field effect transistor in its most elementary form is composed of two parallel oppositely directed planar p–n junctions [3,4]. (See Figure 8.2a for a schematic of this configuration.) In this layered geometry, a p–n–p transistor has a slab of n-material as the middle layer between two slabs of p-material, whereas an n–p–n transistor has a slab of p-material as the middle layer between two slabs of n-material. The resulting system of three layers is electrically configured in a circuit so that a current flowing in the middle layer (i.e., parallel to the slab surfaces) can be modulated by electrical potentials applied to the top and bottom layers of the sandwich of semiconductor layers.

The basic principles of operation of these two types of transistors are similar [3,4]. A current flowing within the middle slab parallel to the slab interfaces is regulated by adjusting the electric potentials of the top and bottom slabs. In this modulation, the electric potentials of the top and bottom layers are chosen to be the same and are configured so that both the junctions formed at the top and bottom interfaces of the middle layer are reverse biased. This arrangement aids in confining the current flow of the middle layer to the middle layer. The current flow through the middle layer is itself the result of a potential difference applied to the middle slab in the direction of the current flow in the slab. This configuration provides a mechanism for interactions between different branches of a circuit, allowing one branch to control the potential of the upper and lower slabs, while another branch supplies the current flow in the middle slab parallel to the slab surfaces.

8.2.1 p–n–p Transistor

Specifically, in the following considerations, first the operation of a *p–n–p* transistor is discussed [3,4]. This is a layering of three slabs of semiconductor, i.e., a slab of *n*-material layered between two slabs of the same type of *p*-materials. (See Figure 8.2a.) Usually in the *p–n–p* transistor, the impurity density of the acceptor impurities in the *p*-material, n_-, is chosen to be much greater than the impurity density of the donor impurities in the *n*-material, n_+. Consequently, since the conductivity

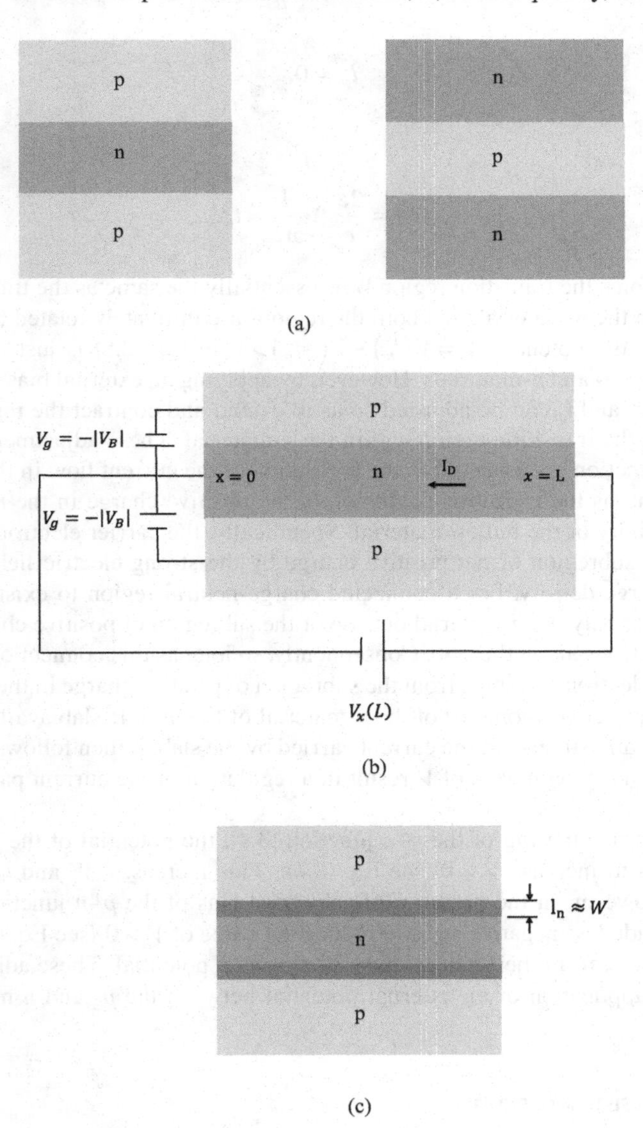

FIGURE 8.2 Schematics for: (a) The three infinite slab layers of the *p–n–p* (left) and *n–p–n* (right) transistors. (b) The basic circuit operating a *p–n–p* transistor as a control of the flow of current through the layer of *n*-type material. The gate potentials $V_g = -|V_B|$ where $|V_B|$ is the battery potential are applied in reverse bias across the two *p–n* junctions, and the potential $V_x(x)$ from the battery $V_x(L)$ is along the center of the slab of *n*-type material. The current flow I_D flows in the slab of *n*-type material due to the potential difference $V_D = V_x(L)$ applied between the $x = 0$ and $x = L$ terminals of the slab of *n*-type material. (c) Schematic of a *p–n–p* transistor with zero $V_D = V_x(L)$ bias for a system with $n_p \gg n_n$. Here, the ionized regions of lengths l_n in the *n*-type material are indicated in dark gray.

is proportional to the carrier density, the p-material has a greater carrier density than the n-materials. As a result of these factors, the p-material then tends to have a much larger conductivity than the n-material, and the p-material maintains a constant potential over its volume.

From Eq. (8.3), it then follows that for $n_- \gg n_+$ at the p–n interfaces [3,4]

$$l_n^2 = \frac{2\varepsilon}{e}\mathcal{V}\frac{1}{n_+}, \tag{8.8a}$$

$$l_p^2 \approx 0, \tag{8.8b}$$

and

$$W^2 = \frac{2\varepsilon}{e}\mathcal{V}\frac{1}{n_+} \approx l_n^2. \tag{8.8c}$$

Under these conditions, the transition region W is essentially the same as the transition subregion l_n of positive charge in the n-material, and both the regions are intimately related to $\mathcal{V}$. In the absence of an external bias, the potential $\mathcal{V} = V(l_n) - V(-l_p) > 0$ in Eq. (8.8) is just the positive contact potential between the n- and p-materials. However, by applying an external bias across the junction, $\mathcal{V}$ and the lengths W and l_n can be adjusted so as to extend and contract the region of net positive charged material in the transition region within the n-material. These adjustments to the transition region of the p–n junction are essential to the regulation of the current flow in the n-material.

This is important, as the transition subregion of net positive charge in the n-material does not exhibit the conductivity of the bulk n-material. Specifically, the carrier electrons of the n-material are swept from the subregion of net positive charge by the strong electric fields created through the net positive charge density. For a connected charge-neutral region to exist across the slab of n-material, therefore, only the n-material outside of the subregion of positive charge participates in the transport of electrons along the slab. Consequently, so long as this connection exists, there is no contribution to the electron transport from the subregion of positive charge in the slab of n-material, and changes in the effective geometry of the n-material of the middle slab available to carry a current will show up as adjustments in the current carried by the slab. It then follows that modifications of the effective geometry by means of $\mathcal{V}$ result in a regulation of the current passed by the middle slab.

In the case of reverse biasing of the p–n junction [3,4], the potential of the p-material is made more negative so as to increase $\mathcal{V} > 0$ (see Eq. (8.7)). This increases W and l_n over their contact potential values. However, in the presence of a forward bias of the p–n junction, the potential of the p-material is made less negative so as to reduce the value of $\mathcal{V} > 0$ (see Eq. (8.6)). This leads to a decrease of W and l_n from their values due to the contact potential. These adjustments are made solely through the application of an external potential between the p- and n-materials as in Eqs. (8.6) and (8.7).

8.2.2 Basic Transistor Circuit

The basic transistor circuit [3,4] for the operation of a p–n–p transistor is now shown in Figure 8.2b. This is the simplest circuit exhibiting most of the features of transistor operation, and all more complex generalizations follow from the basic operation of this circuit. In the following discussions only the system in Figure 8.2b is considered so that from the basic operation of this circuit for the details of more complex transistor designs and networks the reader is referred to the literature [3,4].

In the system of Figure 8.2b, a potential difference is applied along the x-axis, across the length of the n-material slab, causing a current I_D, known as the drain current, to flow through the length

L of this region. The applied potential is dependent on position along the x-axis and along the center of the channel is denoted by $V_x(x)$. The current in the n-material outside the transition regions of the two junctions is an Ohmic current that depends on the geometry and resistivity of the bulk n-material outside the transition regions of the transistor. Consequently, at a general position x in the Ohmic region of the n-material, the potential is given by $V_x(x)$.

In addition, a so-called gate potential V_g is applied to the highly conducting p-material and appears across both of the p–n junctions. This potential adjusts the geometry of the current-carrying region of the bulk n-materials by changing the geometry of the transition regions of net positive charge in the n-material. In this regard, the layers of p-material at each p–n junction are set at the same potential difference relative to the n-material. This assures the reflection symmetry of the n-material about the center of its channel [3].

Next, we discuss the changes in the effective geometry of the Ohmic region of the middle slab of n-material. These changes are caused by the contact potential, and the applied biases $V_x(x)$ and V_g. Under the stated set of model conditions, the regulation of the current through the middle slab as modulated by the potentials applied to the p-material slabs are computed. In this regard, the regulation of the current is seen from Figure 8.2b to be accomplished by the manipulation of $V_x(x = L)$ (also known as the drain potential $V_D \equiv V_x(x = L)$ applied across the n-material slab and V_g applied between the p–n junction interfaces. The theory of these manipulations now presented is a solution from a model approximation to the system and is made for pedagogical purposes. It represents basically what goes on under the operation of the transistor in Figure 8.2b, but involves a number of approximations.

8.2.3 Geometry of the Subregion of Net Positive Charge within the n-Material Layer

In the following, the geometry of the regions of net positive charge within the n-material slab is discussed within the context of a simple approximate model [3,4]. (See Figure 8.2c for a schematic representation of the regions present in the transistor.) As we have seen earlier, this geometry forms the foundation of the mechanism used to control the flow of current through the middle slab of the transistor.

Specifically, within our model the two regions of p-material (known as the gate) are maintained at a constant potential V_g throughout their volumes, and the volume electric potential $V_x(x)$ of the bulk n-material outside the transition region of net positive charge depends only on x. Note that here x is measured in the middle horizontal plane bisecting the slab of n-material. From a consideration of these, the effective geometry of the n-material slab is determined as a function of x.

It is assumed that both the slabs of p-material are maintained at the constant potential V_g measured relative to the $x = 0$ edge in the middle plane of the slab of charge-neutral n-type material. It is also assumed that $V_x(x)$ in the neutral n-material outside the region of net positive charge exists over the horizontal length of the slab defined by $0 \le x \le L$. The zero of electric potential is then defined by $V_x(x = 0) = 0$ and coincides with the zeros for the measurement of both V_g and $V_x(x)$.

The potential difference between the gate and the neutral n-materials outside the transition region, $\mathcal{V}(x)$, is now a function of x, which is obtained from Eq. (8.7) and written as [3,4]

$$\mathcal{V}(x) = V_0 + V_x(x) - V_g, \tag{8.9}$$

where $V_g = -|V_B| < 0$ is the reversed bias configuration of the p–n junction, and $V_x(x)$ is the potential that increases along the axis of the transistor. In this geometry (see Figure 8.2b and c), the electrons travel from the low electric potential $V_x(x = 0) = 0$ left side of the transistor (known as the source) to the high electric potential

$$V_x(x = L) = V_D \tag{8.10}$$

right side of the transistor (known as the drain). Here, V_D is the electric potential of the drain. Consequently, as the electrons move toward the drain, the electric potential across the junction, $\mathcal{V}(x)$, is increased because of the variation of $V_x(x)$ along the mid-plane of the n-material.

The system outlined above provides a reasonable model for the potentials operating within the slab of n-material. In fact, for many practical applications the contact potentials in Eq. (8.9) can also be ignored in a model representing transistor behavior, as it is generally small compared to both V_g and $V_x(x)$. This provides a further simplification of the problem and will be discussed later. For the present, however, we continue with the model as initially described.

From Eq. (8.8) the transition region of net positive charge at the interface is now seen to be dependent on x. Specifically, the length of the extension of this region into the bulk n-material is given by [3,4]

$$W^2(x) \approx l_n^2(x) = \frac{2\varepsilon}{e}\frac{1}{n_+}\mathcal{V}(x), \tag{8.11}$$

where the length $W(x) \approx l_n(x)$ and its x-dependence allows us to characterize the geometry of the n-material slab and its regions of net positive charge, as they are configured between the chemical interfaces of the two p–n junctions. In turn, the charge flow properties of the n-material slab follow directly from these considerations, as under most considerations the current is transported only through the region outside those of net positive charge density [3]. The properties of this region are now determined as a function of x.

If s is half the distance between the chemical interfaces of the two p–n junctions, then half the distance between the regions of net positive charge of the two p–n junctions in the n-material slab is given by [3,4]

$$h(x) = s - l_n(x) = s - \sqrt{\frac{2\varepsilon}{en_+}\mathcal{V}(x)} = s - \sqrt{\frac{2\varepsilon}{en_+}\left[V_0 + V_x(x) - V_g\right]}. \tag{8.12}$$

Here the sole position dependence enters from $V_x(x)$, which causes the current to flow from the drain to the source. Since it is an increasing function of x, it increases the potential difference between the neutral n-material and the gate, as the carrier electrons move toward the drain. This creates an increase in the reverse bias along the channel in the direction of the electron flow in the n-material slab.

The position-dependent separation of the two regions of net positive charge is an important limitation on the effective geometry of n-material carrying the current between the source and the drain. It is a consequence of the fact that, under conditions of reverse bias at the junction interface, the region of net positive charge does not transport charge into the region of p-material. In addition, later it is shown that under a loss of connectivity of the charge-neutral region between the source and drain, a large potential $\mathcal{V}(L)$ will cause a flow of elections from the neutral n-material to the drain through the region of net positive charge. This loss of connectivity occurs when the two regions of net positive charge touch one another, closing the channel of charge-neutral n-material between the source and the drain.

First, we continue our considerations under the assumption that the region between the source and drain are connected by a charge-neutral region of n-material. This is followed by a treatment of a disconnected charge-neutral region between the source and drain.

8.2.4 Connected Region of Charge-Neutral *N*-Material between the Source and Drain

The current through the connected region of bulk n-material outside the subregion of net positive charge is a constant denoted I_D, i.e., it is eventually just the current from the drain [3]. The nature of its motion through the n-material slab is now determined from Ohm's law.

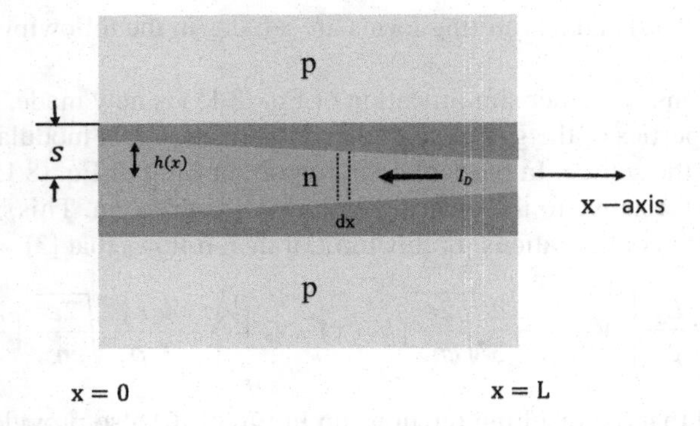

FIGURE 8.3 A p–n–p transistor with an applied potential $V_D = V_x(L) > 0$ showing the spatial measures s, dx, $h(x)$. The length $L_\perp$ is perpendicular to the page, and the drain current flows from right to left in the charge-neutral region of n-type material.

Consider an infinitesimal vertical slice of the slab of n-material having a thickness dx and which is made perpendicular to the horizontal flow of current. (See the schematic in Figure 8.3.) From Ohm's law, it then follows that the potential difference $dV_x(x)$ across the infinitesimal section of the charge-neutral n-material outside the region of net positive charge density is given by [3]

$$dV_x(x) = I_D dR(x) = I_D \frac{\rho_n}{A(x)} dx. \tag{8.13a}$$

Here, from the basic relationship between the resistance and the resistivity, the resistance of the section has been written as

$$dR(x) = \frac{\rho_n}{A(x)} dx. \tag{8.13b}$$

In Eq. (8.13), ρ_n is the resistivity of the bulk n-type material, $A(x) = L_\perp 2h(x)$ is the cross-sectional area perpendicular to the current flow in the region of current-carrying n-material, and $L_\perp$ is the length of the slab perpendicular to the direction of the current flow and to $h(x)$.

From Eqs. (8.12) and (8.13), it then follows that the potential difference $dV_x(x)$ across the slice dx is related to the thickness of the slice by [3]

$$2\frac{L_\perp}{\rho_n}\left\{ s - \sqrt{\frac{2\varepsilon}{en_+}\left[V_0 + V_x(x) - V_g\right]} \right\} dV_x(x) = I_D dx. \tag{8.14}$$

This is a first-order differential equation that is ultimately integrated from the source at $x = 0$ where the electric potential is $V_x(x = 0) = 0$ to the drain at $x = L$ where $V_x(x = L) = V_D$ or to some intermediary point $0 \le x < L$ where $V_x(x)$. It allows for the determination of the potential within the slab as a function of x.

Upon integrating Eq. (8.14) from $x = 0$ to a general $0 \le x \le L$, we find that [3]

$$I_D x = 2\frac{L_\perp}{\rho_n}\left\{ sV_x(x) - \frac{2}{3}\sqrt{\frac{2\varepsilon}{en_+}}\left[V_0 + V_x(x) - V_g\right]^{\frac{3}{2}} \right\} + \frac{4}{3}\frac{L_\perp}{\rho_n}\sqrt{\frac{2\varepsilon}{en_+}}\left[V_0 - V_g\right]^{\frac{3}{2}} \tag{8.15}$$

relates x and $V_x(x)$ subject to the condition of a constant current flow I_D from the drain to the source. Specifically, it allows for the determination of $V_x(x)$ within the slab as a function of x for a fixed I_D. As such, this result is an important transport property of the slab system for technological applications, and Eq. (8.15) and its limiting forms are a focus in the following development of transistor properties.

For many applications, a further simplification of Eq. (8.15) is now made, which is useful in the discussions of the properties of the gate potential and its function as a modulator of the flow of current from the drain to the source. In particular, an important limit of Eq. (8.15) is that in which the contact potential $V_0 \approx 0$ relative to the other potentials in the problem. This, in fact, is found to be the case for most device configurations. In this limit, it then follows that [3]

$$I_D x = 2\frac{L_\perp}{\rho_n}\left\{sV_x(x) - \frac{2}{3}\sqrt{\frac{2\varepsilon}{en_+}}\left[V_x(x) - V_g\right]^{\frac{3}{2}}\right\} + \frac{4}{3}\frac{L_\perp}{\rho_n}\sqrt{\frac{2\varepsilon}{en_+}}\left[-V_g\right]^{\frac{3}{2}}. \tag{8.16}$$

Later, it will be found that the resulting relationship in Eq. (8.16) also provides an important focus in a discussion of transistor switches and amplifiers. As an additional point of interest on the limiting forms of the electrical potentials in the system, note that in the further limit that $V_x(x) \to 0$, the righthand sides of both Eqs. (8.15) and (8.16) go to zero so that $x = 0$. This represents a reduction of both forms to one of the initial boundary conditions for the integration of Eq. (8.14).

Considering the drain end of the slab of n-material (i.e., the case that $x = L$), the expressions in Eqs. (8.15) and (8.16) relate the current flow in the slab I_D to the potential of the drain $V_x(L)$. Specifically, in this region, Eq. (8.15) is written in the simplified form

$$I_D = G_0 V_P\left\{\frac{V_x(L)}{V_P} - \frac{2}{3}\left[\frac{V_0 + V_x(L) - V_g}{V_P}\right]^{\frac{3}{2}} + \frac{2}{3}\left[\frac{V_0 - V_g}{V_P}\right]^{\frac{3}{2}}\right\}. \tag{8.17a}$$

where

$$G_0 = \frac{2L_\perp s}{L\rho_n} \tag{8.17b}$$

and

$$V_P = \frac{en_+}{2\varepsilon}s^2. \tag{8.17c}$$

Similarly, the expression in Eq. (8.16) evaluated at $x = L$ is found to give the form [3]

$$I_D = G_0 V_P\left\{\frac{V_x(L)}{V_P} - \frac{2}{3}\left[\frac{V_x(L) - V_g}{V_P}\right]^{\frac{3}{2}} + \frac{2}{3}\left[\frac{-V_g}{V_P}\right]^{\frac{3}{2}}\right\}. \tag{8.18}$$

Both of these relationships provide a fundamental linkage of the drain current and drain voltage, as they are modulated by the gate potential. Specifically, for a fixed applied $V_x(L)$ between the source and drain, the drain current I_D is regulated by the gate potential V_g. At this level, the slab system can be treated as a black box, yielding a current flow I_D controlled by the parameters characterizing the drain and gate potentials.

The results in Eqs. (8.15)–(8.18) hold when there is a connected region of charge-neutral n-material linking the source to the drain so that the conditions for this to be the case need to be discussed. Specifically, limitations on the connected region of charge-neutral material are obtained from a consideration Eqs. (8.11) and (8.12). This is done by determining the configuration of the system under which the two regions of net positive charge touch each other. When these two regions touch, they close the connected path from the source to the drain in the charge-neutral n-material, and the region becomes disconnected.

Consequently, the requirement that the regions of net positive charge of the upper and lower junctions initially touch in the region $0 \le x \le L$ is given from Eqs. (8.11) and (8.12) by [3]

$$0 = s - l_n(x) = s - \sqrt{\frac{2\varepsilon}{en_+}\,\mathcal{V}(x)}\ . \tag{8.19a}$$

This is essentially the statement that the half-width of the slab of n-material must equal the length $l_n(x)$ of the extension of the region of net positive charge into the n-material slab. From Eq. (8.19a), it then follows that the condition for closure at general x takes the form

$$1 = \sqrt{\frac{V_0 + V_x(x) - V_g}{V_P}} \approx \sqrt{\frac{V_x(x) - V_g}{V_P}}. \tag{8.19b}$$

The resulting relationship in Eq. (8.19b) sets a fundamental restriction on the range of validity of Eqs. (8.15)–(8.18) for general x, and, in particular, for the case that $x = L$. In this regard, a connected channel of neutral n-material exists between the source and the drain when [3]

$$1 \ge \sqrt{\frac{V_0 + V_x(L) - V_g}{V_P}} \approx \sqrt{\frac{V_x(L) - V_g}{V_P}}. \tag{8.19c}$$

Under the restriction given in Eq. (8.19c), it follows that Eqs. (8.17) and (8.18) evaluated at $x = L$ yield valid relationships for I_D and $V_x(L)$ in a connected path between the source and drain.

In the case that Eq. (8.19c) is not satisfied, the problem of the relationship between $V_x(L)$, I_D, V_g must be readdressed. This limit of the problem is now discussed.

8.2.5 Disconnected Region of Charge-Neutral N-Material between the Source and Drain

The case in which the two regions of net positive charge touch is more difficult to treat. For this configuration a path from the source to the drain is no longer contained solely within charge-neutral n-material, and the problem of carrier motion in the region of net positive charge needs to be addressed [3,4]. A treatment of the theory of this type of configuration is now developed based on some simplifying assumptions on the charge conductivity properties of the regions of net charge in our system. These assumptions are consistent with results from computer simulation studies that have been performed on the junction.

Specifically, the reformulated treatment is made by assuming that free charge carriers entering the regions of net electrical charge are easily swept from these charged regions by the high-intensity electric fields generated within them. In our earlier discussions of the connected charge-neutral path between the source and the drain, this kind of argument allowed us to assume that the electrons in the middle slab only travel from the source to the drain in the charge-neutral n-material. Within this view, the charge carriers entering the region of net positive charge are force back into the region of neutral n-material by the strong fields and the reverse bias of the p–n junction. Now, this type of argument is applied to describe the motion of the carrier electrons from the disconnected region of neutral n-material, through the region of net positive charge, and on into the drain.

In our model, we now assume that in a disconnected channel the electron current is swept out of the region of net positive charge to the drain by the electric fields created by the net positive charge and the high positive potential at the drain [3]. The potential of the drain forces the electron carriers to flow into the drain, just as the reverse bias of the junction pushed the electron carriers into the charge-neutral n-material. Basically, it is assumed that in the region of net positive charge between the disconnected neutral n-material and the drain, the current I_D remains constant independent of an increase in the potential $V_x(L)$. In this view then, once the two regions of positive charge touch

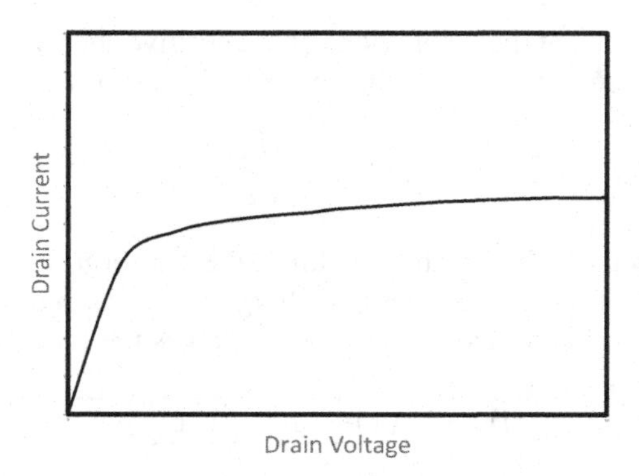

FIGURE 8.4 A schematic plot indicating the typical behavior of the Drain Current I_D versus the drain voltage V_D for a fixed gate potential $V_g<0$ is shown. Note that on the left is an initial linear increase in I_D versus V_D, while a region of constant I_D versus V_D is approximated on the right of the figure.

the current flow remains constant and is little affected by further increases in the potential of the drain. The current is said to be saturated. As an illustration, a plot of the typical drain current versus $V_D = V_x(L)$ under these assumptions is shown in Figure 8.4.

To understand how Figure 8.4 is arrived at, it is necessary to determine the value of the current in the n-material for the configuration in which the upper and lower junction transition regions have just touched at $x = L$ and disconnected the conduction path of neutral n-material from the drain [3]. This is an important configuration of the system as following this disconnection or path closure any further increase in $V_x(L)$ has little effect on I_D, which is approximated as being constant. Consequently, once closure has been made of the neutral n-material conduction path of the system, the current is constant and independent of further increases in $V_x(L)$.

The configuration of $V_x(L)$ and I_D at the moment of channel closure can be obtained from Eqs. (8.15) and (8.16) evaluated at $x = L$. Here Eq. (8.15) includes the contact potential of the junction, and the closure condition for Eq. (8.15) is from Eq. (8.19) given by [3]

$$1 = \sqrt{\frac{V_0 + V_x(L) - V_g}{V_P}}. \tag{8.20}$$

Similarly, Eq. (8.16) neglects the contact potential so that at closure of the channel

$$1 = \sqrt{\frac{V_x(L) - V_g}{V_P}}. \tag{8.21}$$

Evaluating Eqs. (8.15) and (8.17a) for the condition in Eq. (8.20), it then follows that for $1 \le \sqrt{\frac{V_0 + V_x(L) - V_g}{V_P}}$ (i.e., the region of a disconnected region of neutral n-material), the drain current is given by the constant

$$I_D = G_0 V_P \left\{ \frac{V_P + V_g - V_0}{V_P} - \frac{2}{3} + \frac{2}{3}\left[\frac{V_0 - V_g}{V_P}\right]^{\frac{3}{2}} \right\} \tag{8.22}$$

relating the drain current to the potentials of the problem other than $V_x(L)$. In addition, in the case of $V_0 = 0$, for $1 \le \sqrt{\frac{V_x(L) - V_g}{V_P}}$ Eqs. (8.16) and (8.18) reduces to [3]

$$I_D = G_0 V_P \left\{ \frac{V_P + V_g}{V_P} - \frac{2}{3} + \frac{2}{3}\left[-\frac{V_g}{V_P} \right]^{\frac{3}{2}} \right\} \tag{8.23}$$

so that I_D is determined by V_p and V_g. The forms in Eqs. (8.20)–(8.23) are then used in their respective regions of validity to generate the plot in Figure 8.4.

The basic qualitative features of the current to drain voltage exhibited in Figure 8.4 are confirmed by detailed computer simulations of the three-slab system. As long as the neutral n-material path is connected, a fairly rapid increase in drain current with increasing $V_x(L)$ is observed. (See Figure 8.4.) After the connection has been pinched off by the touching of the two regions of net positive charge, however, the variation of the drain current with increasing $V_x(L)$ exhibits a region of slow change, i.e., it is approximately constant. (See Figure 8.4) These two behaviors enter into the operation of the three-layered transistor system, making it a device designed to act as a switch or as an amplifier.

The forms in Eqs. (8.16)–(8.23) are used next to introduce the idea of the gate as a means to modulate the flow of current through the slab of n-material. First the conditions required on the system so that zero drain current is present in the system are discussed. This is then followed by a treatment of general drain current modulations made through adjustments of the gate potential and the application of these adjustments to provide mechanisms for signal switching and amplification.

8.2.6 Conditions of Zero Drain Current

In our first discussion, the two forms in Eqs. (8.17) and (8.18) are used to study the conditions under which the current in the n-material slab is shut off by the gate potential, i.e., the $V_x(L)$ and V_G needed to make $I_D = 0$ are determined [3]. In this configuration, no electrons pass from the source to the drain, and the system behaves as if a gate were closed to their passage through the n-material slab. As shall be seen later, this property of the gate to limit the flow of electrons from the source to the drain may enter the design of electrical switches and amplifiers.

Specifically, from Eq. (8.17a) the condition for $I_D = 0$ is that $V_x(L)$ is a solution of

$$\frac{V_x(L)}{V_P} = \frac{2}{3}\left[\frac{V_0 + V_x(L) - V_g}{V_P} \right]^{\frac{3}{2}} - \frac{2}{3}\left[\frac{V_0 - V_g}{V_P} \right]^{\frac{3}{2}}. \tag{8.24a}$$

Similarly, in the limit that the contact potential can be ignored (i.e., $V_0 = 0$), it follows from Eq. (8.18) that

$$\frac{V_x(L)}{V_P} = \frac{2}{3}\left[\frac{V_x(L) - V_g}{V_P} \right]^{\frac{3}{2}} - \frac{2}{3}\left[-\frac{V_g}{V_P} \right]^{\frac{3}{2}}. \tag{8.24b}$$

These both yield solutions for $V_x(L)$ as a function of the applied gate potential, giving a continuous set of such conditions. It is also seen that independent of V_0 and V_g, $V_x(L) = 0$ is always a solution of these equations, i.e., no current flows in the absence of a potential difference between the source and the drain. Similar relationship can also be obtained from Eqs. (8.22) to (8.23) for the disconnect channel.

As a further simplification an interesting limit is that in which $V_x(L) \gg |V_g|, V_0$. For this configuration, both Eqs. (8.24a) and (8.24b) reduce to the single form [3]

$$V_x(L) = \frac{9}{4} V_P. \tag{8.25}$$

Note that the condition in Eq. (8.25) provides a simple approximate form for the $I_D = 0$ configuration of the transistor. It determines the region of positive charge in the n-material needed to choke the flow of current through the n-material slab, when the potential difference between the source and drain is the primary controlling mechanism of the system.

Similarly, in the case that $|V_g| \gg, V_x(L), V_0$, both Eqs. (8.24a) and (8.24b) are reduced to the form

$$V_g = -V_P. \tag{8.26}$$

For this limit, Eq. (8.26) determines the gate potential required to close the gate on the electron flow through the n-material slab. In this case, the primary control is the dominance of the gate potential.

Equations (8.25) and (8.26) yield two interesting limiting cases of gate closure. For a more complete solution of the problem, however, the complex forms in Eq. (8.24) must be evaluated numerically for general $V_x(L)$, V_0, and V_g.

The conditions considered here have only treated the current flow arising from potentials applied directly to the transistor and have neglected other loads that may enter into more complex systems. An interesting extension of these ideas is that in which the transistor is connected in circuit with a load, and is made to turn on and off the current flowing through the load. In this operation, the transistor acts as a switch. We next addressed this more complex behavior of operating a transistor as a switch regulating the flow of current through a specific load.

8.2.7 Switches and Amplifier Circuits

The closing of the channel for current flow in the n-type material of a p–n–p transistor by means of a gate potential has an application in the design of electrical switches. In addition, lesser modifications of the channel current through the modulation of the gate potential have applications in the design of amplifying systems. A discussion is now given of these applications in the formulation of switches and amplifiers. In the following, only a consideration of the simplest circuits that provide a switching or amplification application are treated [3,4].

First consider the circuit in Figure 8.5 composed of a transistor with the source and drain connected to a load-resistor-battery current source, and with a reverse bias potential applied between the gate and source. The proposed circuit can be made to act either as an on–off switch for the current flowing through the load or, in the case that a small AC source potential is introduced in addition to the DC gate potential, as a basic amplifier circuit. In our first considerations, the configuration is made to operate as a switch for turning on and off the flow of current through the load. For simplicity, the load is represented as a resistance in a DC system but may be a variety of other

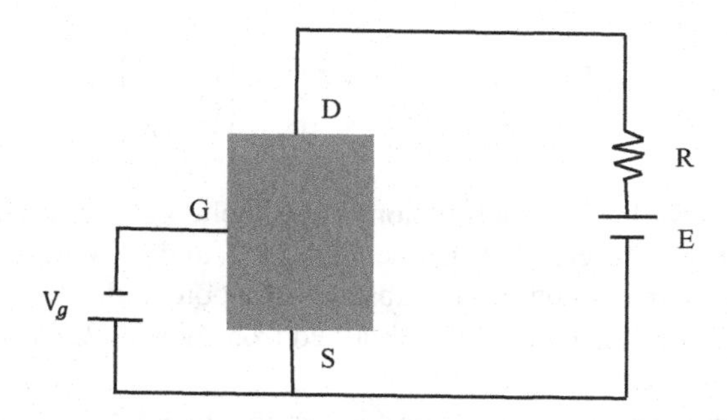

FIGURE 8.5 A circuit for the illustration of the switching and amplification properties of a field effect transistor. S is the source terminal, D is the drain terminal, G is the gate, V_g is the gate potential, R is the load, and E is an applied potential across the drain–source–load circuit. The transistor is indicated on the figure by the shaded rectangle.

devices requiring a DC current for their operation. The focus of the discussions is on illustrating the basic ideas of a transistor switch. In practice, for systems formulated to meet engineering design requirements, more complex discussions may be needed. To obtain a more advanced treatment of such systems, the reader is referred to the literature [3–5].

If the transistor in Figure 8.5 is represented as a device in which the current between the drain and the source I_D is related to the potential applied to the transistor across the source and drain V_D and the gate potential V_g by a nonlinear function $V_D = f(V_g, I_D)$, then the Kirchhoff's law for the circuit in Figure 8.5 containing the load resistance becomes [3]

$$E = f(V_g, I_D) + I_D R, \tag{8.27a}$$

where R is the load resistance, and E is the battery emf. (Note from our earlier discussions that $V_D = f(V_g, I_D)$ for a fixed gate potential V_g relates the potential across the transistor to the current through the transistor. Again, as per our earlier discussion, this has the general form illustrated in Figure 8.4.) Here the emf drops across both the resistor and the transistor path between the source and the drain. The resulting Eq. (8.27a) is a nonlinear equation that can be solved by a variety of numerical methods or through a graphical representation shown schematically as in Figure 8.6. This is now discussed for switching applications.

As an example, in Figure 8.6a, Eq. (8.27a) is rewritten in the form [3]

$$E - I_D R = f(V_g, I_D) = V_D, \tag{8.27b}$$

and both the right- and left-hand sides of Eq. (8.27b) are independently plotted as functions of I_D. The solution of Eq. (8.27) appears as the intersection of the two different curves obtained by plotting the left-hand side and the middle term in Eq. (8.27b). Specifically, in Figure 8.6a, a schematic plot of $f(V_g, I_D)$ for two different values of the gate potential V_{g1} and V_{g2} is shown. In addition, a schematic of the plot of the linear form $E - I_D R$ is presented. The solutions of interest are found at the intersections of the linear and two nonlinear forms.

In Figure 8.6a, the solution at the intersection (I_D, V_D) for the gate potential V_{g1} is seen to occur at a much greater current I_D than the solution of (I_D, V_D) for the gate potential V_{g2}. The difference of these two solutions can be used to mimic the behavior of a current switch operating on the current I_D passing through R. In this operation, the current is considered to be on for the applied gate potential V_{g1} and is considered to effectively be off for the applied gate potential V_{g2}. Consequently, the flow of current in the load is turned on and off by changing the gate potential. It is then an engineering problem to improve on the design of the system just described in order to give the closest approximation to the current flow and no current flow configurations of an ideal switch.

By adjusting the switching behavior just described, an amplifier can be formulated. To see this, consider a system in which the DC gate potential in Figure 8.5 has an additional small AC potential added on top of it. As a result, the gate potential is modulated between the potential extrema V_{g1} and V_{g2}. This situation is shown in Figure 8.6b for a graphical solution of Eq. (8.27b). In Figure 8.6b, the separations of the V_{g1} and V_{g2} curves are now less pronounced than those in Figure 8.6a.

For a fixed potential E applied to the current source part of the circuit in Figure 8.5, the periodic variation of the gate potential between V_{g1} and V_{g2} will show up as a modulation of the drain currents and potentials between their respective solutions (I_{D1}, V_{D1}) and (I_{D2}, V_{D2}). By choosing the configuration correctly, a small variation $|V_{g1} - V_{g2}|$ in gate potential can show up as an amplified modulation of the drain potential $|V_{D1} - V_{D2}|$ and currents $|I_{D1} - I_{D2}|$. In this way a larger current variation $|I_{D1} - I_{D2}|$ is generated in the load containing part of the circuit than would otherwise be generated by a direct application of $|V_{g1} - V_{g2}|$. This provides the basic operation of an electronic amplifier (i.e., the generation of a strong signal output from a weak signal input) and is illustrated schematically in Figure 8.6b.

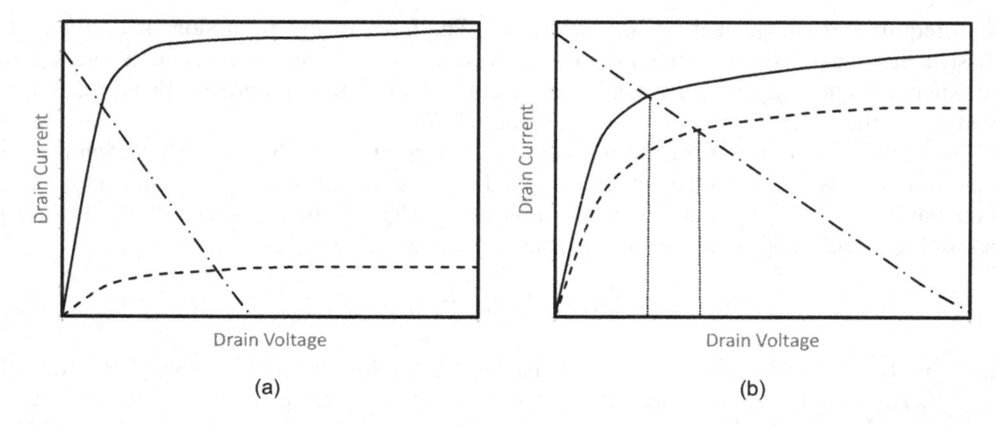

FIGURE 8.6 Schematic for (a) The operation of the circuit in Figure 8.5 as a switch. The solid line is the plot of I_D versus V_D for the gate potential V_{g1}, and the dashed line is the plot of I_D versus V_D for the gate potential V_{g2}. The dot–dashed line is a plot of $E=I_DR+V_D$. By switching between V_{g1} and $V_{g2,}$ the on–off configurations of a switch acting on the current through R are approximated. (b) The operation of the circuit in Figure 8.5 as an amplifier. A small AC voltage is introduced as a signal modulation to the DC gate bias. The solid line is the plot of I_D versus V_D for the gate potential extremum $V_{g1,}$ and the dashed line is the plot of I_D versus V_D for the gate potential extremum V_{g2}. The dot–dashed line is a plot of $E=I_DR+V_D$. For the small variation in gate voltage between V_{g1} and V_{g2}, the two weak doted vertical lines marked on the drain voltage axis give a large variation of the drain voltage for a small difference | $V_{g2}-V_{g2}$|. This is an amplification effect of a small gate voltage modulation transformed to a large drain voltage and current modulation.

8.2.8 *N–P–N* Transistors

The operations just described for the *p–n–p* field effect transistor can also be reformulated for an *n–p–n* sandwich of layers [3,4]. In this formulation, a slab of *p*-material is layered between two slabs of *n*-material. For the *n–p–n* system, now the *p*-material acts to carry a current that is modulated by a gate potential applied across the two parallel and oppositely directed *p–n* junctions. In this arrangement, the gate potentials are configured to confine the current flowing in the *p*-material layer and to adjust the magnitude of the current flow carried by the *p*-material. The confinement is now made by regulating the regions of net negative charge generated in the *p*-material at the junction interfaces. The extent of this region of net negative charge is changed by application of appropriate gate potentials to open and close the charge-neutral channel region of *p*-material that carries the hole current.

It should be noted that in addition to the transistors formed from *p–n* semiconductor junctions, which are created by doping an originally pure intrinsic semiconductor, various other types of metal–semiconductor and semiconductor (A)–semiconductor (B) junctions can be used in transistor designs [3,4]. In addition, only the very basic elements of transistor operation have been treated here, and there are a variety of other interesting physical characteristics of these devices available to technology. The read can find discussions of these in the literature [2–5].

We next turn to a treatment of systems known as bipolar transistors [3,4]. These involve a new set of operating principles that ultimately allows a current in one circuit to be modulated by changes in the current of another circuit. This requires a completely different set of operating procedures from those of the field effect transistor.

8.3 BIPOLAR TRANSISTOR

The other major variant of semiconductor junction transistor is the bipolar junction transistor [3,4]. As in the case of field effect transistors, a bipolar transistor is again formed of three different semiconductor slabs that are interfaced with one another, involving either a *p–n–p* or *n–p–n* layering, in which each layer is connected to external sources by one of the three terminals. As in the field effect geometry, the p–n junctions at the layer interfaces are parallel to one another, but now their properties are organized differently than in the field effect system. Specifically, in the bipolar systems, the current flows and potentials applied across the junctions are essentially different from those found in the field effect transistor. In the bipolar transistor, the current modulated by the transistor flows perpendicular to the junction interfaces, passing though all the three semiconductor layers, rather than parallel to the interfaces and within the center layer. The current flowing through the transistor is now regulated by a modulation of the middle layer rather than in the outer layers.

In the bipolar transistor, the system is configured so that a large current flows between the two p layers of the p–n–p system through the separating n layer or between the two n layers of the n–p–n system through the separating p layer [3]. This large current can then be modulated by a small current injected into the separating middle layer of the transistor. The configuration differs from that of the field effect transistor, which modulates a current flowing in the middle slab of the layering of three slabs by means of a gate potential applied between the center slab and the two outer layers it separates. Consequently, whereas the operation of the field effect transistor was characterized by the properties of the source, drain, and gate transistor terminals, in the bipolar transistors the transistor terminals are now referred to as the emitter, collector, and base. This acknowledges the difference of function in which the system is configured for operation. In the following presentation, the focus will be on details of the operation of a p–n–p bipolar transistor, which is, in principle, somewhat similar to that of the n–p–n transistor.

A schematic of the p–n–p bipolar transistor in its basic operating circuit is given in Figure 8.7. The p-material on the left is denoted p^+, as the carrier concentration of holes in this layer is designed to be much greater than that of the electrons in the layer of n-material or of the holes in the p-material in the layer on the right [3]. This arrangement is made so that a large hole current can be generated and passed through the left p–n junction, which is forward biased for a large flow of holes into the layer of n-material. Eventually most of this large current flow will be passed into the p-material of

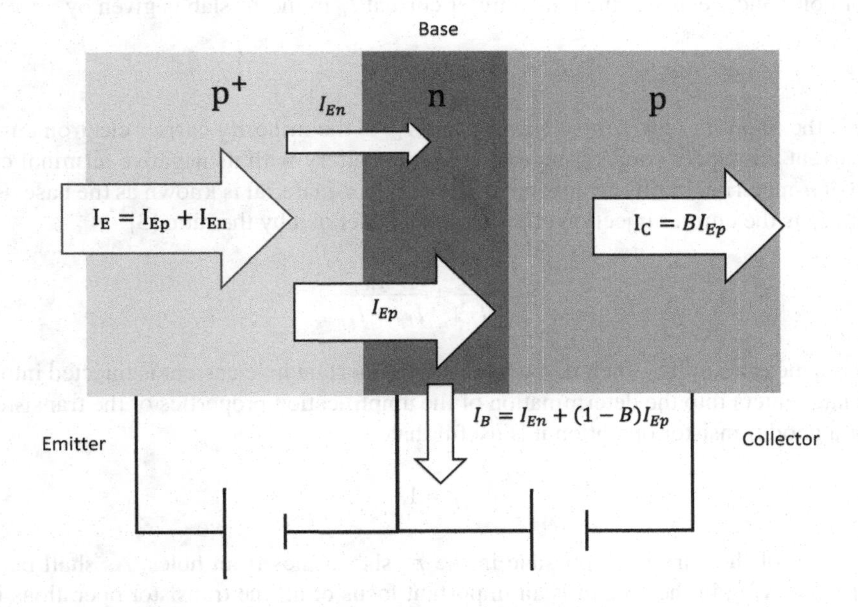

FIGURE 8.7 Current flows in a bipolar p–n–p transistor. On the left side, the emitter current $I_E = I_{Ep} + I_{En}$ is composed of a majority hole current I_{Ep} and a minority electron current I_{En}. On the right side, a fraction B of the hole current I_{Ep} forms the collector current I_C. The base current $I_B = I_{En} + (1-B)\ I_{Ep}$ consists of electrons that pass into the emitter forming the current I_{En} and a current of electrons $(1-B)\ I_{Ep}$ that enter the base in order to neutralize some of the holes transiting the base.

the right layer of the p–n–p transistor. In addition to the hole current in the left slab, there is a small current of minority electron carriers that originates from a flow of electrons into the n-material. This electron current ultimately flows from the terminal connection of the n-material with the battery on the left side of the figure.

The large forward bias at the left junction pushes a large current of holes through the thin layer of n-material, where some of the holes combine with electrons in the charge-neutral region of n-material [3,4]. For this flow, the system is arranged so that most of the holes pass through the second junction on the right and enter the p-material on the right side of the junction. This transfer of holes between the two regions of p-materials is the origin of the large current flowing from the left to the right junction. The small number of holes that combine with the electron carriers in the n-material generate light, while removing both the combining electron and hole participants from the conduction process of the junction. These so-called recombination processes act to remove a small number of both of the two sets of oppositely charged carriers from the conduction processes. The n-material electrons entering into the recombination processes are injected into the n-material from the batteries. In this regard, the p–n junction to the right of the layer of n-material is reversed biased, so that the flow of electrons between the left slab of n-material and the right slab of p-materials is very small and is ignored here. In addition, in order to make the recombination process small, the width of the slab of n-material between the two junctions is made thin so that the transit time of the holes through the n layer is adjusted to be as small as possible.

A consideration of the basic transport properties of the currents flowing through the system allows for an understanding of the amplification properties of the p–n–p transistor. The focus in the following is on a quantitative accounting of the various carrier transport processes in the system and a numerical determination of the signal amplification properties following from the operation of the bipolar transistor. This is now discussed.

Begin by considering the current flow in the p^+ slab of material [3]. This slab is called the emitter and is the origin of the hole current projected into the middle slab of the transistor. Considering both the flow of holes and electrons, the total emitter current I_E in the p^+ slab is given by

$$I_E = I_{Ep} + I_{En}, \tag{8.28}$$

where I_{Ep} is the majority carrier hole current, and I_{En} is the minority carrier electron current. The electron current ultimately comes primarily from the battery with its negative terminal connected to the slab of n-material. In this arrangement, the slab of n-material is known as the base. Related to the current, I_E is the emitter injection efficiency that is defined by the ratio [3]

$$\gamma = \frac{I_{Ep}}{I_E} = \frac{I_{Ep}}{I_{Ep} + I_{En}}. \tag{8.29a}$$

This ratio is a measure of the efficiency at which the important hole current is injected into the transistor and later enters into the determination of the amplification properties of the transistor. In this regard, for a good transistor operation it is useful that

$$\gamma \approx 1, \tag{8.29b}$$

so that as much of the current as possible in the p^+ slab comes from holes. As shall be seen, the amount of I_{Ep} flowing in the system is an important focus of all the transistor operations in p–n–p bipolar systems.

The flow of current I_C in the right slab of p-material (known as the collector) is related to the hole current in the emitter by [3]

$$I_c = BI_{Ep} \tag{8.30}$$

where B is the fraction of I_{Ep} making it to the collector following the hole transit in the thin layer of n-material. Again, for a good operation $B \approx 1$ so that as many holes as possible originating in the emitter are passed into the collector. In this regard, the righthand junction is also reversed biased so that the flow of electrons in the collector is negligible compared to that in the p^+ and n slabs.

Combining Eqs. (8.28) and (8.30), it then follows that [3]

$$\frac{I_C}{I_E} = \frac{BI_{Ep}}{I_{Ep} + I_{En}} = B\gamma = \alpha, \tag{8.31}$$

where α is the current transfer ratio relating the magnitudes of the collector and emitter currents. This provides a measure of the large current flow between the left and right slabs of the transistor, and should be close to unity. The amplification properties of the system are intimately related to the efficient transfer of current between the emitter and collector.

The properties of the electron current flow in the terminal connecting the slab of n-material (known as the base) to the batteries can now be determined in terms of α, γ, and B. In this regard, the current in the base I_B originating from the batteries is given by

$$I_B = I_{En} + (1 - B)I_{Ep}. \tag{8.32}$$

Here I_{En} are electrons from the batteries that pass the left junction and show up as minority carriers in the p^+ slab, and the second term in the sum accounts for the electrons drawn from the battery that go into neutralizing holes in the process of recombination within the n-material.

The amplification properties of the p–n–p transistor can now be determined from a comparison of the base and collector currents. In this regard, from Eqs. (8.30) to (8.32), it follows that [3]

$$\frac{I_C}{I_B} = \frac{BI_{Ep}}{I_{En} + (1 - B)I_{Ep}} = \frac{B\gamma}{1 - B\gamma} = \frac{\alpha}{1 - \alpha} = \beta, \tag{8.33}$$

and for our system designed so that $B \approx 1$, $\gamma \approx 1$, it then follows that $\alpha \approx 1$ and $\beta \gg 1$. As a consequence, from Eq. (8.33), it is seen that a small change in I_B shows up as a large change in I_C. This is a basis for an application of the transistor in signal amplification.

Considering the material composing the system, the ratio in Eq. (8.33) can be related to some of the microscopic properties of the n-material. If the lifetime of the holes in the base for recombination to occur is τ_p, and the transit time of holes in the slab of n-material is τ_t, then in the time τ_p for an electron from the base to combine with a hole, the number of holes passing through the n-material is $\frac{\tau_t}{\tau_p}$. Consequently, the ratio of the base current to the collector current is roughly related by the condition [3]

$$I_B\tau_p = I_C\tau_t \tag{8.34a}$$

so that

$$\frac{\tau_p}{\tau_t} = \beta = \frac{I_C}{I_B}. \tag{8.34b}$$

The potential of the system to act as a signal amplifier is then intimately related to the microscopic characteristics of its conduction properties [1–5]. Therefore, for an optimum operation, the slab of n-type material should be thin in order for the hole transit time τ_t to be small, and an n-type material that has a large recombination time τ_p should be chosen.

By applying a signal to the n-type material slab, in addition to the DC biases shown in Figure 8.7, a small variation of I_B is transformed to a collector current variation by $I_c = \beta I_B$ where $\beta \gg 1$. This results in a net amplification of the output signal generated in the collector.

REFERENCES

1. Kittle, C. 1996. *Introduction to solid state physics, 7th Edition*, 393–410. New York: John Wiley & Sons, Inc.
2. Marder, M. P. 2000. *Condensed matter physics*. New York: John Wiley & Sons, Inc.
3. Streetman, B. G. and S. Banerjee 2000. *Solid state electronic devices, 5th Edition*. New Jersey: Prentice Hall.
4. Schilling, D. L. and C. Belove 1968. *Electronic circuits: Discrete and integrated.* New York: McGraw-Hill Book Company.
5. Datta, S. 2005. *Quantum transport: Atom to transistor*. Cambridge: Cambridge University Press.

9 Toward Single-Electron Transistors

Coulomb Blockade

In this chapter, discussions are directed toward nanoscale systems that function as transistors for processing the flow of single electrons in nanoscale circuits [1–6]. Such circuits are generally composed of material features that allow for the tunneling of electrons between one another in order to store and/or direct the flow of individual electrons. The types of functions exhibited by these circuits provide the basis for the manipulation of single electrons in various logic gates or other circuit components based on controlling the motion of a small number of individual electrons. Unlike macroscopic transistors [6], where the electron–electron interactions in the electrical current are averaged out, nanoscale systems focus on the motions of small numbers of uncompensated electrons in the electron system and their detailed electron–electron interactions [1]. The single-electron dynamics of nanoscale circuits then operates in regions in which the circuits are not charge neutral on average, and Coulomb interactions between individual electrons become important components of the circuit dynamics. In this regard, to be a successful processor of individual charges, this type of nanoscale circuit requires low temperatures and circuit features of low capacitance for the storage of individual electronic charges and their capacitive energies [1–6]. These conditions aid in the observation of the strong electron correlations needed to control the motion of single electrons transport in nanocircuits [7–14].

In this regard, if an electron of charge $-e$ is stored on a nanoscale capacitance C, the characteristic capacitive energy involved in the storage is of order $\frac{1}{2}\frac{e^2}{C}$. For the stability of such a charged capacitor in a nanocircuit, its capacitive energy must be greater than the characteristic thermal fluctuations that, from the Boltzmann distribution, are of order k_BT. If this is not the case, the charge can be knocked off the capacitor by an inelastic event with the large thermal fluctuations, and the flow of electrons in the system is no longer a matter of the motion of individual electrons [1]. The satisfaction of this charging condition is accomplished by operating at low temperatures where the thermal fluctuations are absent and/or by designing low-capacitance (high capacitive energy) nanosystem features for the storage and handling of charge. In addition, for the discussion here, we consider only systems formed from particulates (metal or semiconductor) in which the discrete quantum energy levels of the electrons in the system have small energy separations. In such systems, the electrostatic properties of the electrons dominate the discrete quantum nature of the electron states. This is done by making nanostructures of low capacitance, which are also large enough to support quasi-continuous electron energy states. For such systems, the highly correlated electron motion can be treated in a quasi-classical approach.

In the discussions presented here, a number of simple devices are considered, ultimately leading to the illustration of the fundamental ideas involved in the development of a single-electron transistor [7–11]. The focus is on basic nanoscale systems that are not necessarily charge neutral but may contain a net number of uncompensated electrons, and the treatment emphasizes the theoretical ideas involved, rather than the experimental implementation of these ideas. The Coulomb interactions between the uncompensated stored electrons on proposed circuit devices will be seen to allow the system to develop a so-called Coulomb blockade [1–5]. The idea of a Coulomb blockade is to develop a configuration of net charge on a nanocircuit, which inhibits the flow of individual

DOI: 10.1201/9781003031987-9

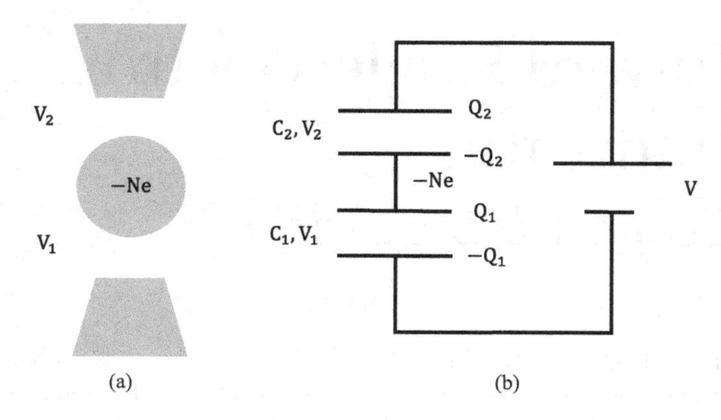

FIGURE 9.1 Schematic plots of: (a) A metallic nanoisland (circular) containing a net charge of uncompensated electrons *–Ne* interacting with two tunneling junction probes (parallelepipeds). (b) The equivalent electronic circuit problem in which the tunneling junctions are represented by capacitors and the potential differences across the junctions are supplied by batteries.

electrons in the system due to the electrostatic energy of individual electrons isolated in the system. The phenomenon of Coulomb blockade will be seen to allow for the manipulation of the transport of individual electrons involving simple tunneling processes developed between nanoparticles forming a nanoelectrical system. The essential mechanisms of electron dynamics in such systems are the electron tunneling between nanoparticles and the capacitive energies associated with the development of a net uncompensated charge on the individual nanoparticles of the system.

In a first consideration, a single island device is treated [1]. (See Figure 9.1 for a schematic of this system.) This is formed as a single nanoparticle known as the island that interacts with the outside world through two tunnel junctions. Upon applying a potential difference across the island–junction system, individual uncompensated electrons can be developed on the island so that the island carries a nonzero total charge. The uncompensated charges tunnel through the junctions onto the island due to the potential difference developed across the tunnel junctions, and, in general, the tunneling rate is chosen to be small. Each of the tunnel junctions develop both a potential and charge difference across it. In this regard, the tunnel junction acts as a kind of capacitor. Specifically, it is a type of leaky capacitor having a capacitance and a large resistance allowing for a weak current flow of single electrons onto the island [1]. A focus in the study of this system is to understand how the net uncompensated electron charges on the island can be manipulated by the potential applied from the outside world.

To study the single island problem, a model electronic circuit system is developed [1]. This exhibits all of the essential properties of the configuration of nanoparticles and its interaction with the outside world. The circuit model involves two series capacitors (each representing a tunnel junction), a region between the two series capacitors that represents the island, and an applied potential across the two junction–island system (representing the potential from the world outside the island). Unlike the standard circuit model of two series capacitors in series with a battery, the present model allows the region between the two capacitors to develop a net uncompensated charge of individual electrons. The properties exhibited by the circuit are shown to form an essential component for the development of a single-electron transistor. In this regard, the single-electron transistor is the next system considered.

The model of the island interacting with two junctions is extended to that of a single-electron transistor by introducing a third tunnel junction into the considerations [1]. (See Figure 9.2 for a schematic of this system.) Now an outside potential is applied, as discussed earlier, across two junctions. These junctions cause electrons to tunnel on and off the island. Similar to the field-effect transistor, the low potential side of the applied potential is known as the source, and the high

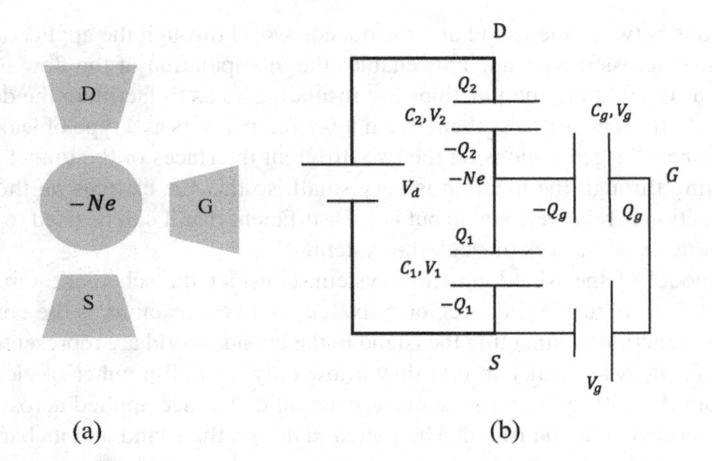

FIGURE 9.2 Schematic plots of: (a) A metallic nanoisland (circular) containing a net charge of uncompensated electrons –*Ne* interacting with three tunneling junction probes (parallelepipeds) denoted as *S* (source), *D* (drain), and *G* (gate). (b) The equivalent electronic circuit problem in which the tunneling junctions are represented by capacitors, and the potential differences across the junctions are supplied by batteries. The *S* (source), *D* (drain), and *G* (gate) branches of the circuit are labeled in the figure.

potential side of the applied potential is known as the drain. The electrons flow between the source and drain in a tunneling current that is dependent on the applied potential from the outside world. In addition, a third junction is introduced to interact with the island that does not pass charge but only acts as a capacitance, which modulates the potential of the island. This third junction is known as the gate, and the flow of charge between the source and drain is regulated by manipulating the gate potential. This regulated flow of electrons through the island forms the basis of transistor control of the electrons in the device.

The electrical circuit analogy for the single-electron transistor is composed of three capacitors that are connected to each other at a single circuit branch. This coupling represents the island of the system, which may contain a net uncompensated charge, and the capacitors are the three junctions interacting with the outside sources of potential. Only two of the capacitors (representing the source and drain junctions) are leaky, so that single electrons can tunnel through them, and the capacitor representing the gate does not leak charge [1]. A potential difference applied across the source and drain helps charge to flow through the system, and another potential difference is applied across the source and gate to regulate the charge flow. This arrangement of capacitors and applied potentials is shown to exhibit features that control the flow of single electrons through the island. The various manipulations of charge transfer in this circuit represent the properties of the single-electron transistor.

In the following, first the two-junction system is discussed [1]. After this, a treatment of the single-electron transistor is given along with discussions of some properties and applications of the single-electron transistor [1]. These properties include applications as logic gates and data storage units, as well as in processes requiring the manipulation of single-electron motions. The focus in the presentation is on the most basic models illustrating the ideas of the nanocircuits, and the reader is referred to the literature for detailed discussion of technical complications involved in the realization of nanocircuits and their remedies [1–5].

9.1 A SINGLE ISLAND DEVICE

As a first device, consider a nanoscale-conducting island that is connected to an outside potential difference through two weak tunnel junctions [1]. The junctions allow the tunneling of a small number of electrons between the island and the outside world through the application of an outside potential difference across the island. This enables the manipulation of the flow of single charges through the island. In addition, the junctions are insulating so as to facilitate the development of a sustained potential differences across them. Each junction then acts as a type of leaky capacitor that supports opposite net charged regions on the two different interfaces of the tunnel junction. In this sense, the tunneling through the junction is very small, so that the charges on the two capacitors formed by the junction is relatively stable but is still sufficient that it can be used to control the passage of a small number of charges through the system.

As a circuit model of the island–junction system, consider the schematic circuit diagram [1] shown in Figure 9.1. In Figure 9.1, the region between the two capacitors is the conducting island, and the two tunnel junctions connecting the island to the outside world are represented by the capacitors. The two capacitors are leaky in that they allow only a small number of electrons to tunnel through them from the battery that represents a potential difference applied across the conducting island from the world outside the island. The potential across the island and its barrier junctions is denoted V, and the barrier capacitors denoted C_1 and C_2 are charged with respective charges Q_1 and Q_2 and to respective potentials V_1 and V_2. As a result of the small tunneling probabilities of the capacitors, a small net charge $-Ne$ is developed on the island where $e > 0$ and $N = 0, \pm 1, \pm 2, \pm 3, \ldots.$ are the net number of electrons on the island. In this regard, the island has a net charge $Q_1 - Q_2 = -Ne$, and, in the case that $N = 0$, the system reduces to the standard problem of two capacitors placed in a series circuit with the external potential V.

The electrostatic equations governing the system in Figure 9.1 are obtained from elementary circuit analysis as follows [1]:

a. The two capacitor equations relating the charges on the upper capacitor plates in the figure to the potential applied across each are given by

$$Q_1 = C_1 V_1 \tag{9.1a}$$

$$Q_2 = C_2 V_2. \tag{9.1b}$$

b. The net charge on the island is the sum of the charge on the lower plate of C_2 with that on the upper plate of C_1. This yields

$$Q_1 - Q_2 = -Ne = C_1 V_1 - C_2 V_2. \tag{9.1c}$$

c. The potential of the battery is the sum of the two potentials across C_1 and C_2, i.e.,

$$V = V_1 + V_2. \tag{9.1d}$$

The solution of these equations for Q_1, Q_2, V_1, V_2 is then obtained in terms of V, N, C_1, C_2. Specifically, using Eqs. (9.1c) and (9.1d), it follows that

$$V_1 = \frac{1}{C_1 + C_2}(C_2 V - Ne) \tag{9.2a}$$

and

$$V_2 = \frac{1}{C_1 + C_2}(C_1 V + Ne) \tag{9.2b}$$

so that from the capacitor formulae in Eqs. (9.1a) and (9.1b)

$$Q_1 = \frac{C_1}{C_1 + C_2}(C_2 V - Ne) \tag{9.2c}$$

and

$$Q_2 = \frac{C_2}{C_1 + C_2}(C_1 V + Ne). \tag{9.2d}$$

Knowing the charges and potentials on each of the capacitors allows for a determination of the energies stored on them and the work done by the battering in configuring the system.

The energy contained in the capacitors modeling the island interactions with an outside potential is given by [1]

$$E_C = \frac{1}{2}C_1 V_1^2 + \frac{1}{2}C_2 V_2^2 = \frac{1}{2}\frac{1}{C_1}Q_1^2 + \frac{1}{2}\frac{1}{C_2}Q_2^2, \tag{9.3a}$$

and, upon substituting, the results in Eq. (9.2) yields

$$E_C(N,V) = \frac{1}{2}\frac{C_1 C_2}{C_1 + C_2}V^2 + \frac{1}{2}\frac{(Ne)^2}{C_1 + C_2}. \tag{9.3b}$$

This expresses the energy stored on the barrier capacitors in terms of N, V. (Note that for $N = 0$ Eq. (9.3) reduces to the standard result for two capacitors in series with the battery.) The origin of the $N \neq 0$ charges on the junctions and the island come from their interactions with the outside world, as represented by the battery V and the ability of individual charges to exhibit a small tunneling probability through the capacitors.

The total system, however, is composed of the capacitors (junctions and island) and the charging battery (outside world). In this scheme, the battery does an amount of work on the capacitors to configure them in their final state. An approximation for this work is given as the energy required from the battery V to place the final charge Q_2 on the plate of C_2 connected to the positive terminal of the battery. In this way, for a small net charge $|-Ne| \ll |Q_1|, |Q_2|$ leaked onto the island, the work done on the capacitors and island by the battery is given by [1]

$$E_B = Q_2 V. \tag{9.4}$$

This charging energy represents a drain on the energy contained on the battery, which along with the two capacitors of the junction form a closed system.

Consequently, the change in energy of the system composed of the battery and capacitors at the end of the charging process is given from Eqs. (9.3) and (9.4) by [1]

$$E(N,V) = \frac{1}{2}\frac{C_1 C_2}{C_1 + C_2}V^2 + \frac{1}{2}\frac{(Ne)^2}{C_1 + C_2} - Q_2 V = -\frac{1}{2}\frac{C_1 C_2}{C_1 + C_2}V^2 + \frac{1}{2}\frac{(Ne)^2}{C_1 + C_2} - \frac{C_1 C_2}{C_1 + C_2}\frac{NeV}{C_1}. \tag{9.5}$$

At zero temperature or for small-enough systems that the capacitive energy is very large compared to thermal energy fluctuations the state of the system is obtained by minimizing Eq. (9.5). In this regard, temperature effects are ignored in the later treatment.

Ignoring temperature effects, at a constant V, the charge configurations N leading to an energy minimum of Eq. (9.5) is given by

$$\frac{\partial E(N,V)}{\partial N} = 0 = \frac{Ne}{C_1 + C_2} - \frac{C_1 C_2}{C_1 + C_2}\frac{V}{C_1}. \tag{9.6}$$

From which it follows that

$$Ne = C_2V, \tag{9.7a}$$

where the energy minima are given as the integer solutions of Eq. (9.7a) for N. Consequently, by changing V one passes through a series of energy minima at the various integer solutions

$$N = \frac{C_2V}{e}. \tag{9.7b}$$

These are then the charge configurations of the minimum energy supported by the system subject to a constant applied V.

Next consider the energy associated with changes in the number N of net electron charges in the system for the transition $N \to N+1$. Returning to Eq. (9.5), it is seen that the energy change of the system at fixed V is given by [1]

$$E(N+1,V) - E(N,V) = \frac{e}{C_1 + C_2}\left[\left(N + \frac{1}{2}\right)e - C_2V\right]. \tag{9.8}$$

From this energy change, the relative stability of the states of N and $N+1$ are obtained. Specifically, in the case that

$$E(N+1,V) - E(N,V) > 0, \tag{9.9a}$$

the system is more stable in the state of N, while in the case that

$$E(N+1,V) - E(N,V) < 0, \tag{9.9b}$$

the system is more stable in the state of $N+1$. Consequently, the state of the system must be an integer solution of Eq. (9.7), which cannot be transitioned out of under the conditions in Eq. (9.9).

As an illustration of these results, in Figure 9.3, a plot is presented for the stable N solutions of Eqs. (9.7)–(9.9) versus V of the system represented in Figure 9.1. The plot shows a series of plateaus that are initiated at $V = \frac{2n+1}{2}\frac{e}{C_2}$ for successive $n = 0, 1, 2, 3, \ldots\ldots$ In this way, the plateaus of constant N in the figure represent the stable configurations over a range of V of the charge on the island.

9.2 SINGLE-ELECTRON TRANSISTOR

The next more complex device is a nanoscale-conducting island, which is connected to two outside potential differences through three weak tunnel junctions [1–5,12–14]. The junctions allow the tunneling of a small number of electrons onto the island from one outside source of potential difference and for some modulation of this tunneling by a second potential difference. In this regard, the terminals are denoted by the terminology of the junctions of a field-effect transistor [6] and are termed the source, drain, and gate. The interactions of the system are modeled electronically by the schematic circuit diagram shown in Figure 9.2. In Figure 9.2, the circuit between the three capacitors is the conducting island and the capacitors represent three junctions, which only allow a small number of electrons to pass from the batteries, which represent potential differences applied across the conducting island from the world outside the island. The focus in the following is on tunneling events between the source, the island, and the drain under the regulation of the gate potential. In this regard, the tunneling rate through the gate is assumed to be much smaller than that through the source and drain junctions.

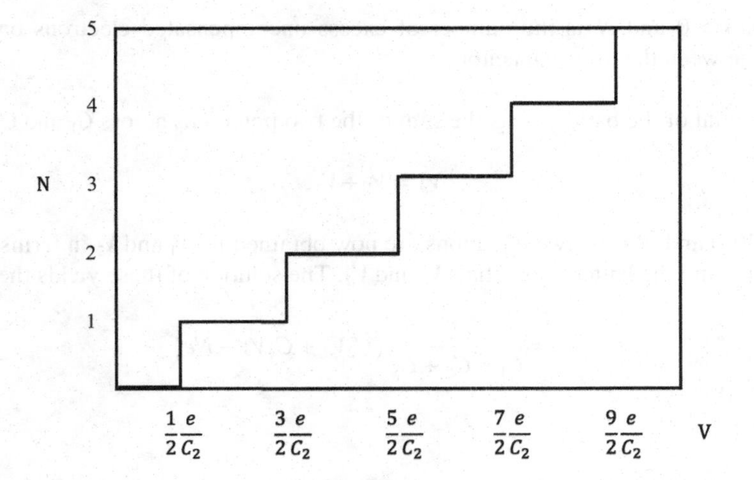

FIGURE 9.3 Plot for the system in Figure 9.1b of the number of electrons on the island N versus the applied potential V. The plateaus of constant N are the stable configurations in V of the charge on the island.

For this configuration, the potential across the island and its barriers between the source and drain terminals is denoted by V_d and is applied across the capacitors denoted by C_1 and C_2, which are charged with respective charges Q_1 and Q_2, and respectively support potentials V_1 and V_2. In addition, a second potential V_g is applied to the island and functions as a gate potential. In regards of the island tunneling, the tunneling of the electrons in the system are primarily through C_1 and C_2, and C_g does not allow tunneling. As a result of the small but nonzero tunneling probabilities of the leaky capacitors C_1 and C_2 a small net uncompensated charge $-Ne$ is developed on the island where $e > 0$ and $N = 0, \pm 1, \pm 2, \pm 3, \ldots$.. are the number of excess (uncompensated) electrons on the island [1]. Consequently, the island has an overall charge $-Ne$ and in the case that $N = 0$ the system is charge neutral, reducing to the standard problem of three capacitors placed in a branched circuit with external potentials V_d and V_g.

From electrostatics, the equations governing the system in Figure 9.2 are given by:

a. The two capacitor equations relating the charges on the upper plates of C_1 and C_2 in the figure to the potential applied across each are

$$Q_1 = C_1 V_1 \tag{9.10a}$$

$$Q_2 = C_2 V_2 \tag{9.10b}$$

and the charge on the gate capacitor C_g (specifically the righthand plate in the figure) is then

$$Q_g = C_g \left(V_g - V_1\right), \tag{9.10c}$$

where $V_g - V_1$ is the potential across C_g.

b. The net charge on the island is the sum of the charges on the lower plate of C_2 with that on the upper plate of C_1 and the left-hand plate of C_g. This is given by the relationship

$$-Q_g - Q_2 + Q_1 = -Ne = -C_g\left(V_g - V_1\right) - C_2 V_2 + C_1 V_1, \tag{9.10d}$$

where $e > 0$ and N is the number of excess uncompensated electrons on the island formed between the three capacitors.

c. The potential of the battery V_d is the sum of the two potentials across C_1 and C_2

$$V_d = V_1 + V_2. \tag{9.10e}$$

From Eqs. (9.10d) and (9.10e), two equations are now obtained for V_1 and V_2 in terms of the capacitances C_1, C_2, C_g and the battery potentials V_g and V_d. The solution of these yields the forms [1]

$$V_1 = \frac{1}{C_1 + C_2 + C_g}\left(C_2V_d + C_gV_g - Ne\right) \tag{9.11a}$$

and

$$V_2 = \frac{1}{C_1 + C_2 + C_g}\left(\left(C_1 + C_g\right)V_d - C_gV_g + Ne\right) \tag{9.11b}$$

so that from Eqs. (9.10a) and (9.10b)

$$Q_1 = \frac{C_1}{C_1 + C_2 + C_g}\left(C_2V_d + C_gV_g - Ne\right) \tag{9.11c}$$

and

$$Q_2 = \frac{C_2}{C_1 + C_2 + C_g}\left(\left(C_1 + C_g\right)V_d - C_gV_g + Ne\right). \tag{9.11d}$$

These formulae relate V_1, V_2, Q_1, Q_2 to the external variables V_d, V_g, N and the island capacitances.

As in the earlier example in Section 9.1 for the two junction–island system, the energy contained in the three capacitors in the current system is given by

$$E_C = \frac{1}{2}\left[Q_g\left(V_g - V_1\right) + Q_1V_1 + Q_2V_2\right], \tag{9.12}$$

and, upon substituting from Eq. (9.11), it follows that [1]

$$E_C\left(N, V_d, V_g\right) = \frac{1}{2}\frac{1}{C_1 + C_2 + C_g}\left[C_1C_2V_d^2 + C_1C_gV_g^2 + C_2C_g\left(V_g - V_d\right)^2 + (Ne)^2\right]. \tag{9.13}$$

It is seen from Eq. (9.13) that for fixed V_d and V_g, the energy in the capacitors has a quadratic dependence on N. Again, as in the earlier discussions in Section 9.1, the energy in Eq. (9.13) is only that of the capacitor part of the system, and the energy changes of the batteries V_g and V_d in charging the capacitors must also be accounted for in computing the total energy of the capacitor–battery system. These changes represent the interaction of the island with the outside world.

In the following the energy of the system composed of the island and barrier potentials is computed and then studied for changes in the net charge configuration N on the island. Specifically, the energy changes of the system are considered at fixed V_g and V_d for the case in which the net uncompensated electrons on the island changes by $N \to N+1$ or $N \to N-1$. This allows us to determine the lowest energy configuration in N of the system and how to make alterations in the applied potentials in order to manipulate the flow of electrons through the system. The manipulation of the movement of electrons in the system is the basis for the technological applications of the single-electron transistor [7–14].

9.2.1 Electron Transitions between the Source and the Island

In a first treatment, we consider a transition of the system in which the total number of uncompensated electrons on the island is changed by making electron exchanges with the source. This is done by passing an electron from the negative terminal of the battery V_d onto the island (i.e., on the island $N \rightarrow N+1$) or by passing an electron from the island onto the negative terminal of the battery V_d (i.e., on the island $N \rightarrow N-1$). In the course of these transitions, it should be remembered that the total system is composed of the capacitors and the charging batteries, and the batteries do an amount of work on the capacitors to reach their final charged states. During these changes, the potentials V_d and V_g remain constant, and the transition is made to the lowest energy of the system of batteries and capacitors for a zero-temperature system.

The work done by the batteries during the transitions $N \rightarrow N+1$ and $N \rightarrow N-1$ of the island net uncompensated charge can be approximated by considering the energy required of the batteries to place the charge Q_2 on the plate connected to the positive terminal of the battery V_d and the charge $Q_g = C_g\left(V_g - V_1\right)$ placed on the plate connected to the positive terminal of the battery V_g. This work is given by [1]

$$E_B\left(N,V_d,V_g\right) = Q_g V_g + Q_2 V_d = \frac{Ne}{C_1 + C_2 + C_g}\left(C_g V_g + C_2 V_d\right) + S, \tag{9.14}$$

where on the far-right side of Eq. (9.14), the terms dependent on N have been separated out for consideration. These are the terms of importance to the following discussions, and the remaining terms not involving N are represented by S.

The energy in Eq. (9.14) represents a drain on the chemical energy contained within the battery and is observed to depend linearly on the net charge on the island N. Consequently, the change in energy of the system composed of the battery and capacitors at the end of the charging process is given by [1]

$$E\left(N,V_d,V_g\right) = \frac{1}{2}\frac{1}{C_1 + C_2 + C_g}\left[C_1 C_2 V_d^2 + C_1 C_g V_g^2 + C_2 C_g\left(V_g - V_d\right)^2 + (Ne)^2\right]$$

$$-\frac{Ne}{C_1 + C_2 + C_g}\left(C_g V_g + C_2 V_d\right) - S, \tag{9.15}$$

where the first term on the right represents an increase of the capacitor energy as they are charged, and, the negative sign of the second and third terms represent the loss of energy of the batteries, as they supply energy to the island.

Ignoring temperature effects, at a constant V_d and V_g, the charge configuration N leading to an energy minimum is obtained from the derivative of Eq. (9.15). In this way, the minimum is given by the condition

$$\frac{\partial E\left(N,V_d,V_g\right)}{\partial N} = 0 = \frac{1}{C_1 + C_2 + C_g}\left[Ne - \left(C_g V_g + C_2 V_d\right)\right] \tag{9.16}$$

from which it follows that

$$Ne = C_2 V_d + C_g V_g. \tag{9.17}$$

Consequently, the energy minima of the capacitor–battery system are the integer N solutions of Eq. (9.17), and, by changing V_d and/or V_g, one passes through a series of minima at the integers values $N = \frac{C_2 V_d + C_g V_g}{e}$.

The first process to be studied is the transition of the island charge from N to $N+1$, i.e., an additional electron is tunneled or leaked onto the island from the source. Returning to Eq. (9.15), it is seen that the energy change in the transition $N \to N+1$ is given by

$$E(N+1,V_d,V_g)-E(N,V_d,V_g)=\frac{e}{C_1+C_2+C_g}\left[\left(N+\frac{1}{2}\right)e-C_2V_d-C_gV_g\right]. \tag{9.18}$$

Here, the term in $C_2V_d+C_gV_g$ represents the decrease in energy of the battery during the process, whereas the positive energy term in $\left(N+\frac{1}{2}\right)e$ comes from the change in capacitor energy arising from the change in the charge state of the capacitors.

From the energy difference in Eq. (9.18), it is found that in the case that [1]

$$E(N+1,V_d,V_g)-E(N,V_d,V_g)>0 \tag{9.19a}$$

the system is more stable in the state of N net uncompensated electrons on the island, while in the case that

$$E(N+1,V_d,V_g)-E(N,V_d,V_g)<0, \tag{9.19b}$$

the system is more stable in the state of $N+1$ net uncompensated electrons on the island. Consequently, these inequalities determine the stability of the island under the transition processes involving the addition of an extra electron to the island from the source.

The second type of exchange involving the source is that in which the net charge on the island changes from N to $N-1$, i.e., a single electron is absorbed from the island into the source. Returning to Eq. (9.15), it is seen that the energy change for the transition $N \to N-1$ is given by [1]

$$E(N-1,V_d,V_g)-E(N,V_d,V_g)=\frac{e}{C_1+C_2+C_g}\left[-\left(N-\frac{1}{2}\right)e+C_2V_d+C_gV_g\right]. \tag{9.20}$$

Here the term in $C_2V_d+C_gV_g$ represents the increase in energy of the battery during the process whereas the energy term in $-\left(N-\frac{1}{2}\right)e$ comes from the change in energy arising from altering the uncompensated charge state of the capacitors. For the case that

$$E(N-1,V_d,V_g)-E(N,V_d,V_g)>0, \tag{9.21a}$$

the system is more stable in the state of N, while for the case that

$$E(N-1,V_d,V_g)-E(N,V_d,V_g)<0, \tag{9.21b}$$

the system is more stable in the state of $N-1$.

Consequently, it is seen that the conditions in Eqs. (9.19) and (9.21) govern the exchange of single electrons between the island and the source. In this respect, they are only a part of the description of the state of the source–island–drain system, and an addition consideration is needed to determine the complete state of the drain–island configurations. Once these additional transition conditions are arrived at, the complex motions of single charges through the entire system can be made.

Next, we consider the case of a system in which the island exchanges charge with the drain. Again, a first case of this motion is that in which the state with N net uncompensated electrons on the island transitions to a state of $N-1$ net uncompensated island electrons. Specifically, this transition is made by allowing an electron to pass from the island to the positive V_d battery terminals.

A second case is that in which N electrons on the island transition to a state of $N+1$ island electrons. This transition is made by allowing an electron to pass from the positive V_d battery terminal to the island. After these results are obtained, it will be shown how successively applying both types of transitions (source to island and drain to island) allows, for example, the study of the conditions under which a single electron is made to pass completely from the source to the drain.

9.2.2 Electron Transitions between the Drain and the Island

In a first consideration of the exchange of electrons between the drain and the island, it follows from Eq. (9.13) that the change of capacitor energy for the transition $N \to N-1$ at fixed V_d and V_g is given by [1]

$$E_C\left(N-1,V_d,V_g\right)-E_C\left(N,V_d,V_g\right)=-\frac{\left(N-\frac{1}{2}\right)e^2}{C_1+C_2+C_g}. \tag{9.22}$$

In addition, an estimate of the work done by the batteries for this same transition between the drain and the island is from Eq. (9.14)

$$\begin{aligned} E_B\left(N-1,V_d,V_g\right)-E_B\left(N,V_d,V_g\right) &= \Delta Q_g V_g+\Delta Q_2 V_d+eV_d \\ &= \frac{e}{C_1+C_2+C_g}\left[\left(C_1+C_g\right)V_d-C_g V_g\right], \end{aligned} \tag{9.23}$$

where $\Delta Q_g = Q_g(N-1)-Q_g(N)$ and $\Delta Q_2 = Q_2(N-1)-Q_2(N)$, and eV_d is the work to pass an electron from the negative to the positive terminal of the drain battery. These two energy changes both contribute to the total energy change when an electron is passed from the island to the drain.

In this regard, it is interesting to take a short break in the discussions of the $N \to N-1$ transition, and note in these considerations that introducing a single electron onto the island from the negative terminal of V_d (see Eq. (9.14)) results in the battery doing work $\frac{e}{C_1+C_2+C_g}\left(C_g V_g+C_2 V_d\right)$, while promoting that electron from the island to the positive terminal of V_d requires from Eq. (9.23) a work from the battery of $\frac{e}{C_1+C_2+C_g}\left[\left(C_1+C_g\right)V_d-C_g V_g\right]$. (Note that both of these expressions are independent of N, as they only involve the motion of a single-electron charge.) In addition, the sum of these two energies is eV_d, which is the work required of the battery to completely move the electron between the negative terminal to the positive terminal of the battery V_d. Consequently, after the passage of one electron through the island, the system is returned to its initial charge configuration before the complete passage. It is seen that during the complete transition, only some energy has been lost from the system, and the capacitors are left ready to electrostatically manipulate another electron between the source and drain reservoirs.

Now continuing from Eqs. (9.22) and (9.23), the analysis of the $N \to N-1$ transition will be completed. From Eqs. (9.13), (9.22), and (9.23), it then follows that during the transition $N \to N-1$, the total energy of the system $E\left(N,V_d,V_g\right)=E_C\left(N,V_d,V_g\right)-E_B\left(N,V_d,V_g\right)$ as a function of N has the minima in N given by [1]

$$\frac{\partial E\left(N,V_d,V_g\right)}{\partial N}=0=\frac{1}{C_1+C_2+C_g}\left[Ne-\left(C_g V_g-\left(C_1+C_g\right)V_d\right)\right] \tag{9.24}$$

Note that here the capacitor energy $E_C\left(N,V_d,V_g\right)$ is obtained from Eq. (9.13) for the processes between the island and drain, and the work done by the battery

$E_B(N,V_d,V_g)=\frac{eN}{C_1+C_2+C_g}\left[C_gV_g-(C_1-C_g)V_g\right]+S$ in agreement with the result in Eq. (9.23). In addition, for the $N\to N-1$ transition, the energy change is

$$E(N-1,V_d,V_g)-E(N,V_d,V_g)=\frac{e}{C_1+C_2+C_g}\left[-\left(N-\frac{1}{2}\right)e-(C_1+C_g)V_d+C_gV_g\right]. \quad (9.25)$$

From Eq. (9.24), the minima of the energy in N are the integer solutions of

$$N=\frac{C_gV_g-(C_1+C_g)V_d}{e}, \quad (9.26)$$

and a transition between $N\to N-1$ is not allowed if

$$E(N-1,V_d,V_g)-E(N,V_d,V_g)>0, \quad (9.27a)$$

but is allowed if

$$E(N-1,V_d,V_g)-E(N,V_d,V_g)<0. \quad (9.27b)$$

The considerations of the transitions between the drain and island are next extended to both the case $N\to N-1$ and the case $N\to N+1$, just as in our considerations of the charge transitions between the source and the island. Proceeding by similar argument for the transition $N\to N-1$, it then follows that

$$E(N+1,V_d,V_g)-E(N,V_d,V_g)=\frac{e}{C_1+C_2+C_g}\left[\left(N+\frac{1}{2}\right)e+(C_1+C_g)V_d-C_gV_g\right], \quad (9.28)$$

and a transition between $N\to N+1$ is not allowed if [1]

$$E(N+1,V_d,V_g)-E(N,V_d,V_g)>0, \quad (9.29a)$$

but is allowed if

$$E(N+1,V_d,V_g)-E(N,V_d,V_g)<0. \quad (9.29b)$$

The two conditions in Eq. (9.29) finish the conditions needed to treat all of the various transitions between the island, source, and drain.

Now, the variety of passages through the island as mediated by the drain and gate potentials are handled within the context of Eqs. (9.19), (9.21), (9.27), and (9.29). In this respect, it is of interest to know when a net uncompensated charge can be stably developed on the island, and what voltage changes are needed to move the charges onto and off the island.

The next step in understanding the stable equilibrium charge configurations of the island is made from a study of Eqs. (9.19), (9.21), (9.27), and (9.29). Specifically, the focus will be on understanding the conditions on V_d and V_g so that a net number of electrons N is stably confined on the island.

9.2.3 Stability of N Net Uncompensated Electrons on the Island

The condition that the island contains a stable configuration of N net uncompensated electrons is obtained when Eqs. (9.19a), (9.21a), (9.27a), and (9.29a) are simultaneously satisfied. This precisely defines a finite bounded region of parameters of the system. To understand the nature of the equilibrium, consider the set of these relationships represented graphically in the $V_g - V_d$ plane where the capacitances of the system are fixed. In this scheme, the region of stable N configurations is composed as a diamond area that is centered about the line $V_d = 0$. This is shown schematically in Figure 9.4 by the shaded region.

To understand Figure 9.4 consider the following: From Eqs. (9.19a), (9.21a), (9.27a), and (9.29a), the stable N configurations of the island are contained as a finite area enclosed within the four lines defined by [1]

$$V_d = \left[N + \frac{1}{2}\right]\frac{e}{C_2} - \frac{C_g}{C_2}V_g, \tag{9.30a}$$

$$V_d = \left[N - \frac{1}{2}\right]\frac{e}{C_2} - \frac{C_g}{C_2}V_g, \tag{9.30b}$$

$$V_d = -\left[N - \frac{1}{2}\right]\frac{e}{C_1 + C_g} + \frac{C_g}{C_1 + C_g}V_g, \tag{9.30c}$$

and

$$V_d = -\left[N + \frac{1}{2}\right]\frac{e}{C_1 + C_g} + \frac{C_g}{C_1 + C_g}V_g. \tag{9.30d}$$

Note that in the case that $V_d = 0$, the four equations define a finite segment of the V_g – axis between the end points $V_g = \left[N + \frac{1}{2}\right]\frac{e}{C_g}$ and $V_g = \left[N - \frac{1}{2}\right]\frac{e}{C_g}$, which are equidistance from the center point $V_g = N\frac{e}{C_g}$ of the line segment they define in the region of stable N configurations. The center point $V_g = N\frac{e}{C_g}$ is obtained from Eqs. (9.17) to (9.26) evaluated at $V_d = 0$ for the center of the region of

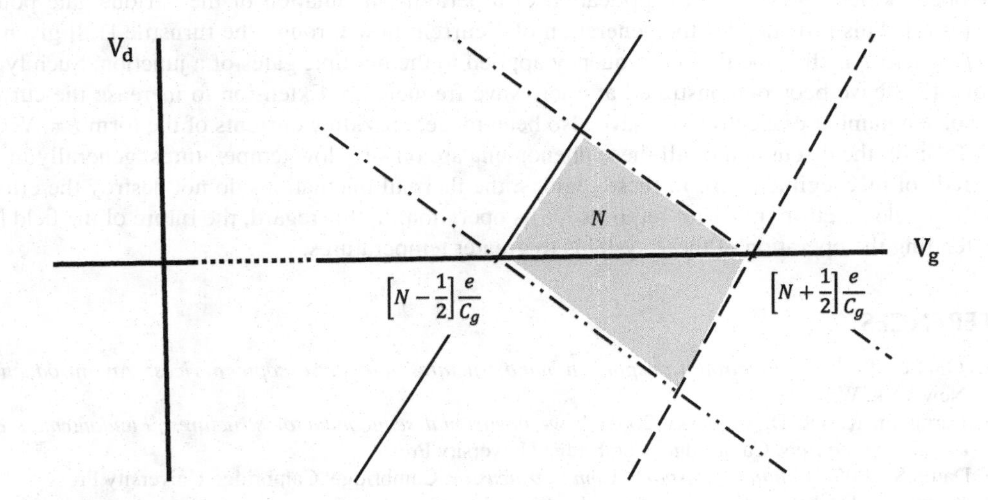

FIGURE 9.4 Schematic plot of V_d versus V_g with a focus on the solution of Eq. (9.30) for the region in which a charge $-Ne$ is stable on the island. The four lines are the stability bounds of the system, and the region of total charge stability is the shaded region.

stability of N. The lines given by Eqs. (9.30b) and (9.30c) define the left sides of the diamond area, while the lines given by Eqs. (9.30a) and (9.30d) define the right side of the diamond area. In this way, the states of stable N configuration of the system are indicated by the shaded area in Figure 9.4. Outside this region, the system of N uncompensated net uncompensated island electrons is unstable and decays to another charge configuration of the island. Consequently, by manipulating either V_g or V_d, the charge on the island is made to remain stable at N or to transition to another change configuration of the island. Note that all the stable N configurations are formed for small $|V_d|$, while increasing $|V_g|$ tends to favor stable configurations of increasing $|N|$.

9.3 APPLICATIONS OF SINGLE-ELECTRON TRANSISTORS

Single-electron devices are a focus in developing low-energy systems, which function in schemes of data storage units and computer logic gates [1–6,12–14]. Of course, in addition to offering low-energy operation, they also provide the element of miniaturization of electronic circuits. This is another important element in the design of computer systems with high-processing capabilities and arises directly from the ability to control the motion of individual electrons. More recently, the manipulations of single electrons have been extended to include magnetic effects related to the electron spin. These controlled motions arise from the ideas of the relatively new field of spintronics [15], which focuses on technological applications of phenomena involving both the electrical and magnetic manipulation of electrons.

In this regard, some developments have been made on NOT and OR gates, which function on the basis of single-electron manipulations and some discussion given of dynamic random access memories [1–5,12–14]. In addition, formulations based on the introduction of more islands and more junctions in the device formulation have also been made. The introduction of more islands helps reduce some of the problems arising from events involving the tunneling of electrons directly from source to drain by a single tunneling process rather that a process involving non-virtual stops on the island. These direct source–drain tunneling events can become important at low temperatures. In addition, multiple gates can be introduced to the formulation of island devices. By the introduction of multiple gate tunnel junctions into the problem, further control of the motion of electrons through the island have been achieved.

As an example of a multiple gate device, recently a so-called single-electron turnstile has been developed, which allows for the application of a periodic modulation of the various gate potentials [7–11]. This provides for the generation of a current flow through the turnstile [1,8] given by $I = ef$, where f is the modulation frequency applied to the multiple gates of a junction. Such types of turnstiles have been demonstrated at microwave frequencies. Extension to increase the current to involve a number of electrons N have also been made, providing currents of the form $I = Nef$.

A focus in the discussion of all these phenomena are on very low temperatures, generally in the hundreds of mK regime [1–5]. In these regions, the thermal fluctuations do not destroy the effects that the single-electron transistor requires for its operation. In this regard, the future of the field lies in extending the operation of these systems to greater temperatures.

REFERENCES

1. Ouisse, T. 2008. *Electron transport in nanostructures and mesoscopic devices: An introduction.* New York: Wiley.
2. Barnham, K. and D. Vedensky 2001. *Low-dimensional semiconductor structures: Fundamentals and device applications.* Cambridge: Cambridge University Press.
3. Datta, S. 2005. *Quantum transport: Atom to transistor.* Cambridge: Cambridge University Press.
4. Heitmann, D. 2010. *Quantum materials: Lateral semiconductor nanostructures, hybrid systems, and nanocrystals.* Heidelberg: Springer.
5. Ryndyk, D. A. 2016. *Theory of quantum transport at nanoscale: An introduction.* Cham: Springer.

6. Streetman, B. G. and S. Banerjee 2000. *Solid state electronic devices, 5th Edition*. New Jersey: Prentice Hall.
7. Takahashi, Y., Ono, Y., Fujiwara, A., et al. 2002. Silicon single-electron devices. *Journal of Physics: Condensed Matter* 14: R995–R1033.
8. Geerling, L. J., Anderegg, V. F., Holweg, P. A. M., et al. 1990. Frequency-turnstile device for single electrons. *Physical Review Letters* 64: 2691–2694.
9. Fulton, T. A. and G. J. Dolan 1987. Observation of single-electron charging effects in small tunnel junction. *Physical Review Letters* 59: 109–112.
10. Zeller, H. R. and I. Giaever 1969. Tunneling, zero-bias anomalies, and small superconductors. *Physical Review* 181: 789–799.
11. Elabd, A. A., El-Sayed, M., El-Rabaie M., et al. 2016. Analysis of co-tunneling effect in single-electronics simulation. *Journal of Computational Electronics* 15: 1351–1360.
12. Datta, S. 2018. How we proposed the spin transistor. *Nature Electronics* 1: 604.
13. Khursheed, R. C. and F. Z. Haque 2019. Review on single electron transistor (SET): Device in nanotechnology. *Austin Journal of Nanomedicine & Nanotechnology* 7: 1055–1066.
14. Goyal, S. and A. Tonk 2015. A review towards single electron transistor (SET). *International Journal of Advanced Research in Computer and Communication Engineering* 4: 36–39.
15. Ehrmann. A. and T. Blachowicz 2019. *Spintronics: Theory, modeling, devices*. Dusseldorf: De Gruyter, Inc.

10 Quantum Hall Effect

An interesting phenomenon found in two-dimensional electron gases at liquid helium temperatures is the quantum Hall effect [1–10]. This involves certain unusual features observed in the magnetoresistance of the gas, which are a direct result of the quantum mechanical nature of the system, and which contradict the expectations for these properties based on a classical electrodynamic description. It is another manifestation in low-temperature physics of the quantum nature of matter, leading to novel properties that are of interest to a variety of technologies. In the case of the quantum Hall effect, the traditional Hall geometry is studied but now for a two-dimensional electron gas. (See Figure 10.1 for the standard Hall effect geometry.) This is a configuration in which the currents and electric potentials of the gas are determined in the presence of a magnetic induction applied perpendicular to the plane of the gas. In general, such two-dimensional electron gases are realized experimentally in a variety of field effect transistors and heterojunction devices [2–10].

The quantum Hall effect is observed in a disordered two-dimensional electron gas when the disorder is such that the Landau levels of the gas are still well defined [1–12]. In general, this requires large magnetic fields and low temperatures. The resulting quantum phenomenon differs considerably from the qualitative properties of the classical electrodynamics of the gas, and the new features of the system allow for important applications in a number of technologies of high-precision metrology, device design, and material science development. The properties of the electron gas focused upon are primarily the conductivities and resistivities that in the absence of the applied magnetic field are those of normal metals or semiconductors. In this regard, due to the two-dimensional nature of the disordered system, Anderson localized [13,14] states are also present in the gas. Some of these localized states, however, have large localization lengths compared to the dimensions of the samples studied. While the localization properties of the states are to an extent diminished by the applied magnetic fields, which violate the time reversal symmetry of the system, localization still forms an important component in understand the physics of quantum Hall systems. Both localization and the dynamics of electrons in an external magnetic field, therefore, enter into the following discussions.

In particular, in the classical electrodynamics picture of a two-dimensional electron gas defined in the $x-y$ plane, and subject to external electric and magnetic fields, the longitudinal components of the resistivity tensor (i.e., ρ_{xx} and ρ_{yy}) are field independent, and the transverse components of the

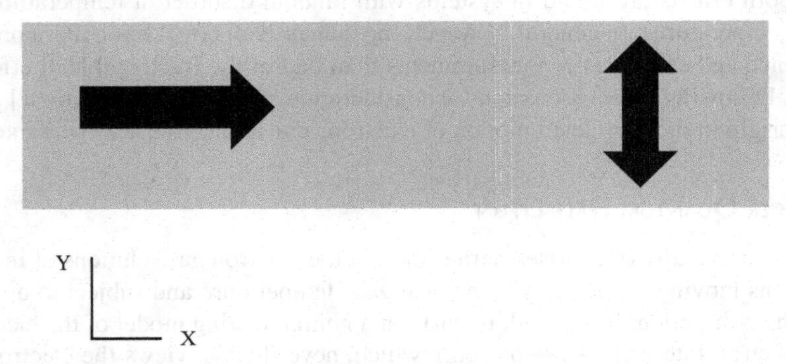

FIGURE 10.1 Standard Hall geometry. A strip of two-dimensional electron gas in the plane of the page is subject to an applied magnetic field perpendicular to the page. Current can flow in the horizontal direction (indicated by the arrow), and a Hall potential is developed vertically across the edges of the strip (indicated by the double arrow). The resistivity of the electron gas is composed of both longitudinal $\rho_{xx}=\rho_{yy}$ and transverse $\rho_{xy}=-\rho_{yx}$ components.

DOI: 10.1201/9781003031987-10

resistivity tensor (i.e., ρ_{xy} and ρ_{yx}) are proportional to the intensity of the applied magnetic induction perpendicular to the $x-y$ plane [1,2]. (See Figure 10.2a for a schematic representation of these properties.) This is a completely different behavior to that found in the quantum system. In the quantum system, the transverse components of the resistivity tensor (i.e., ρ_{xy} and ρ_{yx}) are proportional to the reciprocal of an integer (integer quantum Hall effect) or to the reciprocal of a rational fractional factor (fractional quantum Hall effect). These integers or rational fractions are, in turn, related to the intensity of the applied magnetic induction in the z-direction and to its flux in the $x-y$ plane. Similarly, the longitudinal components of the quantum resistivity tensor (i.e., ρ_{xx} and ρ_{yy}) are no longer constant but are related in a complex way to the magnetic induction and the behavior of the transverse components of the resistivity [1,2].

10.1 TWO QUANTUM HALL EFFECTS

In the integer quantum Hall effect, the transverse quantum resistivity as a function of increasing applied magnetic field exhibits a series of plateaus of constant resistivity [2,7,8]. (See. Figure 10.2b for a schematic representation of this behavior.) On the other hand, the longitudinal quantum resistivities of the integer quantum Hall effect as a function of increasing applied magnetic field exhibits a series of zeros or very small minima over the plateau regions of the transverse resistivity. (See. Figure 10.2b for a schematic representation of this behavior.) These plateaus and minima are intimately related to the quantum numbers of the Landau levels of the electron orbits.

A similar but smaller effect is the fractional quantum Hall effect [2,9,10]. Again, a series of subplateaus are observed in the transverse resistivity at extremely high magnetic fields. These are accompanied in most cases by zeros or minima in the longitudinal resistivity. The plateaus and minima occur as a substructure on some of the plateau and minima features of the integer Hall effect. The origin of the fractional Hall effect is completely different from that of the origin of the integer Hall effect. While the integer Hall effect is completely explained by a model of noninteracting electrons, the fractional quantum Hall effect arises from the electron–electron Coulomb interactions in the two-dimensional system.

The study of the quantum Hall effect then separates into the treatment of the integer quantum Hall effect and the fractional quantum Hall effect [1–10]. These are distinctly different effects in the magnetoresistance of materials, and originate from different physical mechanisms. While the integer quantum Hall effect arises for the most part from the independent motion of single electrons in the absence of electron–electron interactions, the fractional quantum Hall effect involves the correlated motions of electrons in a Fermi liquid [2–10] arising from their electron–electron Coulomb interactions. Both effects are found in systems with random disorder at temperatures of order of liquid helium temperatures. In general, however, the integer Hall effect has a more prominent presence in resistance and conductance measurements than that of the fractional Hall effect. Its theory is treated first. Following these discussions, a consideration is given of the fractional quantum Hall effect and its origin in the correlated motion of electrons due to electron–electron interactions.

10.1.1 INTEGER QUANTUM HALL EFFECT

The integer quantum Hall effect arises in the study of the electron gas solutions of individual independent electrons moving in the $x-y$ plane near zero temperature and subject to an applied magnetic field in the z-direction. It is based, in part, on a noninteracting model of the electron gas (i.e., no electron–electron interactions are present), which, nevertheless, views the electrons as moving in a randomly disordered potential representing impurity scattering in the material [11,12,15]. This scattering causes the electrons to attain a steady state motion described by resistivity and conductivity transport coefficients. In this regard, the bulk of the electrons are grouped into distinct independent Landau levels of the single-particle Landau solutions [2,11,12] of the gas. These solutions were discussed in an earlier chapter (See Chapter 2.), where it was shown that each Landau level

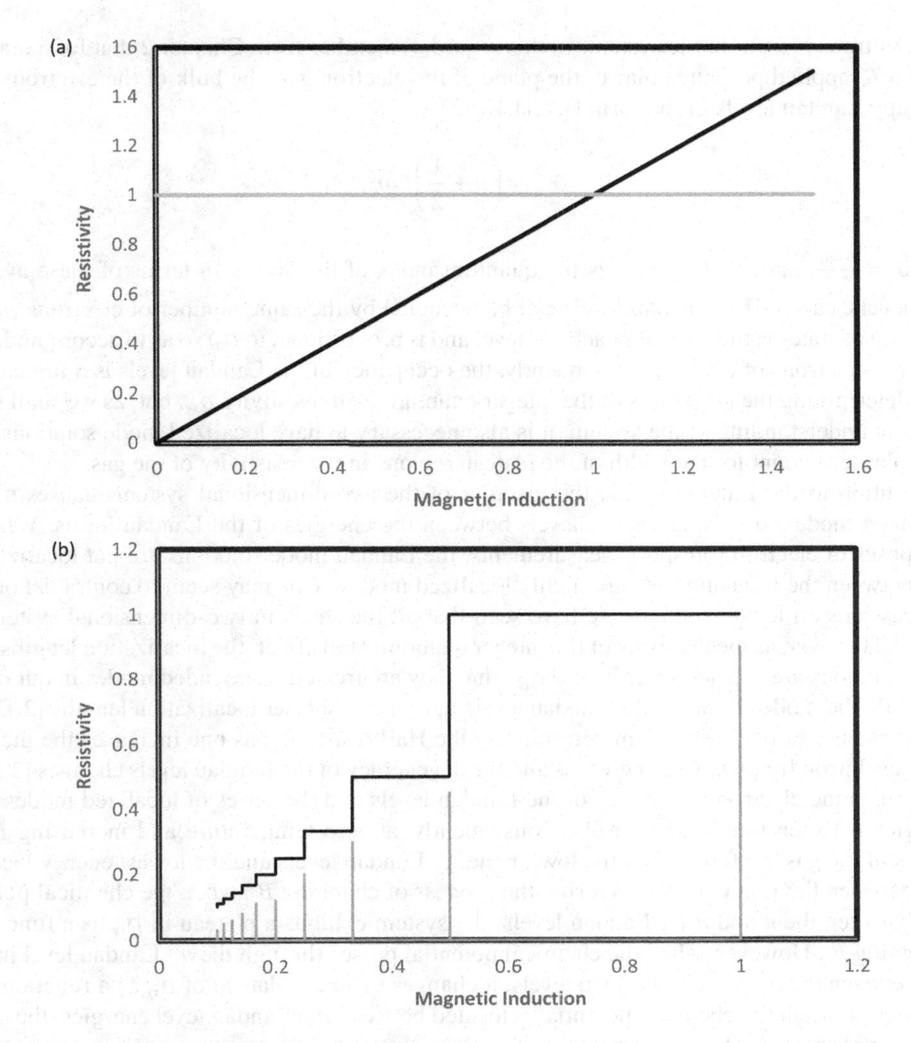

FIGURE 10.2 Highly stylized schematic plot of: (a) Plot of resistivity $(n_e e^2 \tau/m)\, \rho_{xy}$ (black line) or $(n_e e^2 \tau)/m\, \rho_{xx}$ (gray line) versus magnetic induction $e\tau/mc\, B_0$ for the Hall effect of the two-dimensional electron gas in classical electrodynamics. (b) Plot of resistivity $(e^2/2\pi\hbar)\, \rho_{xy}$ (dark stepped line) and $(e^2/2\pi\hbar)\, \rho_{xx}$ (gray spikes) versus $e/(2\pi\hbar c n_e)\, B_0$ for an idealized two-dimensional quantum electron gas. In a real system, the edges of the steps are rounded, and the spiked gray lines are broadened. With the present normalization, $e^2/2\pi\hbar\, \rho_{xy} = 1/n_{\text{total}}$ and $e/(2\pi\hbar c n_e)\, B_0 = 1/n_{\text{total}}$, where $n_e = N_{\text{total}}/(L_x L_y) = n_{\text{total}}\, (eB_0)/2\pi\hbar c$, N_{total} is the total number of electrons in the Landau levels, and n_{total} is the number of completely filled Landau levels in the gas. The plot in (b) is made over the interval $0.1 < e/(2\pi\hbar c n_e)\, B_0 < 1.0$.

contained the same number of electron modes. In addition, the impurity scattering of the gas is strong enough that localized states are formed at energy levels between those of the Landau levels. Both of these types of states (i.e., transport and localized states) enter into the explanation of the observed resistivities and conductivities of the gas.

The integer Hall effect tends to be the dominant effect in the quantum Hall system. It is an effect in which ρ_{xy} exhibits a series of stepped plateaus with increasing magnetic field [2]. (See. Figure 10.2b for a schematic representation of this behavior.) These plateaus depend on the energies of both the Landau and localized energy levels available for electron occupancy in the system and the value of the chemical potential, which sets the occupancy of the energy levels in accommodating all the electrons present in the system. In this regard, remember from Chapter 2 that for a magnetic induction B_0 applied perpendicular to the plane of the electron gas, the bulk of the electrons in the gas occupy Landau levels of the form [1,2,11,12,15]

$$\epsilon_n = \left(n + \frac{1}{2}\right)\hbar\omega_c, \tag{10.1}$$

where $\omega_c = \frac{eB_0}{mc}$, and $n = 0, 1, 2,\ldots$ is the quantum index of the levels. In terms of these available energy levels, each of the Landau levels can be occupied by the same number of electrons (i.e., the degeneracy of states is the same for each $n-$ level and is proportional to B_0) so as to accommodate the bulk of the electrons of the gas. Consequently, the occupancy of the Landau levels is a fundamental feature determining the properties of the integer quantum Hall resistivity ρ_{xy}, but, as we shall see for a complete understanding of the system, it is also necessary to have localized mode solutions in the system. These account for the width of the plateau regions in the resistivity of the gas.

In addition to the Landau levels, the disorder of the two-dimensional systems causes a series of localized modes to exist at energy levels between the energies of the Landau levels. While for the purposes of electron transport measurements, the Landau mode solutions are not localized, the modes between the Landau levels are highly localized modes. This may seem to contradict our earlier discussions on localization, as we have seen that all the states in two-dimensional systems are localized. However, as mechanisms of the integer quantum Hall effect, the localization lengths of the Landau solutions are considered to be so large that they are treated as extended modes in our discussions, while the modes between the Landau levels have much shorter localization lengths [2,13,14].

In the picture of the mechanism determining the Hall resistivity, as one increases the magnetic field the cyclotron frequency ω_c increases and the degeneracy of the Landau levels changes [2,13,14]. This disrupts the electron occupancy of the Landau levels and the series of localized modes found at energies between the Landau levels. Consequently, at zero temperature and increasing B_0, the electrons of the gas are forced into the lowest energy Landau levels and the lowest energy localized modes between the Landau levels. During the process of changing B_0, when the chemical potential occurs between the n and $n+1$ Landau levels, the system exhibits a plateau in ρ_{xy} as a function of the changing B_0. However, when the chemical potential passes through the nth Landau level into the region between the $n-1$ and n Landau levels, it changes to a new plateau of ρ_{xy} as a function of B_0. In this regard, when the chemical potential is located between the Landau level energies, the system exhibits a plateau in ρ_{xy} for changing B_0.

A plateau in ρ_{xy}, consequently, remains independent of the increasing magnetic field, until its electrons begin dropping into lower energy localized mode states below its Landau level. As the localized mode states below the Landau level drop into the next lower Landau level, a new plateau of increased ρ_{xy} is then started, and the plateau is maintained with increasing magnetic field until the next lower Landau level begins to lose electrons. The process is repeated over and again as the system changes its occupancy of the Landau and localized modes, until only a single Landau level remains in the system. During these processes, the longitudinal resistivity ρ_{xx} remains zero while ρ_{xy} is in a plateau but changes with the increasing magnetic field in a rapid spike at the boundary between two different plateaus. The changes in the longitudinal resistivity ρ_{xx} is a related phenomenon to the integer Hall effect known as the Shubnikov-de Hass effect [16] and will form part of the discussions to follow. This, then, is the progression of events found in the integer quantum Hall effect.

10.1.2 Fractional Quantum Hall Effects

The fractional quantum Hall effect is a smaller effect than that of the integer quantum Hall effect [2–10]. Like the integer quantum Hall effect, it is again observed in two-dimensional electron system near zero temperature subject to a perpendicular applied external magnetic field. But it is found in the presence of weaker random disorders than those typically found in the earlier studies of the integer quantum Hall effect and at higher applied magnetic fields than used in those studies. In this regard, the applied magnetic fields are usually large enough that primarily only the lowest (i.e., the $n = 0$) Landau level of electrons are occupied or in some cases the next higher (i.e., the $n = 1$) Landau level. Similar to the integer quantum Hall effect, the fractional quantum Hall effect involves plateaus in the transverse resistivity and occasional peaks in the longitudinal resistivity considered as functions of the applied magnetic field. These features, however, now arise at fractional occupancies of the Landau levels and are shown to be intimately related to the angular momentum states of the electron states in a correlated electron liquid.

Unlike the noninteracting electron model for the integer quantum Hall effect, the fractional quantum Hall effect is based on an essential consideration of the Coulomb interactions between the electrons in the system. The electron–electron interactions cause them to exhibit a correlated motion. The correlations in the electron motions show up in the new plateau features exhibited by the fractional quantum Hall effect in the resistivity versus applied magnetic field behavior of the two-dimensional system.

In the following, we begin with a review of the classical electrodynamic treatment of the classical Hall effect. This is followed by a discussion of the integer quantum Hall effect and finally by a discussion of the fractional quantum Hall effect.

10.2 CLASSICAL MODEL OF THE HALL EFFECT

Begin by reviewing the Hall effect for a two-dimensional electron gas, as treated in the classical electrodynamic Drude model [1,2]. In this model, the individual electron–electron interactions are ignored, so that the electrons only interact with external fields and with a viscous medium representing impurity scattering. In this approach, the two-dimensional system consists of a strip of metal with infinite extent in the x-direction and a finite width L_y in the y-direction. (See Figure 10.3 for a schematic of this geometry.) An external magnetic induction B_0 is applied to the strip in the z-direction, an electric field E_x is applied in the x-direction, and a Hall field E_y is generated in the y-direction. The object of the study is to determine the current in the x-direction in terms of the Hall field, or its potential difference across the width of the strip and the relationship of the current in the x-direction to the field or potential applied in the x-direction.

The dynamics of the individual electrons in the strip is described by the classical single-particle Lorentz force modified to include the viscous drag of a random scattering medium. This is given by [1]

$$m\left(\frac{d}{dt}+\frac{1}{\tau}\right)\vec{v} = -e\left(\vec{E}+\frac{1}{c}\vec{v}\times B_0\hat{k}\right), \tag{10.2}$$

where $\vec{v} = v_x\hat{i} + v_y\hat{j}$ is the velocity of an electron, $e > 0$ is the unit of charge, m is the electron mass, τ is the Drude relaxation time characterizing the viscous drag, and $\vec{E} = E_x\hat{i} + E_y\hat{j}$ is the electric field in the strip. The electrons in the strip move independent of each other, traveling in a random scattering medium. Consequently, their motion in the medium is a steady state, constant-velocity motion.

In the steady state configuration of the strip $\frac{d}{dt}\vec{v} = 0$, and the problem in Eq. (10.2) reduces to a linear algebra, which is solved for the electron drift velocities $\vec{v}$. From these, the current density of the system $\vec{J} = J_x\hat{i} + J_y\hat{j}$ is given in terms of the fields $\vec{E} = E_x\hat{i} + E_y\hat{j}$ and $\vec{B} = B_0\hat{k}$.

The current density of the strip is defined by [1,2]

$$\vec{J} = -n_e e\vec{v}, \tag{10.3}$$

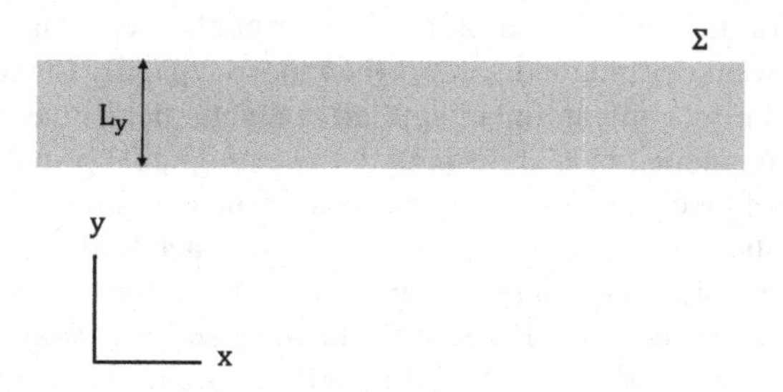

FIGURE 10.3 Schematic plot of the Hall geometry for the electron gas. The infinite strip of the two-dimensional electron gas is shaded, and the magnetic induction B_0 is out of the page. The infinite shaded region of the gas has an area $A \to \infty$, with a surrounding perimeter denoted by Σ.

where n_e is the carrier density of the electrons and $e > 0$. Solving Eq. (10.2) for the electron drift velocities $\vec{v}$ and applying the definition of the current density in Eq. (10.3) gives

$$\begin{vmatrix} J_x \\ J_y \end{vmatrix} = \frac{n_e e^2}{\left(\frac{m}{\tau}\right)^2 + \left(\frac{eB_0}{c}\right)^2} \begin{vmatrix} \frac{m}{\tau} & -\frac{eB_0}{c} \\ \frac{eB_0}{c} & \frac{m}{\tau} \end{vmatrix} \begin{vmatrix} E_x \\ E_y \end{vmatrix} = \begin{vmatrix} \sigma_{xx} & \sigma_{xy} \\ \sigma_{yx} & \sigma_{yy} \end{vmatrix} \begin{vmatrix} E_x \\ E_y \end{vmatrix} \tag{10.4a}$$

where the magnetoconductivities $\sigma_{xx}, \sigma_{xy}, \sigma_{yx}, \sigma_{yy}$ are the components of the two-dimensional antisymmetric conductivity tensor. Alternatively, the results can be expressed in terms of the resistivities of the system. This gives

$$\begin{vmatrix} E_x \\ E_y \end{vmatrix} = \begin{vmatrix} \frac{m}{n_e e^2 \tau} & \frac{B_0}{n_e ec} \\ -\frac{B_0}{n_e ec} & \frac{m}{n_e e^2 \tau} \end{vmatrix} \begin{vmatrix} J_x \\ J_y \end{vmatrix} = \begin{vmatrix} \rho_{xx} & \rho_{xy} \\ \rho_{yx} & \rho_{yy} \end{vmatrix} \begin{vmatrix} J_x \\ J_y \end{vmatrix}, \tag{10.4b}$$

where the magnetoresistivities $\rho_{xx}, \rho_{xy}, \rho_{yx}, \rho_{yy}$ are the components of the antisymmetric resistivity tensor. These two results are used to compute the classical Hall coefficient and strip resistivities that characterize the Hall system.

From the results in Eqs. (10.3) and (10.4), we obtain the Hall coefficient defined in terms of E_y, J_x, B_0, or ρ_{xy} and the elements of the conductivity tensors. In this way, it is found that the Hall coefficient R_H is reduced to an expression in terms of the carrier density n_e, being written in the form

$$R_H \equiv \frac{E_y}{J_x B_0} = -\frac{\rho_{xy}}{B_0} = \frac{\sigma_{xy}}{B_0\left(\sigma_{xx}\sigma_{yy} - \sigma_{xy}\sigma_{xy}\right)} = -\frac{1}{n_e ec}. \tag{10.5}$$

Note that from Eq. (10.5), the Hall coefficient is directly determined by the ρ_{xy} component of the resistivity tensor of the gas or alternatively by the study of its conductivity tensor. This will be a focus in our later considerations of the full quantum mechanical system. As an additional point for later considerations, it follows that the diagonal components of the resistivity are field-independent constants of the form

$$\rho_{xx} = \rho_{yy} = \frac{m}{n_e e^2 \tau}, \tag{10.6}$$

where the relaxation time τ increases with the increasing purity of the system. Consequently, the values of ρ_{xx} and ρ_{yy} decrease as the system is purified. Later, ρ_{xx} and ρ_{yy} of the classical theory are shown to exhibit a difference in behavior from their results in the quantum system.

A schematic of the general behavior of ρ_{xy} and ρ_{xx} as a function of B_0 is shown in Figure 10.2a. In this plot, the longitudinal resistivity ρ_{xx} is represented as a constant independent of B_0, while the transverse resistance ρ_{xy} increases linearly with an increase field B_0. These simplistic behaviors are not found in the full quantum mechanical treatment.

10.3 THEORY OF THE INTEGER QUANTUM HALL EFFECT

The focus in this section is on developing an understanding of the basic conductivity properties of the integer quantum Hall effect [2,7,8]. In this regard, the discussions to follow center on a treatment of a model of spinless two-dimensional electrons treated as independent particles. A consideration of the spin of the electrons is a complication in the problem, which is not necessary to understand the mechanisms at play in the quantum Hall effect. Consequently, it is ignored here. In addition, the classical treatment just given makes no account of the electron intrinsic magnetic moments and is easiest compared to the results for a spinless quantum mechanical model of electrons. Later some discussion of spin effects is made.

In the quantum model, the electrons are constrained to move in the $x-y$ plane in a strip of sides $L_x \to \infty$, $L_y < \infty$ subject to a uniform magnetic field $B_0\hat{k}$ in the $z-$ direction and a uniform electric field $E_0\hat{j}$ in the y-direction [15]. (See Figure 10.3 for a schematic of this geometry.) The single-particle electron modes in the quantum mechanical approach are now treated in the context of the Landau solution for a single electron moving in an external field. This solution was discussed in an earlier chapter of the book but will be briefly reviewed here.

The Hamiltonian for the single-electron orbits in the system is given by [1,2,11,12,15]

$$H = \frac{1}{2m}\left[\vec{p} + \frac{e}{c}\vec{A}\right]^2 + eE_0 y, \tag{10.7}$$

where $\vec{A} = -B_0 y\hat{i}$, $B_0\hat{k} = \nabla \times \vec{A}$, and $e > 0$. This is essentially the electron Hamiltonian of the Landau problem with an added electric field E_0 in the y-direction. The additional electric field term is not a difficulty, as we shall see that it can be transformed away by a change of y-coordinate. From Eq. (10.7), the resulting quantum mechanical eigenvalue problem becomes

$$H\psi_{\vec{k},n}(x,y) = \frac{1}{2m}\left\{\left[\vec{p} + \frac{e}{c}\vec{A}\right]^2 + eE_0 y\right\}\psi_{\vec{k},n}(x,y) = E_{\vec{k},n}\psi_{\vec{k},n}(x,y). \tag{10.8}$$

A solution of Eq. (10.8) is obtained by taking a wavefunction of the general form

$$\psi_{\vec{k},n}(x,y) = e^{ikx}\phi_{k,n}(y - y_0) \tag{10.9}$$

with

$$y_0 = \frac{1}{\omega_c}\left[\frac{\hbar k}{m} - \frac{cE_0}{B_0}\right], \tag{10.10a}$$

$$\omega_c = \frac{eB_0}{mc}, \tag{10.10b}$$

and substituting it in Eq. (10.8). Following some mathematics, it is found that $\phi_{k,n}(y-y_0)$ in Eq. (10.9) is generated from the solutions of

$$\frac{1}{2m}\left[p_y^2+\left(\frac{eB_0}{c}\right)^2 y^2\right]\phi_{k,n}(y)=\left(n+\frac{1}{2}\right)\hbar\omega_c\phi_{k,n}(y), \tag{10.11}$$

which is in the form of a one-dimensional harmonic oscillator, and $n = 0, 1, 2,..$ represents the oscillator modes. In terms of the solutions of Eq. (10.11), the eigenmodes of Eq. (10.8) have wavefunctions given by Eq. (10.9), which are characterized by the quantum numbers k, n and correspond to eigenvalues of the form

$$E_{\vec{k},n}=\left(n+\frac{1}{2}\right)\hbar\omega_c+\frac{1}{2}m\left(\frac{cE_0}{B_0}\right)^2+eE_0\,y_0. \tag{10.12}$$

Note that the functions $\phi_{k,n}(y)$ are those of the harmonic oscillator in Eq. (10.11), and in Eq. (10.9) they are shifted along the y-axis by y_0.

The boundary conditions on the wavefunction of Eq. (10.8) are that in the x-direction periodic boundary conditions are applied to e^{ikx}, and in the y-direction the electrons are contained within the strip. In the x-direction, this yields the restrictions [11,12,15]

$$k=\frac{2\pi}{L_x}n_x, \tag{10.13a}$$

where n_x is an integer and $L_x \to \infty$. These planewave forms are chosen to aid in treating the transport properties of the strip in the x-direction. In the y-direction, it is seen that for the solutions to be contained within the strip y_0 must be located within the strip so that

$$0 \le y_0 \le L_y, \tag{10.13b}$$

or alternatively [2–10]

$$\frac{m}{\hbar}\frac{cE_0}{B_0}\le k\le\frac{m}{\hbar}\left[\frac{cE_0}{B_0}+\omega_c L_y\right] \tag{10.13c}$$

These two sets of conditions then completely specify the wavefunctions in Eq. (10.8).

Applying the results in Eq. (10.13), it follows that the number of states in each harmonics oscillator (Landau) energy level n can be computed. Specifically, for the Landau level with index n the number of possible electron states is given by

$$N_n=\int\frac{L_x dk}{2\pi}, \tag{10.14a}$$

where N_n is the number of modes in the level n, and the integration is over the region defined by (10.13c). In this way, it is found that

$$N_n=\frac{L_xL_y}{2\pi}\frac{eB_0}{\hbar c}. \tag{10.14b}$$

From the result in Eq. (10.14b), it is then seen that the number of modes in each Landau level increases linearly with increasing B_0, and each Landau level n contains the same number of modes. Consequently, the density of electrons in the filled Landau levels per area n_e is given by

$$n_e = n_{\text{total}} \frac{eB_0}{2\pi\hbar c}, \tag{10.14c}$$

where n_{total} is the number of completely filled Landau levels in the gas.

These results for the modal occupancy are essential to understanding the transverse conductivity. This is now addressed.

10.3.1 Transverse Conductivity

In order to determine the Hall conductivity, we study the σ_{xy} component of the conductivity tensor [2]. This relates the current flow in the x-direction to the field applied in the y-direction. Specifically, for the electrons in the nth Landau level, the current in the x-direction is given by

$$I_x = -e\frac{1}{L_x}\int \frac{L_x dk}{2\pi} v_k(k), \tag{10.15}$$

where $v_k(k)$ is the velocity in the x-direction of the electron with wavevector k, and the region of integration is again defined in Eq. (10.13c).

The integral in Eq. (10.15) can be changed to an integral over y_0 by noting that from Eq. (10.10a)

$$dy_0 = \frac{\hbar}{m\omega_c} dk, \tag{10.16}$$

and $v_k(k)$ in Eq. (10.15) is obtained from the dispersion relation given in Eq. (10.12) by a standard relationship

$$v_k(k) = \frac{1}{\hbar}\frac{\partial E_n}{\partial k} = \frac{1}{\hbar}\frac{\partial eE_0 y_0}{\partial k} = \frac{cE_0}{B_0}. \tag{10.17}$$

(Note: The result in Eq. (10.17) is readily seen to be true for the free electron gas where $\frac{1}{\hbar}\frac{\partial}{\partial k}\frac{(\hbar k)^2}{2m} = \frac{\hbar k}{m} = v_k(k)$.) Using Eqs. (10.16) and (10.17) in Eq. (10.15) then gives

$$I_x = j_x L_y = -e\int \frac{1}{2\pi}\frac{m\omega_c}{\hbar} dy_0 \frac{cE_0}{B_0} = -\frac{e^2}{2\pi\hbar}(E_0 L_y), \tag{10.18a}$$

relating the current down the x-axis of the strip to the potential across the strip. Consequently,

$$j_x = -\frac{e^2}{2\pi\hbar} E_0 = \sigma_{xy}^n E_0 \tag{10.18b}$$

where σ_{xy}^n is the transverse conductivity of the nth Landau level. Specifically, Eq. (10.18b) is the contribution from the electrons in the nth Landau level, and the total contribution of the electron gas is obtained by summing over all the fully occupied Landau levels in the gas. The total conductivity of all the filled Landau levels then is given by

$$\sigma_{xy} = -n_{\text{total}} \frac{e^2}{2\pi\hbar}, \tag{10.18c}$$

where n_{total} is the total number of occupied Landau levels in the gas.

As a final point of interest in the approach given in Eqs. (10.7)–(10.18), a generalization in the second term on the righthand side of Eq. (10.7) to treat a less-specific form of the potential across the width of the slab can be made, which also leads to the result in Eq. (10.18c) [6]. Specifically, the second term in Eq. (10.7) given by the form eE_0y can be replaced by a general confining potential $U(y)$. (This is made, for example, in the spirit of the harmonic term $\frac{1}{2}m\omega_0^2y^2$ used in the treatment of the problem of the confined system in Eq. (2.32) of Chapter 2.) The general confining potential offers a little more realistic approach to how the electrons interact with the edges of the slab. In this regard, $U(y)$ is considered as large and positive at the edges of the slab but approximately constant within the bulk of the slab away from its edges. Assuming that the potential changes slowly over the extent of the electron orbital wavefunction, a Taylor series can be made giving

$$U(y) \approx U(y_0) + (y - y_0)\frac{dU(y)}{dy}\bigg|_{y=y_0} = U(y_0) - y_0\frac{dU(y)}{dy}\bigg|_{y=y_0} + y\frac{dU(y)}{dy}\bigg|_{y=y_0} = C + y\frac{dU(y)}{dy}\bigg|_{y=y_0}$$

where y_0 is the center of the orbital wavefunction, and C is a constant that is absorbed into the eigenenergy of the mode. Consequently, taking $eE_0 = \frac{dU(y)}{dy}\Big|_{y=y_0}$ in Eq. (10.7) and shifting the eigenenergy by $-C$, the above calculations in Eqs. (10.8)–(10.18) can be repeated. The result of these then follows by replacing $eE_0 = \frac{dU(y)}{dy}\Big|_{y=y_0}$ in Eq. (10.18a), which now gives the current in the x-direction as $I_x = -\int \frac{1}{2\pi}\frac{m\omega_c}{\hbar}dy_0\frac{c}{B_0}\frac{dU(y)}{dy}\Big|_{y=y_0} = -\frac{1}{2\pi\hbar}e\left[U(L_y) - U(0)\right] = -\frac{e^2}{2\pi\hbar}V_H$, where V_H is the Hall voltage across the slab. From this, the result in Eq. (10.18c) then follows.

10.3.1.1 Another Approach to the Conductivity

An interesting point is that the result in Eq. (10.18) can be obtained using another method, which involves the direct application of Faraday's law to the strip [2]. To see this, we apply Faraday's law in its integral form to the strip in Figure 10.3.

Written in integral form the Faraday's law becomes

$$\oint \vec{dl} \cdot \vec{E} = -\frac{1}{c}\frac{d\Phi}{dt}. \tag{10.19a}$$

Here the line integral is taken around the contour in Figure 10.3, which is the perimeter of the entire strip, and

$$\Phi = B_0 A \tag{10.19b}$$

is the flux of the applied field through the contour. Note that in Eq. (10.19) we consider a rectangular portion (shaded) of the strip in Figure 10.3 of area A and a contour Σ, which is its total perimeter.

For the geometry in Figure 10.3, the current density in the strip flows along the x-direction and is related to the Faraday electric field generated in the y-direction by

$$J_x = \sigma_{xy}E_y. \tag{10.20a}$$

If the current density is integrated around the contour Σ, it follows from Eq. (10.19) that

$$\oint \vec{dl} \cdot \vec{E} = \frac{1}{\sigma_{xy}}\oint \vec{dl} \cdot J_x\hat{j} = -\frac{1}{c}\frac{d\Phi}{dt}. \tag{10.20b}$$

This integral can be rewritten using the charge–current conservation relation (i.e., the continuity equation $\nabla \cdot \vec{J} + \frac{\partial \rho}{\partial t} = 0$) and put into the form

$$\frac{\partial Q}{\partial t} = \sigma_{xy} \frac{1}{c} \frac{d\Phi}{dt}, \tag{10.21}$$

where Q is the total charge of the electrons in the Landau levels contained within the shaded rectangle in Figure 10.3. In this view, a charge dQ is developed in the shaded region by a change of flux $d\Phi$ through the region. Consequently, it then follows from Eq. (10.21) that

$$\sigma_{xy} = c \frac{dQ}{d\Phi}. \tag{10.22}$$

Note, in addition, that from Eq. (10.14b), the total number of electrons in the nth Landau level is given by

$$N_n = \frac{L_x L_y}{2\pi} \frac{eB_0}{\hbar c} = \frac{1}{2\pi} \frac{e\Phi}{\hbar c} = \frac{\Phi}{\Phi_e} \tag{10.23}$$

with

$$\Phi_e = \frac{2\pi\hbar c}{e}, \tag{10.24}$$

defined as a unit of quantum flux. This relates the total number of electrons in the nth Landau level to the flux through the system. Consequently, the total electronic charge contained in all the Landau levels is then from Eqs. (10.23) and (10.24) given by

$$Q = -en_{\text{total}} \frac{\Phi}{\Phi_e}, \tag{10.25}$$

where n_{total} is again the number of completely occupied Landau levels in the gas so that

$$\frac{dQ}{d\Phi} = -\frac{en_{\text{total}}}{\Phi_e} = -n_{\text{total}} \frac{e^2}{2\pi\hbar c}. \tag{10.26}$$

We then have from Eqs. (10.22) to (10.26) that the transverse conductivity of the gas is given by

$$\sigma_{xy} = -n_{\text{total}} \frac{e^2}{2\pi\hbar}, \tag{10.27}$$

and by applying similar considerations to the other off-diagonal component $\sigma_{yx} = -\sigma_{xy}$.

We shall soon see that at the conductivity represented by Eq. (10.27), the resistivity of the gas is given by $\rho_{xy} = -\frac{1}{\sigma_{xy}}$ and $\rho_{yx} = -\frac{1}{\sigma_{yx}}$, and these resistivities also form plateau regions as functions of B_0. The plateau levels are obtained from Eq. (10.27) and the length in B_0 of each resistivity plateau is determined by the range of B_0 over which the chemical potential remains between two neighboring Landau levels. Also, we shall see that in a plateau where ρ_{yx} is constant, the longitudinal resistivities $\rho_{xx} = \rho_{yy} = 0$. These properties are a result of the Shubnikov-de Hass effect. This effect is now discussed.

10.3.2 Shubnikov-de Hass Effect

The Shubnikov-de Hass effect involves the treatment of the longitudinal magnetoresistivity of the two-dimensional electron gas [2,16]. It is closely related to the integer quantum Hall effect and, in fact, is required in part for the explanation of the existence of the resistivity plateaus in the integer quantum Hall effect. These plateaus have been described earlier and are fundamental experimental behaviors observed in the quantum Hall experiment. We now briefly introduce the Shubnikov-de Hass effect in the context of obtaining an explanation for the Hall resistivity plateaus.

First consider the relationship between the conductivities and resistivities of a general two-dimensional medium. The Hall conductivity of the system is an antisymmetric tensor of the form

$$\ddot{\sigma} = \begin{vmatrix} \sigma_{xx} & \sigma_{xy} \\ \sigma_{yx} & \sigma_{yy} \end{vmatrix} = \begin{vmatrix} \sigma_{xx} & \sigma_{xy} \\ -\sigma_{xy} & \sigma_{yy} \end{vmatrix}, \tag{10.28a}$$

which is related to the antisymmetric resistivity tensor $\ddot{\rho}$ by

$$\ddot{\rho} = \begin{vmatrix} \rho_{xx} & \rho_{xy} \\ \rho_{yx} & \rho_{yy} \end{vmatrix} = \left[\ddot{\sigma}\right]^{-1} = \frac{1}{\sigma_{xx}\sigma_{yy} + \sigma_{xy}\sigma_{xy}} \begin{vmatrix} \sigma_{yy} & -\sigma_{xy} \\ \sigma_{xy} & \sigma_{xx} \end{vmatrix}. \tag{10.28b}$$

Consequently, the transverse Hall resistivity is given from the conductivity results by

$$\rho_{xy} = -\frac{\sigma_{xy}}{\sigma_{xx}\sigma_{yy} + \sigma_{xy}\sigma_{xy}} \tag{10.29}$$

and requires a knowledge of all the components of the conductivity tensor for its complete understanding. In addition to the result in Eq. (10.27), a knowledge of σ_{xx} and σ_{yy} is, therefore, required.

In this regard, experimentally we know that when ρ_{xy} as a function of B_0 is in a constant plateau region, $\sigma_{xx} = \sigma_{yy} = 0$. The theoretical reason for this is that ρ_{xy} exhibits a plateau when the chemical potential of the electrons is located within a region of localized, nonconductive modes. Specifically, these modes occur at energies between two neighboring Landau levels of the gas. For this configuration of the chemical potential, all the lower energy Landau level states are completely occupied, and the electron Fermi energy is positioned within a band of localized electron modes. Consequently, no longitudinal transport exists in the electron gas, either from the completely occupied modes of the Landau level or from the localized modes. This result is essentially the Shubnikov-de Hass effect [2].

As an interesting point in this respect, note that an unusual property of the system arises in the context of these discussions. This comes from Eq. (10.28), where it is seen that, when the diagonal components of either the resistivity or conductivity tensors are zero, the diagonal components of the other tensor are also zero. This property holds of course provided that the determinants of the matrices are nonzero. Consequently, under these circumstances, the system can exist in a state in which both the longitudinal conductivities and resistivities are zero. In terms of its longitudinal properties, then the material has no resistance and no conductance.

From Eqs. (10.27) to (10.29), it is then found for $\sigma_{xx} = \sigma_{yy} = 0$ that [2,16]

$$\rho_{xy} = -\frac{\sigma_{xy}}{\sigma_{xx}\sigma_{yy} + \sigma_{xy}\sigma_{xy}} = -\frac{1}{\sigma_{xy}} = \frac{1}{n_{total}}\frac{2\pi\hbar}{e^2}, \tag{10.30}$$

where $n_{\text{total}} = 1, 2, 3, \ldots$ are the number of the of totally occupied Landau levels in the system. These are the plateaus in the transverse resistivities and conductivities, illustrated earlier in Figure 10.2, and their related zero longitudinal conductivities and resistivities.

TABLE 10.1
Summary of Conductivity and Resistivity for Classical and Quantum Gases

	Classical	Integer Quantum
$\sigma_{xx} = \sigma_{yy}$	$\dfrac{n_e e^2}{\left(\dfrac{m}{\tau}\right)^2 + \left(\dfrac{eB_0}{c}\right)^2} \dfrac{m}{\tau}$	0
$\sigma_{xy} = -\sigma_{yx}$	$-\dfrac{n_e e^2}{\left(\dfrac{m}{\tau}\right)^2 + \left(\dfrac{eB_0}{c}\right)^2} \dfrac{eB_0}{c}$	$-n_{total} \dfrac{e^2}{2\pi\hbar}$
$\rho_{xx} = \rho_{yy}$	$\dfrac{m}{n_e e^2 \tau}$	0
$\rho_{xy} = -\rho_{yx}$	$\dfrac{B_0}{n_e ec}$	$\dfrac{1}{n_{total}} \dfrac{2\pi\hbar}{e^2}$

As a general remark on the filling of the Landau levels, it follows from Eq. (10.14c) that for a set of completely filled Landau levels up to and including the nth level, $n_e = (n+1)\dfrac{eB_n}{2\pi\hbar c}$, where B_n is the magnetic field for the filling. Similarly, for completely filled Landau levels up to and including the $(n+1)$th level, $n_e = (n+2)\dfrac{eB_{n+1}}{2\pi\hbar c}$, where B_{n+1} is the magnetic field for the filling. From these two results it is found that for a constant n_e the field at the two filling are related by $\dfrac{1}{n_e} = \dfrac{2\pi\hbar c}{e}\left(\dfrac{1}{B_{n+1}} - \dfrac{1}{B_n}\right)$. The consequences of this filling condition are expected to be observed not only in the resistivity but also in the other physical properties that depend on the occupancies of the Landau levels.

From our earlier discussions and for the convenience of the reader, a summary of the analytical results of the discussions of the forms for the conductivity and resistivity tensors of the classical and quantum gas is now presented in Table 10.1. These summarize the basic results and classical comparisons for the integer Hall effects, and are now briefly discussed.

As a conclusion of the presentations on the integer quantum Hall effect, a brief review and discussion are given of the qualitative features of the results in Figure 10.2b obtained from the forms in Table 10.1. In Figure 10.2b a plot is presented of the resistivity $\dfrac{e^2}{2\pi\hbar}\rho_{xy}$ (dark stepped line) and $\dfrac{e^2}{2\pi\hbar}\rho_{xx}$ (gray spikes) versus $\dfrac{e}{2\pi\hbar c n_e}B_0$ for an idealized two-dimensional electron gas. Here the results in Eqs. (10.14c) and (10.27) are used in normalizing the horizontal and vertical axes in a dimensionless way. In this regard, $\dfrac{e^2}{2\pi\hbar}\rho_{xy} = \dfrac{1}{n_{\text{total}}}$ is plotted with respect to $\dfrac{e}{2\pi\hbar c n_e}B_0$. Our plot in Figure 10.2b is highly idealized, and in a real system the edges of the steps are rounded, and the spiked gray lines are broadened. The rounding and broadening in real systems depends on the disorder charactering the system. In addition, note that in the $B_0 = 0$ limit the longitudinal resistivity becomes that of the quantum electron gas (i.e., $\rho_{xx} = \dfrac{m}{n_e e^2 \tau}$), while the transverse resistivities become zero.

We now turn to a treatment of the fractional quantum Hall effect. In this effect, additional resistivity plateaus are found in the system for which n_{total} in Eq. (10.27) is replaced by some rational fractions $f \leq 1$. The fractional quantum Hall effect is realized in some somewhat smaller stepped plateau features that occur on the plateaus of the integer quantum Hall effect.

10.4 FRACTIONAL QUANTUM HALL EFFECT

The fractional quantum Hall effect occurs in systems with weak random disorder and for extremely high magnetic fields [2–10]. In this regard, the magnetic fields are high enough so that all or most of the modes of the electrons in the system are in the lowest energy set of oscillator modes in the system (i.e., states of the $n = 0$ Landau electron level). It is a smaller effect than the integer quantum Hall effect, and is again observed in types of plateau behavior in the transverse resistivity and types of peak behaviors in the longitudinal resistivity. However, these features now occur at rational fractions of the unit of conductivity rather than at integer multiples of the unit conductivity. Like the integer quantum Hall effect, the fractional quantum Hall effect occurs at extremely low temperatures. Consequently, we will only consider a zero-temperature system.

As we shall see, the new features of the fractional quantum Hall effect arise in the system from the correlated motion of electrons subject to weak residual electron–electron interactions, i.e., the electrons are no longer regarded as a Fermi gas of noninteracting particles. The correlated motion originates in the electron–electron interaction effects of the Coulomb forces between the individual electrons and causes them to behave as a Fermi liquid. For the fractional quantum Hall effect, the residual interactions are treated as small, so that there is only a small correlation of the electron motions in the liquid. Ultimately, however, in the limit of the extreme dominance of the Coulomb electron–electron interactions, the system transforms into a Wigner solid. The Wigner solid is a crystal condensate of electrons formed on a lattice. It is not further discussed here.

Before addressing the theory of the fractional quantum Hall effect, it is necessary to revisit the problem of noninteracting electrons moving in a uniform magnetic field. This problem is now developed in a new set of modes, which are different from those used to treat the integer Hall effect. In the integer Hall effects modes appropriate to handle the effects of weaker magnetic fields on the electron gas were considered. Now modes appropriate to much higher magnetic fields are considered. These are more appropriate to develop an understanding of the origins of electron correlations in the interacting electron system of the fractional quantum Hall effect.

10.4.1 FREE ELECTRONS IN HIGH FIELDS

To understand the fractional quantum hall effect, we begin by discussing the free-particle Schrodinger equation in the limit of a high-intensity uniform external magnetic field and again treat the problem involving spinless electrons, i.e., electron spin effects are ignored. This is done in the context of a different gauge representation of the vector potential than that used in our previous treatments of the integer quantum Hall effect. Specifically, we go back to Eq. (10.8) and consider the uniform magnetic field $\vec{B} = -B_0\hat{k}$ expressed in terms of the vector potential $\vec{A} = -\frac{1}{2}B_0\left[-y\hat{i} + x\hat{j}\right]$. (Note for a mathematical convenience, the uniform magnetic field is taken along the negative z-axis.) In this representation, the vector potential is now written in terms of both x and y coordinates. Within this context, a set of noninteracting electron modes can be developed, which turn out to be very effective in the discussion of high applied fields for systems in which the electrons occupy only the lowest $n = 0$ Landau level. In the following, the focus is directed to systems in which the electron occupancy involves only this lowest Landau level.

For simplicity, in the following only the discussion of the $\frac{1}{3}$ quantum Hall modes are focused upon. This was the first set of fractional modes to be explained. The reader is referred to the literature for other factional states [2–12,15]. Some of these are still in the process of a treatment.

From Eq. (10.8) considered in the limit that the electric field $E_0 = 0$, we then have in terms of the new vector potential the single-electron problem [2]

$$\frac{1}{2m}\left\{\left(\frac{\hbar}{i}\frac{\partial}{\partial x} + \frac{1}{2}\frac{e}{c}yB_0\right)^2 + \left(\frac{\hbar}{i}\frac{\partial}{\partial y} - \frac{1}{2}\frac{e}{c}xB_0\right)^2\right\}\psi_{\vec{k},n}(x,y) = E_{\vec{k},n}\psi_{\vec{k},n}(x,y), \quad (10.31)$$

in which the vector potential components enter into both the x and y derivatives. To facilitate the following discussions, the notation in Eq. (10.31) can be simplified by making a change to new position variables $(\bar{x},\bar{y})$ defined by the relationships

$$x = l_B \bar{x} \tag{10.32a}$$

and

$$y = l_B \bar{y} \tag{10.32a}$$

for $l_B = \sqrt{\frac{2\hbar c}{eB_0}}$. Rewriting the Schrodinger equation in Eq. (10.31) in terms of these new coordinates, it then follows that

$$\frac{\hbar\omega_c}{4}\left\{\left(\frac{1}{i}\frac{\partial}{\partial\bar{x}}+\bar{y}\right)^2+\left(\frac{1}{i}\frac{\partial}{\partial\bar{y}}-\bar{x}\right)^2\right\}\psi_{\bar{k},n}(\bar{x},\bar{y}) = E_{\bar{k},n}\psi_{\bar{k},n}(\bar{x},\bar{y}) \tag{10.33}$$

where $\omega_c = \frac{eB_0}{mc}$ is the cyclotron frequency. The focus is now on the discussion of the modes of the lowest Landau level written in terms of the solutions of Eq. (10.33).

Consider the degenerate solutions of Eq. (10.33) for the lowest $n = 0$ Landau level of electron oscillator states. These are the modes corresponding to the eigenvalue $E_{\bar{k},0} = \frac{\hbar\omega_c}{2}$ and which are the solutions of the resulting differential equation

$$\frac{\hbar\omega_c}{4}\left\{\left(\frac{1}{i}\frac{\partial}{\partial\bar{x}}+\bar{y}\right)^2+\left(\frac{1}{i}\frac{\partial}{\partial\bar{y}}-\bar{x}\right)^2\right\}\psi_m(\bar{x},\bar{y}) = \frac{\hbar\omega_c}{2}\psi_m(\bar{x},\bar{y}). \tag{10.34}$$

From standard considerations of differential equations, it is shown that the general form of the solutions of Eq. (10.34) is given by

$$\psi_m(\bar{x},\bar{y}) = (\bar{x}+i\bar{y})^m e^{-\frac{1}{2}(\bar{x}^2+\bar{y}^2)}, \tag{10.35}$$

where $m \geq 0$ is an integer. The solutions in Eq. (10.35) then form an infinite, linearly independent set of states corresponding to the eigenvalue $\frac{\hbar\omega_c}{2}$. They are a different degenerate set of basis functions than those discussed in Eqs. (10.9)–(10.12) for the integer quantum Hall effect. However, they give just as complete a representation of the properties of the noninteracting Fermi gas as those studied earlier in the context of the integer quantum Hall effect.

Note that the set of polynomials in Eq. (10.35)

$$\left\{(\bar{x}+i\bar{y})^m\right\} \tag{10.36}$$

pre-multiplying the exponential in Eq. (10.35) are an infinite set composed of the linear independent powers of $\bar{x}+i\bar{y}$. This set forms a convenient basis from which a Taylor series for a general analytic function of the form $f(\bar{x}+i\bar{y})$ can be generated. In terms of these analytic functions of $\bar{x}+i\bar{y}$, then, a general solution of Eq. (10.34) is given by the form

$$f(\bar{x}+i\bar{y})e^{-\frac{1}{2}(\bar{x}^2+\bar{y}^2)}. \tag{10.37}$$

It consequently follows that, by an application to the set in Eq. (10.36) of the Gram–Schmidt orthogonalization procedure, a complete set of orthonormal basis states for the eigenvalue $\frac{\hbar\omega_c}{2}$ can be developed as linear combinations of the basis functions in Eq. (10.35).

The linearly independent solutions in Eq. (10.35) have an additional interesting physical property that enters into developing an understanding of the fractional quantum Hall effect. Specifically, they represent single-particle eigenstates of the z-component of angular momentum. This can be seen from a direct application of the angular momentum operator to the states represented in Eq. (10.35). To see this, apply the angular momentum operator in the form [2–10]

$$L_z = -i\hbar\left(x\frac{\partial}{\partial y} - y\frac{\partial}{\partial x}\right) = -i\hbar\left(\bar{x}\frac{\partial}{\partial \bar{y}} - \bar{y}\frac{\partial}{\partial \bar{x}}\right) \tag{10.38}$$

to the wave function in Eq. (10.35). It then follows that

$$L_z = -i\hbar\left(\bar{x}\frac{\partial}{\partial \bar{y}} - \bar{y}\frac{\partial}{\partial \bar{x}}\right)\psi_m(\bar{x},\bar{y})$$

$$= -i\hbar\left(\bar{x}\frac{\partial}{\partial \bar{y}} - \bar{y}\frac{\partial}{\partial \bar{x}}\right)\left[(\bar{x}+i\bar{y})^m e^{-\frac{1}{2}(\bar{x}^2+\bar{y}^2)}\right] = m\hbar\,\psi_m(\bar{x},\bar{y}), \tag{10.39}$$

and $\psi_m(\bar{x},\bar{y})$ is found to be an eigenstate of z-component of angular momentum with eigenvalue $m\hbar$. Note, as a result, that $\psi_m(\bar{x},\bar{y})$ must consequently be composed of states with total angular momenta $\hbar l(l+1) \geq \hbar m(m+1)$. An important point regarding the solutions in Eq. (10.35) can now be made by considering the spatial distribution of these basis functions within the two-dimensional $x-y$ plane.

In this regard, it is of interest to determine the spatial extent or covering in space of each of the basis solutions in Eq. (10.35). To this end, let us compute the position of the maximum in the probability distribution of the wave function

$$\left|\psi_m(\bar{x},\bar{y})\right|^2 = r^{2m}e^{-r^2}, \tag{10.40}$$

where here we use $(\bar{x}+i\bar{y})(\bar{x}-i\bar{y}) = \bar{x}^2+\bar{y}^2 = r^2$. The probability distribution is found to have circular symmetry in the $\bar{x}-\bar{y}$ plane, and the maximum in its spatial distribution is given by the condition

$$\frac{d}{dr}\left|\psi_m(\bar{x},\bar{y})\right|^2 = 2r^{2m}e^{-r^2}\left(\frac{m}{r} - r\right) = 0. \tag{10.41}$$

From Eq. (10.41), it then follows that the maxima of the wave function occur at the radius of the wave function, and this radius is given by

$$r^2 = m. \tag{10.42}$$

From Eq. (10.42), we then find that as the z-component of angular momentum m increases, the diameter in space of the wave function probability distribution increases. This has important consequences for the representation of the spatial distribution of the electron states of the system.

The circular area in the $\bar{x}-\bar{y}$ plane enclosed by the ring of maxima of the m state solution in Eq. (10.40) has an area

$$\bar{A}_m = \pi r^2 = \pi m, \tag{10.43a}$$

where we have used the result in Eq. (10.42). From the length scaling in Eq. (10.32), this translates in the $x - y$ plane to an area given by

$$A_m = \pi(rl_B)^2 = \pi ml_B^2 = \pi m\frac{2\hbar c}{eB_0}. \tag{10.43b}$$

It then follows that

$$m = \frac{eB_0A_m}{2\pi\hbar c} = \frac{B_0A_m}{\Phi_e}, \tag{10.44}$$

where B_0A_m is the flux through the circular ring of maxima of the m state, and $\Phi_e = \frac{2\pi\hbar c}{e}$ is the unit of magnetic flux. (Note that the unit of flux Φ_e earlier entered into our discussions of the integer quantum Hall effect in Eqs. (10.23)–(10.27) and will similarly enter the discussions presented here.) These two fluxes are, consequently, found to be related to one another through the integer m. This has important consequences in regards of the boundary conditions of the solutions in the $x - y$ plane, which are now addressed.

If the electron solution in Eq. (10.35) is contained within an electron gas in the $x - y$ plane, which has an area denoted by A, then A must be the upper limit of the area of the solution's circular ring of maxima. Specifically, in considering the solution m in Eq. (10.35), it is required that

$$A \geq A_m \tag{10.45}$$

for the solution to exist in the gas of area A. Consequently, if

$$\psi_N(\overline{x},\overline{y}) = (\overline{x} + i\overline{y})^N e^{-\frac{1}{2}(\overline{x}^2+\overline{y}^2)} \tag{10.46}$$

is the solution of the highest $m = N$ with an area $A_N = A$ in the $x - y$ plane, it follows from Eq. (10.44), (10.46), and the condition $A_N = A$ that

$$N = \frac{eB_0A_N}{2\pi\hbar c} = \frac{B_0A_N}{\Phi_e} = \frac{\Phi_N}{\Phi_e} = \frac{B_0A}{\Phi_e} = \frac{\Phi}{\Phi_e} \tag{10.47}$$

where Φ_N is the flux through the area of the Nth solution in Eq. (10.46), and the flux through the entire electron gas is $\Phi = B_0A = \Phi_N$. The largest value of m that enters the description of the noninteracting gas is then seen to be given by $N = \frac{\Phi}{\Phi_e}$, and the gas can only be composed as linear combinations of the modes in Eq. (10.35) with $m \leq N$.

The results in Eq. (10.47) for the number of modes in the completely filled Landau level agree with our earlier discussions for the number of modes in the nth Landau level obtained in Eqs. (10.23) and (10.24), and which are generated in the context of the integer quantum Hall effect. Note, in addition, that the results in Eqs. (10.23) and (10.24) are based on the wave functions in Eqs. (10.9) and (10.10). This indicates that these two different sets of wave function bases are consistent in their complete representation and treatment of the properties of the 0th Landau level. Either basis can be used to study the properties of the electron gas.

The modes in Eq. (10.35) with $m \leq N$ are now used to study the many-body problem of the two-dimensional system of electrons. In our discussions, first the properties of the noninteracting electron gas are developed. This is followed by a treatment of a system of electrons with electron–electron interactions and a presentation of a full discussion of the fractional quantum Hall effect.

10.4.2 Wave Function for the Fractional Quantum Hall Effect

Consider the integer quantum Hall effect in which all of the states of the $n = 0$ Landau level are completely occupied by a set of M electrons, i.e. $N = M$ in Eq. (10.46). In the absence of electron–electron interactions, the general form of the many-body wave function for M electrons in the representation based on the solutions in Eq. (10.35) is given by [2,9,10]

$$\Psi(z_0, z_1, \ldots\ldots, z_{M-1}) = f(z_0, z_1, \ldots\ldots, z_{M-1}) e^{-\frac{1}{2}\left[|z_0|^2 + |z_1|^2 + \ldots\ldots + |z_{M-1}|^2\right]}, \tag{10.48}$$

where the position coordinates of the particles in the gas are $z_j = x_j + iy_j$ with $j = 0, 1, 2, \ldots, M-1$. This is a direct generalization of the form in Eq. (10.37), where $f(z_0, z_1, \ldots\ldots, z_{M-1})$ is an analytic function. In addition, however, now the many-body function $f(z_0, z_1, \ldots\ldots, z_{M-1})$ must obey Fermi–Dirac statistics, being antisymmetric under an interchange of particle coordinates. This last property can be accomplished by writing $f(z_0, z_1, \ldots\ldots, z_{M-1})$ as a Slater determinate of independent-particle eigenstates.

For the linearly independent modes in Eq. (10.35), Eq. (10.48) then takes the form [2]

$$\Psi(z_0, z_1, \ldots\ldots, z_{M-1}) = \frac{1}{\sqrt{Z}} \begin{vmatrix} 1 & 1 & \cdot & \cdot & 1 \\ z_0 & z_1 & \cdot & \cdot & z_{M-1} \\ \cdot & \cdot & \cdot & \cdot & \cdot \\ \cdot & \cdot & \cdot & \cdot & \cdot \\ z_0^{M-1} & z_1^{M-1} & \cdot & \cdot & z_{M-1}^{M-1} \end{vmatrix} e^{-\frac{1}{2}\left[|z_0|^2 + |z_1|^2 + \ldots\ldots + |z_{M-1}|^2\right]}, \tag{10.49}$$

where Z is the normalizing factor for the wave function, and the antisymmetry of the wave function follows from the properties of the determinant. Equation (10.49) can alternatively be rewritten as

$$\Psi(z_0, z_1, \ldots\ldots, z_{M-1}) = \frac{1}{\sqrt{Z}} \prod_{0 \le l < l' \le M-1} (z_l - z_{l'}) e^{-\frac{1}{2}\left[|z_0|^2 + |z_1|^2 + \ldots\ldots + |z_{M-1}|^2\right]}, \tag{10.50}$$

which is obtained by writing out the determinant [2–12,15]. The wave functions in Eqs. (10.49) and (10.50) represent the ground state of the noninteracting particle system and can also be used as a Hartree–Fock wave function for an estimate of the ground state energy of the electron system with weak electron–electron interactions. In this regard, it has been shown to give a good representation of the $n = 0$ Landau level electrons.

For these treatments, an important feature of the wave functions is that their Fermi–Dirac statistics introduces spatial separations between particles arising from the factors of $(z_l - z_{l'})$ multiplying the exponential factor. In this regard, these factors introduce a correlation between the particles characterized by a pair correlation function in the electron density given by [2–12,15]

$$g(r) = \frac{M(M-1)}{n_e^2} \int d^2 z_2 \ldots\ldots \int d^2 z_{M-1} \left|\Psi(0, r, z_2, \ldots\ldots, z_{M-1})\right|^2 = 1 - e^{-\frac{1}{2}r^2}, \tag{10.51}$$

where n_e is the density of electrons. Note that as $r \to 0$ the correlation function goes to zero, whereas it uniformly approaches 1 with increasing r. Consequently, the correlations between the particles arise from their avoidance of one another.

The above discussion considered a filled Landau level of M electrons, i.e., for our discussions here we take a totally filled Landau level to contain the maximum $N = M$ electrons. A partially filled Landau level of $P < M$ states can be formed by considering wave functions composed of a reduced set of states involving the basis solutions in Eq. (10.35). For example, consider the wave function of the form

$$\Psi(z_0, z_1, \ldots\ldots, z_{P-1}) = \frac{1}{\sqrt{Z}} \begin{vmatrix} 1 & 1 & \cdot & \cdot & 1 \\ z_0^q & z_1^q & \cdot & \cdot & z_{P-1}^q \\ \cdot & \cdot & \cdot & \cdot & \cdot \\ \cdot & \cdot & \cdot & \cdot & \cdot \\ \cdot & \cdot & \cdot & \cdot & \cdot \\ z_0^{q(P-1)} & z_1^{q(P-1)} & \cdot & \cdot & z_{P-1}^{q(P-1)} \end{vmatrix} e^{-\frac{1}{2}\left[|z_0|^2 + |z_1|^2 + \ldots\ldots + |z_{P-1}|^2\right]}, \tag{10.52}$$

for $q > 1$ an odd integer and where Z is a normalization. (Note here that only odd integers q are consistent with Fermi–Dirac statistics of the wave function, and not all of the modes in Eq. (10.35) are included in Eq. (10.52).) Here for $P \gg 1$, the largest wave function mode of Eq. (10.35) that can fit into the area A of the gas has $m = q(P-1) \approx qP$, so that from our earlier arguments in Eqs. (10.45)–(10.47), by replacing N in these equations with qP, it follows that

$$qP = \frac{eB_0 A_{qP}}{2\pi\hbar c} = \frac{B_0 A_{qP}}{\Phi_e} = \frac{\Phi_{qP}}{\Phi_e} = \frac{B_0 A}{\Phi_e} = \frac{\Phi}{\Phi_e}, \tag{10.53a}$$

and the number of states in the wave function is given by

$$P = \frac{1}{q}\frac{\Phi}{\Phi_e} = \frac{1}{q}M. \tag{10.53b}$$

We see that there are $1/q$ less electrons than in the totally filled Landau level. Later when including electron–electron interactions in our system, the form developed in Eq. (10.52) will be helpful in understanding the states for the fractional quantum Hall effect.

In the context of the integer quantum Hall effect, the electron–electron interactions were largely ignored. These interactions, however, are fundamental to the physics of the fractional quantum Hall effect. Specifically, in the following, we will study the fractional 1/3 quantum Hall effect, which is a case in which the 1/3 filling of the $n = 0$ Landau level is considered. The focus will be on a set of electrons that weakly interact with each other through their residual Coulomb interactions.

For the study of the fractional 1/3 quantum Hall effect, the lowest Landau level is only partially occupied. In the treatment of this partially occupied level, Laughlin [8–10] proposed a new type of variational wave function to calculate the properties of the partially filled Landau level based on the features found in the wave functions of Eqs. (10.50) and (10.52). The idea is to make a direct generalization of Eq. (10.52) to treat the 1/3 filled $q = 3$ ground state of weakly interacting particles. In the generalization, an attempt is also made to introduce correlations into the wavefunction, so that the particles tend to avoid being in proximity of one another. In this regard, the ground state wave function for a Hartree–Fock-like treatment is taken to have the form

$$\Psi(z_0, z_1, \ldots\ldots, z_{P-1}) = \frac{1}{\sqrt{Z}} \prod_{0 \le l < l' \le P-1} (z_l - z_{l'})^3 e^{-\frac{1}{2}\left[|z_0|^2 + |z_1|^2 + \ldots\ldots + |z_{P-1}|^2\right]}, \tag{10.54}$$

where Z is the normalization of the wave function, and $P = \frac{1}{3}M$ for a 1/3 occupancy of the Landau level. Consequently, our considerations of Eq. (10.52) for the case $q = 3$ are a motivating factor for a representation of the 1/3 occupied Landau level, and the factors of $(z_l - z_{l'})^3$ are introduced in analogy with Eq. (10.50), as they tend to spatially separate the particles.

The mode in Eq. (10.54) forms the basis of a Hartree–Fock type of variational treatment for the ground state of the 1/3 fractional quantum Hall effect. Before we discuss this, we first look at some of the properties of the wavefunction in Eq. (10.54). Specifically, a consideration of its partial filling of the $n = 0$ Landau level is given.

The wave function in Eq. (10.54) is a generalization of Eqs. (10.49) and (10.50), in which the highest power of z_j entering the prefactor multiplying the exponential is $z_j^{3(P-1)}$. The circular area in the $\bar{x} - \bar{y}$ plane enclosed by the ring of maxima of the $z_j^{3(P-1)}$ solution from Eq. (10.40) has an area [2]

$$\bar{A}_{3P} = \pi r^2 = \pi 3(P-1) \approx \pi 3P, \tag{10.55a}$$

where we have used the result in Eq. (10.42). From the spatial scaling in Eq. (10.32), this translates in the $x - y$ plane to an area given by

$$A_{3P} = \pi(r l_B)^2 = \pi 3 P l_B^2 = \pi 3P \frac{2\hbar c}{eB_0}. \tag{10.55b}$$

It then follows that

$$3P = \frac{eB_0 A_{3P}}{2\pi\hbar c} = \frac{B_0 A_{3P}}{\Phi_e}, \tag{10.56}$$

where $B_0 A_{3P}$ is the flux through the circular ring of maxima of the $3P$ state, and $\Phi_e = \frac{2\pi\hbar c}{e}$ is the unit of magnetic flux. This largest angular momentum state must fit into the area A of the gas, so that $A_{3P} = A$ and the flux through A is

$$\Phi = B_0 A = B_0 A_{3P}. \tag{10.57}$$

Consequently, from Eqs. (10.56) and (10.57), it follows that the wave function represents a filling

$$P = \frac{1}{3}\frac{\Phi}{\Phi_e} = \frac{1}{3}M, \tag{10.58}$$

where for the completely filled Landau level $M = N = \frac{\Phi}{\Phi_e}$, as determined in Eq. (10.47).

10.4.3 The Electron–Electron Interactions

Applying the wavefunction form in Eq. (10.54), a computation of the ground state energy of the 1/3 occupied $n = 0$ Landau level is made using variational techniques in the presence of electron–electron interactions [2–10]. Further excited states of the system in the presence of electron–electron interactions are then obtained as various types of excited modes existing at energies greater than that of the ground state. In this regard, it can be shown that the excited states of the system are separated by energy gaps from the ground state. There are, consequently, a variety of excited energy states involving charge and chargeless excitations introduced to the system, which must be tested for excitation energy gaps. This determination is made using analytic and computer simulation methods [2–12,15].

In addition to these excited states in the absence of a random disorder, in the presence of a weak random disorder localized modes are introduced in the energy gaps between these modes [2–10]. This situation is then much like that observed earlier in the integer quantum Hall effect, where energy levels of localized modes were found in the energy gaps between the Landau levels, and which arise from similar mechanisms of random disorder. The function of these localized modes in the fractional quantum Hall effect is similar to that of the localized modes in the integer quantum Hall effect.

The ground state of the 1/3 occupied $n = 0$ Landau level is then found to be stable relative to the types of excitations just discussed [2–10]. As a result, a current flow is developed in the 1/3 occupied $n = 0$ Landau level but is absent from the localized energy modes. Due to these characteristics of the modes of the system, the current of the fractional quantum Hall effect exhibits a characteristic ρ_{xy} resistivity plateau and an associated region of $\rho_{xx} = 0$ as functions of the field B_0. This plateau and zeros of the fractional quantum Hall effect are present arising from the same mechanisms involving localized modes as have been discussed earlier in the context of the integer quantum Hall effect [2–12,15].

10.4.4 Conductivity and Resistivity of the $\frac{1}{3}$ Fractional Hall State

To see how the fractional resistivity is accounted for in terms of the ground state wave function, we revisit the discussion of the transverse resistivity for the integer quantum Hall effect now applying it to the fractional quantum Hall effect. For these considerations again, we treat the geometry in Figure 10.3 in which the current density in the strip flows along the x-direction and is related to the Faraday electric field generated in the y-direction by

$$J_x = \sigma_{xy} E_y. \tag{10.59}$$

We use arguments that are closely based on those for the integer quantum Hall effect. Applying Faraday's law and the continuity equation around the contour Σ, it again follows that

$$\frac{\partial Q}{\partial t} = \sigma_{xy} \frac{1}{c} \frac{d\Phi}{dt}, \tag{10.60}$$

where Q is the total charge of the electrons in the 1/3 occupied $n = 0$ Landau level ground state contained within the shaded rectangle in Figure 10.3. From Eq. (10.60), it then follows that the transverse conductivity is related to the change in flux through the area by

$$\sigma_{xy} = c \frac{dQ}{d\Phi}. \tag{10.61}$$

From our earlier discussions of the fractional Hall effect wave function in Eq. (10.58), the total number of electrons in the 1/3 occupied $n = 0$ Landau level ground is given by

$$P = \frac{1}{3} \frac{\Phi}{\Phi_e}. \tag{10.62}$$

The total electronic charge contained in the 1/3 occupied $n = 0$ Landau level ground state is then obtained from Eq. (10.62) as

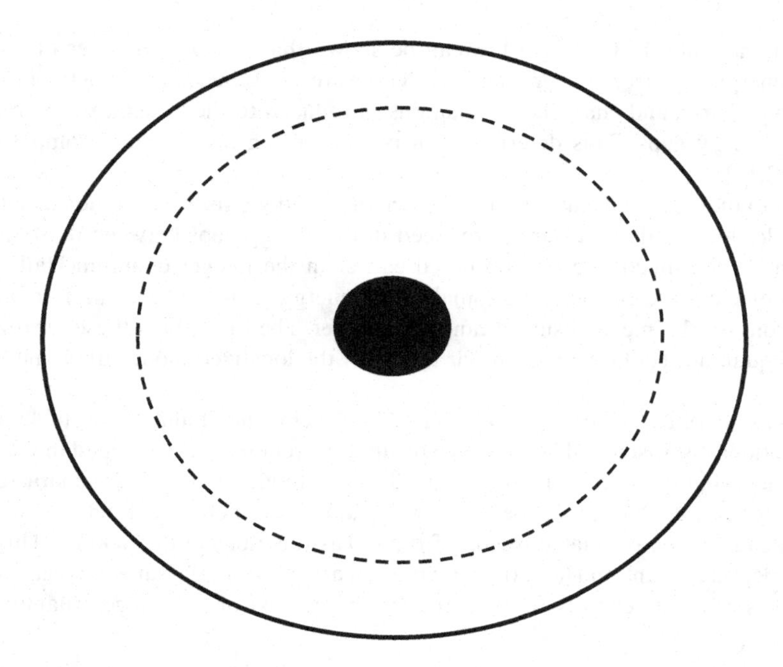

FIGURE 10.4 Schematic for determining the transverse resistivity. The black disc at the center is a solenoid containing a region of uniform magnetic field. Outside the solenoid is a disc of two-dimensional interacting Fermi liquid. By changing the magnetic flux in the solenoid, an azimuthal electric field is generated. This creates a radial current arising from the transverse Hall resistivity that passes charge radially through the dashed surface.

$$Q = -eP = -e\frac{1}{3}\frac{\Phi}{\Phi_e} = -\frac{1}{3}\frac{e^2}{2\pi\hbar c}\Phi. \tag{10.63}$$

Using this result in Eq. (10.61), it follows that

$$\sigma_{xy} = -\frac{1}{3}\frac{e^2}{2\pi\hbar}, \tag{10.64}$$

where now the transverse conductivity involves the fraction 1/3 in place of the integer coefficient found in the integer quantum Hall effect.

Another common variation of the argument just given involves the geometry presented in Figure 10.4. Here, an infinite solenoid containing a uniform magnetic field B perpendicularly pierces a two-dimensional disc of the electron system [2–12,15]. The current flow through the dashed surface within the disc is studied using Eq. (10.60), where Φ is now the flux carried by the solenoid, and $\frac{\partial Q}{\partial t}$ is the current flow out of the dashed region. By a consideration of the charge passed through the surface by the changing flux, again the result in Eq. (10.64) is obtained.

In addition, an elegant interpretation of these later results in terms of factionally charged particles can be made [2–6]. This contains some semblance of the ideas found in the theory of quarks. The reader is referred to the literature for the details [2–10].

10.4.5 Conclusions

The results presented here were a consideration of the fractional 1 / 3 quantum Hall effect. This was the first fractional quantum Hall property to be discovered and explained. Since then, a variety of other factional effects have been discovered. These include, for example, fractional $\frac{1}{3}, \frac{2}{5}, \frac{3}{7}$, etc.

and $\frac{2}{3}, \frac{3}{5}, \frac{4}{7}$, etc. Hall effects. Some of these levels have been explained, but there remain others that have not [2,5]. The discussions regarding these other levels are not treated here, and the reader is referred to the literature for their discussions [3–10].

As another point, the discussions given here of both the integer and factional quantum Hall effects were made for zero temperatures systems. More recent considerations have treated the temperature-dependent behaviors of the quantum Hall effects into regions significantly above liquid helium temperatures. For these recent undertakings, the reader is referred to the literature [17–19]. Such developments at higher temperatures would certainly facilitate the application of quantum Hall effects to device technology. An additional complication to the results presented in the earlier treatments involves the presence of spin on the electrons. This has been ignored in our spinless Fermi discussions. However, the inclusion of the additional energy of the interaction of the electron spins with the applied magnetic field can be included in the discussions. This is done as a modification of the spinless fermion results to include the bookkeeping associated with the shifting energy arising from the spin states of the electrons [2–12,15].

To finish the chapter, some additional brief considerations on the effects of spin on spinless fermion model are made in the context of the quantum Hall effect as well as the Landau diamagnetism treated earlier in Section 2.3.2. This is followed by a summary of some interesting recent work in optics on optical analogies of the quantum Hall effect.

10.5 SPIN

The earlier considerations were based on the spinless fermion model of the electron gas. This model removes the spin energy from the treatment, so that in the real world each particle occurring in the spinless theory must be replaced by either a spin up or a spin down particle. For the treatment of conductivity and resistivity, this is not too much of a difficulty, as the spin energy splits each Landau energy level of the spinless system into a level with all spin up and a level with all spin down electrons [6]. In the determination of the low-temperature conductivity, the transport properties of the system are then determined predominately by the positioning of the chemical potential within the highly degenerate energy levels of the electron levels, which include the splitting due to spin energy. Consequently, as the magnetic field is varied, the change in conductivity in the system is described by the passage of a single energy level through the chemical potential. The resulting change provides a characterization of the essentially quantized nature of the conductivity and resistivity entering the quantum Hall effect [6].

On the other hand, in the context of the Landau diamagnetism discussed in Section 2.3.2, the situation is a little more difficult. In diamagnetism, the focus is on field dependence of the total energy of the conduction electrons of the system, and this involves all of the Landau levels, which in the real world system are split between spin up and spin down states. The paramagnetism and diamagnetism of the gas are then intimately connected to one another. In this picture, an approximation for the diamagnetic contribution can be made by treating two spinless solutions: one for the spin up electrons and one for the spin down electrons. The total diamagnetic contribution of the system is then formed as a sum of the two effects. This approximation is made based on the spinless fermion results in Section 2.3.2.

In regard to the spin interactions of the Fermi gas, the coupling of the electrons to the externally applied magnetic field through the spins are highly dependent on the band structure of the material being approximated [1,2,6]. The strength of the splitting of the Landau levels due to spin effects are large or small, depending on the chemical nature of the system being treated. As well, the Landau level splitting is highly dependent on the effective mass of the material studied. These band structure considerations ultimately explain many of the deviations of observed material properties from the idealized models studied earlier.

10.6 SPIN HALL EFFECT AND THE SPIN HALL EFFECT OF LIGHT

The original theory of the classical Hall effect [1,2] has a number of modifications other than those of the quantum Hall effects. Most of these involve a closer consideration of the details of the interaction of the conduction electrons with their environment in the materials transporting them. In regard to these modifications, shortly after the original discovery of the Hall effect, an anomalous Hall effect [20,21] was found in ferromagnetic materials. In addition, more recently so-called topological Hall effects have been described in the literature [21] and have been a focus of considerable current research. These phenomena generally result from a variety of scattering processes between the conduction electrons and impurities in the system that arise from the interactions of the spin of the electrons with their own orbital angular moment or the orbital angular momentum of the impurity ions in the media. As an added consideration, certain factors originating from the Barry phase associated with the electron motion can enter into the theory of the scattering processes forming the basis of these effects. All these mechanisms account for various aspects of charge separation within a conducting medium.

In addition, an interesting variant of these Hall effects that shall be a focus for us presently is known as the spin Hall effect [22,23]. This involves a transverse spin separation within the media in response to a longitudinal electric field. The resulting spin separation occurs in the absence of the usual Hall charge separation found in a strip of conducting material. The spin Hall effect also has an interesting counterpart in the study of optical systems known as the spin Hall effect of light [24,25]. This optical analogy will be described at the close of this section.

An early extension of the classical Hall effect was the anomalous Hall effect [20,21,23]. This occurs in ferromagnetic material such as *Ni, Co, Fe* , which have a permanent magnetization M arising from the conduction electrons and results from the scattering of the conduction electron spins with impurities. In these materials, the impurity scattering involves the electron spins and can be anisotropic in space so that electrons with different spins are preferentially sent to different (spin-dependent) directions relative to their incident trajectories. A consequence of this for the classical Hall effect is that the transverse resistivity in Eq. (10.14b) is modified by the addition of a magnetization-dependent term. Specifically, it is found that the transverse resistivity is changed to have the form [23]

$$\rho_{xy}(B,M) = R_B B + R_M M, \tag{10.65}$$

where R_B is the coefficient describing the original Hall effect in the $M = 0$ limit, and R_M is a coefficient for the magnetization contribution to the anomalous Hall effect. In this regard, the magnitude of the magnetization term can be much greater than that arising from the term involving only the magnetic induction.

From Eq. (10.65), it is also found that in these systems of permanent magnetization, the Hall effect exists even in the limit that $B = 0$, and results in a transverse Hall electric field and charge separation arising solely from the application of a longitudinal electric field [20–23]. In this limit, the transverse electron separation and resulting Hall field in a slab of finite width are due to the anisotropy in space of the spin-dependent electron scattering. This then accounts for the anomalous Hall effect, but the same discussions enter into an interesting spin separation, which occurs in the systems in the limit that B, $M = 0$. This is now considered.

Related to the anomalous Hall effect is the spin Hall effect [22,23]. The spin Hall effect arises from the limit in which both $B,M \to 0$ in the anomalous Hall systems. In this limit $\rho_{xy} = 0$, so that there is no transverse or Hall electric field and no charge separation. However, the anisotropic spin-dependent scattering is still present in the system so that there is development of a transverse separation of the electron spins [23]. Specifically, for a thin film of width W in the presence of an applied longitudinal electric field along the length of the film, a net concentration of one spin type will occur on the upper edge of the film, and an opposite spin type will occur on the lower edge of

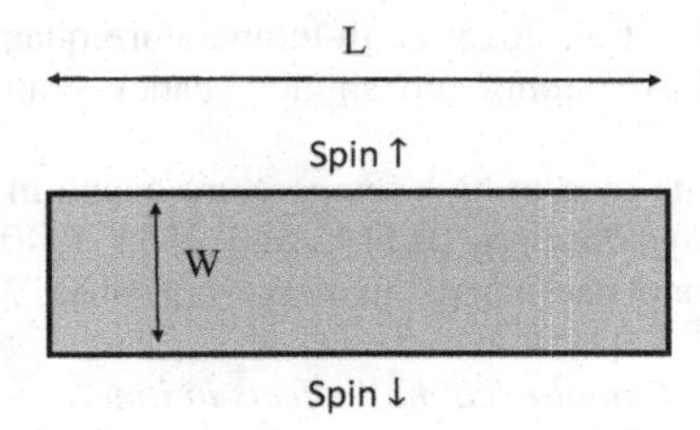

FIGURE 10.5 A thin film of conducting ferromagnetic material considered in the B, $M \to 0$ limit. The film is of transverse width W and length L. Upon application of a longitudinal electric field along the length L of film, a net opposite spin density is developed across the upper and lower edges of the strip, so that on the upper edge a net spin up is developed, and along the lower edge a net spin down is developed.

the film. (See Figure 10.5.) Consequently, in the spin Hall effect, there is a spin density separation across the width of the slab in the absence of a charge separation across the slab.

Recently, an optical analogy of the spin separation has been found in optical systems involving surface reflections and transmission. It is known as the spin Hall effect of light [24,25], and it is related to the difference in surface scattering of different polarizations of light. Specifically, in an interesting experiment, a Gaussian beam of linear polarized light has been reflected from a thin film and has been shown to be reflected into two oppositely circular polarized beams of light that are shifted upward or downward from the plane of incidence dependent on the left- or right-handed nature of their circular polarization [25]. This is the basis of a variety of effects known as the spin Hall effect of light [24,25].

REFERENCES

1. Kittle, C. 1996. *Introduction to solid state physics, 7th Edition*, 393–410. New York: John Wiley & Sons, Inc.
2. Marder, M. P. 2000. *Condensed matter physics*. New York: John Wiley & Sons, Inc.
3. Prange, R. E. and S. M. Girvi 1987. *The quantum Hall effect*. New York: Springer.
4. Avron, J. E., Osadchy, D., and R. Seiler 2003. A topological look at the quantum Hall effect. *Physics Today* 56: 38–42.
5. Shrivastava, K. N. 2002. *Introduction to quantum Hall effect*. New York: Nova Science Publishers, Inc.
6. Tong, D. 2016. Quantum Hall effect. arXiv:1606.06687.
7. von Klitzing, K., Dorda, G., and M. Pepper 1980. New method for high-accuracy determination of the fine-structure constant based on quantum Hall resistance. *Physical Review Letters* 45: 1545–1547.
8. von Klitzing, K. 1986. The quantum Hall effect. *Reviews of Modern Physics* 58: 519–531.
9. Laughlin, R. B. 1981. Quantized Hall conductivity in two dimensions. *Physical Review B* 23: 5632–5633.
10. Laughlin, R. B. 1983. Anoumalous quantum Hall effect: An incompressible quantum fluid with fractional charged excitations. *Physical Review Letters* 50: 1395–1398.
11. Landau, L. D. 1930. Diamagnetism in metals. *Zeitschrift fur Physik* 64: 629–637.
12. Landau, L.D. and E. M. Lifshitz 1977. *Quantum mechanics, 3rd Edition*. Oxford: Pergamon Press.
13. Mott, N. F. and E. A. Davies 1979. *Electronic processes in non-crystalline materials*. Oxford: Clarendon Press.
14. McGurn, A. R. 2021. *Introduction to nonlinear optics of photonic crystals and metamaterials, 2nd Edition*. Bristol: IOP Publishing Ltd.
15. Kroemer, H. 1994. *Quantum mechanics for engineering, material science, and applied physics*. New Jersey: Prentice Hall.
16. Fujita, S. and A. Suzuki 2014. Theory of Shubnikov-De Haas and quantum Hall oscillations in graphene under bias gate voltage. *Global Journal of Science Frontier Research: A Physics and Space Science* 14: 19–32.
17. Matthews, J. and M. E. Cage 2005. Temperature dependence of the Hall and longitudinal resistances in a quantum Hall resistance standard. *Journal of Research of the National Institute of Standards and Technology* 100: 497–510.

18. Han, Y., Wan, J.-G., Ge, G.-X., et al. 2015. High-temperature quantum anomalous Hall effect in honeycomb bilayer consisting of Au atoms and single-vacancy graphene. *Science Reports*. https://doi.org/10.1038/srep16843
19. He, K and Q.-K. Xue 2019. The road to high temperature quantum anomalous Hall effect in magnetic topological insulators. *Spin*. https://doi.org/10.1142/S2010324719400162
20. Hall, E. H. 1881 On the "rotational coefficient" in nickel and cobalt". *Proceedings of the Physical Society of London* 12: 157–172.
21. Shiomi, Y. 2013. *Anomalous and topological hall effects in itinerant magnets*. Tokyo: Springer.
22. Dyakonov, M. I. and V. I. Perel 1971. Current-induced spin orientation of electrons in semiconductors. *Physics Letters A* 35: 459–460.
23. Hirsch, J. E. Spin Hall effect. *Physical Review Letters* 83: 1834–1832.
24. Ling, X., Zhou X., Huang K., et al. 2017. Recent advances in the spin Hall effect of light. *Reports on Progress in Physics* 80: 066401.
25. Ren, J., Li, Y., Lin, Y., et al. (2012) Spin Hall effect of light reflected from a magnetic thin film. *Applied Physics Letters* 101: 171103.

11 Resonance Properties

Frequency resonance is an important feature found in many physical systems. It involves the rapid change in physical properties of the system over a narrow frequency band of an external stimulus, and usually is associated with an underlying instability or transition between energy levels or an interaction with a structural feature of the system defined over characteristic length scales. Due to their anomalous properties, when an externally applied stimulus is applied to them, resonant systems display an enhanced response over that otherwise found in the system at nonresonant frequencies, and these enhanced properties often lead to important new technology [1]. While resonant properties are traditionally of interest in their own right, in recent years the properties of resonance have become increasingly important in material science applications as a means of designing new materials with unusual physical properties. An important example of such applications is in the engineering of optical and acoustical media with novel refractive properties for the design of perfect lenses, cloaking devices, and for applications in surface physics [1–6].

In this regard, as we shall see, the nature of the resonant responses can occur in a variety of functional frequency forms, and such forms can be shown to have a natural classification. Common examples of resonance frequency response are observed in acoustic, optical, and electronic materials and devices [1–4], yielding a variety of different useful frequency behaviors. From an engineering perspective, these anomalous changes and frequency behaviors provide for a number of design possibilities for device technology and in the development of new engineering applications [1–6]. As an illustration of some of the interesting aspect of these properties, a discussion of the nature and origins of Fano resonances is presented.

As an important example, electron transitions between atomic energy levels and transitions between vibrational states offer mechanisms for optical resonance. Such transitions between energy levels depend on the natural frequency response characteristics of the atomic electron and vibrational modes being studied, and the coupling of the modes to the optical stimuli. On the other hand, the spatial characteristics of, for example, Fabry–Perot resonators used in laser designs provides for resonant frequency responses of the light generated within the resonator [7]. These later types of resonator responses are examples in which the geometry of the system, rather than its dynamical frequency response, determine the frequency interactions with optical radiation [8,9]. Both forms of resonance are seen to exhibit optical behaviors, which enter into the design of important optical systems, and similar resonant behaviors are available in acoustic and electronic systems. For example, in electron conduction and tunneling systems, a variety of resonant transitions occurring in the resonant interactions of electrons with material features are observed and can have important applications. And in the context of an illustration from acoustics, acoustic resonators can be arrayed to engineer a medium with specific vibrational resonances [1]. These properties in electron and acoustic systems too will be a focus in the presentation to follow.

In this chapter, we shall look at aspects of these two types of mechanisms of resonant response. First a treatment of resonant response commonly applied in the design of metamaterials is considered [1]. In this regard, a metamaterial is a material that is designed as an array of nanoscopic frequency resonators and formulated to act as a homogeneous material in its interactions with frequencies of radiation with which it is engineered to provide a modulation. The idea for this system is similar to that found in crystalline materials, which are formed of atoms and molecules but appear as homogeneous media to optical frequencies of radiation with which they interact. The metamaterial is then a case in which the frequency response of the resonator is a focus. The second topic is that of Fano resonances, which are engineered responses in optical and electron systems based on the spatial characteristics of device structures. These systems are formulated using the spatial properties of engineered materials to influence the functional characteristics in frequency of the

DOI: 10.1201/9781003031987-11

resonant response. In the case of Fano resonances, a variety of frequency-dependent resonant forms exist, and allow for the classification and design of resonance characteristics. Both optical and electron resonance forms will be a consideration in this regard.

11.1 METAMATERIAL RESPONSES AND SIMPLE SPATIAL RESONANCES

In the design of metamaterials, we first consider a single nanostructured resonator unit that does not depend on the spatial characteristics of the radiation it interacts with, i.e., the resonator only supplies a frequency response [1–6]. The metamaterial is then formulated as a spatial array of these types of resonator units and is designed to give a tailored frequency response to an externally applied plane wave of radiation. Common radiations that have been manipulated in these meta-systems are electromagnetic and acoustical fields [1], which involve material designs focused, respectively, on the resonant properties of dielectric and acoustical engineered arrays of resonator units. In this regard, such metamaterials enter into a variety of problems, involving the use of new materials with unusual properties in the refraction and reflection of light or sound [1].

An optical metamaterial that has been of much current interest is that in which the nanostructure resonator unit is a so-called split ring resonator. In its simplest form, a single-split ring resonator is formed as a ring of conducting material, which has a gap cut in it so that the conducting path is broken [1–4]. (See Figure 11.1a for a schematic of the split ring oscillator.) This device is modeled as a basic LRC circuit in which the inductor L is the ring of conductor, the capacitance C of the ring is the gap in the ring, and the ring resistance R is the resistance of the LRC circuit ring. In addition, the split ring resonator can couple through Faraday's law to a time varying applied external magnetic induction $\vec{B}$, which is applied perpendicular to the plane of the ring. By means of this coupling, the magnetic field induces an emf in the circuit, which then acts as a driving potential in the circuit so that the system forms a forced oscillator. (See Figure 11.1b for the resulting forced harmonic oscillator circuit.) It should be emphasized here that this is a highly simplified model, which, in practice, is modified to treat a number of engineering problems. Nevertheless, it is representative of the basic function of the split ring resonator [1–4].

For nanoscale split ring resonators and electromagnetic waves with wavelengths large compared to the resonator dimensions, the circuit in Figure 11.1b then represents the physics of the coupling of the ring to electromagnetic waves with magnetic induction perpendicular to the plane of the ring. In this model, the field of the electromagnetic radiation drives the system as a forced LRC oscillator, where the driving AC potential ε models the Faraday potential generated by the applied alternating magnetic field. In addition, the split ring resonators of the metamaterial are formulated in an array, so that they are coupled together by mutual induction, which then enters as part of the resonant response of the metamaterial [1–4].

The equation of motion for the current in the single-split ring resonator unit in Figure 11.1b as driven by the Faraday emf from the coupling with the electromagnetic field can be modeled by the LRC circuit equation [1–4]

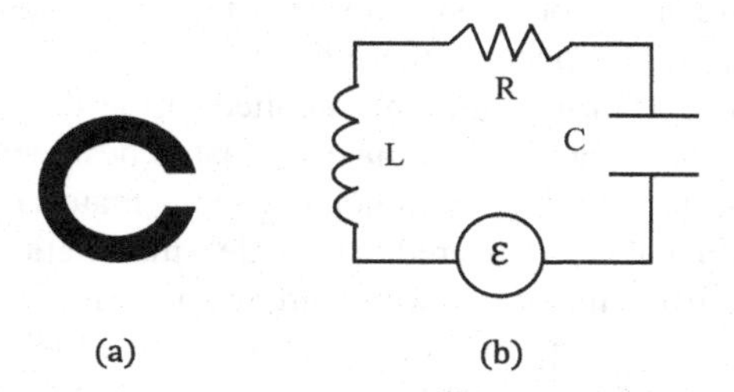

FIGURE 11.1 Schematic of: (a) split ring resonator formed as a ring with a gap and an applied magnetic induction perpendicular to the plane of the ring, (b) an LRC-forced harmonic oscillator circuit driven by an AC potential source labeled $\mathcal{E}$. Here, the forcing emf arises through Faradays law from a time-dependent magnetic induction passing through the split ring resonator.

$$L\frac{d^2I}{dt^2}+R\frac{dI}{dt}+\frac{I}{C}=\frac{\partial \varepsilon}{\partial t}, \tag{11.1}$$

where ε is an applied emf arising from the coupling to the magnetic induction of the driving electromagnetic plane wave. For a harmonic current $Q = Q_0 e^{-i\omega t}$ and an applied emf $\varepsilon = \varepsilon_0 e^{-i\omega t}$, it then follows that

$$Q_0 \propto \frac{\omega_0^2-\omega^2-i\omega\frac{R}{L}}{\left(\omega_0^2-\omega^2\right)^2+\left(\omega\frac{R}{L}\right)^2}\varepsilon_0, \tag{11.2}$$

where $\omega_0 = \frac{1}{\sqrt{LC}}$ is the natural frequency of the oscillator in the absence of the driving field. Note that this is the standard from of a frequency resonance that occurs in a forced harmonic oscillator driven by a frequency source. In this regard, the spatial dimensions of the oscillator do not enter into the resonance properties in Eq. (11.2), but the resonance is characterized by the minimum value of the denominator that occurs at $\omega = \omega_0$, and the resonance frequency width is set by $\omega\frac{R}{L}$. Within the frequency band characterized by ω_0 and $\omega\frac{R}{L}$, the split ring resonator exhibits a rapid change in its response to the electromagnetic wave.

These ideas of the resonant response of the single-split ring resonator to an applied electromagnetic field can be extended to treat a planar array of resonators, which are coupled to each other by mutual inductive responses. The response of the resonator array to an applied electromagnetic wave with wavelength greater than the dimensions of the individual split ring resonators ultimately is characterized by an effective permittivity and permeability, representing the dielectric response of the array to the electromagnetic wave. This allows for the ability to engineer media with specific dielectric properties required for applications. By setting up a three-dimensional array of intersecting planar arrays of resonators, a fully three-dimensional dielectric material (i.e., a metamaterial) can then be formulated [1–6].

The idea of an engineered response based on arrays formed of resonator units has not just been applied to electromagnetic media but has also been used in acoustics to engineer the acoustic response of materials, which are termed as acoustic metamaterials [1]. In this regard, an important application of these techniques is in the design of materials exhibiting negative index of refraction [1]. Of particular interest is an application to electrodynamics in which it is shown that materials that simultaneously exhibit negative permittivity and negative permeability display a negative index of refraction. Similarly, acoustic media have also been shown to exhibit negative index refraction properties based on acoustic metamaterials. For the details of these discussions in both optical and acoustic media, the reader is referred to the literature [1].

Here, we only mention that it is known in electrodynamics that systems with both negative permittivity and permeability exhibit a negative index of refraction [1–6]. This is an interesting situation in which the phase and group velocities of an electromagnetic plane wave are antiparallel. This leads to a variety of interesting consequences that extends the refractive properties of optical systems. In addition, negative index contributes to the development of the perfect lens, electromagnetic cloaking, and to other unusual optical properties involving the characteristics of radiation [1]. For a treatment of these, the reader is referred to the literature [1–6].

The ideas of resonance we have just discussed are all based only on the frequency response of systems. Resonance effects are also important when they arise from the spatial characteristics of a system interacting with an applied radiation. A common example of such a spatial resonance is that found in the optics of a thin dielectric slab in air. For light that is incident at near normal to the surface of the slab, the condition for the maximum reflection of light of wavelength λ in vacuum from the slab is that [7,10]

$$2nt = \left(m + \frac{1}{2}\right)\lambda, \tag{11.3a}$$

where t is the slab thickness, $n > 1$ is the index of refraction of the slab, and m is a nonnegative integer. Alternatively, destructive interference is then observed when [7,10]

$$2nt = m\lambda \tag{11.3b}$$

for m a nonnegative integer. This enhanced reflection is seen to be largely dependent on the sample geometry rather that the material science of the thin film.

In the following sections, further examples of spatial interference and energy effects are discussed. Specifically, resonant effects related to the scattering from features found in layered media are treated. These effects will involve scattering resonances arising from bound-state energy levels that are within layers of the system and feature modifications to the energy-dependent resonant scattering functions known as Fano resonances. These modifications will be shown to introduce interesting modulation effects into the shape of the scattering resonances as a function of frequency.

11.2 STANDARD RESONANCE INVOLVING QUANTUM WELLS

An important type of resonance, commonly arising from the interaction between a continuum of states and a single discrete energy level, is described by a Lorentzian form [1–4,11,12]. Such a type of interaction is often observed in layered media when a plane wave is scattered from layers containing highly localized, nearly bound, modes. The resonant scattering observed from these types of materials is often described by energy-dependent properties represented as peaked resonant functions of energy characterized by a resonant amplitude and a resonant width. In this section, a simple model of such a resonant scattering as exemplified in a one-dimensional system is presented, and its fundamental properties are discussed. The treatment is meant to illustrate the basic features of a Lorentzian resonance in a system of interest to nanoscience studies.

Consider a Schrodinger particle moving in the one-dimensional step potential [11] shown in Figure 11.2a. This particular potential profile provides a simple model illustrating many of the basic properties of Lorentzian resonances. Specifically, the particle moves along the x-axis with a position-dependent potential energy given by

$$\begin{aligned} &V(x) = 0 \text{ in the regions } x < -b, \\ &V(x) = v(x) \text{ in the region } -a < x < a, \text{and} \\ &V(x) = V_0 > 0 \text{ otherwise,} \end{aligned} \tag{11.4}$$

where $-b < -a$, $0 < v(x) < V_0$ is an even function of x, and the potential $v(x)$ is specifically designed to support at least a single resonant bound state of energy ε_i located within the interval $0 < \varepsilon_i < V_0$. For the development of the ideas of resonance in the context of the model potential in Eq. (11.4), the focus in the following discussions is on the dynamical solutions of a particle propagating with a kinetic energy less than V_0 for the case that $-b \ll -a$. Consequently, the solutions involve quantum mechanical tunneling in which the left region of $V(x) = 0$ is only weakly coupled to the $-a < x < a$ region of $V(x) = v(x)$.

Two basic types of solutions arise in the context of the proposed model. One involves the scattering of a particle incident on the barrier from the far left (i.e., from the region $x < -b$), and the other involves the decay of a particle in the $-a < x < a$ region of $V(x) = v(x)$ into the region to the far left of the barrier. Both of these two different solutions are set by the nature of the initial boundary conditions on the particle.

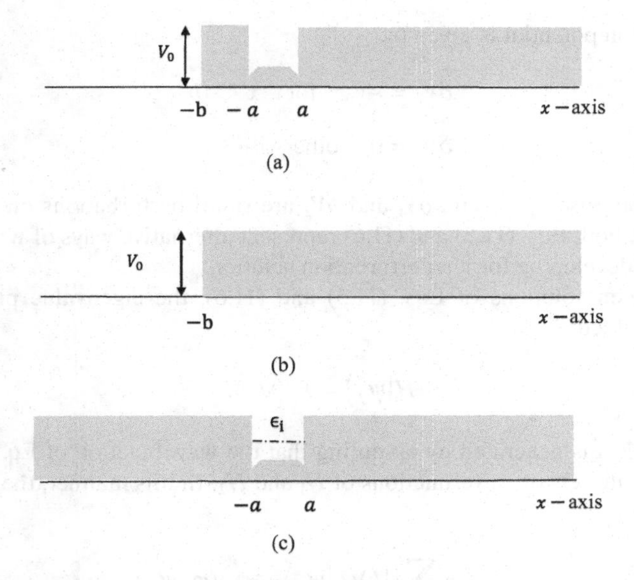

FIGURE 11.2 Plot of the potential energy versus x for: (a) the problem involving a planewave incident from the left on a potential energy step barrier that contains a well supporting a resonant-bound-state mode, (b) the auxiliary problem of a planewave incident from the left on a semi-infinite potential energy step barrier, and (c) the auxiliary problem of an infinite region of constant potential energy V_0 containing a well that has a single bound state of energy $0<\epsilon_i<V_0$. In all of the plots, the zero of potential energy is set at the x-axis.

For both types of solutions, the barrier problem in Figure 11.2a can be studied in terms of a weak perturbation treatment between the solutions of two different auxiliary problems. The first auxiliary problem involves the barrier system in Figure 11.2b, where $V(x)=0$ in the region $x<-b$ and $V(x)=V_0>0$, otherwise. The second auxiliary problem involves the bound-state system in Figure 11.2c where $V(x)=v(x)$ in the region $-a<x<a$ and $V(x)=V_0$, otherwise. The solutions of these two problems (See Figure 11.2b and c) can then be used to obtain the solutions of the problem in Figure 11.2a in the context of a perturbation treatment [11].

In this perturbation view, the Hamiltonian for the dynamics of the problem of interest in Figure 11.2a is denoted by H. The Hamiltonians of the two auxiliary problems used to generate the solutions of H (shown in Figure 11.2b and c) are, respectively, denoted by H_1 and H_2. These auxiliary Hamiltonians characterize the problem of the scattering of a particle wave from a step barrier H_1 and a bound-state problem H_2. For the systems in both Figure 11.2a and b in the common regions, $x<-b$, the particle is a free particle, and the solutions of interest are planewaves forms with energies $\epsilon_k<V_0$ and wavenumbers k. For the well problem of Figure 11.2c, however, the lowest energy-bound-state solution is of energy $0<\epsilon_i<V_0$ and is an even function of x.

In the perturbation approach, the Hamiltonian H can be decomposed into two parts in two different ways. In the first decomposition [11]

$$H = H_1 + \delta V_1 \tag{11.5a}$$

where the perturbation is of the form

$$\begin{aligned} \delta V_1 &= -V_0 + v(x) \quad \text{for} \quad -a<x<a \\ \delta V_1 &= 0, \quad \text{otherwise} \end{aligned} \tag{11.5b}$$

In the second decomposition

$$H = H_2 + \delta V_2, \tag{11.6a}$$

where the perturbation potential is given by

$$\begin{aligned} \delta V_2 &= -V_0 \quad \text{for} \quad x < -b \\ \delta V_2 &= 0, \quad \text{otherwise} \end{aligned} \tag{11.6b}$$

In both cases, for the posed problem, δV_1 and δV_2 are small perturbations on H_1 and H_2, respectively. In this regard, both Eqs. (11.5) and (11.6) represent alternative ways of writing the same total Hamiltonian H and developing for it a perturbation solution.

To obtain the eigen solutions of Eqs. (11.5) and (11.6), the eigenvalue problem of the total Hamiltonian is expressed as

$$H|\psi_f\rangle = E_f|\psi_f\rangle. \tag{11.7}$$

A solution of Eq. (11.7) is generated by assuming that the wavefunction of Eq. (11.7) is composed in terms of the separate sets of wavefunctions of H_1 and H_2. In this manner, the general eigenstates are written as

$$|\psi_f\rangle = \sum_k |k\rangle\langle k|\psi_f\rangle + |\varphi_i\rangle\langle\varphi_i|\psi_f\rangle. \tag{11.8}$$

Here $|k\rangle$ is the planewave-based eigenstate from

$$H_1|k\rangle = \epsilon_k|k\rangle \tag{11.9}$$

of the scattering states of the step potential, and $|\varphi_i\rangle$ is the lowest-energy eigenstate of the bound-state problem

$$H_2|\varphi_i\rangle = \epsilon_i|\varphi_i\rangle. \tag{11.10}$$

Substituting Eq. (11.8) into Eq. (11.7) and projecting out the state $|k\rangle$ gives an algebraic equation [11]

$$\begin{aligned} \langle k|H|\psi_f\rangle &= \langle k|H\sum_{k'}|k'\rangle\langle k'|\psi_f\rangle + \langle k|H|\varphi_i\rangle\langle\varphi_i|\psi_f\rangle = E_f\langle k|\psi_f\rangle \\ &\approx \epsilon_k\langle k|\psi_f\rangle + \alpha(k,i)\langle\varphi_i|\psi_f\rangle \end{aligned}, \tag{11.11}$$

where

$$\langle k|H|k'\rangle = \langle k|H_1 + \delta V_1|k'\rangle = \epsilon_k\delta_{k,k'} + \langle k|\delta V_1|k'\rangle \approx \epsilon_k\delta_{k,k'} \tag{11.12a}$$

$$\langle k|H|\varphi_i\rangle = \langle k|H_2 + \delta V_2|\varphi_i\rangle = \epsilon_i\langle k|\varphi_i\rangle + \langle k|\delta V_2|\varphi_i\rangle = \alpha(k,i). \tag{11.12b}$$

Here the approximation in Eq. (11.12a) follows, as δV_1 is only nonzero over the interval $-a < x < a$, and in this region both $|k\rangle$ and $|k'\rangle$ are small so that $\langle k|\delta V_1|k'\rangle \ll \epsilon_k$. Similarly, substituting Eq. (11.8) into Eq. (11.7), and projecting out the state $|\varphi_i\rangle$ gives the algebraic equation

$$\begin{aligned} \langle\varphi_i|H|\psi_f\rangle &= \langle\varphi_i|H\sum_{k'}|k'\rangle\langle k'||\psi_f\rangle + \langle\varphi_i|H|\varphi_i\rangle\langle\varphi_i|\psi_f\rangle = E_f\langle\varphi_i|\psi_f\rangle \\ &\approx \epsilon_i\langle\varphi_i|\psi_f\rangle + \sum_{k'}\alpha(k',i)^*\langle k'|\psi_f\rangle \end{aligned}, \tag{11.13}$$

where

$$\langle \varphi_i | H | k' \rangle = \langle \varphi_i | H_2 + \delta V_2 | k' \rangle = \epsilon_i \langle \varphi_i | k' \rangle + \langle \varphi_i | \delta V_2 | k' \rangle = \alpha^* (k', i) \approx \alpha(k, i) \tag{11.14a}$$

$$\langle \varphi_i | H | \varphi_i \rangle = \langle \varphi_i | H_2 + \delta V_2 | \varphi_i \rangle = \epsilon_i + \langle \varphi_i | \delta V_2 | \varphi_i \rangle \approx \epsilon_i. \tag{11.14b}$$

Here the approximation in Eq. (11.14a) follows, as the bound state $|\varphi_i\rangle$ is a real even parity function and has little spatial overlap with $|k'\rangle$. Note that the approximation in Eq. (11.14b) follows, as δV_2 is only nonzero over the interval $x < -b \ll -a$, and in this spatial interval $|\varphi_i\rangle$ is small so that $\langle \varphi_i | \delta V_2 | \varphi_i \rangle \ll \epsilon_i$. In addition, it is assumed that $\alpha(k,i) \approx \alpha(k',i)$ has a weak dependence on k over the energy region of the resonance [11].

From Eq. (11.11), it then follows that the amplitudes $\langle k | \psi_f \rangle$ and $\langle \varphi_i | \psi_f \rangle$ are related by

$$\langle k | \psi_f \rangle = \frac{\alpha(k,i)}{E_f - \epsilon_k} \langle \varphi_i | \psi_f \rangle, \tag{11.15}$$

and applying this in Eq. (11.13) gives [11]

$$E_f - \epsilon_i - \sum_k \frac{\alpha^2(k,i)}{E_f - \epsilon_k} = 0. \tag{11.16}$$

The relationship in Eq. (11.16) is now used to determine the eigenvalues E_f, while Eq. (11.15) along with the condition of wavefunction normalization (i.e., $\sum_k |\langle k | \psi_f \rangle|^2 + |\langle \varphi_i | \psi_f \rangle|^2 = 1$.) determine $\langle k | \psi_f \rangle$ and $\langle \varphi_i | \psi_f \rangle$.

From the normalization condition on $|\psi_f\rangle$, it then follows from Eqs. (11.8) to (11.15) that

$$|\langle \varphi_i | \psi_f \rangle|^2 = \frac{1}{1 + \sum_k \frac{\alpha^2(k,i)}{(E_f - \epsilon_k)^2}} \tag{11.17}$$

and

$$|\langle k | \psi_f \rangle|^2 = \frac{\alpha^2(k,i)}{(E_f - \epsilon_k)^2} \frac{1}{1 + \sum_k \frac{\alpha^2(k,i)}{(E_f - \epsilon_k)^2}}. \tag{11.18}$$

These represent the distribution of the wavefunction in the energy space E_f of the eigen energies of the system and can be ultimately expressed as function of $E_f - \epsilon_i$.

In order to eliminate the factors of ϵ_k in Eqs. (11.17) and (11.18), it is helpful to work on a simplification of Eq. (11.16). Specifically, consider that the summation terms in Eq. (11.16) are largest for terms in which $E_f - \epsilon_k \approx 0$. Under this condition if the system is near a resonant interaction of the continuum with the bound state (i.e., $E_f \approx \epsilon_k$), the three energies of the dominant terms of the series have $\epsilon_k \approx E_f \approx \epsilon_i$. In this region, the wavevector k can be approximated by a linear form so that the dispersion relation ϵ_k of the continuum states of H_1 has the approximate form [11]

$$\begin{aligned}\epsilon_k &\approx \epsilon_i + \left.\frac{d\epsilon_k}{dk}\right|_{k_i}(k-k_i) \\ &= \epsilon_0 + \left.\frac{d\epsilon_k}{dk}\right|_{k_i} k\end{aligned} \tag{11.19}$$

where $\epsilon_0 = \epsilon_i - \left.\frac{d\epsilon_k}{dk}\right|_{k_i} k_i$ and k_i is the wavevector of the continuum state with energy $\epsilon_k \approx \epsilon_i$ of the bound-state mode.

The continuum wavevector states of H_1 have wavevectors that are of the general form $k = k_0 n$, where n is an integer, and the constant k_0 arises from the boundary conditions of the continuum modes. Consequently, it follows if it is assumed that $\alpha^2(k,i) = \alpha^2$ is a constant over the region of the pole that [11]

$$\sum_k \frac{\alpha^2(k,i)}{E_f - \epsilon_k} = \alpha^2 \sum_k \frac{1}{E_f - \epsilon_k} \approx \frac{\alpha^2}{\left.\frac{d\epsilon_k}{dk}\right|_{k_i} k_0} \sum_{n=-\infty}^{\infty} \frac{1}{\left[\frac{E_f - \epsilon_0}{\left.\frac{d\epsilon_k}{dk}\right|_{k_i} k_0}\right] - n}. \tag{11.20}$$

Applying the identity [13]

$$\sum_{n=-\infty}^{\infty} \frac{1}{z-n} = \pi \cot \pi z \tag{11.21}$$

to Eq. (11.20) then gives

$$\begin{aligned}\sum_k \frac{\alpha^2(k,i)}{E_f - \epsilon_k} &\approx \frac{\alpha^2}{\left.\frac{d\epsilon_k}{dk}\right|_{k_i} k_0} \sum_{n=-\infty}^{\infty} \frac{1}{\frac{E_f - \epsilon_0}{\left.\frac{d\epsilon_k}{dk}\right|_{k_i} k_0} - n} \\ &= \frac{\pi\alpha^2}{\left.\frac{d\epsilon_k}{dk}\right|_{k_i} k_0} \cot\left(\pi \frac{E_f - \epsilon_0}{\left.\frac{d\epsilon_k}{dk}\right|_{k_i} k_0}\right)\end{aligned} \tag{11.22}$$

and using this in Eq. (11.16) we find

$$E_f - \epsilon_i = \frac{\pi\alpha^2}{\left.\frac{d\epsilon_k}{dk}\right|_{k_i} k_0} \cot\left(\pi \frac{E_f - \epsilon_0}{\left.\frac{d\epsilon_k}{dk}\right|_{k_i} k_0}\right) \tag{11.23}$$

as a transcendental equation for the eigen energies E_f.

A second identity obtained from Eq. (11.21) is

$$\sum_{n=-\infty}^{\infty}\frac{1}{(z-n)^2}=\pi^2\left[1+\cot^2\pi z\right]. \tag{11.24}$$

From this it follows, proceeding similarly to the discussions leading to Eq. (11.22), that

$$\sum_k \frac{\alpha^2(k,i)}{(E_f-\epsilon_k)^2}\approx\alpha^2\sum_k\frac{1}{(E_f-\epsilon_k)^2}=\frac{\pi^2\alpha^2}{\left(\left.\frac{d\epsilon_k}{dk}\right|_{k_i}k_0\right)^2}\left[1+\cot^2\left(\pi\frac{E_f-\epsilon_0}{\left.\frac{d\epsilon_k}{dk}\right|_{k_i}k_0}\right)\right], \tag{11.25}$$

$$=\frac{\Gamma^2}{4\alpha^2}+\frac{(E_f-\epsilon_i)^2}{\alpha^2}$$

where Eq. (11.23) is applied in the last line, and $\Gamma=\frac{2\pi\alpha^2}{\left.\frac{d\epsilon_k}{dk}\right|_{k_i}k_0}$. Applying Eq. (11.25) in Eq. (11.17), it is found that

$$\left|\langle\varphi_i|\psi_f\rangle\right|^2=\frac{\alpha^2}{\frac{\Gamma^2}{4}+(E_f-\epsilon_i)^2}, \tag{11.26}$$

where it is assumed that $\alpha^2\ll\frac{\Gamma^2}{4}$, and from Eq. (11.18) it then follows that [11]

$$\left|\langle k|\psi_f\rangle\right|^2=\frac{\alpha^2}{(E_f-\epsilon_k)^2}\frac{\alpha^2}{\frac{\Gamma^2}{4}+(E_f-\epsilon_i)^2}. \tag{11.27}$$

The expression for $\left|\langle\varphi_i|\psi_f\rangle\right|^2$ in Eq. (11.26) represents the Lorentzian distribution of energy states in the $|\varphi_i\rangle$-bound state of the system. Specifically, consider the case of boundary conditions in which a state is originally localized within the well at $-a<x<a$ and transitions to $|k\rangle$ planewave like states propagating away from the well. This is one of the two possible boundary conditions imposed on the eigenmodes. Under this condition, it can be shown, in the time representation which complements the E_f energy space representation, that the decay of the particle out of the well is described by the time-dependent term $e^{-\frac{\Gamma}{2}t}$. This time dependence is typical of the decay of particles through a barrier. On the other hand, for scattering boundary conditions representing a particle incident into the well region, $e^{-\frac{\Gamma}{2}t}$, relates to the trapping time of a particle within the well.

11.3 FANO RESONANCE INVOLVING QUANTUM WELLS

A more complex type of resonance is the Fano resonance [11–13]. This involves the interaction of the $|k\rangle$ continuum states with more than one discrete energy level. Under these conditions, there are additional scattering paths of the particle, as it interacts with either of the two discrete states, and the two discrete states in turn interact with one another. These multiple scattering paths generate a phase-coherent output from the total interference of the correlated scattering. From this interference a complex variety of energy-dependent resonant scattering effects are observed in the system properties.

To understand the origins of Fano resonance in a highly simplistic model, consider the model of resonance scattering treated in the previous section but with the addition of a second new weakly coupled discrete resonant interaction. The new resonance may arise from the interaction with an addition bound state present in the interval $-a < x < a$ or in another region of space that is weakly coupled to the $|\psi_f\rangle$ continuum states of Eq. (11.8) and to the $|\varphi_i\rangle$ bound state. The new resonant bound state in our present considerations is denoted $|\vartheta_j\rangle$ and has an energy $\epsilon_j \approx \epsilon_i \approx \epsilon_k$. As with the resonance state $|\varphi_i\rangle$ treated earlier, there is a bound-state Hamiltonian and eigenvalue problem associated with the considerations of the state $|\vartheta_j\rangle$. In the case that the new resonance is not a state of H_2, the auxiliary Hamiltonian associated with it is denoted H_3 and forms the eigenvalue problem given by [11]

$$H_3|\vartheta_j\rangle = \epsilon_j|\vartheta_j\rangle. \tag{11.28}$$

Consequently, the bound-state problem in Eq. (11.28) is of a similar nature to that for the bound state of H_2 in Eq. (11.10).

The eigenvalue problem of the of the modified Hamiltonian, H_{new}, of the new system including the $|\vartheta_j\rangle$ state is given by [11]

$$H_{new}|\psi_{f,new}\rangle = E_{f,new}|\psi_{f,new}\rangle, \tag{11.29}$$

where now the eigenfunctions are assumed to have the form

$$|\psi_{f,new}\rangle = |\psi_f\rangle\langle\psi_f|\psi_{f,new}\rangle + |\vartheta_j\rangle\langle\vartheta_j|\psi_{f,new}\rangle. \tag{11.30a}$$

In the spirit of the wavefunction designed in Eq. (11.8) for the earlier resonance treatment, this represents a combination of the solutions of H and of the new bound-state wavefunction from H_3. Consequently, as per the discussions in the earlier section, here

$$H_{new} = H + \delta V_3 \tag{11.30b}$$

for H defined in Eq. (11.7) and δV_3 representing the scattering potential acting between the $|\psi_f\rangle$ modes and the $|\vartheta_j\rangle$ mode. For simplicity in the following, it is also assumed that the magnitude of the coupling δV_3 to the new resonance-bound state $|\vartheta_j\rangle$ is much weaker than the coupling to the resonance state $|\varphi_i\rangle$ discussed in the earlier section. This assures that the resonance associated with $|\varphi_i\rangle$ is little changed from that associated with $|\vartheta_j\rangle$.

Next, let us focus on the scattering transitions between the states $|\psi_f\rangle$ and $|\vartheta_j\rangle$, as mediated by the system described by Eqs. (11.29) and (11.30). This involves a study of the transition matrix of the form $\langle\vartheta_j|\delta V_3|\psi_f\rangle$, which ultimately forms part of a Fermi golden rule treatment of the scattering between these two parts of the system. Written in the form $\langle\vartheta_j|\delta V_3|\psi_f\rangle$, the transition matrix is now shown to be a resonant function of the energy difference $E_f - \epsilon_i \approx E_f - \epsilon_j$ between the unperturbed states $|\psi_f\rangle$ and $|\varphi_i\rangle$.

In order to determine the resonant structure associated with the new factor of $|\vartheta_j\rangle$, the matrix element generated from Eq. (11.30b) is considered. Specifically, applying Eq. (11.8) for the $|\psi_f\rangle$ wavefunction, it follows that [11]

$$\begin{aligned}\langle\vartheta_j|\delta V_3|\psi_f\rangle &= \sum_k \langle\vartheta_j|\delta V_3|k\rangle\langle k|\psi_f\rangle + \langle\vartheta_j|\delta V_3|\varphi_i\rangle\langle\varphi_i|\psi_f\rangle \\ &\approx \sum_k W\langle k|\psi_f\rangle + w_j\langle\varphi_i|\psi_f\rangle\end{aligned} \tag{11.31}$$

is the transition matrix between the initial and final states of a scattering process involving the new bound state. A simplification is obtained in Eq. (11.31) by assuming that the matrix elements in Eq. (11.31) are of the approximate form

$$W = \langle \vartheta_j | \delta V_3 | k \rangle \tag{11.32a}$$

$$w_j = \langle \vartheta_j | \delta V_3 | \varphi_i \rangle, \tag{11.32b}$$

and under the conditions of interest here can be represented as constants.

For a final determination of the energy structure of Eq. (11.31), the general form of the coefficients $\langle k | \psi_f \rangle$ and $\langle \varphi_i | \psi_f \rangle$ obtained in the previous section as functions of $E_f - \epsilon_i$ is now applied in Eq. (11.31). To this end, from Eqs. (11.15), (11.26) and (11.27) it follows that

$$\langle \varphi_i | \psi_f \rangle = \frac{\alpha}{\sqrt{\frac{\Gamma^2}{4} + \left(E_f - \epsilon_i\right)^2}} \tag{11.33a}$$

and

$$\langle k | \psi_f \rangle = \frac{\alpha}{E_f - \epsilon_k} \langle \varphi_i | \psi_f \rangle = \frac{1}{E_f - \epsilon_k} \frac{\alpha^2}{\sqrt{\frac{\Gamma^2}{4} + \left(E_f - \epsilon_i\right)^2}}. \tag{11.33b}$$

Combining Eq. (11.33) with Eq. (11.31), it then follows that

$$\langle \vartheta_j | \delta V_3 | \psi_f \rangle = \left[W \sum_k \frac{\alpha^2}{E_f - \epsilon_k} + w_j \alpha \right] \frac{1}{\sqrt{\frac{\Gamma^2}{4} + \left(E_f - \epsilon_i\right)^2}}. \tag{11.34}$$

Applying Eq. (11.16) to evaluate the sum in Eq. (11.34) yields the general form [11]

$$\langle \vartheta_j | \delta V_3 | \psi_f \rangle = \frac{2W}{\Gamma} \frac{\frac{w_j}{W}\alpha + E_f - \epsilon_i}{\sqrt{1 + \frac{4}{\Gamma^2}\left(E_f - \epsilon_i\right)^2}}, \tag{11.35}$$

which now exhibits both a numerator and denominator dependent on $E_f - \epsilon_i$. This can be rewritten in the form

$$\langle \vartheta_j | \delta V_3 | \psi_f \rangle = W \frac{\epsilon + q}{\sqrt{1 + \epsilon^2}} \tag{11.36}$$

where $\epsilon = \frac{2}{\Gamma}\left(E_f - \epsilon_i\right)$ and $q = \frac{w_j}{W}\frac{2\alpha}{\Gamma}$.

The general Fermi Golden Rule transmission matrix is then given by

$$\left| \langle \vartheta_j | \delta V_3 | \psi_f \rangle \right|^2 \propto \frac{\left(\epsilon + q\right)^2}{1 + \epsilon^2}, \tag{11.37}$$

indicating the rate of transitions as a function of the energy variable ϵ. The resonance form in Eq. (11.37) has a number of interesting limiting cases. In the case that $q = 0$, the form in Eq. (11.37) reduces to [11]

$$\left|\langle \vartheta_j | \delta V_3 | \psi_f \rangle\right|^2 \propto \frac{\epsilon^2}{1+\epsilon^2} \tag{11.38a}$$

and exhibits a single minimum at the $\epsilon = 0$ resonance so that the system exhibits a dip centered on the resonant condition. In the case that $q \gg 1$, the form reduces to [11]

$$\left|\langle \vartheta_j | \delta V_3 | \psi_f \rangle\right|^2 \propto \frac{q^2}{1+\epsilon^2} \tag{11.38b}$$

so that the transition rate as a function of the energy variable is a Lorentzian peak centered about the $\epsilon = 0$ resonance condition. In the case that $q = 1$, the form in Eq. (11.37) reduces to

$$\left|\langle \vartheta_j | \delta V_3 | \psi_f \rangle\right|^2 \propto \frac{(\epsilon+1)^2}{1+\epsilon^2}. \tag{11.38c}$$

Under these conditions, the resonant structure exhibits a single minimum at the $\epsilon = -1$ for which $\left|\langle \vartheta_j | \delta V_3 | \psi_f \rangle\right|^2 = 0$ and a single maximum resonance peak at $\epsilon = 1$. As $\epsilon \to \pm\infty$, it is found that $\frac{(\epsilon+1)^2}{1+\epsilon^2} \to 1$. For the convenience of the reader, a schematic comparison of these three different resonant forms [11] as function of ϵ is made in Figure 11.3.

From the three different forms in Eq. (11.38), it is found possible for the resonant behavior of a system to exhibit a variety of energy dependencies differing from those of the simple Lorentzian resonances discussed earlier in the chapter. In this regard, depending on the value of q resonant structures exhibit a certain amount of leeway in their design, and this can be used to tailor the functional type of their frequency response. This has been the case, for example, in the flexibilities of the design of metamaterials, and in their application in creating materials with negative index of

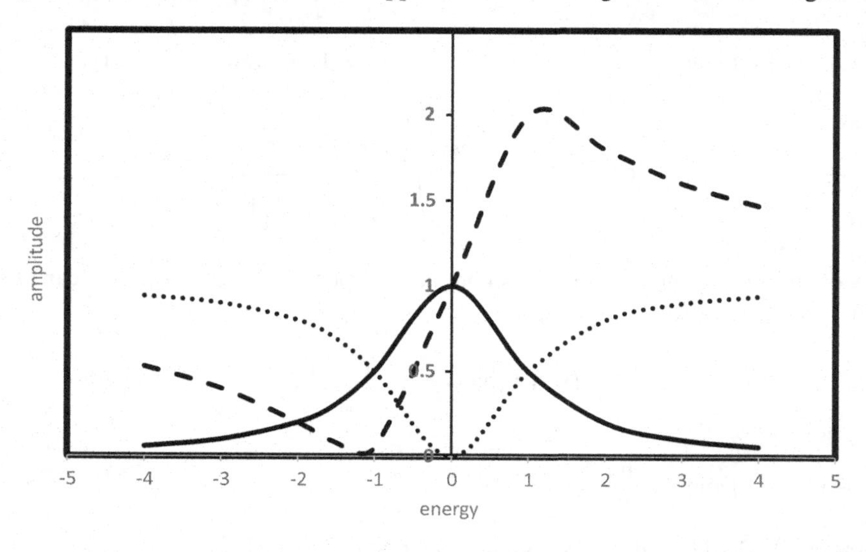

FIGURE 11.3 Schematic representation of the Lorentzian resonance form $1/(1+\epsilon^2)$ (solid line), the resonance form $\epsilon^2/(1+\epsilon^2)$ (dotted line), and the resonance form $(\epsilon+1)^2/(1+\epsilon^2)$ (dashed line), all presented versus ϵ. In the plot of these functions, their amplitudes are scaled relative to the Lorentzian form (solid line) on a linear scale, so that at $\epsilon=0$ and the Lorentzian resonance has unit amplitude.

refraction and electromagnetic cloaking properties. Another example is the use of photonic crystals as a basis for formulating high Q, low loss, cavity resonators for laser and other frequency-specific applications [1–4]. Applications of resonance also arise in the design of waveguides and waveguide couplings, and these are considerations often associated with switching, sensor, and other modulation effects [1–4].

The resonant structures that have been treated here are associated with both elastic and inelastic events, resulting in a rapid change of scattering properties with an approach to a resonant frequency. In this regard, it should be noted that the resonant response often also involves energy loss in the system, and these losses tend to be the greatest at the resonant frequency. This, for example, is found to be the case in the response represented in Eq. (11.2) and is a fundamental limitation on the design of metamaterials. Consequently, a problem of much recent focus has been the development of ways of overcoming such losses to metamaterial systems [1–6].

As a final topical note, a recently developed means of designing resonances within engineered systems has been focused on types of so-called topological excitations [14]. Topological excitations arise from the features that can be engineered into a system that modify its basic geometry and as such greatly change its resonant properties. They provide, in some instances, for the formulation of resonant structures that meet important engineering criteria, being stable to small disorder in the system, and easily realized in the laboratory. As a conclusion, we shall turn to a brief discussion of the ideas of topological features and a simple example of their application [14].

11.4 TOPOLOGICAL EXCITATIONS

An interesting feature of some impurity systems is the presence in them of topological excitations associated with topological impurities [14]. These are systems in which the basic geometry of the system is locally modified, giving rise to impurity modes. An example of such a topological modification in a one-dimensional system is shown in Figure 11.4, consisting as an interface mismatch between two semi-infinite chain segments. This example will be discussed in the following. Other types of similar locally formed topological impurities are associated with surface states in higher dimensional systems, but here a focus will be on a simple one-dimensional system and the interface it displays.

In Figure 11.4a, a layered photonic crystal composed of the periodic arrangement of three basic slabs labeled A, B, and C is shown. Since C always occurs between the A and B layers, the photonic crystal can be described as an ...A, B, A, B, A, B, ... layering. Due to the periodic arrangement of the slabs, an electromagnetic band structure composed as a series of stop and pass bands is created for electromagnetic plane wave radiation propagating along the axis perpendicular to the slab surfaces.

A topological-bound state can now be created in this structure by introducing the layering shown in Figure 11.4b. In Figure 11.4b, the layering is seen to be ...A, B, A, B, B, A, B, ... Now, at the center of the layering, a B, B dislocation of two B slabs separated by a C slab has been introduced. Otherwise, the layering in Figure 11.4a is maintained in the regions to the left and right of the dislocation [14].

The B, B dislocation changes the geometry of the layering of the slabs, and it is found that the dielectric properties of the layers are easily adjusted so that a highly localized bound state can be located near the center of one of the stop bands of the structure. This localized mode is bound to the B, B dislocation and is known as a topological-bound state. In this regard, to the left and right of the B, B dislocation, the two sides of the layering have the same frequency band structure, and the B, B dislocation now binds to it a highly localized state with a frequency positioned near the center of one of the stop bands of the band structure of the one-dimensional photonic crystal.

These types of spatially localized modes have been of great current interest. One recent application of these ideas in a one-dimensional system has been made in the form of the coupling between two bus waveguides in which the topological system exhibits a variety of Fano resonances. These resonances have been observed in the resonant transmission characteristics of the system of joined waveguides.

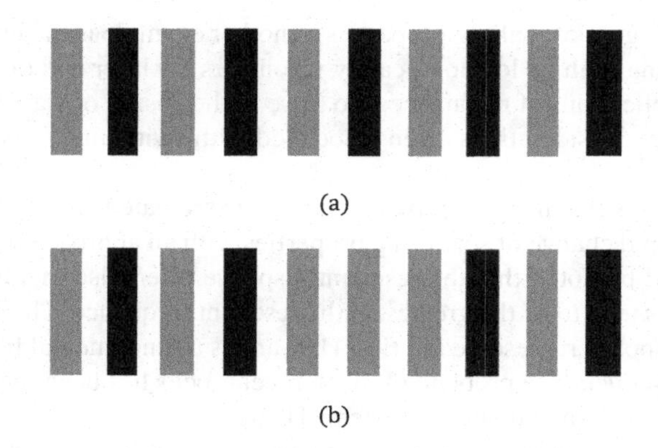

FIGURE 11.4 Schematic of: (a) one-dimensional photonic crystal layering of slabs A (black), B (gray), and C (white) for a shown layering of …BABABABABA…, (b) topological defect at the center of the crystal with two B layers forming the defect for a shown layering …BABABBBABA…. The topological defect is the BB at the center of the layers.

REFERENCES

1. McGurn, A. R. 2020. *Introduction to photonic and phononic crystals and metamaterials.* San Rafael: Morgan & Claypool Publishers.
2. McGurn, A. R. 2021. *Introduction to nonlinear optics of photonic crystals and metamaterials, 2nd Edition.* Bristol:IOP Publishing Ltd.
3. McGurn, A. R. 2015. *Nonlinear optics of photonic crystals and meta-materials.* San Rafael: Morgan & Claypool Publishers.
4. McGurn, A. R. 2018. *Nanophotonics.* Cham: Springer.
5. Pendry, J. B. 2000. Negative refraction makes a perfect lens. *Physical Review Letters* 85: 3966–3969.
6. Pendry, J. B., Schurig, D., Smith D. R., et al. Controlling electromagnetic fields. *Science* 312: 1780–1782.
7. Chartier, G. 2005. *Introduction to optics.* New York: Springer.
8. Kittle, C. 1996. *Introduction to solid state physics, 7th Edition*, 393–410. New York: John Wiley & Sons, Inc.
9. Marder, M. P. 2000. *Condensed matter physics.* New York: John Wiley & Sons, Inc.
10. Serway, R. A. and J. W. Jewett 2010. *Physics for scientists and engineers, 8th Edition.* Belmont: Brooks/Cole Cengage Learning.
11. Ouisse, T. 2008. *Electron transport in nanostructures and mesoscopic devices: An introduction.* New York: Wiley.
12. Kamenetskii, E., Sadreev, A. and A. Miroshnicenko 2018. *Fano resonances in optics and microwaves.* Cham: Springer.
13. Franklin, P. 1958, *Functions of complex variables.* Englewood Cliffs, NJ: Prentice Hall Inc.
14. Gu, L., Wang, B., Yuan, G., et al. 2021. Fano resonance from a one-dimensional topological photonic crystal, *APL Photon* 6: 086105.

12 Josephson Junction Properties and Basic Applications

In this chapter, the study of Josephson junctions is introduced along with a variety of basic solutions found in device applications involving the Josephson effect [1–12]. These include a discussion of the space- and time-dependent features of the Josephson effect itself and a variety of types of interference phenomena that are associated with junction geometries and elementary junction circuit configurations. Following these treatments, a presentation is given of the characteristics of Type I and Type II superconducting materials [1–8] and how these materials are classified in terms of the parameters in the Ginzburg–Landau theory. The chapter concludes with a discussion of the modification of the Ginzburg–Landau theory required to treat the properties of high-temperature superconductors [8–11].

These treatments focus on applications of the Ginzburg–Landau theory, which was developed in an earlier chapter (See Chapter 2.) [1–12]. The Ginzburg–Landau formulation is a reliable approach to the theoretical understanding of many properties important in device applications and has been shown to be obtained as a limit of the microscopic BCS theory. In this regard, in the last part of this chapter, a discussion of the extension of the Ginzburg–Landau theory to treat the properties of high-temperature superconductors is given.

We begin the discussions with an introduction to the ideas of the time-dependent Ginzburg–Landau free energy [12–14], and some treatment of the gauge symmetry requirements for the considerations of applied electromagnetic fields [15]. This leads naturally to the Josephson effect found in the interaction of two superconductors separated by a barrier of normal material [16–18].

12.1 TIME-DEPENDENT GINZBURG–LANDAU FREE ENERGY

The theory of electrodynamics is characterized to an important extent by its properties of gauge symmetry, and this symmetry is found to govern the forms and types of interactions the electromagnetic fields can have when interacting with other dynamical systems [3–8,15]. To understand the origins of the gauge symmetry, consider the Maxwell equations written in the form [15]

$$\nabla \cdot \vec{D} = 4\pi\rho, \tag{12.1a}$$

$$\nabla \cdot \vec{B} = 0, \tag{12.1b}$$

$$\nabla \times \vec{H} = \frac{4\pi}{c}\vec{J} + \frac{1}{c}\frac{\partial \vec{D}}{\partial t}, \tag{12.1c}$$

and

$$\nabla \times \vec{E} + \frac{1}{c}\vec{B} = 0. \tag{12.1d}$$

DOI: 10.1201/9781003031987-12

A simplification of these equations in the consideration of some problems is made by relating the electric and magnetic fields to a set of new fields given by a scalar potential ϕ and vector potential $\vec{A}$. This is done by defining [15]

$$\vec{B} = \nabla \times \vec{A} \tag{12.2a}$$

$$\vec{E} = -\nabla\phi - \frac{1}{c}\frac{\partial \vec{A}}{\partial t}. \tag{12.2b}$$

which in linear systems are, in turn, related to the magnetic field $\vec{B} = \mu\vec{H}$ and the electric displacement $\vec{D} = \varepsilon\vec{E}$ through the permeability μ and permittivity ε. Consequently, all the four fields $\vec{E}$, $\vec{B}$, $\vec{D}$, and $\vec{H}$ are expressed in terms of the scalar potential ϕ and vector potential $\vec{A}$, as are the Maxwell equations in Eq. (12.1). In addition, the two Maxwell equations $\nabla \cdot \vec{B} = 0$ and Faradays law are automatically satisfied by Eq. (12.2).

The gauge symmetry is a symmetry focused on the relationships in Eq. (12.2) between the fields and the potentials. It arises because the forms in Eq. (12.2) can give the same fields $\vec{E}$, $\vec{B}$ under the application of a particular set of transformations made on ϕ and $\vec{A}$. Specifically, these transformations come from the realization that Eq. (12.2) are invariant in form when the functions $\vec{A}$ and ϕ are replaced by the functions [3–8,15]

$$\vec{A} - \nabla\chi \tag{12.3a}$$

and

$$\phi + \frac{1}{c}\frac{\partial \chi}{\partial t}, \tag{12.3b}$$

where χ is a differentiable scalar function. In addition, as $\vec{D}$ and $\vec{H}$ are related to $\vec{E}$ and $\vec{B}$ through the constitutive equations of the material, under these transformations of the potentials the field equations in Eq. (12.1) are also invariant in their form. The set of all transformations given by Eq. (12.3) form a symmetry group known as the gauge symmetries.

The electromagnetic field must exhibit gauge symmetry, but this symmetry must also extend to its interactions with other dynamical systems. Consequently, any theory of matter interacting with the electric and magnetic fields must be invariant under the set of gauge transformations. As a result, the field gauge symmetry would then influence the set of properties exhibited by the system. These effects on the properties of the systems interacted with are now discussed for the electrodynamics of a charged particle, and this is followed by a generalization of gauge symmetry to the treatment of the Ginzburg–Landau theory.

12.1.1 Schrodinger Gauge Symmetry

To begin, consider the Schrodinger equation for a single electron in the semi-classical theory. In the presence of electromagnetic fields, it takes the gauge invariant form [2–8,15]

$$-\frac{\hbar^2}{2m}\left(\nabla + \frac{ie}{\hbar c}\vec{A}\right)^2 \psi - e\phi\psi = i\hbar\frac{\partial}{\partial t}\psi, \tag{12.4}$$

where under the gauge transformation in Eq. (12.3), we require that ψ is replaced by

$$\psi \exp\left(i\frac{e}{\hbar c}\chi\right), \tag{12.5}$$

and where $e > 0$ is the unit of electric charge. It is seen by substitution that under the changes in Eqs. (12.3) and (12.5), the Schrodinger equation in Eq. (12.4) retains its form, and the dynamics of the electron interacting with the electric fields is invariant under the same gauge transformations governing the field dynamics. The symmetry group of the electromagnetic gauge symmetries therefore also applies to the Schrodinger equation in Eq. (12.4). In this regard, the results in Eqs. (12.4) and (12.5) are standard from quantum mechanics. They will now be directly generalized to treat the Ginzburg–Landau free energy form for superconductivity.

In the following, the gauge invariance of the energy equation in Eq. (12.4) will be seen to be directly applicable to develop a gauge invariant theory of the interaction of the Ginzburg–Landau free energy with an electromagnetic field. This allows us to introduce a leading order space and time variation into the dynamics of the Ginzburg–Landau free energy form considered in our earlier discussion of superconductivity.

12.1.2 Ginzburg–Landau Form

In particular, in Eq. (2.148), it was shown that the function $\psi(\vec{r})$ characterizing the state of the superconductor in a static magnetic field is given as a solution of the Ginzburg–Landau equation. For time-independent systems, this has the form [1–3]

$$\frac{1}{2m_s}\left(-i\hbar\nabla - \frac{q}{c}\vec{A}\right)^2 \psi + \alpha\psi + \beta|\psi|^2\psi = 0, \tag{12.6}$$

where $q = -2e$ for $e > 0$, $\alpha = \alpha_0 \dfrac{T - T_c}{T_c}$, $\alpha_0 > 0$, T is the temperature of the system, T_c is the critical temperature, $\beta > 0$, and $\vec{A}(\vec{r})$ is the magnetic vector potential. The solution of Eq. (12.6) and its interpretation in terms of physical variables generates a reasonable approximation to the static field-dependent behaviors of superconducting systems. These include, for example, the Meissner effect, the superconducting current density, and the flux quantization behaviors.

Continuing this treatment to include time-dependent properties of $\psi(\vec{r},t)$ requires a gauge invariant generalization of Eq. (12.6) involving both space- and time-dependent electromagnetic fields. This can be accomplished to leading order in the space–time variables by analogy with the space–time treatment of the gauge invariant Schrodinger in Eq. (12.4). In this way, a simple generalization is offered by the form [12–14]

$$\frac{1}{2m_s}\left(-i\hbar\nabla - \frac{q}{c}\vec{A}\right)^2 \psi + \alpha\psi + \beta|\psi|^2\psi = \frac{1}{2\gamma}\left(\frac{\partial}{\partial t} + iq\tilde{\phi}\right)\psi, \tag{12.7a}$$

in which [12–14]

$$\tilde{\phi} = \phi + \frac{2\mu}{q}, \tag{12.7b}$$

where ϕ is the electric potential, $\gamma < 0$ is a parameter related to the rate of relaxation of ψ, and μ is the chemical potential of the medium. The resulting extension in Eq. (12.7) is then found to be gauge invariant under a transformation involving the replacement of ψ by

$$\psi \exp\left(-i\frac{q}{\hbar c}\chi\right), \tag{12.8}$$

and at the same time replacing $\vec{A}(\vec{r},t)$ and $\phi(\vec{r},t)$ by the forms

$$\vec{A}-\nabla\chi \tag{12.9a}$$

and

$$\phi+\frac{1}{c}\frac{\partial\chi}{\partial t}, \tag{12.9b}$$

where $\chi(\vec{r},t)$ is a differentiable scalar space–time function. Under this transformation, both the Maxwell equations and the Ginzburg–Landau free energy form remain the same. Consequently, the introduction of χ to the problem does not affect the physics that is represented by the solutions of Eq. (12.7). In addition, the set of all $\chi(\vec{r},t)$ constitute a symmetry group that leaves the physics of the problem unchanged.

As developed in this manner, the free energy form in Eq. (12.7) represents the simplest generalization of the basic static Ginzburg–Landau free energy polynomial to a theory involving a first-order time derivative and a second-order space derivative. The resultant theory should then provide the lowest order corrections in space- and time dependence of the superconducting system. This holds, in particular, in the vicinity of the phase transition.

12.1.2.1 An Example

As an example of the gauge symmetry applied to a time-dependent problem, consider the space-independent form of Eq. (12.7) in the absence of electromagnetic fields. In this limit, Eq. (12.7) reduces to the time-dependent form [12–14]

$$\alpha\psi+\beta|\psi|^2\psi=\frac{1}{2\gamma}\left(\frac{\partial}{\partial t}+2i\mu\right)\psi, \tag{12.10}$$

where $\gamma<0$. This can be used to study the relaxation of a non-equilibrium system to its final equilibrium configuration.

To study the relaxation in Eq. (12.10), assume that $\psi=|\psi|e^{i\theta}$. Substituting this general form of ψ into Eq. (12.10) gives two equations: one for the real part and one for the imaginary part of ψ. These are represented by

$$\frac{\partial|\psi|}{\partial t}=2\gamma\left(\alpha|\psi|+\beta|\psi|^3\right) \tag{12.11a}$$

and

$$\frac{\partial\theta}{\partial t}+2\mu=0. \tag{12.11b}$$

The equilibrium configuration state of the system is obtained as $\psi_{eq}=|\psi_{eq}|e^{i\theta_{eq}}$, which is characterized by $\frac{\partial|\psi_{eq}|}{\partial t}=0$ and a time-dependent solution of Eq. (12.11b) for θ_{eq}. Note, in this regard that, to the order of approximation represented in Eq. (12.11), θ does not depend on the real amplitude $|\psi_{eq}|$, so that the time dependence in Eqs. (12.11a) and (12.11b) are decoupled from one another.

First consider the equilibrium configuration of the system. From Eq. (12.11a), the equilibrium state is obtained from the solution of the time-independent form for the amplitude

$$\alpha\psi_{eq}+\beta|\psi_{eq}|^2\psi_{eq}=0 \tag{12.12a}$$

and the time-dependent phase equation

$$\frac{\partial \theta_{eq}}{\partial t} + 2\mu = 0. \tag{12.12b}$$

Solutions of Eq. (12.12) are then given as

$$|\psi_{eq}|^2 = -\frac{\alpha}{\beta} = -\frac{\alpha_0}{\beta}\frac{T - T_c}{T_c}, \tag{12.13a}$$

and

$$\theta_{eq} = -2\mu t + \theta_0, \tag{12.13b}$$

where θ_0 is an integration constant set by boundary conditions. In this regard, the equilibrium configuration $\psi_{eq} = |\psi_{eq}|e^{i\theta_{eq}}$ has a constant amplitude and a periodic time dependence determined by the constant chemical potential of the system.

To obtain the relaxation of a non-equilibrium state ψ to the equilibrium configuration ψ_{eq}, consider that the system begins in a non-equilibrium state represented by the form $\psi = |\psi|e^{i\theta}$. Writing the amplitude of the non-equilibrium state as $|\psi| = |\psi_{eq}| + \delta\psi$, where $\delta\psi$ is a small, real parameter characterizing the deviation of the system from equilibrium [12–14] and substituting in Eq. (12.11a) gives to order $\delta\psi$

$$\frac{\partial \delta\psi}{\partial t} = -4\gamma\alpha\ \delta\psi. \tag{12.14a}$$

Solving Eq. (12.14a), the relaxation of $\delta\psi$ is of an exponential nature given by [12]

$$\delta\psi = \delta\psi_0 e^{-4\gamma\alpha t} = \delta\psi_0 \exp\left(-4\gamma\alpha_0\frac{T - T_c}{T_c}t\right) = \delta\psi_0 \exp\left(-4\left|\gamma\alpha_0\frac{T - T_c}{T_c}\right|t\right), \tag{12.14b}$$

where $\delta\psi_0$ characterizes the initial non-equilibrium configuration. In addition, from Eq. (12.11b), the argument of the complex exponential form remains unchanged and is given by [12–14]

$$\theta = -2\mu t + \theta_0, \tag{12.14c}$$

where θ_0 represent the phase of the initial configuration.

From Eq. (12.14b), the amplitude of the order parameter is seen to exponentially decay to that of the equilibrium configuration, and the rate of decay slows as the critical temperature is approached from below. To the order of approximation used here, the phase of the order parameter is not affected by the relaxation.

12.2 JOSEPHSON JUNCTION

Next, we consider the problem of the tunneling of superconducting electron pairs between separate superconducting materials [1–11,16–18]. For simplicity, we shall treat the system at zero temperature and adopt a discussion based on the Schrodinger equation of the system. Specifically, a thin planar region of non-superconducting, insulating, material separates two semi-infinite regions of superconducting media. See Figure 12.1 for a schematic of this geometry. To the left of the barrier, the superconductor is described by the eigenfunction [1,2]

$$\psi_1 = |\psi_1|e^{i\theta_1} = \sqrt{n_1}e^{i\theta_1}, \tag{12.15a}$$

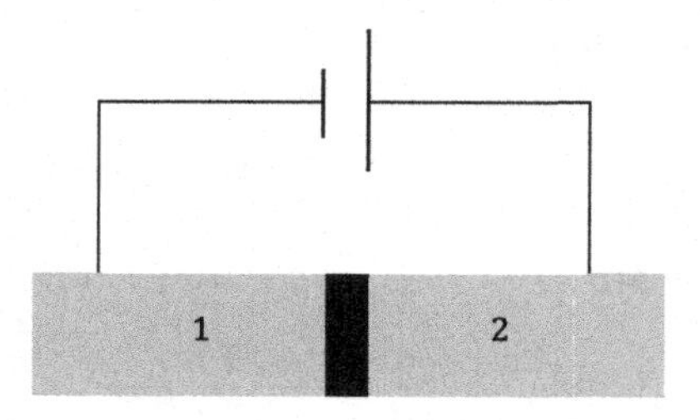

FIGURE 12.1 Biased Josephson junction composed as a superconductor 1 (gray region labeled 1) with ϵ_1, n_1, θ_1 and a superconductor 2 (gray region labeled 2) with ϵ_2, n_2, θ_2. The black region between them is a weak link formed of normal metal.

and to the right of the barrier

$$\psi_2 = |\psi_2| e^{i\theta_2} = \sqrt{n_2} e^{i\theta_2}, \tag{12.15a}$$

where n_1 and n_2 are the Cooper pair densities in the two respective superconducting regions.

When the barrier separating the two superconductors is infinitely thick, so that the wavefunctions of the two superconductors do not interact, the two eigenfunctions of the noninteracting systems satisfy [1–4]

$$i\hbar \frac{\partial \psi_1}{\partial t} = H_1 \psi_1 = \epsilon_1 \psi_1 \tag{12.16a}$$

and

$$i\hbar \frac{\partial \psi_2}{\partial t} = H_2 \psi_2 = \epsilon_2 \psi_2. \tag{12.16a}$$

Here, H_1 and H_2 are the Hamiltonians of the two superconducting regions, and ϵ_1 and ϵ_2 are the eigenenergies of the two superconductors. Now, as the barrier is made, finite electron pairs can tunnel between the two superconductors in a manner similar to the tunneling of electrons between two metals separated by an insulating barrier.

In the case in which the barrier between the superconductors is thin, one must therefore include the possibility of tunneling between the two superconductors. With this possibility, the Hamiltonians in Eq. (12.16) are coupled by terms related to the rate of transition between the two superconductors. In this way, assuming the simplest possible forms for weakly coupled systems, we assume that [1–4]

$$i\hbar \frac{\partial \psi_1}{\partial t} = \epsilon_1 \psi_1 + \epsilon \psi_2 \tag{12.17a}$$

and

$$i\hbar \frac{\partial \psi_2}{\partial t} = \epsilon_2 \psi_2 + \epsilon \psi_1. \tag{12.17a}$$

Here ϵ is related to the transition rate of the tunneling between the two separate superconductors, and ϵ_1 and ϵ_2 are the eigenenergies of the two separate superconductors. Note, in this regard, that the form of Eq. (12.17) is reminiscent of both the electron tunneling problem and the two-level states problem. Specifically, the rate of tunneling into one of the superconductors is proportional to the wavefunction of the other superconductor.

To solve the tunneling problem, first consider the time derivatives in Eq. (12.17). Given wavefunctions of the form [1–4]

$$\psi_l = \sqrt{n_l}\,e^{i\theta_l}, \tag{12.18a}$$

where $l = 1, 2$ label the states on the respective sides of the insulating barrier, it follows that

$$\frac{\partial \psi_l}{\partial t} = \frac{\partial}{\partial t}\left(\sqrt{n_l}\,e^{i\theta_l}\right) = \frac{1}{2}\frac{\dot{n}_l}{\sqrt{n_l}}e^{i\theta_l} + i\dot{\theta}_l\sqrt{n_l}\,e^{i\theta_l}. \tag{12.18b}$$

Substituting these derivatives into Eq. (12.17), we obtain four equations generated from the real and imaginary parts of the two equations. These are

$$\dot{n}_1 = 2\frac{\epsilon}{\hbar}\sqrt{n_1 n_2}\sin(\theta_2 - \theta_1) = -\dot{n}_2, \tag{12.19a}$$

$$\dot{\theta}_1 = -\frac{\epsilon_1}{\hbar} - \frac{\epsilon}{\hbar}\sqrt{\frac{n_2}{n_1}}\cos(\theta_2 - \theta_1), \tag{12.19b}$$

and

$$\dot{\theta}_2 = -\frac{\epsilon_2}{\hbar} - \frac{\epsilon}{\hbar}\sqrt{\frac{n_1}{n_2}}\cos(\theta_1 - \theta_2). \tag{12.19c}$$

For an additional simplification, in the treatment of these equations to follow, it will be assumed that the superconductors on the left and right of the barrier are made of the same material. The generalization to different materials is left to the reader.

It then follows from Eqs. (12.19b) and (12.19c) that, for the case in which the superconductors on both sides of the barrier are of the same material, $n_1 = n_2 = n$ and

$$\dot{\theta}_2 - \dot{\theta}_1 = -\frac{\epsilon_2 - \epsilon_1}{\hbar}, \tag{12.20}$$

so that the rate of change of the relative phases of the two superconductors depends only on the energy difference of the two uncoupled superconductors. Consequently, if the superconductors to the left and right of the barrier are at the same electric potential, $\epsilon_2 - \epsilon_1 = 0$ so that $\theta_2 - \theta_1$ is a constant. On the other hand, if a constant electric potential difference V is applied between the two sides of the barrier, then $\epsilon_2 - \epsilon_1 = qV \neq 0$, and from Eq. (12.20) the phase is given by the linear time dependence [1–4]

$$\theta_2 - \theta_1 = -\frac{\epsilon_2 - \epsilon_1}{\hbar}t + \delta\theta_0, \tag{12.21}$$

where $\delta\theta_0$ is an integration constant. See Figure 12.2 for a schematic of these two cases.

Next, consider the current flow between the left and right sides of the barrier. From Eq. (12.19a), it is found that the rate of change of the charge density $\dot{\rho}_1$ in superconductor ψ_1 resulting from the pair tunneling is given by [2]

$$\dot{\rho}_1 = q\dot{n}_1 = 2q\frac{\epsilon}{\hbar}n\sin(\theta_2 - \theta_1). \tag{12.22a}$$

Since the change in charge is caused by the charge transfer between the two superconductors, it follows that [1–3]

$$J_{1\to 2} \propto 2q\frac{\epsilon}{\hbar}n\sin(\theta_2 - \theta_1). \tag{12.22b}$$

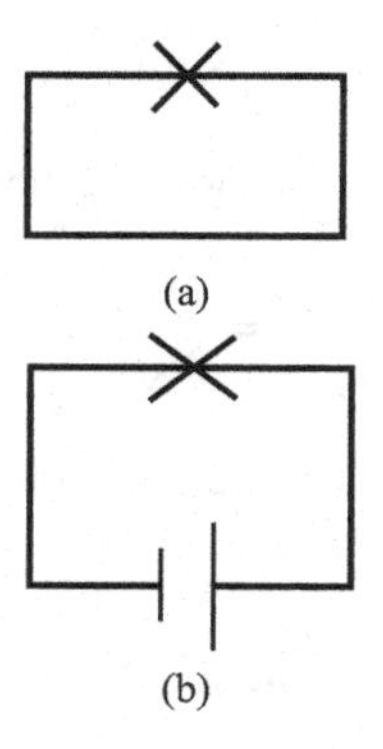

FIGURE 12.2 Schematic of (a) an isolated Josephson junction and (b) a Josephson junction with a battery bias.

Specifically, the current density $J_{1\rightarrow 2}$ for charge flow from ψ_1 into ψ_2 is proportional to the rate of change of the net charge in ψ_1.

Summarizing, from Eqs. (12.19) to (12.22), we then find that

$$J_{1\rightarrow 2} = J_0 \sin(\theta_2 - \theta_1), \tag{12.23}$$

where J_0 is the current amplitude, and the current flow is ultimately seen to depend on the relative phase difference between the superconductors. The case in which $\epsilon_2 - \epsilon_1 = 0$ so that $\theta_2 - \theta_1$ is a constant is referred to as the DC Josephson effect. This is an interesting configuration for which a current flows through the barrier in the absence of an applied electric potential difference. On the other hand, the case in which $\epsilon_2 - \epsilon_1 = qV \neq 0$, as seen in Eqs. (12.21) and (12.23), gives a time-dependent phase and is known as the AC Josephson effect. In this configuration, an AC current is generated across the barrier.

12.3 SPATIAL DEPENDENT EFFECTS: MAGNETIC FIELDS

The Josephson effect exhibits a variety of interesting space-dependent properties coming from its descriptions in terms of the solutions of a nonlinear wave equation [1–3]. (In this regard, see the formulations in Eqs. (12.6) and (12.7).) Since the solution ψ of the wave equation can also interact with electromagnetic fields, the additional requirements of gauge invariance of the interactive field terms are basic to understanding many of the wave interference effects observed in superconductors. We now turn to a discussion of the origin of gauge invariance in the Josephson effect, and its consequences on the wavefunctions and current flow in superconducting junction devices.

To handle spatial field effects in the context of the Ginzburg–Landau theory, it is necessary to return to a consideration of the gauge invariant form [2,12–14]

$$\frac{1}{2m_s}\left(-i\hbar\nabla - \frac{q}{c}\vec{A}\right)^2 \psi + \alpha\psi + \beta|\psi|^2 \psi = \frac{1}{2\gamma}\left(\frac{\partial}{\partial t} + iq\tilde{\phi}\right)\psi, \tag{12.24}$$

given earlier in Eq. (12.7a). Specifically, it is noted that the space dependence of ψ arises in Eq. (12.24) largely through the $\left(-i\hbar\nabla - \frac{q}{c}\vec{A}\right)^2 \psi$ term involving the spatial derivative and the vector potential. In the following, we shall see that a more precise understanding of the effects of the vector potential on the system wavefunction and the nature of its phase coherence are attained by making a simple transformation on Eq. (12.24). This transformation leads to an understanding of the origins of a variety of spatial junction properties based on phase interference.

In this regard, consider the transformation made by rewriting $\psi(\vec{r})$ in the form [1–4,18]

$$\psi(\vec{r}) = \exp\left[\frac{iq}{\hbar c}\int^{\vec{r}} \vec{A}(\vec{r})\cdot d\vec{r}\right]\psi_0(\vec{r}), \tag{12.25}$$

which separates out a phase generated by the vector potential from a function $\psi_0(\vec{r})$ that is independent of the vector potential. Substituting, Eq. (12.25) into Eq. (12.24), it then follows that $\psi_0(\vec{r})$ must be a solution of

$$\frac{1}{2m_s}(-i\hbar\nabla)^2\psi_0 + \alpha\psi_0 + \beta|\psi_0|^2\psi_0 = \frac{1}{2\gamma}\left(\frac{\partial}{\partial t} + iq\tilde{\phi}\right)\psi_0, \tag{12.26}$$

which does not involve the vector potential. Consequently, the net effect of the vector potential is to add a position-dependent phase to the wavefunction of the total system in Eq. (12.25). This added phase gives rise to a variety of important coherence effects that are now treated.

Returning to the earlier discussion of the Josephson junction in Eqs. (12.15)–(12.23), in the presence of a static magnetic field described by a vector potential $\vec{A}(\vec{r})$ the superconductor wavefunctions in Eq. (12.15) must now include a contribution to the phase from the vector potential. This changes the basic forms of the wavefunctions by adding a spatial variation to their phases. In this regard, for the superconductor on the left of the barrier, it follows that

$$\psi_1 = |\psi_1|e^{i\varphi_1} = \sqrt{n_1}e^{i\varphi_1} \tag{12.27a}$$

where n_1 is the pair density and [1–4,18]

$$\varphi_1 = \theta_1 + \frac{q}{\hbar c}\int_{\vec{r}_0}^{\vec{r}_1} \vec{A}(\vec{r})\cdot d\vec{r}, \tag{12.27b}$$

where θ_1 is the zero-field phase, and $\vec{r}_1$ is the position at which θ_1 is measured. In the superconductor to the right of the barrier, similarly,

$$\psi_2 = |\psi_2|e^{i\varphi_2} = \sqrt{n_2}e^{i\varphi_2} \tag{12.27c}$$

where n_2 is the pair density and [2]

$$\varphi_2 = \theta_2 + \frac{q}{\hbar c}\int_{\vec{r}_0}^{\vec{r}_2} \vec{A}(\vec{r})\cdot d\vec{r}, \tag{12.27d}$$

where θ_2 is the zero-field phase of the wavefunction and $\vec{r}_2$ is the position at which θ_2 is measured. In the expressions for both the order parameters (i.e., ψ_i for $i = 1$ or 2), the line integral is performed relative to a fixed position $\vec{r}_0$.

12.4 JOSEPHSON JUNCTION IN A STATIC MAGNETIC FIELD

The earlier treatment leading to Eqs. (12.22) and (12.23) can now be reworked for the wavefunctions in Eq. (12.27), in the case of a static magnetic field applied to the system. The modified derivation involves the vector potential contributions to the phases of the wavefunctions, which introduces a space dependence to their phase difference. Specifically, this enters as a modification of the phase difference $\theta_2 - \theta_1$ in Eqs. (12.22) and (12.23) through the introduction from Eq. (12.27) of a term in the vector potential. Nevertheless, the treatment of the phase and current densities parallels that for the system in the absence of the vector potential.

Proceeding in this manner, we then see that Eq. (12.23) becomes [1–7,18]

$$J_{1\to 2} = J_0 \sin\left(\theta_2 - \theta_1 + \frac{q}{\hbar c}\int_{\vec{r}_1}^{\vec{r}_2} \vec{A}(\vec{r})\cdot d\vec{r}\right), \tag{12.28}$$

where J_0 is again the current amplitude. The current flow is ultimately found to depend on the relative phase difference between the superconductors, which now includes a shifting effect from the position-dependent vector potential. In addition, the time dependence of the position-dependent phase is from Eq. (12.21), given by the generalized form [2]

$$\theta_2 - \theta_1 + \frac{q}{\hbar c}\int_{\vec{r}_1}^{\vec{r}_2} \vec{A}(\vec{r})\cdot d\vec{r} = -\frac{\epsilon_2 - \epsilon_1}{\hbar}t + \delta\theta_0, \tag{12.29}$$

where the line integral of the vector potential is evaluated between the positions of the θ_2 and θ_1 phase measurements.

12.5 REAL-WORLD VERSUS IDEAL JOSEPHSON JUNCTION

The earlier discussions treated an ideal Josephson junction. (See the circuit schematic in Figure 12.3a.) In the idealized junction, the barrier only acts as a tunneling mechanism, yielding a phase-dependent current crossing the barrier. In the real world, however, the barrier also acts as a source of resistance R for the current flowing through it, and, in addition, as a capacitor C for the charges developed across the barrier [2–4]. These additional aspects of the junction can be introduced by representing the real-world junction by an equivalent circuit involving an ideal Josephson junction, and the junction resistance and capacitance. (See the circuit schematic in Figure 12.3b.) This equivalent circuit is the analog of the equivalent circuit for the internal resistance of a real-world battery, written in terms of an ideal battery and its internal resistance. We now make a brief aside to remark on the real-world corrections to the ideal Josephson junction before proceeding with our ideal-world discussions.

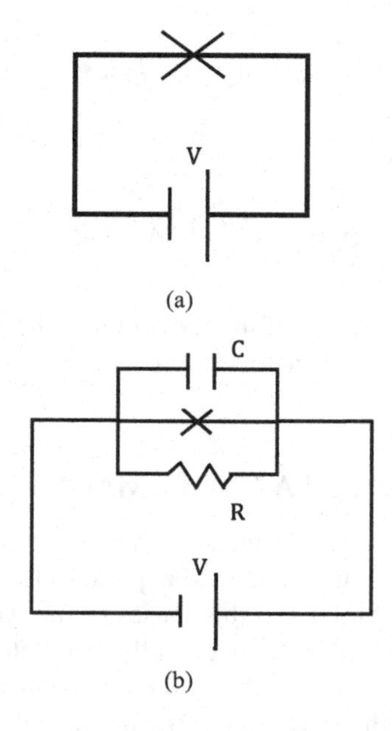

FIGURE 12.3 Schematic of (a) an ideal Josephson junction with a battery bias and (b) a circuit for a real Josephson junction with a battery bias.

In this regard, we have seen that the ideal junction in the absence of magnetic fields is characterized by relationships of the form

$$I_{1\to 2} = I_0 \sin(\theta_2 - \theta_1) \tag{12.30a}$$

and

$$\dot{\theta}_2 - \dot{\theta}_1 = -\frac{qV}{\hbar}, \tag{12.30b}$$

where $I_{1\to 2}$ is the junction current, I_0 is the current amplitude, and V is the electric potential applied across the barrier. In the real world, these relationships must now be generalized to include a consideration of the barrier resistance and capacitance. Under these considerations the current through the barrier takes the form given by [2,3]

$$I_{1\to 2} = \frac{V}{R} + I_0 \sin(\theta_2 - \theta_1) + C\dot{V} \tag{12.31a}$$

and

$$\dot{\theta}_2 - \dot{\theta}_1 = -\frac{qV}{\hbar}. \tag{12.31b}$$

These forms arise directly from a consideration of the real-world circuit in Figure 12.3b. Here R in the first term on the right of Eq. (12.31a) is the junction resistance, C in the last term on the right of Eq. (12.31a) is the junction capacitance, and $V = \frac{\epsilon_2 - \epsilon_1}{q}$ is the electric potential applied across the barrier.

By making a change of variables, Eq. (12.31) can be written in a more convenient from. Specifically, defining $\tau = \frac{|q|}{\hbar} I_0 R t$ and $\beta_C = \frac{|q|}{\hbar} I_0 R^2 C$, $\delta\theta = \theta_2 - \theta_1$, $i = \frac{I_{1\to 2}}{I_0} = \frac{I}{I_0}$, Eq. (12.31a) becomes [2]

$$i = \sin(\delta\theta) + \delta\dot{\theta} + \beta_C \delta\ddot{\theta}, \tag{12.32}$$

where $\delta\dot{\theta} = \frac{d\delta\theta}{d\tau}$ and $\delta\ddot{\theta} = \frac{d\delta\dot{\theta}}{d\tau}$. In general, the solution of this equation requires a numerical approach. However, for $\beta_C = 0$, Eq. (12.32) can be integrated. As an illustration, this limit is now considered.

For $\beta_C = 0$, Eq. (12.32) can be rewritten in the form [2]

$$\int^{\delta\theta} \frac{d(\delta\theta)}{i - \sin\delta\theta} = \tau + a, \tag{12.33}$$

where a is an integration constant. In the case that $i^2 > 1$, it follows on evaluating the integral that

$$\frac{2}{\sqrt{i^2 - 1}} tan^{-1} \left[\frac{\text{i} \tan\frac{\delta\theta}{2} - 1}{\sqrt{i^2 - 1}} \right] = \tau + a, \tag{12.34a}$$

which is then rewritten in the form

$$\delta\theta = 2 \tan^{-1} \left\{ \frac{\sqrt{i^2 - 1}}{i} \tan\left[\frac{\sqrt{i^2 - 1}(\tau + a)}{2} \right] + \frac{1}{i} \right\}. \tag{12.34b}$$

Note that the tangent function in Eq. (12.34b) is periodic in integer multiples of π. Consequently, the period of the motion of $\delta\theta$ in the time variable τ, denoted by Γ, is given by

$$\Gamma = \frac{2\pi}{\sqrt{i^2 - 1}}, \tag{12.35a}$$

or alternatively in our original variable t, the period is

$$T = \frac{\hbar}{|q| I_0 R}\Gamma = \frac{2\pi\hbar}{|q| R\sqrt{I^2 - I_0^2}} = \frac{\pi\hbar}{eR\sqrt{I^2 - I_0^2}}. \tag{12.35b}$$

It then follows from the relationship $V = \frac{\hbar}{|q|}\frac{d(\delta\theta)}{dt} \approx \frac{\hbar}{2e}\frac{2\pi}{T}$ in Eq. (12.31b) that the average of the potential over a period of the motion gives

$$V = R\sqrt{I^2 - I_0^2}. \tag{12.36}$$

A similar procedure also applies to the case $I^2 < I_0^2$.

The results for the system are that on time average

$$V = 0 \text{ for } I^2 < I_0^2, \tag{12.37a}$$

which is expected from the AC Josephson effect and

$$V = \frac{I}{|I|}\sqrt{(IR)^2 - (I_0 R)^2} \text{ for } I^2 > I_0^2 \tag{12.37b}$$

from our earlier result. For the limit that $I^2 \gg I_0^2$, the system is seen to exhibit Ohmic behavior, as the resistance of the junction dominates the behavior of the system [2,3].

Though Eq. (12.31) characterizes the real-world junction, in the treatments to follow we shall continue the discussions in the ideal world. This leads to a simplification of the considerations, while illustrating much of the basic physics involved in Josephson junctions. We shall now return to a consideration of some basic properties of Josephson junctions in magnetic fields with a discussion of junctions having finite cross-sectional areas.

12.6 JOSEPHSON JUNCTIONS OF FINITE CROSS-SECTIONAL AREA

Consider a junction [2–8,18] between two superconductors characterized by ψ_1 and ψ_2 that are interfaced in the $y-z$ plane with a rectangular interface area of sides L_y, L_z. (See Figure 12.4.) A constant magnetic field $B\hat{k}$ exists in the barrier region between the two superconductors, which is of length d and has within it a vector potential $\vec{A} = -By\hat{i}$.

The current density through the barrier is obtained as a function of the y coordinate in the junction from Eq. (12.28) written as [2–4,18]

$$J_{1\to 2}(y) = J_0 \sin\left(\theta_2 - \theta_1 + \frac{q}{\hbar c}\int_{\vec{r}_1}^{\vec{r}_2} \vec{A}(\vec{r})\cdot d\vec{r}\right), \tag{12.38}$$

where $\vec{r}_2 = x_2\hat{i} + y\hat{j}$ and $\vec{r}_1 = x_1\hat{i} + y\hat{j}$ for x_2 and x_1 in the bulk of ψ_2 and ψ_1 where there is no magnetic induction. The integration path in Eq. (12.38) is along the line passing through $\vec{r}_2 = x_2\hat{i} + y\hat{j}$ and $\vec{r}_1 = x_1\hat{i} + y\hat{j}$. It then follows that

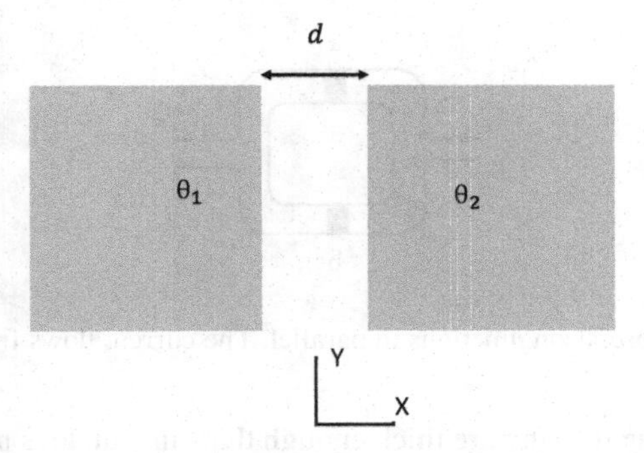

FIGURE 12.4 Junction between two superconductor with the uniform applied magnetic field out of the page.

$$\frac{q}{\hbar c}\int_{\vec{r}_1}^{\vec{r}_2}\vec{A}(\vec{r})\cdot d\vec{r}=-\frac{q}{\hbar c}Byd \tag{12.39}$$

so, consequently,

$$J_{1\to 2}(y)=J_0\sin\left(\theta_2-\theta_1-\frac{q}{\hbar c}Byd\right). \tag{12.40}$$

The total current through the junction is then given by [2–4,18]

$$I=J_0L_z\int_{-L_y/2}^{L_y/2}\sin\left(\theta_2-\theta_1-\frac{q}{\hbar c}Byd\right)dy, \tag{12.41a}$$

which when evaluated has the form

$$I=\frac{J_0L_z\hbar c}{qBd}\left\{\cos\left[-\frac{q}{\hbar c}Bd\frac{L_y}{2}+\theta_2-\theta_1\right]-\cos\left[\frac{q}{\hbar c}Bd\frac{L_y}{2}+\theta_2-\theta_1\right]\right\} \tag{12.41b}$$

Simplifying Eq. (12.41) then yields [18]

$$I=I_0\frac{\sin\pi\frac{\Phi}{\Phi_{2e}}}{\pi\frac{\Phi}{\Phi_{2e}}}\left|\sin(\theta_2-\theta_1)\right|, \tag{12.42}$$

where $\Phi_{2e}=\frac{\pi\hbar c}{e}$, $\Phi=\mathrm{Bd}L$, and $I_0=J_0L_yL_z$.

It is interesting to note that a similar result as that in Eq. (12.42) is found in the theory of the diffraction of light from an aperture. Consequently, similar interference effects are present in the two systems. The results in both the systems then arise from the same types of phase-coherent processes.

12.6.1 Two Junctions in Parallel

Another configuration based on phase interference effects and which is of interest in devices that make measurement of magnetic fields is composed of two Josephson junctions in a parallel circuit [1–3]. (See Figure 12.5.) In this circuit, a current is input at the left terminal labeled 1; splits into the two currents I_a and I_b that, respectively, pass through junctions a and b; and exits at the right output terminal labeled 2. Outside the two junctions, the circuit ring is composed of superconductor, and

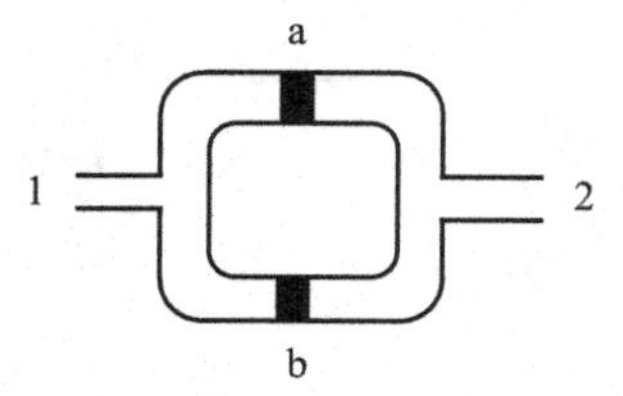

FIGURE 12.5 Plot of two Josephson junctions in parallel. The current flows from 1 to 2 through either a or b branches.

the superconductors forming the ring are thick enough that current does not flow in the center of the superconductors (located away from the superconductor surfaces), but it only flows near superconducting surfaces of the ring. The object in the following is to determine the nature of the Josephson current flow in the system and its dependence on the magnetic flux enclosed by the circuit. The result has an important technological application.

To start with, remember, from our discussions in Chapter 2, that the current in the system is given by [2–4]

$$\vec{J} = q\psi^*\vec{v}\psi = q\psi^*\frac{1}{m}\left(\vec{p}-\frac{q}{c}\vec{A}\right)\psi = q\psi^*\frac{1}{m}\left(-i\hbar\nabla-\frac{q}{c}\vec{A}\right)\psi = \frac{n}{m}\left(\hbar\nabla\theta-\frac{q}{c}\vec{A}\right) \quad (12.43)$$

Here $\vec{v} = \frac{1}{m}\left(\vec{p}-\frac{q}{c}\vec{A}\right)$ is the velocity operator, $\psi = \sqrt{n}e^{i\theta}$ is the superconducting pair wavefunction, m is the mass of the charge carriers, $q = -2e < 0$, and n is the superconducting pair density, which is considered a constant in our discussions. A closed circuit path C shown in Figure 12.5 is taken through the center of the superconductor ring. The path runs through a region well away from the surfaces of the ring so that no current flows along the path, i.e., $\vec{J} = 0$ over C. The closed circuit starts at 1 and passes through a proceeding to 2. From 2, it passes through b and returns to 1. As a result, from Eq. (12.43) and the $\vec{J} = 0$ condition, it follows that along this path

$$\nabla\theta = \frac{q}{\hbar c}\vec{A}. \quad (12.44)$$

In the discussion to follow, we focus on a study of the currents flowing from 1 to 2 through each Josephson junction. To this end, we begin by calculating the flux through the current loop.

The Josephson current from 1 to 2 depends on the phase difference between 1 and 2 going through a, denoted by δ_a, and on the phase difference between 1 and 2 going through b, denoted by δ_b. In terms of these two phases, it follows that [2–4]

$$I = I_0\left[\sin\delta_a + \sin\delta_b\right], \quad (12.45)$$

that is, it is the sum of the currents passing through each of the two parallel junctions. Here each of the sine terms in Eq. (12.45) comes from a transition argument based on that in Eqs. (12.17)–(12.23) for tunneling transitions between two superconductors separated by a barrier region exhibiting a pair tunneling transition rate between them. The focus now is to on how to determine the nature of the phases δ_a and δ_b in terms of the magnetic flux through the circuit.

To determine the flux, consider evaluating the line integral of both sides of Eq. (12.44) over the closed circuit C. This gives [2–4]

$$\oint\nabla\theta\cdot d\vec{l} = \theta^r(a)-\theta^l(a)-\left[\theta^r(b)-\theta^l(b)\right] = \frac{q}{\hbar c}\oint\vec{A}\cdot d\vec{l} = -\frac{2\pi\Phi}{\Phi_{2e}}, \quad (12.46)$$

where $\theta^r(a)$ is the phase difference going from the right edge of junction a to 2, $\theta^l(a)$ is the phase difference going from 1 to the left edge of the junction to a, $\theta^r(b)$ is the phase difference going from the right edge of junction b to 2, and $\theta^l(b)$ is the phase difference going from 1 to the left edge of the junction to b. Specifically, defining $\delta_a = \theta^r(a) - \theta^l(a)$ and $\delta_b = \theta^r(b) - \theta^l(b)$, Eq. (12.46) can be rewritten as

$$\delta_a - \delta_b = -\frac{2\pi\Phi}{\Phi_{2e}}, \tag{12.47}$$

so that the phase around the path is an integer multiple of 2π, where the integer $\frac{\Phi}{\Phi_{2e}}$ is the number of flux units $\Phi_{2e} = \frac{2\pi\hbar c}{2e}$ contained in the total flux Φ through the loop.

Both δ_a and δ_b must be linear functions of $\frac{2\pi\Phi}{\Phi_{2e}}$ with the general linear forms

$$\delta_a = \delta_0 - \frac{\pi\Phi}{\Phi_{2e}} \tag{12.48a}$$

$$\delta_b = \delta_0 + \frac{\pi\Phi}{\Phi_{2e}} \tag{12.48a}$$

for the δ_a and δ_b. Here, δ_0 is a constant that eventually is used to set boundary conditions [1–4].

In terms of δ_a and δ_b, the total current between 1 and 2 from Eqs. (12.46) to (12.48) is then [2–4,18]

$$I = I_0\left[\sin\left(\delta_0 - \frac{\pi\Phi}{\Phi_{2e}}\right) + \sin\left(\delta_0 + \frac{\pi\Phi}{\Phi_{2e}}\right)\right] = 2I_0 \sin\delta_0 \cos\frac{\pi\Phi}{\Phi_{2e}}. \tag{12.49}$$

From Eq. (12.49), we see that δ_0 enters into fixing the current amplitude, and the current is periodic in the total magnetic flux Φ through the loop of the circuit.

Next, we consider two different types of superconducting materials that display different properties dependent on the nature of their flux exclusion properties. These are the so-called Type I and Type II superconductors. All superconductors are classified as either Type I or Type II materials.

12.7 TYPE I AND TYPE II SUPERCONDUCTORS AND INTERFACES WITH NORMAL METALS

In the absence of an external magnetic induction, the difference in the free energy per volume of a bulk superconductor, F_s, and its bulk-normal metal state, F_n, can be written as [1–4]

$$\Delta F = F_n - F_s. \tag{12.50a}$$

It then follows that in the case that $\Delta F > 0$ the superconductor is stable, whereas if $\Delta F < 0$ the normal metal is stable. When the material is stable in the superconducting state, it is often helpful to write Eq. (12.50a) in terms of a magnetic energy density so that it becomes

$$\Delta F = F_n - F_s = \frac{1}{8\pi}H_c^2, \tag{12.50b}$$

and H_c is known as the critical field. The factor $\frac{1}{8\pi}H_c^2$ then represents the free energy per volume needed to be added to the bulk superconductor in order to change it to a bulk-normal metal and is computed as the magnetic work done on the superconductor, as it is positioned in an external magnetic field. In regard of this relationship, note that though the magnetic induction is zero in the superconductor, the magnetic field inside the superconducting bulk is given by $H = B - 4\pi M = -4\pi M$, so that the superconductor is a perfect diamagnet.

In the case of a Type I superconductor, Eq. (12.50b) describes the essential stability of the material in an external magnetic field. Specifically, for magnetic fields $H < H_c$ the material is a superconductor in which $B = 0$, and for magnetic fields $H > H_c$ the material is transformed to a normal metal in which $B \neq 0$. This is observed in a large class of low-temperature superconductors. There is, however, another class of materials known as Type II superconductors in which the transition between the superconductor and its normal metal is a little more complicated. In these materials, the bulk superconductor does not make a direct transition to the normal metal state at an applied magnetic field $H = H_c$. Instead, there is a transition to an intermediate state, which is composed of a mixture of superconducting and normal metal. This occurs at a critical magnetic field $H_{c1} < H_c$ at which the zero temperature system transitions from a bulk superconductor to a mixed superconductor–normal metal state. In this mixed state, the bulk superconductor is penetrated by regions of normal metal that contains a nonzero applied magnetic induction, i.e., $B \neq 0$. As the magnetic field is increased above H_{c1}, the relative amount of normal metal and superconductor in the mixed state increases until a critical field $H_{c2} > H_c$ is reached. Above H_{c2}, all of the material becomes a normal metal [1–3].

The reason for the mixed state between $H_{c1} < H < H_{c2}$ to exist comes from the surface energy associated with the interface of superconducting and normal metal materials. In Type I materials, the interface between the superconducting and normal metal state is energetically unfavorable, whereas in Type II materials the interface between the superconducting and normal metal state is energetically favorable. Consequently, the nature of the surface energy between a superconductor–normal metal interface is key to determining whether or not a material is of Type I or Type II in nature. To understand this difference in surface energies, we must go back to the Ginzburg–Landau treatment of the system near the critical temperature. Specifically, near the critical temperature, we have from Eqs. (2.148a) to (2.149) the two defining equations of the Ginzburg–Landau theory [1–4]

$$\frac{1}{2m_s}\left(-i\hbar\nabla - \frac{q}{c}\vec{A}\right)^2 \psi + \alpha\psi + \beta|\psi|^2\psi = 0 \tag{12.51}$$

and

$$\nabla \times \vec{B} = \frac{4\pi}{c}\left\{\frac{q\hbar}{2m_s}\left[-i\psi^*\nabla\psi + i\hbar\psi\nabla\psi^*\right] - \frac{q^2}{m_s c}|\psi|^2\vec{A}\right\} = \frac{4\pi}{c}\vec{j}_s. \tag{12.52}$$

In these equations, m_s is the effective mass, $q = -2e$ is the charge, $\alpha = \alpha_0 \frac{T - T_c}{T_c}$ where $\alpha_0 > 0$ is a constant, and $\beta > 0$. For the consideration of a bulk superconductor in which $\vec{A} = 0$, $\nabla\psi = 0$ so that from Eq. (12.51), it follows that

$$|\psi|^2 = -\frac{\alpha}{\beta} \tag{12.53}$$

and from Eq. (2.145)

$$F_s(T) = F_N(T) + \alpha|\psi|^2 + \frac{\beta}{2}|\psi|^4 = F_N - \frac{1}{2}\frac{\alpha^2}{\beta}. \tag{12.54a}$$

Consequently, the difference in free energy per volume near the critical temperature is

$$F_N(T) - F_s(T) = \frac{1}{8\pi} H_c^2(T) = \frac{1}{2}\frac{\alpha^2}{\beta} > 0, \tag{12.54b}$$

where now $H_c(T)$ represents the critical field at the temperature T so that

$$H_c(T) = 2|\alpha|\sqrt{\frac{\pi}{\beta}} \tag{12.55}$$

characterizes the fields needed to transform the material from the superconducting to the normal state. In addition, to the critical field in Eq. (12.55), there are two length scales that enter into fixing the properties of the superconductor. These are now considered.

If we let $\vec{A} = 0$, then below the critical temperature Eq. (12.51) takes the form [1–4]

$$\xi^2 \frac{d^2\phi}{dx^2} + \phi + |\phi|^2 \phi = 0, \tag{12.56a}$$

where $\phi = \psi\sqrt{-\frac{\beta}{\alpha}}$ and

$$\xi^2 = \frac{\hbar^2}{2m_s|\alpha|} \tag{12.56b}$$

is the square of the coherence length. In the resulting Eq. (12.56), it is then seen that ξ is a measure of the stiffness of the Ginzburg wave function, i.e., it is a measure of the characteristic length over which the wavefunction changes significantly in value. Similarly, for $\vec{A} \neq 0$ and in a medium with a constant wave function, Eq. (12.52) becomes [1–4]

$$\vec{j}_s = -\frac{q^2}{m_s c}\frac{|\alpha|}{\beta}\vec{A} = -\frac{c}{4\pi\lambda_L{}^2}\vec{A}, \tag{12.57a}$$

where

$$\lambda_L{}^2 = \frac{m_s c^2 \beta}{4\pi q^2 |\alpha|} \tag{12.57b}$$

is the square of the London penetration length. Defined in this way from Eq. (12.57), the London penetration length λ_L provides an indication of the characteristic length in which a magnetic field can penetrate into a superconductor.

These two characteristic lengths (the coherence and the London penetration lengths) are the only lengths entering into the description of the bulk properties of a superconductor. The parameters $H_c(T)$, $\xi(T)$, and $\lambda_L(T)$ give a complete characterization of the properties of the superconduct and from Eqs. (12.55), (12.56b), and (12.57b) are found to be related by

$$\lambda_L(T)\xi(T)H_c(T) = \frac{1}{2\sqrt{2}\pi}\left(\frac{\pi\hbar c}{e}\right) = \frac{1}{2\sqrt{2}\pi}\Phi_{2e}, \tag{12.58}$$

where $\Phi_{2e} = \frac{\pi\hbar c}{e}$ is the unit flux. They are now shown to determine the surface energetics at a superconductor–normal metal interface and, consequently, the difference in the properties between the Type I and Type II superconductors.

We begin by considering the penetration of an applied magnetic induction into an otherwise bulk superconductor. The smallest geometry associated with the penetration of flux into a bulk superconductor is a cylinder with a circular cross section. (See the schematic in Figure 12.6 for a cross-sectional view of such a cylinder.) In the figure, the black circle represents a region of normal metal containing lines of constant magnetic flux, which are perpendicular to the page, and the bulk superconductor is outside. At a sufficient distance outside the cylinder of normal material is a region of bulk superconductor with a wave function of the form

$$\psi = \psi_0 e^{i\theta(\vec{r})}, \tag{12.59}$$

where ψ_0 is a constant wave function amplitude, and $\theta(\vec{r})$ is a position-dependent phase.

Consider this wave function along the circular path indicated by the dashed line in Figure 12.6. The path is well within the bulk of the superconductor. Along this path, we substitute the form in Eq. (12.59) into Eq. (12.52), and take $\vec{j}_s = 0$ in the bulk region where no current exists to obtain [1–4]

$$\hbar c \nabla\theta = -2e\vec{A}. \tag{12.60}$$

Integrating this around the dashed line, it follows that

$$\hbar c \oint \nabla\theta \cdot d\vec{l} = \hbar c 2\pi s = -2e \oint \vec{A} \cdot d\vec{l} = -2e\Phi, \tag{12.61}$$

where $\oint \vec{A} \cdot d\vec{l} = \int_\sigma \vec{B} \cdot d\vec{\sigma} = \Phi$ is the flux through the black circle, and s is an integer. From Eq. (12.61), we see that the smallest nonzero flux through the cylinder is

$$\Phi_{2e} = \frac{2\pi\hbar c}{2e} = \frac{\pi\hbar c}{e}. \tag{12.62}$$

All other flux through the cylinder are integer multiples of the unit of flux defined in Eq. (12.62). This result on flux can now be used to obtain estimates of the critical fields for flux penetration into a Type II superconductor [1–3].

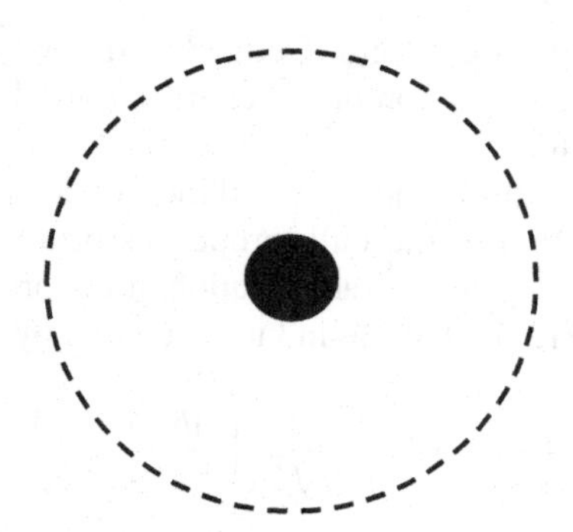

FIGURE 12.6 Circular cross section area of a normal metal flux containing cylinder (black area) surrounded by white area of bulk superconductor. The dashed line is an integration path within the bulk.

Returning to our treatment of the critical fields H_{c1} and H_{c2} of a Type II superconductor, first consider the field H_{c1} at which the mixed state first occurs. At this field, flux is beginning to enter the bulk superconductor, and it is expected that the basic flux cylinder should have a cross-sectional radius equal to the London penetration length, λ_L. The London penetration length measures the characteristic length in which the system can be made to go from the superconducting state to the normal state and, hence, provides a natural measure of the flux cylinder. At the transition to the mixed state, the flux in such a single cylinder of radius λ_L is given by $\pi\lambda_L{}^2H_{c1}$. For the transition to the mixed state to begin, this flux must just become equal to the basic unit of flux Φ_{2e} supported in the superconductor. Consequently, the condition for the transition to begin is that

$$\pi\lambda_L{}^2H_{c1} = \Phi_{2e}. \tag{12.63}$$

On the other hand, at the upper critical field H_{c2} the radii of the flux cylinders should begin to overlap one another, and this overlap is characterized by the other characteristic length of the material. The typical length scale for the superconductor wave function to make a spatial change from the bulk superconductor to the normal metal is governed by the coherence length ξ, and this length should characterize the separation between touching normal regions of neighboring flux cylinders. The flux through a cylinder of radius ξ at the upper critical field is then $\pi\xi^2H_{c2}$, and its smallest nonzero value is Φ_{2e}. Consequently, the condition on H_{c2} is that

$$\pi\xi^2H_{c2} = \Phi_{2e}, \tag{12.64}$$

and this sets the condition for a nonzero H_{c2}. In this regard, from both Eqs. (12.63) and (12.64), it follows that

$$\frac{H_{c2}}{H_{c1}} = \frac{\lambda_L{}^2}{\xi^2}. \tag{12.65}$$

(Note: In this picture, at the upper field transition, the flux cylinders can be thought of as a close packed array of flux cylinders, each of radius ξ.)

Next, consider the energetics of these flux configurations in the mixed state and the consequences to the mixed state stability. The free energy associated with creating a single flux cylinder in the bulk superconductor at H_{c1} increases the superconductor free energy toward that of the normal state. The free energy per length of the magnetic flux cylinder can be written as

$$f_{\text{mag}} = \frac{1}{8\pi}H_{c1}^2\pi\lambda_L{}^2. \tag{12.66a}$$

We see then that f_{mag} is the free energy per length of a single flux cylinder in the bulk superconductor. Similarly, from a consideration of Eqs. (12.64) and (12.50b)

$$f_{\text{core}} = \frac{1}{8\pi}H_c^2\pi\xi^2 \tag{12.66b}$$

is related to the free energy per length associated with converting the superconductor directly to the normal state. Specifically, to see this consider the bulk superconductor as being formed of a close packed array of superconductor cylinders of radii ξ and compute the energy required to change a single cylinder from superconductor to normal metal by application of a magnetic field. In this regard, from Eq. (12.50b) we see that the f_{core} in Eq. (12.66) represents the free energy per length to change a single cylinder of area $\pi\xi^2$ directly from superconductor to normal metal.

Next, we use the results in Eq. (12.66) to create a second relationship for H_{c1}. In this picture, the free energy per length of a cylinder of radii ξ, which contains a single flux vortex cylinder of radius λ_L is then given by combining the contributions in Eq. (12.66) to obtain [1–4]

$$f_{\text{Total}} = f_{\text{core}} - f_{\text{mag}} = \frac{1}{8}\left[H_c^2\xi^2 - H_{c1}^2{\lambda_L}^2\right]. \tag{12.67}$$

The total free energy per length f_{Total} is representative of the free energy of a single fluxoid in the system just before the transition of the material from the bulk superconductor to the mixed state and in which there is a fluxoid with the minimum flux Φ_{2e}. At the transition between the bulk superconductor and the mixed state, we would then expect that $f_{\text{Total}} = 0$. Under this condition, the system is just losing its stability upon acquiring the minimum unit of flux. From Eq. (12.67), it then follows that at the bulk superconductor–mixed state transition

$$\frac{H_{c1}}{H_c} = \frac{\xi}{\lambda_L} \tag{12.68}$$

and from Eqs. (12.63), (12.64), (12.68), and (12.58), it is found that [2]

$$\sqrt{H_{c1}H_{c2}} = \sqrt{2\sqrt{2}}\,H_c \approx H_c \tag{12.69}$$

The form in Eq. (12.69) is seen to relate the three critical fields that characterize the superconductor.

A way of obtaining a more accurate value of H_{c2} can be made by considering Eq. (12.51) in the vicinity of the magnetic transition. In this region $|\psi|^2 \approx 0$, so that Eq. (12.51) becomes [2,3]

$$\frac{1}{2m_s}\left(-i\hbar\nabla - \frac{q}{c}\vec{A}\right)^2 \psi \approx -\alpha\psi, \tag{12.70}$$

and the solution of this eigenvalue problem generates the difference of the normal free energy per volume subtract the superconductor free energy per volume. For a uniform applied magnetic field H_{c2}, the lowest free energy per volume then comes from the smallest eigenvalue of Eq. (12.70). Consequently, the solution of Eq. (12.70) is just the problem of a charged particle moving in a uniform applied magnetic field. Once the lowest $-\alpha$ is obtained, it then follows that

$$\Delta F = F_n - F_s = -\alpha = \frac{1}{2}\hbar\omega_c = \frac{e\hbar}{mc}H_{c2}, \tag{12.71}$$

where $\omega_c = \dfrac{2eH_{c2}}{mc}$ is the cyclotron frequency, and the particle charge is $q = -2e$. Viewing the transition in terms of Eq. (12.54), however, it follows that ΔF can be approximated as

$$\Delta F = F_n - F_s = \frac{1}{2}\frac{\alpha^2}{\beta} = \frac{H_c^2}{8\pi} \tag{12.72}$$

so that

$$H_c^2 = 4\pi\frac{\alpha^2}{\beta} \tag{12.73}$$

Equating Eqs. (12.71)–(12.72), and applying Eqs. (12.56b) and (12.57a) it is then found that

$$\frac{H_{c2}}{H_c} = \sqrt{2}\frac{\lambda_L}{\xi} = \sqrt{2}\kappa. \tag{12.74}$$

For a region of Type II superconductor to exist, it is required that $H_{c1} < H_c < H_{c2}$. This defines the region of the mixed state of superconductor–normal materials and is the characterizing feature of a Type II material. From Eq. (12.74), we have seen that this condition requires that

$$\frac{H_{c2}}{H_c} = \sqrt{2}\frac{\lambda_L}{\xi} = \sqrt{2}\kappa > 1. \tag{12.75}$$

The result in Eq. (12.75) is the standard quoted condition for a material to be Type II. Otherwise, it is Type I.

12.8 HIGH-TEMPERATURE SUPERCONDUCTORS

For most of the twentieth century, physicists were only familiar with low-temperature superconductors [8–11]. These are materials that exhibit superconductivity at temperatures below 25°K and are described microscopically by the BCS theory and macroscopically by the Ginzburg–Landau theory [9]. Near the end of the twentieth century, however, a new class of high-temperature superconductors were found so that superconductivity can now be observed in materials at temperatures of order of 100°K and above. While the mechanism associated with the appearance of low-temperature superconductivity involves the interaction of electrons with phonons the mechanism forming the basis of high-temperature superconductivity is less clear and may, in fact, be associated with magnetic interactions in these materials. Consequently, a microscopic theory of high-temperature superconductors is still a question. However, in some cases, a macroscopic approach based on a modified form of the Ginzburg–Landau theory can be useful to describe many of the features of high-temperature superconductivity [9]. This approach will be briefly introduced here.

An important class of high-temperature superconductors that have dominated the field are the so-called copper-oxide or cuprate materials. Examples of these are $YBa_2Cu_3O_7$ $(T_c = 90°\text{K})$, $Tl_2Ba_2Ca_2Cu_3O_{10}$ $(T_c = 120°\text{K})$, $Hg_2Ba_2Ca_2Cu_3O_8$ $(T_c = 134°\text{K})$. In these systems, the superconductivity arises in parallel planes of copper-oxides, and these planes weakly interact with one another. Consequently, the properties of the systems are highly anisotropic, so that in a Ginzburg–Landau approach the materials are viewed to have highly isotropic properties within the copper-oxide planes but a totally different response in the direction perpendicular to the copper-oxide planes [1–11].

In a crystallographic notation, the copper-oxide planes contain and are characterized by a set of $a-b$ crystal axes, and the crystallographic axis perpendicular to the planes is the c-axis. (Here the $a-b-c$ form a set of orthogonal crystallographic axes.) This will be the notation used in the generalization of the Ginzburg–Landau theory to a layered medium in the following discussions [9]. It should also be noted that, in addition to the copper-oxides, there have also been found high-temperature superconductors involving iron (e.g., $LaFeAs(O,F)$, $FeSe$, $(Ba,K)Fe_2As_2$), magnesium dibromide, and Fulleride superconductors (e.g. Cs_3C_{60}). These later types of systems have been less of a focus than the cuprates, however, and are not a consideration here and in the following.

Specifically, in the anisotropic Ginzburg–Landau approach, Eqs. (12.51) and (12.52) are modified to take the forms [2–4,9]

$$\frac{1}{2m_\gamma}\left(-i\hbar\nabla_\gamma - \frac{q}{c}A_\gamma\right)^2 \psi + \alpha\psi + \beta|\psi|^2\psi = 0 \tag{12.76a}$$

and [9]

$$J_\gamma = \frac{q\hbar}{2m_\gamma}\left[-i\psi^*\nabla_\gamma\psi + i\psi\nabla_\gamma\psi^*\right] - \frac{q^2}{m_\gamma c}A_\gamma. \tag{12.76b}$$

where $\gamma = a, b, c$ labels the crystallographic axis. In these equations, m_γ is the anisotropic effective mass, $q = -2e$ for $e > 0$ is the charge, $\alpha = \alpha_0 \frac{T - T_c}{T_c}$ where $\alpha_0 > 0$ is a constant, T_c is the critical temperature of the superconducting transition, and $\beta > 0$. From Eq. (12.76), it is seen that the system as a whole exhibits a single bulk transition temperature at T_c characterized by the Ginzburg–Landau wave function [9]

$$|\psi|^2 = -\frac{\alpha}{\beta}, \tag{12.77}$$

and that the anisotropy of the material enters primarily through the spatial varying terms and through the terms involved with the coupling to the external magnetic fields. Consequently, below the transition temperature, the system displays an anisotropic response to the flow of current and to the field penetration into the material. For the most part, in the superconducting planes, it is found that $m_a \approx m_b$ so that we will assume $m_a = m_b$. The focus in the following is then on the properties arising from these anisotropies and their semiquantitative behaviors in the Ginzburg–Landau treatment [9].

As in Eq. (12.54b) for the isotropic system, the difference in the free energy per volume of the normal and superconducting states of a material can be written in terms of a critical field H_c so that [2–4,9]

$$F_N(T) - F_s(T) = \frac{1}{8\pi} H_c^2(T) = \frac{1}{2}\frac{\alpha^2}{\beta} > 0. \tag{12.78}$$

Note that this result follows from the temperature transition in Eq. (12.77) for the bulk material and is, consequently, independent of the anisotropy of the material. However, due to the spatial anisotropy of the high T_c materials, the other critical fields characterizing the magnetic transitions of Type II materials are dependent on the spatial anisotropy of the material. The response to these fields depends on the direction in which they are applied to the material, and arises from the position- and field-dependent terms of Eq. (12.76b).

In the limit of a slowly varying $\psi(\vec{r})$, it then follows from Eq. (12.76b) that the first term on the right of the equation can be ignored so that [2–4,9]

$$J_\gamma \approx -\frac{q^2}{m_\gamma c}|\psi|^2 A_\gamma = -\frac{c}{4\pi\lambda_{L\gamma}^2} A_\gamma, \tag{12.79}$$

where [9]

$$\lambda_{L\gamma} = \sqrt{\frac{m_\gamma c^2}{4\pi|\psi|^2 q^2}} = \sqrt{\frac{m_\gamma}{m}}\lambda_L, \tag{12.80a}$$

$$\lambda_L = \sqrt{\frac{mc^2}{4\pi|\psi|^2 q^2}}, \tag{12.80b}$$

and

$$m = \left[m_a m_b m_c\right]^{\frac{1}{3}} = \left[m_a^2 m_c\right]^{\frac{1}{3}}. \tag{12.80c}$$

Similarly, the coherence length of the anisotropic material from Eq. (12.58a) generalizes to [9]

$$\xi_\gamma = \sqrt{\frac{\hbar^2}{2m_\gamma |a|}} = \sqrt{\frac{m}{m_\gamma}}\sqrt{\frac{\hbar^2}{2m|a|}} = \sqrt{\frac{m}{m_\gamma}}\xi. \tag{12.81}$$

In Eqs. (12.80a) and (12.81), the effective lengths λ_L and ξ are defined as an effective London penetration length and coherence length for the characterization of the material on whole.

In terms of these parameters, the condition for Type II superconducting behavior is approximated as

$$\kappa = \frac{\lambda_L}{\xi} > \frac{1}{\sqrt{2}}. \tag{12.82}$$

In fact, in high T_c cuprate superconductors [9], it is generally found that $\kappa = \frac{\lambda_L}{\xi} \gg 1$ so that they are all Type II. A consequence of these particularly large values of κ is that the cylinders of magnetic flux penetrating the mixed state of the Type II material have small circular cross sections. This presents some difficulties in the applications of these materials, as the cylinders of flux can exhibit a mobility in the materials.

For modeling the copper-oxide superconductors [9], it is found that $m_a = m_b \ll m_c$ and, consequently, $\lambda_{La} = \lambda_{Lb} \ll \lambda_{Lc}$ and $\xi_a = \xi_b \gg \xi_c$. (Some specific examples of the ratios of the effective masses are: $YBa_2Cu_3O_7$ with $m_c/m_a = 5.3$, YBCO with $m_c/m_a = 9.0$, Hg-2201 with $m_c/m_a = 27$, and $Ba_2CuTl_2O_{6.1}$ with $m_c/m_a = 117$. An additional example of the resulting London penetration and coherence length of a material is that of YBCO with $\lambda_{La} = \lambda_{Lb} = 150$ nm $\ll \lambda_{Lc} = 800$ nm and $\xi_a = \xi_b = 2$ nm $\gg \xi_c = 0.4$ nm.) As a result, the geometry of the cylinders of flux penetration depends on the field direction relative to the $a-b-c$ crystallographic axes as do the critical fields for the Type II transition.

Specifically, for external applied fields along the c-axis, the three critical fields are given by [9]

$$H_{c1}^c = \frac{\Phi_{2e}}{4\pi\lambda_{La}^2}\ln\kappa_c \le H_c \le H_{c2}^c = \frac{\Phi_{2e}}{2\pi\xi_a^2}, \tag{12.83}$$

where H_{c1}^c is the lower critical field of the mixed state, H_{c2}^c is the upper critical field of the mixed state, and $\Phi_{2e} = \frac{2\pi\hbar c}{2e}$ is the unit of flux. Here [2–4,9]

$$\kappa_c = \frac{\lambda_{La}}{\xi_a} \tag{12.84a}$$

and

$$H_c = \frac{\Phi_{2e}}{2\sqrt{2}\pi}\frac{1}{\xi_a\lambda_a} = \frac{\Phi_{2e}}{2\sqrt{2}\pi}\frac{1}{\xi_c\lambda_c} = \sqrt{4\pi\frac{a^2}{b}}. \tag{12.84b}$$

For external applied fields along the a-axis, the three critical fields are given by [9]

$$H_{c1}^a = \frac{\Phi_{2e}}{4\pi\lambda_{La}\lambda_{Lc}}\ln\kappa_a \le H_c \le H_{c2}^a = \frac{\Phi_{2e}}{2\pi\xi_a\xi_c}, \tag{12.85}$$

where H_{c1}^a is the lower critical field of the mixed state, H_{c2}^a is the upper critical field of the mixed state. Here [9]

$$\kappa_a = \sqrt{\frac{\lambda_{La}\lambda_{Lc}}{\xi_a\xi_c}} = \sqrt{\frac{m_c}{m_a}}\kappa_c, \tag{12.86}$$

and as a general relationship it is found that

$$\frac{H_{c1}^{\alpha}}{H_{c2}^{\alpha}} = \frac{\ln \kappa_{\alpha}}{2\kappa_{\alpha}^{2}} \tag{12.87}$$

for $\alpha = a, c$.

Aside from the critical fields, the anisotropic Ginzburg–Landau approach has been applied to the study of various other thermodynamic properties, i.e., heat capacity, correlations of the wave function, and fluctuations [9]. For these, the reader is referred to the literature [9]. In general, a requirement for the application of the Ginzburg–Landau formulation is that the thermodynamic fluctuations in the system are small. While the Ginzburg–Landau approach has been generally successful in the discussion of the properties of low-temperature superconductors [9], the properties of high-temperature superconductors are studied at temperatures where the fluctuations are greater, and in this regard the use of the theory may become problematic. In addition, the coherence lengths of high-temperature systems should be larger than the lattice parameters of the material being treated. This may be difficult along the c-axis, and, in some cases, it has been necessary to treat the interaction between the $a-b$ planes of the system as a type of Josephson tunneling effect. For the treatment of these considerations, the reader is referred to the literature [1–11].

REFERENCES

1. Kittle, C. 1996. *Introduction to solid state physics, 7th Edition*, 393–410. New York: John Wiley & Sons, Inc.
2. Marder, M. P. 2000. *Condensed matter physics.* New York: John Wiley & Sons, Inc.
3. Fossheim, K. and A. Sudbo 2004. *Superconductivity: Physics and applications*, 141–198. New York: John Wiley & Sons, Inc.
4. Schrieffer, J. R. 1983 *Theory of superconductivity.* Boca Raton: CRC Press
5. Barone, A. and G. Paterno 1982. *Physics and applications of the Josephson effect.* New York: John Wiley & Sons, Inc.
6. Wolf, E. L. 1985. *Principles of electron tunneling spectroscopy, 2nd Edition.* Oxford: Oxford University Press.
7. de Gennes, P. G. 1989. *Superconductivity of metals and alloys.* New York: Addison-Wesley.
8. Saxena, A. K. 2012. *High-temperature superconductors, 2nd Edition.* Heidelberg: Springer.
9. Plakida, N. 2010. *High-temperature cuprate superconductors: Experiment, theory, and applications.* Heidelberg: Springer.
10. Tewordt, L., Wermbter, S. and Th. Wolkhausen 1989. Ginzburg-Landau-Gorkov theory for high-temperature superconductors. *Physical Review* B40: 6878–6883.
11. Banerjee, B., Ramakrishnan, T. V. and C. Dasgupta 2011. Phenomenological Ginzburg-Landau-like theory for superconductivity in the cuprates. *Physical Review* B83: 024510.
12. Schuller, I. K. and K.E. Grey 2006. Time-dependent Ginzburg-Landau: From single particle to collective behavior. *Journal of Superconductivity and Novel Magnetism* 19: 401–407.
13. Schmid, A. 1966. A time dependent Ginzburg-Landau equation and its application to the problem of resistivity in the mixed state. *Physik der Kondensierten Materie* 5: 302–317.
14. Schmid, A. 1968. The approach to equilibrium in a pure superconductor the relaxation of the Cooper pair density. *Physik der Kondensierten Materie* 8: 129–140.
15. Jackson, J. D. 1975. *Classical electrodynamics.* New York: John Wiley & Sons, Inc.
16. Josephson B. D. 1962. Possible new effect in superconductive tunneling. *Physics Letters* 1: 251–253.
17. Josephson B. D. 1974. The discovery of tunneling supercurrents. *Reviews of Modern Physics* 46: 251–254.
18. Jaklevic, R. C., Lambe, J., Silver, A. H., et al. 1964. Quantum interference effects in Josephson tunneling. *Physical Review Letters* 12: 159–160.

13 Scaling and Renormalization

The most interesting parts of a field theory are its singularities [1–7]. These are the anomalous regions where a variety of non-analyticities are found in the macroscopic properties of the thermodynamic system, including infinities, cusps, and discontinuities. Such anomalies are most often associated with phase changes in the systems, e.g., the development of a permanent magnetization in ferromagnets or liquid–gas–solid transitions in fluids, etc. The origins of such features are often linked to collective behaviors associated with phase changes and the development of correlations between individual particles of the material. These transformations of macroscopic properties from one state to another can be driven by temperature changes, pressure changes, the applications of external fields, or the introduction of random disorder. In this chapter, we look at some of the basic approaches that have been developed to understand singular behaviors in the macroscopic properties of materials.

Magnetic systems offer good examples of singular field theories [1,2]. See Figure 13.1 for a schematic of some of the qualitative properties of a ferromagnetic about its critical temperature T_c in the absence of an applied magnetic field. In magnetism, the magnetization $\vec{M}(T,H)$ of a material is studied as a function of temperature T and the applied magnetic field H. The magnetization field is often found to exhibit singularities in the temperature and the applied field. These singularities occur at a specific critical temperature denoted T_c or a specific critical field denoted H_c, and take general forms dependent on the characteristics of the microscopic model of the material being studied. In this regard, at zero applied field in the neighborhood of the critical temperature, the magnitude of a ferromagnetic magnetization is found to exhibit a cusp singularity of the form [1,2]

$$\begin{aligned} M(T,H=0) &\to m_0\left(-\frac{T-T_c}{T_c}\right)^{\beta} && \text{with } \beta \geq 0 \text{ for } T < T_c \\ &\to 0 && \text{for } T > T_c \end{aligned} \tag{13.1a}$$

where m_0 is a constant, and β is a constant exponent that depends on the dimension of the magnetic system (i.e., 1 – d, 2 – d, 3 – d) and the type of model being studied (i.e., Ising model, Heisenberg model, etc.). Other properties of the ferromagnetic field theory are also singular at T_c and $H = 0$, so that in the neighborhood of T_c the magnetic susceptibility

$$\chi(T,H=0) = \left.\frac{\partial M}{\partial H}\right|_{H=0} \to \chi_0\left|\frac{T-T_c}{T_c}\right|^{-\gamma} \quad \text{with } \gamma \geq 0 \tag{13.1b}$$

has an infinity, and the specific heat

$$C(T,H=0) = -\frac{1}{V}T\left.\frac{\partial^2 G}{\partial T^2}\right|_{H=0} \to C_0\left|\frac{T-T_c}{T_c}\right|^{-\alpha} \quad \text{with } \alpha \geq 0, \tag{13.1c}$$

becomes infinite. In addition, for small nonzero applied magnetic fields the magnetization as a function of the applied field at T_c is characterized by the relationship

$$M(T_c,H) \to m_1|H|^{\frac{1}{\delta}} \text{ sign }(H) \text{ with } \delta > 0. \tag{13.1d}$$

DOI: 10.1201/9781003031987-13

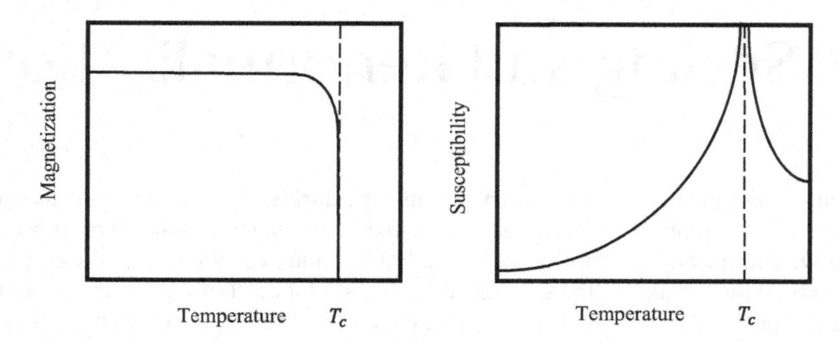

FIGURE 13.1 Schematic of some typical properties of a ferromagnet lattice of spins about its critical temperature T_c: (a) the magnetization versus temperature, and (b) the susceptibility versus temperature. Here, the magnetization exhibits a cusp at T_c, whereas the susceptibility is infinite at T_c. In the figures, the position of T_c is shown by the vertical dashed line and $H=0$.

In these singular forms χ_0, C_0, m_0, m_1, α, β, γ, and δ are model-dependent constants, and in Eq. (13.1c) $G(T,H)$ is the free energy of the ferromagnet with volume V. The exponents are all defined so that α, β, $\gamma,\delta > 0$ characterizes the power law nature of the singular properties in Eq. (13.1), and the constants χ_0 and C_0 may be different above and below T_c.

The notation for the critical exponents defined in Eq. (13.1) is a standard notation, and tables of exponents have been developed listing them for a variety of types of models and in a number of spatial dimensions [1–7]. Under certain broad assumptions on the nature of the origins of the singularities occurring in thermodynamics, a set of scaling relations have been developed that allow for relationships between the critical exponents to be developed. In this regard, an interesting point is that the critical exponents can be shown to be related to one another so that they satisfy [1,2]

$$\alpha + 2\beta + \gamma = 2 \tag{13.2a}$$

and

$$\gamma = \beta(\delta - 1). \tag{13.2b}$$

The power law singularities found in the study of magnetization fields are also observed in the singular nature of field theories developed for other types of physical systems. For example, at the critical point of a fluid, the specific heat of the fluid is singular with a form given by [1,2]

$$C(T,H=0) \to C_0\left|\frac{T_c - T}{T_c}\right|^{-\alpha}, \tag{13.3a}$$

and the density of the fluid ρ behaves as

$$\rho - \rho_c \to \rho_0\left|\frac{T_c - T}{T_c}\right|^{\beta}, \tag{13.3b}$$

where ρ_c is the density of the fluid at the critical point, and ρ_0 is a constant. Here, the fundamental field of the fluid density replaces the magnetization density of the magnetic system and is, nevertheless, characterized by the exponent β. Note that α in Eq. (13.3a) again occurs in the context of the specific heat, and β is used to characterize the temperature dependence of the density field in the neighborhood of the critical temperature. In this regard, there is a certain correlation of the critical properties of the magnetic and fluid systems. This correlation is recognized in the fundamental study of the thermodynamic critical exponents. However, the exponent values in the magnetic system differ numerically from those in the fluid system.

A final critical exponent that occurs in both magnetization and fluid fields shall be consider here. It is that of the correlation length $\xi(T,H)$. The correlation length is a measure of the flexibility in space of the magnetization (magnetic system) or density (fluid system), and in this sense is similar to the coherence length encountered in the Ginzburg–Landau theory of superconductivity [1]. The correlation length measures the characteristic length of the spatial thermodynamic fluctuations of the magnetization or fluid density of the respective systems, and enters as a length scale in the expression for the diffuse elastic optical or neutron scattering from the magnetic or fluid systems. Similar to the superconductor coherence length, at zero field near the critical temperature the correlation length has the form

$$\xi(T,H=0) \to \xi_0 \left| \frac{T-T_c}{T_c} \right|^{-\nu}, \tag{13.4}$$

where $\nu > 0$ is introduced as another critical exponent, and the constant ξ_0 may be different above and below T_c. (Note that in an earlier chapter, it was shown that $\nu = 1$ for the superconductor at the superconductor–normal metal transition.) In terms of the other critical exponents, it is found from scaling studies that the critical exponent ν satisfies the relationship [1,2]

$$d\nu = 2 - \alpha. \tag{13.5}$$

Here d is the dimension of the system, i.e., $d = 1$ for a chain, $d = 2$ for a plane, $d = 3$ for a bulk block of material, and α is the specific heat exponent.

In this chapter, a brief introduction to some of the methods used to study the singularities of the macroscopic properties of materials is presented. First, as an illustration, some preliminary discussion is given of a simple one-dimensional model of a ferromagnet. In one-dimensional systems, in the absence of long-range interactions, a phase transition with its associated singularities cannot occur at a nonzero temperature. However, the phase transition can be regarded as occurring at zero temperature, and certain of the singularities of this system in the neighborhood of zero temperature are studied. This is followed by a treatment of first- and second-order phase transitions within the context of the Landau theory [1–7]. The Landau theory is a mean field treatment of the properties of a system, with a phase transition and a set of exponents α, β, γ, δ, ν obtained in its context. After this, an introduction to the ideas of scaling is present along with a discussion of the origins of the identities given in Eqs. (13.2) and (13.5). The chapter concludes with an introduction to the ideas of the renormalization group treatment of singularities in field theories exhibiting a phase transition.

13.1 ONE-DIMENSIONAL FERROMAGNET

A simple continuum model of a ferromagnet in one-dimension can be formulated in terms of a spin density field, which is position dependent along, for example, the x-axis. This can be represented in units of the lattice constant by the vector function [8]

$$\vec{S}(x) = \left(\sin\theta(x)\cos\varphi(x), \sin\theta(x)\sin\varphi(x), \cos\theta(x)\right), \tag{13.6}$$

where $\vec{S}(x)\cdot\vec{S}(x) = 1$. In this model, each of the atomic spins is also associated with an atomic magnetic moment, causing it to interact with external magnetic fields (i.e., the spin density is associated with a magnetization density). This type of spin model could represent, for example, a polymer material composed as a chain of atoms in which the magnetic moments are associated with some of the atoms composing the polymer. In addition, there are also other types of quasi-one-dimensional materials composed as chains of magnetic atoms, which are weakly coupled together. These type of crystalline materials then resembles stacks of pipes forming a three-dimensional system, which is dominated by one-dimensional properties. The magnetism of such systems then is essentially that of a chain of coupled spins [8].

For these types of materials, a simple continuum model of a ferromagnetic Hamiltonian takes the basic form [8]

$$\mathcal{H} = \int dx\ J\left|\frac{d\vec{S}(x)}{dx}\right|^2, \tag{13.7a}$$

where $J > 0$ is an exchange term favoring the parallel alinement of neighboring parts of the magnetization. In this regard, from Eq. (13.7a), it is seen that the lowest energy configuration of the system is given by the condition that $\left|\frac{d\vec{S}(x)}{dx}\right|^2 = 0$, so in this configuration the spins are completely aligned along the entire chain. In the presence of an external magnetic field applied along, for example, the z-direction—an additional term of the form

$$\mathcal{H}_H = -g\mu\int dx\ HS_z(x) \tag{13.7b}$$

is added to Eq. (13.7a). Here g is a constant g-factor, μ is the dipole moment, and the externally applied field is H. In the following discussion, only the zero-field properties of the system are considered, but the form in Eq. (13.7b) will be needed for the treatment of the zero-field magnetic susceptibility.

In terms of the Hamiltonian in Eq. (13.7a), the partition function of a magnetic chain of length $L \to \infty$ is written as [8]

$$Z = \int d\vec{S}(x)\exp\left(-\frac{\mathcal{H}}{k_BT}\right) = \int d\vec{S}(x)\exp\left(-\frac{1}{k_BT}\int dx\ J\left|\frac{d\vec{S}(x)}{dx}\right|^2\right), \tag{13.8}$$

where k_B is the Boltzmann constant, and T is the temperature. This is a functional integral that is familiar from the path integral formulation of quantum mechanics. The evaluation of the functional integral in Eq. (13.8) can be made in terms of the eigenvalue and eigenvector results from the eigenvalue problem given by [8]

$$\frac{L^2}{4\beta^2 J}|n\rangle = E_n|n\rangle, \tag{13.9}$$

where

$$L^2 = -\frac{1}{\sin\theta}\frac{\partial}{\partial\theta}\sin\theta\frac{\partial}{\partial\theta} - \frac{1}{sin^2\theta}\frac{\partial^2}{\partial\varphi^2} \tag{13.10}$$

is essentially the quantum mechanical total angular momentum operator, and $|n\rangle$ is an eigenvector with eigenvalue E_n. In terms of the eigenmodes of Eq. (13.9), the partition function of the polymer spin chain is obtained as

$$Z = e^{-\beta\Delta EL}\sum_{n=0}^{\infty} e^{-\beta E_n L}, \tag{13.11}$$

where

$$\Delta E = k_B T ln\left(2\beta J\right) - 2J. \tag{13.12}$$

(For a detailed derivation of the relationships in Eqs. (13.8)–(13.12), the reader is referred to the Appendix of this chapter.)

Given the partition function in Eq. (13.11), the thermodynamic energy and heat capacity of the chain are determined from the standard thermodynamic relationships, and they exhibit no singularities at nonzero temperatures for the one-dimensional system [8]. The magnetization can also be determined. For the isotropic system in the absence of an applied field, the magnetization is zero on average at nonzero temperature. However, at zero temperature the spins of the ferromagnetic chain are all parallel oriented in the lowest energy state. Consequently, the chain exhibits a nonzero net magnetization. We regard this as a type of phase transition at zero temperature. Note, however, in a quantum system the quantum fluctuations would destroy such an ordering even at zero temperature. This is due to the zero-point fluctuations.

Looking at the higher order moments of the magnetization, the correlation function between $S_z(x)$ and $S_z(0)$ located at two different positions on the chain is defined by [8]

$$g_{zz}(x) = \langle S_z(x)\, S_z(0)\rangle = \sum_{n=1}^{\infty} \left|\langle n|\cos\theta|0\rangle\right|^2 e^{-\beta x(E_n - E_0)} \approx \left|\langle 1|\cos\theta|0\rangle\right|^2 e^{-x/\xi_{zz}}, \tag{13.13}$$

where

$$\xi_{zz} = \left[\beta(E_1 - E_0)\right]^{-1} \tag{13.14}$$

is the correlation length of the spins along the chain. The correlation length is essentially the length along the chain over which the pointing of two different spins exhibits a degree of correlations, and the correlation function of the chain enters as an essential element in the description of the neutron scattering from the chain.

In terms of the correlation function in Eq. (13.13), the magnetic susceptibility for an applied field along the z-direction in the limit of zero magnetic field is obtained. Specifically, it is written as [8]

$$\chi_{zz} = \frac{g^2\mu^2}{k_B T}\int dx g_{zz}(x) \approx \frac{g^2\mu^2}{k_B T}\frac{2}{3}\xi_{zz}. \tag{13.15}$$

Note that while the specific heat has no singular at zero temperature, the correlation length and the magnetic susceptibility are both singular at zero temperature. In the limit that $H = 0$ and $T \to 0$, it then follows that

$$\xi_{zz} \to \frac{1}{T} \to \infty \tag{13.16a}$$

and

$$\chi_{zz} \to \frac{1}{T^2} \to \infty. \tag{13.16b}$$

Both of these responses of the chain of spins are singular at the $T_c = 0$, with critical exponents given by $\nu = 1$ and $\gamma = 2$.

The one-dimensional chain exhibits some of the rudimentary ideas of the singularities found in a thermodynamics system at a phase transition. Specifically, in this model a transition is made from a magnetically order system at $T = 0$ to a system with a zero magnetization above $T = 0$. Next, we look at systems with $T_c \neq 0$ in the context of a very simple model, proposed by Landau, for the general study of phase transitions. This model is capable of treating two types of phase transitions known, respectively, as first- and second-order phase transitions. We shall first deal with second-order transitions. These are transitions involving, for example, a continuous magnetization, which exhibits a simple cusp singularity at the transition temperature. This is followed by the study of a first-order phase transition, in which the system is described by a field that is discontinuous at the critical temperature.

13.2 SECOND-ORDER PHASE TRANSITIONS

A second-order phase transition is characterized by a field theory in which the thermodynamic behavior of the average field is described by a continuous function of temperature and that, nevertheless, exhibits a singularity at the critical temperature. An example of such a system is the magnetization of a ferromagnetic material [1] or, alternatively, the polarization response of a ferroelectric material [9]. In the case of the magnetic material, the field characterizing the system is the magnetization that below a critical temperature is nonzero on average. As the temperature approaches the critical temperature, the magnetization continuously goes to zero at the critical temperature and remains zero at temperatures above the critical temperature. Its behavior in the neighborhood of the critical temperature is then characterized by the singular forms in Eqs. (13.1) and (13.4). This type of continuous transition with temperature is also found in the polarization exhibited by ferroelectric materials where singular forms similar to those in Eqs. (13.1) and (13.4) are developed to characterize the temperature transition. In this regard, as a final example, earlier in this chapter we have, in addition, seen that in a fluid material the density of the fluid near its critical point is continuous but characterized by the singularities defined in Eqs. (13.3) and (13.4). The thermodynamics of all of these different types of materials are roughly explained by a theory originally proposed by Landau [1,9]. The development of the Landau theory is now the focus of this section.

Consider a ferromagnetic material, which is described by a magnetization field $m(\vec{r})$ defined over space and for simplicity taken to be a scalar function [1,2,9]. The basic qualitative properties of the system in the neighborhood of the critical temperature can often be described by a simple free-energy form that was originally proposed by Landau. The proposed form is found to be generalizable to systems exhibiting a second-order transition and to give a semi-quantitative representation of the system properties about T_c. In this sense, it is developed on certain broadly based ideas involving the general characteristics of second-order phase transitions.

In Landau's approach, the material is studied in the vicinity of the critical temperature T_c, where it is assumed to be described by a specific form written as an expansion in the small parameter of the magnetization field $m(\vec{r})$. This offers a power series representation of the free energy near the transition temperature. In this expansion, the properties of the system near the critical temperature are semi-quantitatively modeled by a simple choice of the temperature dependence of the coefficients of the expansion in the magnetic field.

Specifically, near the critical temperature the magnetization as a function of temperature is becoming small so that it is assumed that the free energy can be written as an expansion in the small parameter $m(\vec{r})$, which for the bulk system is taken to be a constant independent of $\vec{r}$. The expansion then is given by the power series [1,2,9]

$$G(T,h,m) = G(T,h,m=0) + a(T)m^2 + \frac{1}{2}bm^4 - mh, \tag{13.17}$$

where $a(T) = a_0(T - T_c)$, $a_0 > 0$ is a constant, $b > 0$ is a constant, $G(T,h,m=0)$ is approximated as a constant independent of T and h, and $-mh$ represents the coupling of the magnetization with an external magnetic field that is proportional to h. In Eq. (13.17), $G(T,h,m)$ represents the free energy of the system at T and h in the case that it has a magnetization m. The important properties of the phase transition near T_c enter through the change in sign of $a(T)$ at the critical temperature.

In the Landau picture, the state of the magnetization of the system m is obtained by minimizing the free energy $G(T,h,m)$ with respect to m at fixed T and h. Once $m(T,h)$ is determined in this way, the free energy as a function of temperature and field is given by [1,2,9]

$$G(T,h) = G(T,h,m(T,h)). \tag{13.18}$$

The thermodynamics of the field at T and h is then completely contained in $G(T,h)$, where the thermodynamic averages of the macroscopic properties of the material are computed using standard expressions in terms of the free energy $G(T,h)$. Applying this approach, we now determine the properties of the magnetic material described by Eq. (13.17).

Consider minimizing Eq. (13.17) for the case in which $h = 0$. From Eq. (13.17), it then follows in this case that [1,2,9]

$$\frac{\partial G(T,h,m)}{\partial m} = 2a(T)m(T) + 2bm^3(T) = 0. \tag{13.19}$$

This is now solved for the temperature dependence of the magnetization $m(T)$ characterizing the system. For the case in which $T > T_c$ and $a(T) > 0$, the magnetization from Eq. (13.19) is then given by

$$m(T) = 0, \tag{13.20a}$$

and, for the case in which $T < T_c$ and $a(T) < 0$, the magnetization from Eq. (13.19) is given by

$$m^2(T) = -\frac{a}{b} = \frac{a_0}{b}(T_c - T). \tag{13.20b}$$

From Eq. (13.20), the magnetization is found to exhibit a cusp behavior in which it is only nonzero at temperatures below T_c but continuously goes to zero above T_c. In regard to this singular dependence, it follows from Eq. (13.1a) that the critical exponent of the magnetization cusp is $\beta = \frac{1}{2}$. Note, however, that the temperature derivatives of the magnetization will exhibit more serious singularities, and the resulting derivatives of the system are no longer continuous in the temperature.

Next, consider minimizing Eq. (13.17) for $m(T, h)$ in the case in which $h \neq 0$. Again, it follows from Eq. (13.17) that

$$\frac{\partial G(T,h,m)}{\partial m} = 2a(T)m(T,h) + 2bm^3(T,h) - h = 0, \tag{13.21a}$$

so that is found that

$$2a(T)m(T,h) + 2bm^3(T,h) = h. \tag{13.21b}$$

The magnetic susceptibility is then obtained from Eq. (13.21b) by taking its derivative with respect to h. In this way, it follows that

$$\left[2a(T) + 6bm^2(T,h)\right]\frac{\partial m(T,h)}{\partial h} = 1 \tag{13.22}$$

In the case that $T > T_c$ in which $a(T) > 0$ and for $h \to 0$, $m(T,h = 0) = 0$ so that from Eq. (13.22) the magnetic susceptibility is given by

$$\chi(T,h=0) \propto \left.\frac{\partial m(T,h)}{\partial h}\right|_{h=0} = \frac{1}{2a} = \frac{1}{2a_0}\frac{1}{T-T_c}. \tag{13.23a}$$

On the other hand, for the system in which $T < T_c$ with $a(T) < 0$ and for $h \to 0$, $m^2(T,h=0) = -\frac{a}{b}$. From Eq. (13.22) , it then follows again that

$$\chi(T,h=0) \propto \left.\frac{\partial m(T,h)}{\partial h}\right|_{h=0} = -\frac{1}{4a} = \frac{1}{4a_0}\frac{1}{T_c - T}, \tag{13.23b}$$

In regards of Eq. (13.1b), it is seen from Eq. (13.23) that the critical exponent $\gamma = 1$ both above and below the critical temperature.

From Eq. (13.21b) the dependence of the magnetization on the magnetic field can also be obtained at the critical temperature. In this case, $a(T_c) = 0$ so that it follows from Eq. (13.21b) that [1,2,9]

$$m(T_c) = \left|\frac{h}{2b}\right|^{\frac{1}{3}}. \tag{13.24}$$

Applying the definition in Eq. (13.1d), it is found that the critical exponent $\delta = 3$.

The zero-field specific heat of the system is obtained by evaluating Eq. (13.17) at the magnetization obtained in Eq. (13.20). From Eq. (13.17) evaluated in this way, the free energy is

$$\begin{aligned} G(T,h=0,m) - G(T,h=0,m=0) &= -\frac{a_0^2}{2b}(T-T_c)^2 \quad \text{for } T < T_c \\ &= 0 \qquad\qquad\qquad\qquad \text{for } T > T_c. \end{aligned} \tag{13.25}$$

The specific heat then follows from $C = -\frac{1}{V}T\frac{\partial^2 G}{\partial T^2}$. Treating $G(T,h=0,m=0)$ as a constant independent of the temperature gives a discontinuity in the specific heat at the critical temperature represented by the form

$$\Delta C = C(T_c - \epsilon) - C(T_c + \epsilon) = \frac{1}{V}\frac{a_0^2}{b}T_c, \tag{13.26}$$

where $\epsilon > 0$ is an infinitesimal. Note that the theory in its present form only predicts a discontinuous gap in the specific heat at T_c. Our earlier assumptions that $G(T,h=0,m=0)$ and $b > 0$ are temperature-independent constants do not allow us to continue the result into the neighborhood of T_c. To determine the temperature dependence of these terms would require that more information be put into the theory than is currently available. In terms of the critical exponents, the discontinuity in Eq. (13.26) is considered to give $\alpha = 0$ as the critical exponent in Eq. (13.1c).

Going back to our earlier discussions, in Eqs. (13.2) and (13.5) a number of relationships between the critical exponents were given. It is interesting to see how well the exponents of the Landau theory satisfy these relationships. The critical exponents obtained in the Landau theory are summarized as $\alpha = 0$, $\beta = \frac{1}{2}$, $\gamma = 1$, $\delta = 3$. From a consideration of Eq. (13.2a), it follows for these values that [1–7]

$$\alpha + 2\beta + \gamma = 0 + 2\frac{1}{2} + 1 = 2 \tag{13.27a}$$

and from Eq. (13.2b) that

$$\gamma = 1 = \beta(\delta - 1) = \frac{1}{2}(3 - 1) = 1. \tag{13.27b}$$

Both of these forms are satisfied by the Landau theory critical exponents.

One can also compute the correlation length in the context of the Landau treatment. This is done by introducing a gradient term of the form $c|\nabla m(\vec{r})|^2$ into the free energy density of the system in Eq. (13.17) similarly as was done in an earlier treatment in Eq. (2.145) for the Ginzburg–Landau theory of superconductivity [10]. (Specifically, for the Ginzburg–Landau theory in Eq. (2.145), in the $\vec{A}(\vec{r}) = 0$ limit the gradient term introduced into the superconductor free-energy form was given by $\frac{1}{2m_s}|-i\hbar\nabla\psi(\vec{r})|^2$.) The resulting magnetic free-energy model is then used to determine the relaxation of the magnetization at an interface with a nonmagnetic material. In this way, it is found just as in the Ginzburg–Landau treatment that $\nu = \frac{1}{2}$, both above and below the transition at the critical temperature [10].

(The reader should further note here the similarity of our discussions of the magnetic system and our earlier treatment of the Ginzburg–Landau theory of superconductivity. Specifically, the free energy of the magnet material is a generalized form of the Ginzburg–Landau treatment of the superconducting system to handle a real scalar field representing the magnetization of the magnetic material. In this regard, in our earlier discussions of the superconducting coherence length, which is the analogy of the magnetic Landau correlation length, it was found that $\nu = \frac{1}{2}$ for the superconducting coherence length at the superconducting transition temperature. The reader is also referred to the literature for a detailed discussion of this similarity between the two theories [1,2,10].)

Applying the result $\nu = \frac{1}{2}$ in Eq. (13.5), it is found that [1,2,10]

$$d\nu = \frac{d}{2} = 2 - \alpha = 2. \tag{13.28}$$

This relationship is observed to be satisfied for $d = 4$ but not for systems defined in other spatial dimensions. Later, we shall see that the Landau theory gives an exact result for a thermodynamic system in four-dimensional space so that the relationship in Eq. (13.28) is not totally off the mark. Nevertheless, it is not valid for $d \neq 4$.

In the next section, the Landau theory is extended to consider materials exhibiting a type of phase change termed as a first-order phase transition. In these phase changes, the systems exhibit discontinuities in their macroscopic properties, rather than the cusps and infinities of a second-order phase transition.

13.3 FIRST-ORDER PHASE TRANSITIONS

There are a number of ways that the Landau free-energy form can be generalized to handle first-order phase transitions [9,11–13]. Here, we shall consider just one of them. Again, the idea is to study an expansion of the free energy treating $m(\vec{r})$ as a small parameter, with the temperature dependence of the coefficients containing the phase change properties. Similar to the treatment of the second-order transition, the free-energy form of the expansion is minimized in $m(\vec{r})$ to give the free energy $G(T,h)$ of a system as a function of only T and the magnetic field h. Again, to provide an easy comparison, we treat a system involving a magnetization, but the type of model discussed has been also generalized and applied to structural phase transitions and to the related problems of ferroelectric materials. In terms of a magnetic system, the model of a first-order transition given here has been used in the study of surface magnetism [11–13].

Specifically, consider the free-energy form for a bulk magnetization described by the constant m and in which for simplicity we take $h = 0$. This is assumed to be given by [9,11–13]

$$G(T,m) = G(T,m=0) + a(T)m^2 + \frac{1}{2}bm^4 + \frac{1}{3}cm^6, \tag{13.29}$$

where $a(T) = a_0(T - T_o)$, $a_0 > 0$ is a constant, $b < 0$ is a constant, $c > 0$ is a constant, and $G(T,m=0)$ is approximated as a constant independent of T. Note that now there is a m^6 term, $b < 0$, and the temperature T_o at which $a(T)$ changes sign is not necessarily the critical temperature of the magnetization. The origin of the first-order transition will be found to arise from the $b < 0$ coefficient, which in the second-order system entered as a positive constant.

The magnetization at fixed temperature is obtained by minimizing Eq. (13.29) so that [9,11–13]

$$\frac{\partial G(T,m)}{\partial m} = 2a(T)m(T) + 2bm^3(T) + 2cm^5 = 0. \tag{13.30}$$

From Eq. (13.30), it is found that the local minima of the magnetization are at the solutions of

$$m\left[a(T) + bm^2 + cm^4\right] = 0, \tag{13.31}$$

and depending on the coefficients of the polynomial there can be more than one zero. In addition, from Eq. (13.29) the free energy of the magnetized and unmagnetized state become equal at

$$G(T,m) - G(T,m=0) = 0 = a(T)m^2 + \frac{1}{2}bm^4 + \frac{1}{3}cm^6, \tag{13.32}$$

which can be rewritten as

$$m^2\left[a(T) + \frac{1}{2}bm^2 + \frac{1}{3}cm^4\right] = 0. \tag{13.33}$$

Again, in Eq. (13.33), depending on the polynomial coefficients, the free energies of the magnetized and unmagnetized materials can be equal at multiple solutions of the magnetization. Both of these equations must be simultaneously satisfied to define the state of the system at the temperature of the phase transition, and due to $b < 0$ can exhibit multiple phases coexisting together. Now, we look for multiple solutions of Eqs. (13.31) and (13.33).

A solution for both Eqs. (13.31) and (13.33) is $m = 0$. An interesting question to ask is if there is a nonzero magnetization that is also a solution of these equations. If so, it would satisfy

$$a(T) + bm^2 + cm^4 = 0 \tag{13.34a}$$

and

$$a(T) + \frac{1}{2}bm^2 + \frac{1}{3}cm^4 = 0. \tag{13.34b}$$

The system in Eq. (13.34) is solved for $a(T)$ and m^2 to obtain

$$a(T) = a_0(T - T_o) = \frac{3}{16}\frac{b^2}{c} \tag{13.35a}$$

and

$$m^2 = -\frac{3}{4}\frac{b}{c}. \tag{13.35b}$$

It is seen from Eq. (13.35b) that for $b < 0$, a phase of nonzero magnetization occurs simultaneously with a phase with $m^2 \neq 0$, i.e., Eq. (13.35b) is the condition that two phases appear at T. This is recognized as the condition for a first-order phase transition. The temperature at which this condition first occurs is the temperature condition in Eq. (13.35a), i.e., the transition temperature T_c is given by [10,12]

$$a_0(T_c - T_o) = \frac{3}{16}\frac{b^2}{c} \text{ or } T_c = \frac{1}{a_0}\left[\frac{3}{16}\frac{b^2}{c} + a_0 T_0\right]. \tag{13.36}$$

The first-order phase transitions are seen from the Landau treatment to exhibit discontinuities in their thermodynamic properties. Generally, these types of non-analyticities can be handled in realistic models for thermodynamic systems developed from first principles (i.e., many-body models based on the motion of individual gas molecules or the dynamics of systems of magnetic atoms and molecules) by applications of computer simulation methods. They are less difficult to handle in realistic models than the infinities arising at second-order transitions. Consequently, first-order phase transitions are not a further consideration here. In the following, we will go on to discuss the scaling theory treatment of the singularities found in second-order phase transitions and try to generate result for more realistic, fundamental, models than that of the Landau formulation.

13.4 SCALING THEORY

Scaling theory is involved with understanding the origins of the thermodynamic singularities and the relationships between the critical exponents associated with second-order phase transitions [1,2]. In this regard, the nature of the singularities and some of their relationships were first introduced in Eqs. (13.1)–(13.5). We return in this section to the study of these topics with a focus on the nature of the general form of the free-energy singularities near the critical temperature and critical field, and for a development of how these singularities result in the power law forms observed in the general physical properties of the system. The explanation of the origins of these behaviors can be viewed in terms of a question of mathematics or of the development of a physical picture as to how the phase transition develops about the neighborhood of T_c and H_c. In the following, only the mathematical side of the problem is pursued. For an alternative physical approach, the reader is referred to the literature [1–12].

A general type of free-energy function that exhibits the types of singular behaviors observed at a second-order phase transition is represented mathematically by a homogeneous function of temperature and the external field. The mathematical properties of a homogeneous function are described in the most general way how all the power law singularities defined in Eqs. (13.1)–(13.5) can arise mathematically, and how they are related to one another. First, let us consider the properties of a homogeneous function considered as a mathematical structure.

Mathematically, in order to be a homogeneous function of the real variables x and y, a function $f(x,y)$ needs to obey the relationship [1,2]

$$f(x,y) = \lambda f(\lambda^s x, \lambda^r y), \tag{13.37}$$

where λ, s, and r are real numbers. The form in Eq. (13.37) suggests that homogeneous functions need not necessarily be singular, but we shall see later that when they are singular their homogeneous nature is helpful in classifying and relating their singularities and those of their derivatives. Some general examples of homogeneous functions are

$$f(x,y) = x^2 y^6 = \lambda\left[\lambda^1 x\right]^2\left[\lambda^{-\frac{1}{2}} y\right]^6 \tag{13.38a}$$

and

$$f(x,y) = x^2 + y^2 = \lambda\left\{\left[\lambda^{-\frac{1}{2}}x\right]^2 + \left[\lambda^{-\frac{1}{2}}y\right]^2\right\}. \tag{13.38b}$$

Note that these examples illustrate not only that homogeneous functions exist but that they are found to involve both products and sums of terms. This attests to their generality of form.

Now, consider applying the assumption of a homogeneous function form to the free energy of our thermodynamics problem in the neighborhood of the critical temperature and critical field of a second-order transition. This is not so far-fetched an assumption, as there are a number of many-body systems for which the free energy can be solved exactly or at least reasonably approximated, and these are found to have the form of homogeneous functions near their second-order transitions. In this way, we see that a homogeneous form of the free energy for the magnetic material defined in Eqs. (13.1)–(13.5) about a critical temperature T_c and critical field H_c is then written as [1–7]

$$G(t,h) = \lambda G\left(\lambda^s t, \lambda^r h\right), \tag{13.39}$$

where

$$t = \frac{T - T_c}{T_c} \tag{13.40}$$

and

$$h = H - H_c. \tag{13.41}$$

The thermodynamics of the material represented by Eq. (13.39) is found in the usual way by applying the appropriate derivatives of temperature and fields to the form characterized in Eq. (13.39).

In this way, the magnetization of the material of volume V characterized in Eq. (13.39) is obtained as

$$m(t,h) = -\frac{1}{V}\frac{\partial G(t,h)}{\partial h}, \tag{13.42a}$$

so that [1–7]

$$m(t,h) = -\frac{1}{V}\frac{\partial}{\partial h}\left\{\lambda G\left(\lambda^s t, \lambda^r h\right)\right\} = \lambda^{r+1} m\left(\lambda^s t, \lambda^r h\right). \tag{13.42b}$$

Note that by taking $\lambda = |t|^{-\frac{1}{s}}$ and $h = 0$ in Eq. (13.42b), it follows that below the critical temperature

$$m(t,0) = |t|^{-\frac{r+1}{s}} m(-1,0). \tag{13.43}$$

In addition, for $\lambda = |h|^{-\frac{1}{r}}$ and $t = 0$, it follows that in the vicinity of the critical field

$$m(0,h) = |h|^{-\frac{r+1}{r}} m(0,\pm 1). \tag{13.44}$$

These considerations can also be continued to treat the magnetic susceptibility, which is obtained from the relationship

$$\chi(t,h) = \frac{\partial m(t,h)}{\partial h} = \frac{\partial}{\partial h}\lambda^{r+1} m(\lambda^s t, \lambda^r h) = \lambda^{2r+1}\chi(\lambda^s t, \lambda^r h), \tag{13.45a}$$

where Eq. (13.42b) is used for applying the derivative. From Eq. (13.45a), it then follows at $\lambda = |t|^{-\frac{1}{s}}$ and $h = 0$ that

$$\chi(t,h=0) = |t|^{-\frac{2r+1}{s}}\chi(\pm 1,0) \tag{13.46}$$

These represent the magnetic properties associated with the magnetic field.

The specific heat can similarly be obtained from a consideration of the temperature derivatives of Eq. (13.39). Consequently, the specific heat of a system of volume V is obtained from a consideration of the thermodynamic relationship [1–7]

$$C(t,h) = -\frac{1}{V} T \frac{\partial^2 G(t,h)}{\partial T^2}, \tag{13.47a}$$

so that from Eq. (13.39) it follows that

$$C(t,h) = \lambda^{2s+1} C(\lambda^s t, \lambda^r h). \tag{13.47b}$$

Again, taking $\lambda = |t|^{-\frac{1}{s}}$, in the case that $h = 0$, it follows that

$$C(t,h=0) = |t|^{-\frac{2s+1}{s}} C(\pm 1,0). \tag{13.48}$$

It is, consequently, seen that the singularity at the critical temperature is extracted explicitly from the assumption of a homogeneous free-energy form.

The divergent forms obtained in Eqs. (13.39)–(13.48) can be rewritten in terms of the critical exponents defined in Eqs. (13.1)–(13.5). In this way, considering Eqs. (13.43), (13.44), (13.46), and (13.48), we find that the critical exponents are given by [1,2]

$$\alpha = \frac{2s+1}{s}, \tag{13.49a}$$

$$\beta = -\frac{r+1}{s}, \tag{13.49b}$$

$$\gamma = \frac{2r+1}{s}, \tag{13.49c}$$

$$\delta = -\frac{r}{r+1}. \tag{13.49d}$$

Eliminating r and s between the exponents, a variety of identities involving the critical exponents are obtained. Specifically, from the critical exponent forms in Eq. (13.49), it follows that [1,2]

$$\alpha + 2\beta + \gamma = 2 \tag{13.50a}$$

and

$$\gamma = \beta(\delta - 1). \tag{13.50b}$$

These are the relationships given to the reader earlier in Eq. (13.2).

Next, consider the correlation length $\xi(t,h)$ and how, as a homogeneous function, it would scale. In this regard, note that $\xi(t,h)$ has the units of length, so that written as a homogeneous function it looks like [1,2]

$$\xi(t,h) = L\xi\left(L^y t, L^x h\right), \tag{13.51a}$$

where we have used the notation L for the scaling parameter rather than λ. This is done to remind us that L changes the length scale of the system. For $h = 0$, it then follows from Eq. (13.51a) that

$$\xi(t,0) = L\xi\left(L^y t, 0\right). \tag{13.51b}$$

Upon taking $L = |t|^{-\frac{1}{y}}$, it is then found from Eq. (13.51b) that [1,2]

$$\xi(t,0) = |t|^{-\frac{1}{y}}\ \xi(\pm 1,0) = |t|^{-\nu}\ \xi(\pm 1,0), \tag{13.52}$$

where in the last equality we have used the definition of the critical exponent $\nu = \frac{1}{y}$ defined in Eqs. (13.4) and (13.52) for $H_c = 0$.

Consider now the dimension of the material characterized by the free-energy form in Eq. (13.39). In thermodynamics, we generally treat 1-d, 2-d, or 3-d materials. Depending on the dimension d of the system, the free-energy associated with a sample in the form of a cube (in 3-d) of material, square (in 2-d) of material, or length (in 1-d) of material of side l is given by [1,2]

$$l^d g(t,0), \tag{13.53}$$

where $g(t,0)$ is the free-energy density. Consequently, the free energy of a volume of side $\xi(t,0)$ is

$$\xi(t,0)^d\ g(t,0) \tag{13.54}$$

Assuming that the free energy of the volume $\xi(t,0)^d$ of material remains constant under the change of length scale L, it then follows from Eq. (13.54) that

$$\xi(t,0)^d\ g(t,0) = \left[L\xi\left(L^y t,0\right)\right]^d \left[\frac{1}{L^d} g\left(L^y t,0\right)\right]. \tag{13.55}$$

Note, in this regard, that the length scales have changed but the amount of material has not. Under this assumption, the free-energy density scales as

$$g(t,0) = \frac{1}{L^d} g\left(L^y t,0\right), \tag{13.56a}$$

and the correlation length scales as

$$\xi(t,0) = L\xi\left(L^y t,0\right). \tag{13.56b}$$

The scaling form in Eq. (13.56) is now used to obtain the relationship in Eq. (13.5) for the critical exponents of the materials. Specifically, taking $L = |t|^{-\frac{1}{y}}$ in Eq. (13.56), it then follows that [1,2]

$$g(t,0) = |t|^{\frac{d}{y}} g(\pm 1,0). \tag{13.57}$$

Alternatively, from Eq. (13.39), it is seen that

$$G(t,0) = \lambda G(\lambda^s t,0), \tag{13.58}$$

and taking $\lambda = |t|^{-\frac{1}{s}}$ in Eq. (13.58), it then follows that

$$G(t,0) = |t|^{-\frac{1}{s}} G(\pm 1,0). \tag{13.59}$$

From Eqs. (13.49a), (13.52), (13.57), and (13.59), it is seen that the temperature singularities require that

$$-\frac{1}{s} = 2 - \alpha = \frac{d}{y} = d\nu. \tag{13.60}$$

This can be rewritten as

$$d\nu = 2 - \alpha, \tag{13.61}$$

giving the critical exponent relationship in Eq. (13.5).

13.4.1 Examples of the Two-Dimensional Ising Model and Landau Theory

Note that the relationships between the scaling exponents can be tested on a small number of exactly solvable models exhibiting phase transitions. One of the most important of these models in the history of statistical mechanics is the two-dimensional Ising model [1,2]. This is a model of magnetism in which the magnetic atoms of the material have a spin associated with each of them, which carriers the magnetic moment of the atom. The atomic spins and their associated magnetic moments are aligned in space by the chemical bonds between the atoms. This interaction can cause, for example, a permanent ferromagnetic magnetization to arise in the material. For such a material composed as a ferromagnetic array of magnetic atoms, each located on a two-dimensional lattice, the thermodynamics of the material can be solved exactly. Consequently, the scaling properties and the relations between the critical exponents of the material are exactly known.

The two-dimensional Ising model treats a system of spins that can only assume one of the two spin values $S_i = \pm 1$. Here the spins are arranged on a square lattice with the lattice sites labeled by the integer i and have either of the two values, i.e., 1 for spin up and −1 for spin down with corresponding oppositely oriented magnetic moment vectors. The spins on the different lattice sites only interact with their nearest neighbor spins, and the total energy of the system is characterized by the Hamiltonian [1,2]

$$\mathcal{H} = -J \sum_{i,j} S_i S_j, \tag{13.62}$$

where $\langle i, j \rangle$ indicates a sum over the nearest neighbor pairs of spins, and $J > 0$ is the exchange coupling. Note that in the lowest energy configuration, the spins are all either spin up or spin down, and the Hamiltonian models a ferromagnet.

The thermodynamics of Eq. (13.62) is obtained from the partition function

$$Z = Tre^{-\frac{\mathcal{H}}{k_B T}} \tag{13.63}$$

in the usual way. Note that in Eq. (13.63), Tr represents a trace over all possible configurations of the spin in the system, k_B is the Boltzmann constant, and T is the temperature. In this model, it is found from a famous solution that the critical temperature occurs at T_c given by the relationship [1,2]

$$\sinh \frac{2J}{k_B T_c} = 1, \tag{13.64}$$

and the energy and magnetic properties are obtained exactly as functions of the temperature.

The singular nature of the solutions in the neighbor of the critical point are all observed to exhibit the form of homogeneous functions. Specifically, the exact critical exponents from these homogeneous forms [1,2] are also obtained and given as $\alpha = 0$, $\beta = \frac{1}{8}$, $\gamma = \frac{7}{4}$, $\delta = 15$, and $\nu = 1$.

Due to the homogeneity properties of the singularities, the two-dimensional Ising solution exhibits the series of identities for the critical exponents obtained earlier. In this regard, referring back to the identities in Eqs. (13.2a) and (13.5), it then follows for Eq. (13.2) that

$$\alpha + 2\beta + \gamma = 0 + \frac{1}{4} + \frac{7}{4} = 2 \tag{13.65a}$$

and

$$\gamma = \frac{7}{4} = \beta(\delta - 1) = \frac{1}{8}(15 - 1) = \frac{7}{4}. \tag{13.65b}$$

Similarly, for the identity in Eq. (13.5), it is found that

$$d\nu = 2(1) = 2 - \alpha = 2 - 0 = 2 \tag{13.65c}$$

Returning to a consideration of the Landau theory, the form of the Landau free energy as a homogeneous function can be explicitly exhibited. In this way, from Eqs. (13.25) to (13.39), we see that for $h = 0$ and $T < T_c$, the free-energy function scales as

$$G(t, h = 0) = -\lambda \frac{a_0^2}{2b} T_c^2 \left[\lambda^{-\frac{1}{2}} t \right]^2, \tag{13.66}$$

so that $s = -\frac{1}{2}$ from Eq. (13.49a) and the exponent $\alpha = \frac{2s+1}{s} = 0$. In addition, at $T = T_c$ and $h \neq 0$, it follows from Eq. (13.24) that

$$m(t = 0, h) = \left[\frac{h}{2b} \right]^{\frac{1}{3}}, \tag{13.67}$$

so that the free energy in Eqs. (13.17) and (13.39) becomes

$$G(t = 0, h) = \frac{1}{2} b \left[\frac{h}{2b} \right]^{\frac{4}{3}} - \left[\frac{h}{2b} \right]^{\frac{1}{3}} h = -\frac{3}{4} \left[\frac{1}{2b} \right]^{\frac{1}{3}} h^{\frac{4}{3}} \tag{13.68a}$$

The free energy in Eq. (13.68a) is also seen to scale as

$$G(t=0,h) = -\lambda \frac{3}{4}\left[\frac{1}{2b}\right]^{\frac{1}{3}} \left[\lambda^{-\frac{3}{4}} h\right]^{\frac{4}{3}}, \tag{13.68b}$$

so that $r = -\frac{3}{4}$. From our results for r and s and the relationships in Eq. (13.49), it then follows that [1,2]

$$\beta = -\frac{r+1}{s} = \frac{1}{2}, \tag{13.69a}$$

$$\gamma = \frac{2r+1}{s} = 1, \tag{13.69b}$$

$$\delta = -\frac{r}{r+1} = 3. \tag{13.69c}$$

Consequently, the Landau form also represents the properties of a homogeneous function of t and h.

13.4.2 Natural Length Scales

An important point to note about the properties of the scaling forms is that, in the neighborhood of the critical point, they are independent of any natural length scale in the model. To see what is meant by this, consider the example of the homogeneous function of the general real variable x having the form

$$f(x) = \frac{1}{x^s}. \tag{13.70}$$

If x is scaled by us by the constant λ, then

$$f(\lambda x) = \frac{1}{\lambda^s}\frac{1}{x^s} \tag{13.71a}$$

or viewed in terms of homogeneity

$$f(x) = \frac{1}{x^s} = \lambda f\left(\lambda^r x\right) = \lambda \frac{1}{\left(\lambda^r x\right)^s} = \lambda^{1-rs}\frac{1}{x^s}, \tag{13.71b}$$

so that $r = \frac{1}{s}$. From Eq. (13.71), it is found that the amplitude of the function is changed by the factor $\frac{1}{\lambda^s}$, but its fundamental functional dependence on x is not changed, i.e., it is still the case that $f(x) \propto \frac{1}{x^s}$. In general, then the power law singularities of thermodynamic systems indicate the absence of an inherent length scale in the model.

As a further illustration of the nature of natural length scales, two common examples of functions in physics with natural length scales can be given. Specifically, consider the function of the variable x of the form

$$f(x) = \frac{1}{x^s + b}. \tag{13.72a}$$

Here $b^{\frac{1}{s}}$ now enters as a natural length measure associated with x, as defined by Eq. (13.72a). If x is scaled by the constant λ, then it follows that

$$f(\lambda x) = \frac{1}{\lambda^s x^s + b} = \frac{1}{\lambda^s} \frac{1}{x^s + \frac{b}{\lambda^s}}. \tag{13.72b}$$

The amplitude of the function in Eq. (13.72b) is changed by a factor $\frac{1}{\lambda^s}$ but so is the form of the function, i.e., after the scaling it is no longer the case that $f(x) \propto \frac{1}{x^s + b}$. It is seen that the length scale set by $b^{\frac{1}{s}}$ can have a greater or a less effect on the value of the function depending on the ratio $\frac{b}{\lambda^s}$, and that it changes the essential dependence on x of the form of the function. In this regard, b is part of a natural scale introduced into the problem by the system itself, while λ is a scale we introduce.

Note that the form in Eq. (13.72b) is similar to the form of a Lorentzian resonance

$$f(x) = \frac{1}{x^2 + \delta^2}, \tag{13.72c}$$

where the natural length scale is set by δ^2. As $\delta^2 \to 0$, it is also noted that the homogeneity of the form is restored. Another example of a form occurring in physics that has a natural length scale is

$$f(x) = \frac{e^{-\frac{x}{\xi}}}{x}. \tag{13.72d}$$

Here, the natural length scale is set by ξ, and a homogeneous form is returned upon taking the limit $\xi \to \infty$. The functional forms given in Eq. (13.72) frequently occur in the discussion of physical systems with properties characterized by natural length scales.

We next look at the theory of renormalization group, which allows for the systematic calculation of the critical exponents of a many-body model from first principles. It arises from the considerations that are heavily based on the theory of scaling and the representation of the critical properties of a system by a treatment of homogeneous forms.

13.5 RENORMALIZATION GROUP APPROACH

The renormalization group approach arises from the assumption that the critical properties of a medium around its critical points are described by homogeneous functions [1–7]. Specifically of interest are the homogeneity relationships in Eq. (13.56) for the change of the correlation length and of the free energy per volume when the lengths in the material are scaled by the factor L. (Note that, for simplicity, in the following discussions, we only consider the critical properties of the system in the $t \to 0$ limit for $h = 0$. This restriction, however, can be lifted in a straightforward way.) In particular, consider the limit $t \to 0$ of the correlation length [1–7]

$$\xi(t \to 0,0) \to L\xi\left(L^y t \to 0,0\right) \to \infty, \tag{13.73}$$

which becomes infinite and independent of the length scale L. Similarly, upon taking $t \to 0$, it is found that the power law features in Eqs. (13.1) through (13.5) and (13.56) at $t = 0$ also become singular and independent of a length scale change introduced into the system. In this regard, the free energy per volume at the critical temperature exhibits a fix point under a change of scale by the factor of L, and the determination of the critical temperature of the system reduces to the discussion of the fixed points of the system under a change of scale by a factor L.

To see how a renormalization group study works, consider its application to a specific system of considerable importance to thermodynamics. In this regard, a model studied early in the development of renormalization group theory is the two-dimensional Ising model. It is an extremely important model as an exact solution of its thermodynamic properties exists, and it has features that make it easily accessible to a renormalization group treatment. Consequently, the results of the exact solution will enable us to offer a comparison with the approximate results from our renormalization group study, and its transformation properties under a change of length scale are easily generated.

The Hamiltonian of the two-dimensional Ising model of ferromagnetic spins defined on a square lattice is written in the form [1–3]

$$\mathcal{H} = -J\sum_{\langle i,j\rangle}\sigma_i\sigma_j, \tag{13.74a}$$

where $\langle i,j\rangle$ represents a sum over the nearest neighbor pairs, $\sigma_i = \pm 1$ is the spin at the site i, which can be spin up $\sigma_i = 1$ or spin down $\sigma_i = -1$, and $J > 0$ is the ferromagnetic coupling of the system. The schematic in Figure 13.2a shows the square lattice array of lattice sites with the nearest neighbor lattice separation a, each of which is labeled by a site index i so that the ith lattice site is occupied by the spin σ_i.

For this Hamiltonian, the partition function of the system of $N \to \infty$ spins has the form [1–3]

$$Z_N = Tr_{\{\sigma_l\}}e^{-\frac{\mathcal{H}}{k_BT}} = Tr_{\{\sigma_l\}}e^{K\sum_{\langle i,j\rangle}\sigma_i\sigma_j}, \tag{13.74b}$$

where $K = \frac{J}{k_BT}$, $Tr_{\{\sigma_l\}}$ represents a trace over all possible spin configurations of the system, k_B is the Boltzmann constant, T is the temperature, and the free energy is obtained from the logarithm of Eq. (13.74b). The focus in the following is to obtain the critical temperature of the system characterized by Eq. (13.74) by determining the fixed points of the free energy per volume of the system under a change of length scale. In addition, the critical exponent of the correlation length is also determined. Specifically, these calculations require us to determine how Eq. (13.74b) changes under a change of length scale.

To do this, we begin by treating a more general problem than that characterized by the Hamiltonian in Eq. (13.74a). Specifically, consider the Ising Hamiltonian $\mathcal{H}'$ of the form [1–3]

$$-\frac{\mathcal{H}'}{k_BT} = \sum_{i,j}K_{ij}\sigma_i\sigma_j + \sum_{i,j,l,m}K_{ijlm}\sigma_i\sigma_j\sigma_l\sigma_m + \sum_{i,j,l,m,s,t}K_{ijlmst}\sigma_i\sigma_j\sigma_l\sigma_m\sigma_s\sigma_t + \ldots \tag{13.75}$$

Here the first term is a sum over all distinct spin pairs in the system, the second term is the sum over all distinct spin quartets, the third term is the sum over all distinct products of six spins, etc. In this model, we have expanded the possible interactions of the spins on the lattice to treat all the possible combinations of spin interactions. Again, the partition function of Eq. (13.75) for a system of $N \to \infty$ spins is given by

$$Z_N = e^{NK_0}Tr_{\{\sigma_l\}}e^{-\frac{\mathcal{H}'}{k_BT}}, \tag{13.76}$$

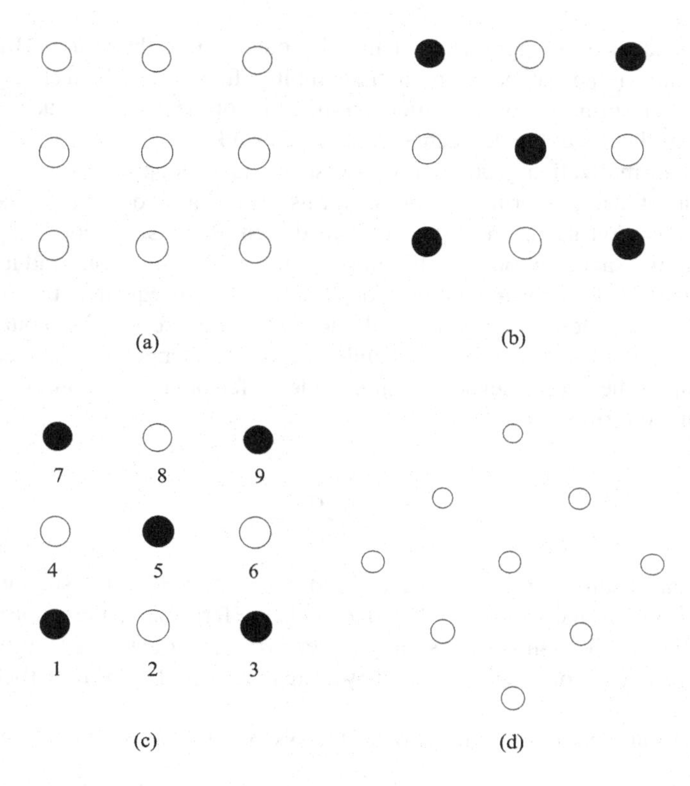

FIGURE 13.2 Schematics for the renormalization group treatment of the two-dimensional Ising model: (a) the two-dimension square lattice of spin sites, as indicated by the open circles, (b) the lattice in (a) divided into two interpenetrating square lattices: one of open circles and the other of closed circles, (c) a portion of the lattice in (b) in which some of the circle sites are numbered, (d) the resulting lattice of open circles that remain upon tracing over the spins on the black circle sites.

where we have allowed for the additional possibility of a constant factor e^{NK_0} occurring outside the trace. For the spins defined on the square lattice of Figure 13.2a, the problem in Eq. (13.76) turns out to be less complex to consider than that in Eq. (13.74), but it can ultimately be easily related to the problem of finding the critical point of Eq. (13.74). This is the subsequent focus of the considerations here.

To accomplish a scale change in Eqs. (13.75) and (13.76), consider the square lattice in Figure 13.2b. It is the same as that in Figure 13.2a with the same spin occupancy, but now half the sites have been colored black and the other half remain white. Consequently, in Figure 13.2b the lattice consists of both black and white sites with the same nearest neighbor separation a as the lattice in Figure 13.2a. Both white and black sites contain the individual spins labeled σ_i with the same i site labels in both Figure 13.2a and b, regardless of whether the site is white or black. To obtain the total partition function of the system, the trace in Eq. (13.76) is then made over all the spins on both the white and black lattice.

Next, consider a new view of the trace over the spins in Figure 13.2b. First, we trace out only the spins on the black sites so that the remaining trace only involves the $\frac{N}{2} \to \infty$ spins on the white lattice [1–3]. The remaining spins on the white lattice also form a square lattice; however, it is rotated by 45°, and the nearest neighbor separation between spins is now $\sqrt{2}a$. The remaining problem on the white lattice is scaled by $\sqrt{2}$ from that of the original lattice in Figure 13.2a.

If in Eq. (13.76), the spins on only the black sites are traced out, it can be shown using the identity $(\sigma_l\sigma_j.....)^2 = 1$ and the Taylor series expansion of factors of the form $e^{A\sigma_l\sigma_j.....}$ that a new problem is obtained on the white lattice sites. It has the form of a new Ising Hamiltonian given by

$$-\frac{\mathcal{H}''}{k_BT} = \sum_{i,j} K'_{ij}\sigma_i\sigma_j + \sum_{i,j,l,m} K'_{ijlm}\sigma_i\sigma_j\sigma_l\sigma_m + \sum_{i,j,l,m,s,t} K'_{ijlmst}\sigma_i\sigma_j\sigma_l\sigma_m\sigma_s\sigma_t + \ldots \quad (13.77)$$

Here the first term is now the sum over all distinct spin pairs on the white lattice, the second term is the sum over all distinct spin quartets on the white lattice, the third term is the sum over all distinct products of six spins on the white lattice, etc. The Hamiltonian in Eq. (13.77) is of the same form as that in Eq. (13.75) but with new primed coefficients. Again, the partition function of Eq. (13.76) for a system of $N \to \infty$ spins is given by [1–3]

$$Z_N = e^{NK_0'} Tr_{\{\sigma_l\}} e^{-\frac{\mathcal{H}''}{k_BT}}, \quad (13.78)$$

where the trace is only over the white lattice spins. Note that by this transformation, the original square lattice Ising model with lattice constant a and coupling constants $\{K_{ij}, K_{ijlm}, K_{ijlmst}, \ldots.\}$ has been mapped onto a new Ising model with lattice constant $\sqrt{2}a$ and coupling constants $\{K_{ij}', K_{ijlm}', K_{ijlmst}', \ldots.\}$. In effect, the resulting transformation represents a scale change by $\sqrt{2}$ of the original system. This change of scale will show up, for example, in the correlation lengths of the new system.

Consequently, the $\{K'\}$ are functionally related to the $\{K\}$ by the transformation, so that

$$K_{ij}' = K_{ij}'(K_{ij}, K_{ijlm}, K_{ijlmst}, \ldots),\ K_{ijlm}' = K_{ijlm}'(K_{ij}, K_{ijlm}, K_{ijlmst}, \ldots), \text{ etc.} \quad (13.79a)$$

From our earlier discussions, we expect that the fixed points of this transformation should contain the critical point of the magnetic phase transition, and that at this point Eq. (13.79a) becomes [1]

$$K_{ij} = K_{ij}'(K_{ij}, K_{ijlm}, K_{ijlmst}, \ldots),\ K_{ijlm} = K_{ijlm}'(K_{ij}, K_{ijlm}, K_{ijlmst}, \ldots), \text{ etc.} \quad (13.79b)$$

As a result, at the fixed point the correlation length singularity in Eq. (13.73) is valid at each iteration of the transform, and at the critical point only the couplings (and not the constant K_0', which is outside the trace) enters directly into the scaling considerations of the infinite correlation length.

To see how the development of the renormalization group transformation of the Ising model works, consider a crude analytic solution for the critical point of the Ising model obtained within the formulation. This example is only meant as a pedagogical illustration of how the transformation works. Nevertheless, our procedure not only illustrates the general method, but it also gives a rough approximation to the exact solution of the problem.

Specifically, consider the square lattice Ising model and initially focus on the patch of spins shown in Figure 13.2c. We will develop an approximate transformation from the nearest and the next nearest coupling on the $\{K\}$ lattice (white and black sites) to the nearest and the next nearest couplings on the $\{K'\}$ lattice (white sites). This is done by tracing out the spins on the black lattice sites so that only the spins on the white sites remain untraced. In developing the transformation, we focus on the nearest and the next nearest couplings, and for the most part ignore the higher order couplings in the systems. This is the source of the simplifying approximation. On the $\{K\}$ lattice, the nearest neighbor coupling is denoted by K, and the next nearest coupling is denoted by P, while on the $\{K'\}$ lattice the nearest neighbor coupling is denoted by K', and the next nearest coupling is denoted by P'. The idea is then to develop a transformation of the form [1–3]

$$K' = K'(K,P) \tag{13.80a}$$

$$P' = P'(K,P) \tag{13.80b}$$

relating the couplings in the primed Hamiltonian to those in the unprimed Hamiltonian.

To generate the transformation, consider the sites in Figure 13.2c, which are white sites labeled 2, 4, 6, 8, and black a site labeled 5. In the partition function, these sites enter as a product of exponential factors of the form [1–3]

$$e^{K(\sigma_5\sigma_2+\sigma_5\sigma_4+\sigma_5\sigma_6+\sigma_5\sigma_8)}e^{P(\sigma_4\sigma_2+\sigma_6\sigma_2+\sigma_6\sigma_8+\sigma_4\sigma_8)} = e^{K\sigma_5(\sigma_2+\sigma_4+\sigma_6+\sigma_8)}e^{P\sigma_2(\sigma_4+\sigma_6)}e^{P\sigma_8(\sigma_6+\sigma_4)}, \tag{13.81}$$

where we ignore the couplings of these sites to sites outside their cluster in Figure 13.2c. Consequently, we shall make the approximation of focusing only on these factors in developing the approximate transformation in Eq. (13.80). To obtain the transformation from the lattice of white and black sites with lattice constant a to the lattice of only white sites of lattice constant $\sqrt{2}a$ in Figure 13.2b and c, the spin σ_5 is traced over. Doing this, it follows that [1,2]

$$\begin{aligned} Tr_{\sigma_5} e^{K(\sigma_5\sigma_2+\sigma_5\sigma_4+\sigma_5\sigma_6+\sigma_5\sigma_8)} &= 2\cosh K(\sigma_2+\sigma_4+\sigma_6+\sigma_8) \\ &= e^{K_0'}e^{\frac{1}{2}K'(\sigma_2\sigma_4+\sigma_2\sigma_6+\sigma_8\sigma_6+\sigma_8\sigma_4)}e^{P'(\sigma_4\sigma_6+\sigma_2\sigma_8)}e^{M'\sigma_2\sigma_4\sigma_6\sigma_8}, \end{aligned} \tag{13.82}$$

where in the last term the $M'\sigma_2\sigma_4\sigma_6\sigma_8$ is a four-particle interaction that shall be handled later. In our approximation, the four-body interaction, however, does not affect the properties of the system and is essentially ignored. Note also that the factor $\frac{1}{2}K'$ in Eq. (13.82) arises due to the fact that a term in $\sigma_2\sigma_4+\sigma_2\sigma_6+\sigma_8\sigma_6+\sigma_8\sigma_4$ enters the problem from the traces of other black sites neighboring σ_5.

Focusing on Eq. (13.82), we then find that [1–3]:

a. For $\sigma_2 = \sigma_4 = \sigma_6 = \sigma_8$

$$2\cosh 4K = e^{K_0'+2K'+2P'+M'} \tag{13.83a}$$

b. For $\sigma_2 = \sigma_4 = \sigma_6 = -\sigma_8$

$$2\cosh 2K = e^{K_0'-M'} \tag{13.83b}$$

c. For $\sigma_2 = -\sigma_4 = -\sigma_6 = \sigma_8$

$$2 = e^{K_0'-2K'+2P'+M'} \tag{13.83c}$$

d. For $\sigma_2 = \sigma_4 = -\sigma_6 = -\sigma_8$

$$2 = e^{K_0' - 2P' + M'}. \tag{13.83d}$$

These are the only four unique relations obtained from Eq. (13.82) that express the primed parameters in terms of those in the unprimed model. Dividing Eq. (13.83a) by Eq. (13.83c) it follows that [2,3]

$$4K' = \ln\cosh 4K, \tag{13.84a}$$

and similarly dividing Eq. (13.83a) by Eq. (13.83d) it follows that

$$2K' + 4P' = \ln\cosh 4K. \tag{13.84b}$$

Using Eq. (13.84a) in Eq. (13.84b) then gives [1,2] a further simplification for P' in the form

$$P' = \frac{1}{8}\ln\cosh 4K. \tag{13.85}$$

The relationships in Eqs. (13.84a) and (13.85) express K' and P' in terms of K, but they need an adjustment that we have left out.

Now, we must remember the exponents $e^{P\sigma_2(\sigma_4+\sigma_6)}e^{P\sigma_8(\sigma_6+\sigma_4)}$ in Eq. (13.81). This factor has not has yet entered our considerations, but it contributes to the nearest neighbor interactions in the primed model. Including this term in our treatment, it follows that Eq. (13.84a) gives [1–3]

$$K' = \frac{1}{4}\ln\cosh 4K + P. \tag{13.86a}$$

From Eq. (13.85), it then follows that

$$P' = \frac{1}{8}\ln\cosh 4K. \tag{13.86b}$$

Within our approximations, Eqs. (13.86a) and (13.86b) are then the two transformations for our system having the required form of Eq. (13.80).

A further simplification is made in the transformation by assuming that K and P can be treated as small. This allows the transformation to be written in the form of polynomial functions. In this approximation, Eq. (13.86) becomes [1–3]

$$K' \approx 2K^2 + P \tag{13.87a}$$

and

$$P' \approx K^2. \tag{13.87b}$$

These equations are easier to work with than those in Eq. (13.86) but still qualitatively contain the physics of the problem. They will be the focus, as we proceed in our study of the mapping of the system in Eqs. (13.75) and (13.76) defined on Figure 13.2a to the system in Eqs. (13.77) and (13.78) defined on Figure 13.2d.

The transformations in Eqs. (13.86) and (13.87) are of a type of recursion relation familiar from nonlinear dynamics and the study of the stability properties of dynamical systems. The rest of the analysis of our system proceeds along the standard lines of approach developed in studying the stability of dynamical systems. Specifically, we first find the fix points of the transformation and then perform the analysis of the stability of these fixed points. This allows for the determination of the critical exponents of our model and the critical points of the system.

First, consider the fixed points of the transformations in Eqs. (13.86) and (13.87). One fixed point of both the transformations is $(K^*, P^*) = (0,0)$. For a model of coupled Ising spins, this corresponds to a system at infinite temperature. In this limit, the spins are totally disordered, and there is no correlation between them. A second fixed point of both transformations is $(K^*, P^*) = (\infty, \infty)$. For a model of coupled Ising spins, this corresponds to the opposite limit of a system at zero temperature. In this limit, the spins are totally ordered so that they are ferromagnetically aligned. Between these two limits is the fixed point corresponding to the critical point of the systems. We now study this critical point in the context of the algebraic form in Eq. (13.87).

The critical fixed point (K^*, P^*) of the ferromagnetic transition obtained from the algebraic transformation in Eq. (13.87) is [1–3]

$$K^* = \frac{1}{3} \tag{13.88a}$$

and

$$P^* = \frac{1}{9}. \tag{13.88b}$$

The transformation in Eq. (13.87) also allows us to extend our study to a consideration of the stability properties of the system in the neighborhood of the critical point. This allows us to obtain, for example, the critical exponents of the model.

In this regard, we make a linearization of the transformation in the neighborhood about its fixed points. This determines the nature of the stability properties of the system under various iterations of the transformation. To do this we write [1–3]

$$K = K^* + k_1, \quad K' = K^* + k_1', \tag{13.89a}$$

and

$$P = P^* + k_2, \quad P' = P^* + k_2', \tag{13.89b}$$

where k_1, k_2, k_1', k_2' are considered small. These are now substituted into Eq. (13.87) to obtain relationships between the k_1, k_2, k_1', k_2'.

Upon substituting Eqs. (13.89) into (13.87) and retaining only the linear terms k_1, k_2, k_1', k_2', we find [1–3]

$$\left| \begin{array}{c} k_1' \\ k_2' \end{array} \right| = \left| \begin{array}{cc} \frac{4}{3} & 1 \\ \frac{2}{3} & 0 \end{array} \right| \left| \begin{array}{c} k_1 \\ k_2 \end{array} \right|. \tag{13.90}$$

For a given deviation of the system from the critical point given by (k_1, k_2), Eq. (13.90) generates the new deviation (k_1', k_2') arising from a scale change introduced into the system by Eq. (13.87). In order to understand the dynamics of this transformation, it is helpful to determine the eigenvectors of the transformation in Eq. (13.90), and express (k_1, k_2) and (k_1', k_2') in terms of this basis of the eigenvalue problem. This is a standard technique in the study of stability theory of dynamical systems described by iterative processes.

Consider the eigenvalue problem [1–3]

$$\lambda \begin{vmatrix} x \\ y \end{vmatrix} = \begin{vmatrix} \frac{4}{3} & 1 \\ \frac{2}{3} & 0 \end{vmatrix} \begin{vmatrix} x \\ y \end{vmatrix} \tag{13.91}$$

involving the dynamical matrix in Eq. (3.90). There are two solutions of this problem:

$$\text{a.} \quad \lambda_1 = \frac{2+\sqrt{10}}{3} > 1, \text{ with } \psi_1 = \begin{vmatrix} 2+\sqrt{10} \\ 2 \end{vmatrix} \tag{13.92a}$$

$$\text{b.} \quad \lambda_2 = \frac{2-\sqrt{10}}{3} < 1, \text{ with } \psi_2 = \begin{vmatrix} 2-\sqrt{10} \\ 2 \end{vmatrix}. \tag{13.92a}$$

Note that one solution has an eigenvalue greater that one, and the other solution has an eigenvalue less than one. This property of the eigenvalues is significant in understanding the iterative properties of the transformation. In addition, the two eigenvectors are complete in (k_1, k_2) -space, so that, in terms of these two eigenvectors, we can express a general deviation from the critical point $(K^*, P^*) = \left(\frac{1}{3}, \frac{1}{9}\right)$ as

$$\begin{vmatrix} k_1 \\ k_2 \end{vmatrix} = a_1\psi_1 + a_2\psi_2. \tag{13.93a}$$

Similarly, by solving for $\begin{vmatrix} a_1 \\ a_2 \end{vmatrix}$ in terms of $\begin{vmatrix} k_1 \\ k_2 \end{vmatrix}$, we also have the inverse relation obtained as [1–3]

$$\begin{vmatrix} a_1 \\ a_2 \end{vmatrix} = \frac{1}{4\sqrt{10}} \begin{vmatrix} 2 & -2+\sqrt{10} \\ -2 & 2+\sqrt{10} \end{vmatrix} \begin{vmatrix} k_1 \\ k_2 \end{vmatrix}. \tag{13.93b}$$

This alternatively gives (a_1, a_2) in terms of a particular proposed (k_1, k_2).

Consider now applying the transformation in Eqs. (13.90)–(13.93a) n times. This generates the transformation

$$\begin{vmatrix} \frac{4}{3} & 1 \\ \frac{2}{3} & 0 \end{vmatrix}^n \begin{vmatrix} k_1 \\ k_2 \end{vmatrix} = \lambda_1^n a_1\psi_1 + \lambda_2^n a_2\psi_2 = \begin{vmatrix} k_1' \\ k_2' \end{vmatrix} \tag{13.94}$$

Note that, in the limit of large $n \gg 1$, since $\lambda_1 > 1 > -\lambda_2 > 0$, it follows that the transformation generates

$$\begin{vmatrix} \frac{4}{3} & 1 \\ \frac{2}{3} & 0 \end{vmatrix}^n \begin{vmatrix} k_1 \\ k_2 \end{vmatrix} \to \lambda_1^n a_1\psi_1, \tag{13.95}$$

and only the term in ψ_1 contributes significantly to the motion of $\begin{vmatrix} k_1' \\ k_2' \end{vmatrix}$, as the transformation proceeds. In addition, for increasing n, the deviations from the critical fixed point at $(K^*, P^*) = \left(\frac{1}{3}, \frac{1}{9}\right)$ increase by the factor

$$\lambda_1^n a_1 \psi_1 = \lambda_1^n a_1 \begin{vmatrix} 2+\sqrt{10} \\ 2 \end{vmatrix}, \tag{13.96}$$

i.e., the system moves farther and farther away from the fixed point at an exponential rate.

On the other hand, if $a_1 = 0$ in Eq. (13.93), then for $n \gg 1$, we see from Eq. (13.94) that only ψ_2 is involved in the motion of the system. Under these conditions, it follows that [1–3]

$$\begin{vmatrix} \frac{4}{3} & 1 \\ \frac{2}{3} & 0 \end{vmatrix}^n \begin{vmatrix} k_1 \\ k_2 \end{vmatrix} \to \lambda_2^n a_2 \psi_2 \to 0 = \begin{vmatrix} k_1' \\ k_2' \end{vmatrix}, \tag{13.97}$$

where $\lambda_2 = \frac{2-\sqrt{10}}{3} < 1$. Consequently, under these transformation iterations, the system always moves only toward the fixed point at $(K^*, P^*) = \left(\frac{1}{3}, \frac{1}{9}\right)$. As a result, from Eq. (13.93b) and the condition $a_1 = 0$, we see that the set of (k_1, k_2) defined by

$$2k_1 - \left(2 - \sqrt{10}\right)k_2 = 0 = a_1 \tag{13.98a}$$

or alternatively

$$k_2 = -\frac{2+\sqrt{10}}{3} k_1 \tag{13.98b}$$

defines a line of points in (k_1, k_2) space. Along this line the renormalization group transformation always acts to move the transformed points only toward the critical fixed point.

This line of points moving only toward $(K^*, P^*) = \left(\frac{1}{3}, \frac{1}{9}\right)$ is seen to cross the $P = 0$ axis at [1–3]

$$(K, P) = \left(K^* + \frac{1}{9}\frac{3}{2+\sqrt{10}}, P^* - \frac{1}{9}\right) = \left(\frac{4+\sqrt{10}}{18}, 0\right) = (0.3979, 0). \tag{13.99}$$

See Figure 13.3 for a schematic representation of the critical point and the line of points in Eq. (13.99), which converge upon the critical point under successive applications of the transformation. Consequently, the critical point of the Ising model in Eq. (13.74) is initially located at $K_c = 0.3979$, and under successive transformations it eventually converges on $(K^*, P^*) = \left(\frac{1}{3}, \frac{1}{9}\right)$. In this motion, couplings are generated between successive higher order many-body terms found in Eq. (13.79), but this does not change the fact that the phase transition in the nearest neighbor only system is given by Eq. (13.99).

In addition, it is found that while all points on the line defined in Eq. (13.98) move toward $(K^*, P^*) = \left(\frac{1}{3}, \frac{1}{9}\right)$, all of the other points in the (k_1, k_2) plane move away from $(K^*, P^*) = \left(\frac{1}{3}, \frac{1}{9}\right)$ under the actions of the renormalization group transformation. Specifically, all of these points off

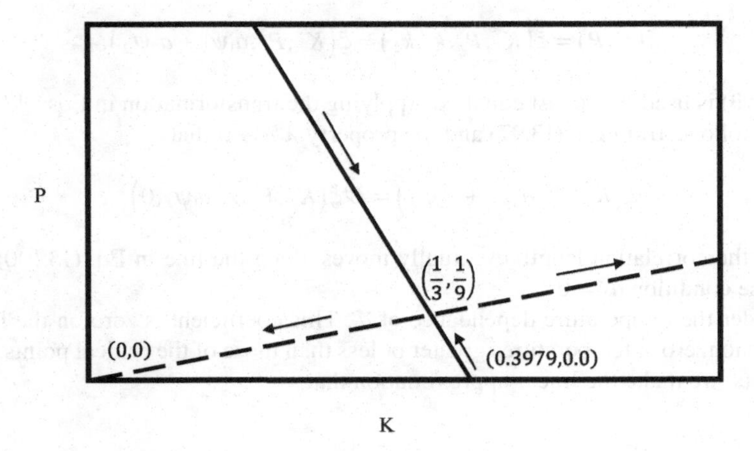

FIGURE 13.3 Schematic for the fixed points and the flows about the fixed points in (K, P)-space from the linearized form of the renormalization group equations of the two-dimensional Ising model given in Eq. (13.87). Here, the fixed point of the mapping is at (1/3, 1/9), and two lines pass through the fixed point. The solid line shows the trajectory of points that travel toward (1/3, 1/9). Note, the arrows show the directions of the motion toward the fixed point. The point (0.3979, 0.0) is the critical point of the two-dimensional Ising model with only the nearest neighbor interactions. Similarly, on the dashed lines the renormalization group mapping moves points away from (1/3, 1/9). This is again shown by the arrows. To the right of the solid line, points in the figure travel off to infinity. To the left of the solid line, points in the figure travel to the fixed point (0, 0).

the line have $a_1 \neq 0$. Of these other points, a point (k_1,k_2) is either on one side or the other of the line of points, which only move toward the critical fixed point. If a point is on the side containing $(0,0)$, it moves toward $(0,0)$. If it is on the side containing (∞,∞), it moves toward (∞,∞). Next, consider how these results are used to determine the critical exponent of the correlation length.

Another interesting line passing through the fixed point is that defined in Eq. (13.93) by [1–3]

$$a_2 = 0 = -2k_1 + \left(2+\sqrt{10}\right)k_2 \tag{13.100}$$

along this line $a_1 \neq 0$ so that Eq. (13.95) applies. In this case, as the transformation is iterated, the points on the line defined by Eq. (13.100) move away from the fixed point. Specifically, if $a_1 > 0$ the points move toward $(K,P) \to (\infty,\infty)$, and if $a_1 < 0$ the points move toward $(K,P) \to (0,0)$. (This behavior is shown schematically in Figure 13.3.) The motions, generated under the transformation, toward $(K,P) \to (\infty,\infty)$ and $(K,P) \to (0,0)$ are important, as they allow us to study the critical exponents of the phase transition. We now consider this in the context of the critical exponent ν, which is associated with the correlation length of the Ising spins.

Consider our transformations in Eq. (3.80) in terms of $\left(K^*,P^*\right)$ and (k_1,k_2). That is, we explicitly write out [1–3]

$$K' = K'(K,P) = K'\left(K^*,P^*,k_1,k_2\right) = K^* + k_1' \tag{13.101a}$$

$$P' = P'(K,P) = P'\left(K^*,P^*,k_1,k_2\right) = P^* + k_2'. \tag{13.101b}$$

In the following, the focus is on applying the notation in Eq. (13.101) to study the transformation in Eq. (13.87) for the limit in which only the terms linear in (k_1,k_2) are retained in the transformation. In terms of this notation, the correlation length of the Ising model is then written in the form

$$\xi(K,P)=\xi\left(K^*,P^*,k_1,k_2\right)=\xi\left(K^*,P^*,a_1\psi_1+a_2\psi_2\right), \tag{13.102}$$

where Eq. (13.93) is used in the last equality. Applying the transformation in Eqs. (13.101)–(13.102) $n \gg 1$ times, it follows from Eq. (13.97) and the property $\lambda_2^n \to 0$ that

$$\xi\left(K^*,P^*,a_1\psi_1+a_2\psi_2\right)=L^n\xi\left(K^*,P^*,\lambda_1^n a_1\psi_1,0\right). \tag{13.103}$$

Consequently, the correlation length eventually moves along the line in Eq. (13.100), i.e., the line governed by the condition $a_2 = 0$.

Next, consider the temperature dependence of a_1. This coefficient is zero on the line of critical points, but it is nonzero at temperatures greater or less than those of the critical points. Near the line of critical points, we make the linear approximation that

$$a_1 = at, \tag{13.104a}$$

where $t = \dfrac{T-T_c}{T_c}$ and a is a coefficient of the linearization. From this it follows from Eq. (13.103) and the power law behavior

$$\xi(a_1) \propto [a_1]^{-\nu} \tag{13.104b}$$

that near the critical point [1,2]

$$a_1^{-\nu} = L^n\left[\lambda_1^n a_1\right]^{-\nu}. \tag{13.105}$$

From Eq. (13.105), we then have the condition

$$1 = L^n\left[\lambda_1^n\right]^{-\nu} = L^n\lambda_1^{-\nu n}. \tag{13.106}$$

Solving Eq. (13.106) for ν and evaluating the result for our model in which $L=\sqrt{2}$ gives [1–3]

$$\nu = \frac{\ln L}{\ln\lambda_1} = \frac{\ln\sqrt{2}}{\ln\left[\dfrac{2+\sqrt{10}}{3}\right]} = 0.6385. \tag{13.107}$$

This represents a rough approximation for the critical exponent ν. For a better result, a better approximate transformation must be developed. Note that by a similar procedure, the other critical exponents $\alpha, \beta, \gamma, \delta$ of the model are also obtained.

13.6 APPLICATION OF THE RENORMALIZATION GROUP TO THE LANDAU FORMULATION

Another way of approaching the renormalization group is to work on the problem of scale changes in the context of wavevector space rather than posed in position space [1–3]. Both the approaches are based on a length rescaling, but the facility of their application depends on the nature of the model being studied. In this section, we look at a renormalization made in wavevector space, presenting it in its application to a particular model with a formulation based on the Landau free-energy form. In the presentation, first the new model is defined, following which the ideas of a wavevector space renormalization are introduced in its context. The result of an application of the approach to determine the fixed points of the newly formulated model is then presented. The application is given as a brief outline so that for the details of the extensive mathematics the reader is referred to the literature [1,3]. Only a summary of the general method is the focus here.

In our earlier discussion of the Landau approach, we treated the free-energy form given in Eq. (13.17). This expression is now rewritten to represent the free energy per volume of a spatially dependent magnetization $m(\vec{r})$. Doing this, the free energy per volume is given by [1–3]

$$g(T,h=0,m) = g(T,h=0,m=0) + c|\nabla m|^2 + a(T)m^2 + \frac{1}{2}bm^4, \tag{13.108}$$

where $a(T) = a_0(T - T_c)$, $a_0 > 0$ is a constant, $b > 0$ is a constant, and $g(T,h=0,m=0)$ is approximated as a constant independent of T. Note that, for simplicity, here we consider only the $h = 0$ system, and since $g(T,h,m)$ represents the free energy per volume, the coefficients a, b are slightly redefined for the per volume system. For this notation, in the standard Landau approach the free energy $G(T,h=0,m)$ is then an integral over space written as

$$G(T,h=0,m) - G(T,h=0,m=0) = \int d^d r \left[c|\nabla m|^2 + a(T)m^2 + \frac{1}{2}bm^4 \right], \tag{13.109a}$$

and the equilibrium state of the system is obtained by minimizing the free energy with respect to $m(\vec{r})$, so that

$$\frac{d}{dm}G(T,h=0,m) = 0 \tag{13.109b}$$

at equilibrium.

Another approach to treating the problem of studying the thermodynamics of the magnetic system is obtained by generalizing the considerations of the problem defined in Eq. (13.109). Specifically, the approach in Eq. (13.109) ignores the thermodynamics fluctuations in the systems, and these fluctuations can be accounted for by reformulating the problem. In the reformulation, $G(T,h=0,m) - G(T,h=0,m=0)$ is regarded as part of a probability distribution representing the probability that a particular $m(\vec{r})$ is the state of the system. In this view, the probability distributions that the system is in the state $m(\vec{r})$ is postulated as being given by a Boltzmann distribution of the form [1,3]

$$\begin{aligned} P\left[m(\vec{r})\right] &= P_0 e^{-\frac{1}{k_B T}\int d^d r\left[c|\nabla m|^2 + a(T)m^2 + \frac{1}{2}bm^4\right]} \\ &= P_0 e^{-\int d^d r\left[\frac{\alpha}{2}|\nabla m|^2 + \frac{u(T)}{2}m^2 + \frac{\lambda}{4!}bm^4\right]}. \end{aligned} \tag{13.110}$$

Here, P_0 is the normalization of the distribution, $\frac{\alpha}{2} = \frac{1}{k_B T}c$, $\frac{u(T)}{2} = \frac{1}{k_B T}a(T)$, and $\frac{\lambda}{4!} = \frac{1}{k_B T}\frac{1}{2}b$. This gives the weight of $m(\vec{r})$ as the state of the system and, consequently, is a natural way of introducing thermal fluctuations into the problem.

Based on the Boltzmann distribution in Eq. (13.110), the partition function of the system is then given by the functional integral [1,3,5,6]

$$Z = \int Dm(\vec{r}) e^{-\int d^d r\left[\frac{\alpha}{2}|\nabla m|^2 + \frac{u(T)}{2}m^2 + \frac{\lambda}{4!}bm^4\right]}, \tag{13.111}$$

where $Dm(\vec{r})$ is the measure of the functional integral, and we have ignored some constants that do not participate in the singularities of interest in the system. Specifically, the integration $\int Dm(\vec{r})$ represents a sum developed over all continuous scalar functions $m(\vec{r})$ for the magnetization subject to a reasonable set of boundary conditions. In addition, note that if in the reformulated problem the approximation is made of representing the state of the system as the most probable configuration of Eq. (13.110), the state of the system reverts to that given by Eq. (13.109). In this sense, the result in Eq. (13.109) may be regarded as a reasonable approximation to the probability distribution of Eq. (13.110).

In the discussions to follow, the renormalization group approach is applied to the system described by Eq. (13.110). The treatment involves a series of scale changes that map the system into itself in such a way that only a state with no natural length scale is left invariant. Consequently, in our discussions of the thermodynamic singularities of the partition in Eq. (13.111), the considerations focus on the fixed points of the scale transformations, as under such conditions the system losses its recognition of lengths within the problem. From our earlier considerations, we recognize this as the signature condition at the phase transition. Next, we describe the general scaling transformation made on the system, and this is followed by a summary of some of the results from the study of Eq. (13.111). The reader is referred to the literature for the details [1,3].

In the transformation applied to the model in Eq. (13.111), the scale changes are made both in position and wavevector space. Specifically, the position space magnetization field $m(\vec{r})$ can also be represented in wavevector space as $\tilde{m}(\vec{k})$, so that the problem can be posed in either of these representations. The two representations of the magnetization are related to one another by the Fourier transform [3]

$$m(\vec{r}) = \int_{-\Lambda}^{\Lambda} d^d k e^{i\vec{k}\cdot\vec{r}} \tilde{m}(\vec{k}), \tag{13.112}$$

where the region of integration is a $d-$ dimensional hypercube centered about the origin in $k-$ space with a side of length 2Λ. Here we introduce a so-called cutoff Λ to the limits of the integration in wavevector space. The cutoff is an acknowledgment that our thermodynamic system is usually defined on a lattice with the smallest separation a between neighboring particles in the medium. In this regards, wavevectors greater that $\frac{2\pi}{a}$ really do not enter the description of the system, as the nature of the material at distance smaller that a is meaningless to our studies. In addition, from a technical standpoint, the cutoff is often found helpful in treating systems that otherwise become singular in the limit $\Lambda \to \infty$. There are standard ways of removing the singularities, and the reader is referred to the literature for these [1,3].

The renormalization group transformations applied to Eq. (13.111) then involve integrating out bands of wavevectors near the edges of the hypercubes in Eq. (13.112). These bands represent the spatially small details of the system, which are not associated with the largescale features developed in the phase transition. Following the integration of these bands, a change of length scales is made, which places the system back to its original covering of wavevector space. This also involves a renormalization of the magnetization amplitude to adjust for the degrees of freedom, which have been integrated out of the system.

The scale changes made on the system are made based on a dimensionless scaling factor $b > 1$. The adjustments to the system are accomplished by a series of specific steps consisting of [3]:

1. The modes of the system are integrated out over a band of wavelengths along each one of the space axes given by $\frac{\Lambda}{b} \le k \le \Lambda$ and $-\Lambda \le k \le -\frac{\Lambda}{b}$, where Λ is the cutoff factor in Eq. (13.112), which enters all the integrals over wavevector. This removes the fine spatial details from the problem.
2. The length scale is rescaled by $\vec{r} \to \vec{r}' = \frac{\vec{r}}{b}$ and is rescaled by $\vec{k} \to \vec{k}' = b\vec{k}$. This puts back the system to contain spatial details.

3. The field is also renormalized by a factor $m(\vec{r}) = b^{-d_\varphi} m'(\vec{r}')$. This accounts for the space dependence of the field amplitude under the rescaling, and the degrees of freedom adjusted by steps 1 and 2. In addition, the exponent d_φ enters into the determination of some of the critical exponents of the model.

The transformations just outlined have been applied to Eq. (13.111) and have successfully given results. These results, however, are primarily of theoretical interest. This is due to the conditions of the solution that are now discussed.

The results obtained for this model are primarily of interest theoretically, as they have a quantitative nature that makes them of limited value experimentally. Specifically, the treatment involves a perturbation theory approach in terms of the integration measure of space, i.e., $\int d^d r$. Usually, $d = 1, 2, 3$ in one- two-, three-dimensional space. In the present approach, however, d has been treated as a real number, so that the definition of the measure is generalized to spatial dimensions d of a real value. Once this is done, the phase transition of the system is studied for $d = 4 - \varepsilon$ in a perturbation series for $|\varepsilon| \ll 1$. Consequently, the study of the model is highly idealized, but the results are, nevertheless, of interest in understanding the origin of the phase transition and its general nature [3].

Applying these transformations to Eq. (13.111), the system can be solved in the neighborhood of $d = 4$ for a set of transformation equations taking (u, λ, α) to a new (u', λ', α'). As in the earlier discussion of the Ising model, the phase transition is obtained from the fixed points of the transition, and the critical exponents are given by the behavior of the points in the neighborhood of the fixed points. In this way, for $d > 4$ the fixed point of the system is given by [3]

$$u^* = 0 \tag{13.113a}$$

$$\lambda^* = 0 \tag{13.113b}$$

$$\alpha^* = 1, \tag{13.113c}$$

independent of the value of ε. For $d < 4$, the fixed point of the system is given by [3]

$$u^* = -\frac{\Lambda^2}{6}\varepsilon \tag{13.114a}$$

$$\lambda^* = \frac{16\pi^2}{3}\varepsilon \tag{13.114b}$$

$$\alpha^* = 1, \tag{13.114c}$$

where now $\varepsilon = 4 - d$. In terms of these critical points, for $d > 4$ the critical exponents of the

system are those of the Landau field theory and are independent of ε. For $d < 4$, on the other hand, the critical exponents of the system are [3]

$$\alpha = \frac{\varepsilon}{6} \tag{13.115a}$$

$$\beta = \frac{1}{2} - \frac{\varepsilon}{6} \tag{13.115b}$$

$$\gamma = 1 + \frac{\varepsilon}{6} \tag{13.115c}$$

$$\delta = 3 + \varepsilon \tag{13.115d}$$

and

$$\nu = \frac{1}{2} + \frac{\varepsilon}{12}. \tag{13.115e}$$

Note as an interesting property that for $d \geq 4$ the model transition temperature and critical exponents are successfully treated by the Landau approximation studied earlier in the text. On the other hand, for $d < 4$ the Landau approximation begins to fail due to the importance of fluctuations in the system.

Consequently, the fluctuations become an important property of the system for $d < 4$. In fact, it is interesting to note that for $d = 1$ it can be shown in the absence of long-range interactions in a system that there are no phase transitions [14]. This is seen from the general form of free energy for one-dimensional materials. In these materials, the entropy dominates the thermodynamic behavior of the system, making them unable to support an ordered phase at nonzero temperatures.

APPENDIX

The path integral in Eq. (13.8) is evaluated using standard methods of functional integrals [15]. Consider the evaluation of the path integral for a spin density defined along the x-axis and expressed as

$$Z = \int d\vec{S}(x) \exp\left(-\frac{1}{k_B T} \int dx \; J \left| \frac{d\vec{S}(x)}{dx} \right|^2 \right). \tag{A.1}$$

The integral in the argument of the exponent can be discretized along the x-axis, so that

$$\int dx \; J \left| \frac{d\vec{S}(x)}{dx} \right|^2 = \sum_{i=1}^{N} \Delta x \; J \left| \frac{\vec{S}_{i+1} - \vec{S}_i}{\Delta x} \right|^2, \tag{A.2}$$

where Δx is an infinitesimal length used to discretize the x-axis, the length of the chain of spins is composed of $N \to \infty$ segments, $\vec{S}_i = \vec{S}(x_i)$ is the spin at the ith segment, and x_i is the position of the ith segment along the spin chain. Periodic boundary conditions are applied to the chain so that it is really an infinite ring of spin density. In terms of this discretized form, Eq. (A.1) then becomes

$$Z = \int \prod_{i=1}^{N \to \infty} \delta\vec{S}_i \exp\left\{ -\frac{1}{k_B T} \sum_{i=1}^{N \to \infty} \Delta x \; J \left| \frac{\vec{S}_{i+1} - \vec{S}_i}{\Delta x} \right|^2 \right\}, \tag{A.3}$$

where $\delta\vec{S}_i = d\vec{S}(x_i)$.

Next, consider the eigenvalue problem based on a consideration of the ith segment [15]

$$\int \delta\vec{S}_i \exp\left\{-\frac{1}{k_B T}\Delta x\ J\left|\frac{\vec{S}_{i+1}-\vec{S}_i}{\Delta x}\right|^2\right\}\psi_n\left(\vec{S}_i\right) = \lambda_n\psi_n\left(\vec{S}_{i+1}\right), \tag{A.4}$$

where λ_n is the eigenvalue and $\psi_n\left(\vec{S}_{i+1}\right)$ is the normalized eigenvector. In terms of the total set of eigenvalues $\{\lambda_n\}$ and their corresponding set of normalized eigenvectors $\left\{\psi_n\left(\vec{S}_i\right)\right\}$

$$\exp\left\{-\frac{1}{k_B T}\Delta x\ J\left|\frac{\vec{S}_{i+1}-\vec{S}_i}{\Delta x}\right|^2\right\} = \sum_{n=0}^{\infty}\psi_n\left(\vec{S}_{i+1}\right)\lambda_n\psi_n\left(\vec{S}_i\right). \tag{A.5}$$

Applying the results in Eqs. (A.4) and (A.5) into Eqs. (A.1) and (A.3) then gives

$$Z = \sum_{n=0}^{\infty}\lambda_n^N, \tag{A.6}$$

and the correlation functions of the spin follow in a similar manner.

REFERENCES

1. Pilschke, M. and B. Bergersen 1989. *Equilibrium statistical physics*. New Jersey: Prentice Hall.
2. Pathria, R. K. 1996. *Statistical Mechanics, 2nd edition*. Oxford: Butterworth Heinemann.
3. McComb, W. D. 2004 *Renormalization methods: A guide for beginners*. Oxford: Clarendon Press.
4. Ma, S.-K. 1978. *Modern theory of critical phenomena*. London: The Benjamin/Cummings Publishing Company.
5. Fisher, M. E. 1983. Scaling, universality and renormalization group theory. In *Critical phenomena: Proceedings, Stellenbosch, South Africa 1982* Ed. F. J. W Hahne, 1–137. Berlin: Springer-Verlag.
6. Wilson, K. G. 1975. The renormalization group: Critical phenomena and the Kondo problem. *Reviews of Modern Physics* 47: 773–837.
7. Wilson, K. G. and J. Kogut 1974. The renormalization group and the ε expansion. *Physics Reports* 12: 75–200.
8. McGurn, A. R. and D. J. Scalapino 1975. One-dimensional ferromagnetic classical-spin-field model. *Physical Review B* 11: 2552–2558.
9. Kittle, C. 1996. *Introduction to solid state physics, 7th Edition*, 393–410. New York: John Wiley & Sons, Inc.
10. Fossheim, K. and A. Sudbo 2004. *Superconductivity: Physics and applications*, 141–198. New York: John Wiley & Sons, Inc.
11. Lipowsky, R. 1987. Surface critical phenomena at first order phase transitions. *Ferroelectrics* 73: 69–81.
12. Ader, J.-P. and A. I. Buzdin 2003. On the theory of surface magnetic first-order phase transition. *Physics Letters A* 319: 360–366.
13. Lipowsky, R. 1984, Surface-induced order and disorder: Critical phenomena at first-order phase transitions. *Journal of Applied Physics* 55: 2485–2490.
14. Landau, L. D. and E. M. Lifshitz 1970. *Statistical physics*, 478. Oxford: Pergamon Press.
15. Feynman, R. P. and A. R. Hibbs 2005. *Quantum mechanics and path integrals*. New York: Dover Publications, Inc.

Index

AC Josephson effect 286–288
Ahronov-Bohm effect 65–66
Airy functions 25
amplifier 168–171, 217–223
Anderson localization 4, 8–9, 40–57, 242–245, 252–253
Anderson transition 5, 8, 41, 57
anomalous Hall effect 264
anti-commutators 91–96
atom lasers 188–189

ballistic transport 6–7, 65–66, 69–73, 77, 80–86, 122
band structure 2–4, 89–91, 118, 126
base 220–224
bipolar transistor 220–224

cavity modes 7–8, 145–147, 151–153, 169–174
chaotic light 162–164
classical Hall effect 12, 245–247
coherence length 62, 299–301, 303–304
coherent states 147–151, 156–157, 170–173
collector 220–224
commutators 21, 143, 146
correlated electrons 10–11, 59, 225, 244–245
correlation length 307–309, 313, 318–324
Coulomb blockade 12–13, 225–230
critical exponents 13–14, 306–307, 312, 317–320, 332, 335
critical fields of superconductors 295–301
critical temperature 13–14, 58–62

DC Josephson effect 286–288
density matrix 176–177
density of states 22–24, 30, 37–39, 108, 182
dispersion relation 12, 21–24, 34, 69, 106–107, 117–118, 122
dissipative transport 17–20
distribution of photons in laser 187–188
distribution of thermal photons 156–157
drain 212–217, 226–227
Drude model 6–7, 17–20, 56

effective mass 112–113
electrical conductivity 17, 22, 41
electromagnetic modes 141–147
electron-electron interactions 12–13, 254
electron gas 17–20, 36–37
electrons 191–195
electron turnstile 238
emitter 220–224
envelope function approximation 113–115
equations of motion 21–22, 25–29
excitons 123–125
extrinsic semiconductors 191–195

Fano resonance 13, 275–279
field effect transistor 208–220
first order phase transitions 313–315
flux quantization
- in Ahronov-Bohm effect 12, 65–66
- in quantum Hall effect 250–251, 257
- in superconductors 63–64

Fock state 147–156, 162
fractional quantum Hall effect 12, 242–245, 254–263
free energy 59–61

GaAs, $Ga_{1-x}As_x$ 111–113
gate 211–215, 226–227
gauge symmetry 281–284
Ginzburg-Landau theory 11, 58–61, 281–285, 295–304, 332–336
graphene 91–108
graphene nanotubes 108–109

Hall coefficient 246
high temperature superconductivity 301–304
holes 6, 191–195
homogeneous functions 315–318

integer quantum Hall effect 12, 242–244, 247–252
interaction picture 176–177
interface between a superconductor and a normal metal 61–63
interference between Josephson junctions 293–295
interference within a Josephson junction 289–290, 292–293
Intrinsic semiconductors 191–195
Ioffe-Regel criterion 56–57

Jaynes-Cummings Model 7–8, 151–157, 174–177
Josephson junction 11–12, 285–288
Josephson junction in a static magnetic field 289–299

Landau diamagnetism 29–34
Landauer-Buttiker formula 77–82
Landauer model 6–7, 12, 69–73, 85–87
Landau free energy 310–315
Landau levels 12, 29–33
Landau solution in the presence of external electric and magnetic fields 247–249
lasers 163–171
Lindbaden formulation 177–183
localization
- and dimension 50–52
- of one-dimensional electrons 41–48
- and phase coherence 48–50

London penetration depth 63, 297–301

magnetic effects 25–40
magnetic field in Josephson Junction of finite cross-sectional area 292–293
magnetization 14, 32–33, 35–40
magnetoconductivity 246
magnetoresistivity 246, 251–254
maser 173–188
metallic and semiconductor photonic crystals 125–134
metamaterials 7, 13, 268–270
minimum metallic conductivity 56–57

Mott transition 8–9, 57–58

n type material 5–6, 191–195

one-dimensional electrons 41–48
one-dimensional magnet 307–310, 336–337
optical analogy of Hall effect 265
optical correlations and coherence 157–163

Pauli paramagnetism 34–36
Periodic wavefunctions 119–121, 126–129
phase coherence in localization 48–50
phase transition 13–14
phononic crystals 7, 134
phononic metamaterials 7, 13
photonic crystals 4,7, 125–134
photonic metamaterials 7, 13, 278–279, 268–269
p-n junction 5–6, 195–202
p-n-p and n-p-n transistors 203–204
p type material 5–6, 191–195

quantum dots 91, 110–111, 121–123
quantum Hall effect 9–10
quantum heterostructures 91, 110–111, 115–121
quantum wells 91, 115–118
quantum wires 121–123

rate equations 169–171
real-world Josephson junction 290–292
rectifiers 204–206
renormalization group theory 13–14
renormalization group theory treatment of Landau free energy 332–336
resonance 13, 270–275
resonance cavity 171–173
Rotating Wave Approximation 154

scaling theory 13–14
scaling theory identities for critical exponents 306–307
second order phase transition 310–313
semiconductors 5–6, 89–91, 191–194
Shubnikov-de Haas effect 252–254
single electron
- in an electric and magnetic field 247–249
- in an electric field 24–25
- in a magnetic field 25–32
- transistor 230–238

source 212–217, 226–227
spasers 188–189
spatial resonances 268–270
specific heat 306
speckle 7, 83, 85
spin Hall effect 263–265
spin Hall effect of light 264–265
spontaneous emission 166–168
statistics of laser light 184–188
stimulated emission 166–168
Stoner ferromagnetism 36–40
superconductivity 10–11, 58–64
switching 217–220

tight binding model 3, 91–107
time-dependent Ginzburg-Landau theory 283–285
topological excitations 279–280
two-dimensional electron gas 17–36
two-dimensional Ising model 319–332
two-dimensional photonic crystals 129–134
two Josephson junctions in parallel 293–295
Type I superconductor 61–64, 295–301
Type II superconductor 61–64, 295–301

universal conductance fluctuations 82–85

variable range hopping 52–55

wavefunctions in periodic media 90
Wigner crystal 58, 254
Wigner solid 9, 17, 41, 254

Youngs experiment 159–161

Printed in the United States
by Baker & Taylor Publisher Services